HANDBOOK OF BIOETHICS

Philosophy and Medicine

VOLUME 78

The titles published in this series are listed at the end of this volume

HANDBOOK OF BIOETHICS:

TAKING STOCK OF THE FIELD FROM A PHILOSOPHICAL PERSPECTIVE

Edited by

GEORGE KHUSHF

Center for Bioethics and Medical Humanities and Department of Philosophy, University of South Carolina, Columbia, SC, U.S.A.

KLUWER ACADEMIC PUBLISHERS

DORDRECHT / BOSTON / LONDON

A C.I.P. Catalogue record for this book is available from the Library of Congress.

ISBN 1-4020-1870-3

Published by Kluwer Academic Publishers,
P.O. Box 17, 3300 AA Dordrecht, The Netherlands.

Sold and distributed in North, Central and South America
by Kluwer Academic Publishers,
101 Philip Drive, Norwell, MA 02061, U.S.A.

In all other countries, sold and distributed
by Kluwer Academic Publishers, Distribution Center,
P.O. Box 322, 3300 AH Dordrecht, The Netherlands.

Printed on acid-free paper

Printed in the Netherlands.

TABLE OF CONTENTS

SECTION IV/ THE PUBLIC POLICY CONTEXT

SECTION V/ FOUNDATIONS OF THE HEALTH PROFESSIONS

GEORGE KHUSHF

INTRODUCTION: TAKING STOCK OF BIOETHICS FROM A PHILOSOPHICAL PERSPECTIVE

I. A TWOFOLD RISK

Bioethics is a large, interdisciplinary field, with contributions from philosophy, theology, literature, history, law, sociology, anthropology, and the diverse health professions. Broadly conceived, there are now well over a dozen major journals exclusively devoted to bioethics, and scores of newsletters and briefs. Medical and nursing journals regularly feature essays on the topic. This huge body of scholarship indicates an active, vital field, and it documents the increasing importance of ethical reflection for all dimensions of science, practice, and policy in the health care arena.

Attending this richness and diversity, there is a twofold risk. First, there is the danger that scholarship in one area of research will lose touch with what is taking place in other areas, and thus insufficiently account for factors that actually have a bearing on the research topic. Now there is simply too large a literature for one person fully to master. We have reached the point where, for example, a bibliography on ethical issues associated with genetics alone could list thousands of references. How does one assimilate all of this and simultaneously keep abreast of all the other areas in bioethics? While the extent of the literature now requires an increasingly specialized focus, the different elements of health care are part of an interconnected network, which cannot be sufficiently addressed in atomistic fashion. It is thus important to have a sense of the larger field, even if the intent is to focus on a narrow element. This, of course, becomes increasingly difficult as the field continues to grow. There is thus a need to provide a general overview of the different elements, and to give a sense of the field at large.

The second danger relates to the disciplinary character of reflection. Bioethicsh as unquestionably benefitted from its interdisciplinarity, and thus from the different canons and criteria that have been brought to bear on its problems and proposed solutions. However, the very strength of this interdisciplinarity is threatened, not by a divisiveness and fragmentation, but rather by the degree to which bioethics now becomes its own discipline, with its own emerging canons, criteria, and orthodoxy. In this emergent discipline, the distinctive voice and

G. Khushf (ed.), Handbook of Bioethics, 1–28.

strengths of each contributing discipline risk being lost, along with the disciplinary identity of the scholars who now see themselves primarily as "bioethicists." It is thus becoming increasingly difficult to identify any salient features distinguishing those coming from philosophy, religious studies, or a general medical humanities program.

In order to respond to this twofold risk (overspecialization and loss of disciplinarity), while simultaneously celebrating the richness and diversity of the field in general, this volume takes stock of bioethics from a philosophical perspective. Twenty-six essays, written by many of the most prominent leaders in the field, provide a survey of the most important theoretical and practical areas of bioethics. Each essay reviews the extant literature on the topic, identifying the important philosophical themes and resources. Each sketches important areas where future research needs to be conducted and where valuable collaboration can take place with those doing more traditional philosophical research on topics such as personal identity, moral theory, or the nature of scientific judgment.

II. THE ROLE OF PHILOSOPHICAL REFLECTION WITHIN THE FIELD OF BIOETHICS

Although this volume attempts to consider bioethics from a philosophical perspective, there is no rigid adherence to disciplinary labels and an attempt has been made to include the full range of western philosophical perspectives. A few of the individuals chosen to contribute to this volume do not have degrees in philosophy, although they all have made important philosophical contributions to the discipline of bioethics. Further, while there are important differences between the task posed here and that found in clinical ethics, one cannot always draw a sharp line, and no attempt is made rigidly to define what counts as philosophy.

While there is no attempt to rigidly define philosophy and exclude those perspectives that do not neatly fit the definition, this does not mean that there is no guiding sense of what philosophy is about and what it brings to the field of bioethics. Philosophers bring an interest in clarity and transparency, simplicity and economy of expression, and systematicity. And perhaps most distinctly, philosophers bring these interests together in order to grasp in thought the essential aspects of a practice, and understand these essentials in the light of previous aspirations (its history) and future goals. Such transparency for thought, simplicity of focus, and systematic reflection on the diverse elements of practice and context provide the vantage point for both criticism (e.g., does the practice live up to its ideals?) and constructive alternatives for improvement (how can practice be reconstructed so that deficiencies are diminished, inconsistencies eliminated, and goals realized?). To this extent, philosophy is constitutive of human life more generally; it is essential to the way knowledge is created in science, and the ways such knowledge is realized in the learned forms of practice. When one refers to the "discipline of philosophy" or "the philosopher," one is simply concerned with these activities in a more disciplined, rigorous, thoroughgoing, and explicit form.

To think of philosophy in this way highlights both what it might include and what it does not. First, regarding what philosophy is not: as a moment of life more generally – albeit an essential moment for the realization of human excellences – it is not the whole thing. As much as philosophers from Plato to Hegel have made pretense to subsuming the whole, or at least what was essential in it (whether as form or concept), and despite the earlier history where "science" (with the exception of theology) generally came under philosophy (as in "natural philosophy" for what we would call "natural science"), today the philosopher must accept a more modest place, especially when discussing the role of philosophy within bioethics and healthcare more generally. Any attempt to define philosophy in a way that encompasses the content of scientific knowledge would make the term useless in our current context – as if a physician-in-training learns the "philosophy" of anatomy upon entering medical school. Whatever philosophy of anatomy might mean, it can't simply be equated with a knowledge of the essentials in that domain.

As a moment of life and not the whole, philosophy presupposes and directly relates to a content that is outside of its own domain; namely it relates to that which is not philosophy, or at least not just philosophy. Philosophy of science relates to science; philosophy of medicine to medicine. The same thing is true for the role of philosophy in bioethics. Although it is not as obvious as in the case of anatomy or medicine, bioethics is not simply philosophical ethics applied to healthcare. Bioethics is now an established practice, with its own history, culture, and norms. Like medicine, it has a body of knowledge and even its own apprenticeships. Bioethics can thus be presupposed by the philosopher in the same way that biology can be presupposed by the philosopher of biology.

All this does not imply that philosophical activity is somehow merely external to bioethics. It is, in fact, constitutive of the practice. But the philosopher cannot construct bioethics *de novo*, as if the norms of the field could in Euclidean fashion be spun out of "pure" philosophical premises. This kind of philosophical reconstruction might serve as a regulative ideal, but I am skeptical that the norms of health care can ever be established in such a manner. (Many philosophers will disagree with me on this point.) Whether or not they can be, we are not at that place; current attempts are at best provisional. This means that the messiness and contingency of the field will be readily apparent in any attempt to take stock of it philosophically. Such messiness will be apparent even in the general divisions and categories used to describe the main areas. It is only with some violence that standard contrasts such as theory vs. practice or policy vs. clinical ethics can be used to organize this volume. In fact, a good case can be made that the next step in bioethics involves a challenge to some of the traditional divisions that have dominated the field to this point (see e.g., Bayertz and Schmidt in this volume).

Positively stated, the philosopher takes the given content of bioethics – that material that now constitutes its practice and conveys its norms – and subjects this to critical analysis, seeking to make explicit what is essential, analyze this in terms of deeper principles, and provide guidance for the next stage in reform. Paraphrasing what Joseph Boyle says about moral reasoning generally: the philosopher seeks to bring the moral life of health care under rational scrutiny. By reflecting on the history, goals and norms of health care, its moral life can be made transparent, allowing for the attainment of goals that are only partially realized. That is what the essays in this volume seek to accomplish.

III. THE EMERGENCE OF BIOETHICS

In his historical introduction to this volume, Albert Jonsen outlines the different strands of reflection that came together in the formation of bioethics (see also Jonsen, 1998). He speaks as an insider, perhaps too modestly passing by his well known role in the formation of the field. From his history we see that the field emerged from a confluence of factors. Among these were:

- The needs of healthcare professionals: bioethics addresses the needs of physicians, biologists, policy makers, and patients who must make decisions. The demands of action, and the questions that attend them, evoke a response. And the actors themselves have played an important role in facilitating that response and determining its substance.

- Theological reflection: Among those trained in ethical reflection, theologians – not philosophers – played the more prominent role in the early days of the field. Informed by their traditions of faith, they took up prominent issues such as death and dying or genetics, and formulated responses that still frame many of the current debates.

- Governmental commissions: While some of the early work was facilitated by those involved in the practice – for example, physicians who needed to make decisions about how limited organs would be distributed – in other cases those involved in practices were blind to important ethical issues. Thus, for example, in areas such as human subjects research, a history of abuse and neglect lead to bioethical commissions such as the one that framed the Belmont Report (see Jonsen in this volume, pp. 38-39).

- Philosophical reflection: At a later stage, but with an enduring influence, philosophers too engaged the ethical issues.

These are representative of the diverse influences. There were many more; for example, there were legal cases, which reframed issues such as abortion and end-of-life care. There were also broad movements, such as those surrounding civil rights or Vietnam, which, in turn, framed the issues that were considered important: issues such as autonomy and open deliberation, as well as criticisms of older notions of paternalism.

Each of these factors by itself cannot be taken as a sufficient cause of the new field. But when they are taken jointly, it is clear that something new emerged. The ethics of health care, and of medicine in particular, was different. No longer was it simply the purview of physicians making paternalistic decisions on behalf of their patients. To this extent, as Edmund Pellegrino (1993) notes in his own review of the history of medical ethics, there was in the late 1960s and early 1970s a break with a two millennia old Hippocratic tradition.

Perhaps most interesting in Albert Jonsen's analysis, however, is the claim that it was not just medical ethics that was transformed. The different disciplines and traditions that merged to form bioethics were themselves changed. In particular, philosophy changed. Or rather, the philosophy that was current – influenced by the sensibilities of logical positivists – was challenged, and, as philosophers came into contact with the concrete practices of healthcare, they were forced to recover ways of doing philosophy that were themselves ancient, with precedent going back to Aristotle and more recent incarnations in the American pragmatists.

Philosophers were challenged to formulate their ideas in a way that actively engages the concrete practices they seek to inform. Jonsen puts this in terms of the classical notions of "invention" and "discovery." Namely, rather than simply apply "honed skills and methods of their discipline," philosophers needed to improvise, applying in creative ways forms of argumentation that were not sufficiently formulated for the health care context. Put another way, a rich dialectic was needed between traditional philosophical reflection, on one hand, and the health care context, on the other. Both were transformed as a result of the encounter.

This creative dialectic is perhaps nowhere more apparent than in the theory that informs bioethical deliberation. Within philosophy, there is a long history of reflection on theory; in fact, together with metaethics, normative theories are the central concern of philosophical ethics. Occasional forays into particular applied problems are often valued more for the way they illustrate theoretical concerns than for their own intrinsic value. However, these systems – whether of deontology, consequentialism, virtue theory, or some more recent variant – were often framed in terms that did not engage those involved in the health care practices themselves. From the experience that many of the early pioneers in the field had on commissions and in the clinical trenches, a form of theoretical reflection was developed that was nurtured by the classical philosophical systems, but recast. After Jonsen's historical introduction, the first main section of this volume considers the more prominent variants of these theories, with many of the chapters written by those who are themselves the pioneers of this new bioethical discourse.

IV. BIOETHICAL THEORY

Several authors in this volume note how in the early days of bioethics, most people focused on "first order concerns" in health care (Clouser and Gert; Beauchamp and DeGrazia); namely, people addressed problems such as end-of-life care, rationing scarce resources, and genetics. Solutions were formed in an ad hoc manner, with a local sense of whether it was adequate to the problem at hand. People in the trenches were not looking at ethical theory, and few at that time (especially among the philosophers) would have thought that theoretical work could or should be framed in a way that was specifically wed to the health care arena. "Theory," largely associated with utilitarian and deontological systems, was almost by definition framed in general terms, applicable to all domains of practice. Early attempts to bridge the theory and first order concerns involved a juxtaposition of the classics of philosophy (e.g., Kant) with the specific problems in the health care arena (see e.g., Gorovitz, Jameton, and Macklin, 1976; also Jonsen in this volume, p. 38). The work of mediating these domains was left to the reader. Beauchamp and DeGrazia nicely summarize this when they state that "[n]o bridge had been built to connect work in philosophical ethical theory with the problems of biomedical ethics, and professional ethics was not conceptualized in biomedicine as an interdisciplinary field" (p. 55).

As the first-order, problem-oriented work continued, those involved in the nuts-and-bolts of bioethics practice found that they could make some generalizations about the nature of ethical deliberation in the health care arena. Such generalization – the first flowering of a full-blown bioethical theory – did not simply arise from the bottom upward, as an inductive move from the particulars of deliberation in healthcare. Those involved – people like Tom Beauchamp and James Childress (1979), Albert Jonsen and Stephen Toulmin (1988), H. Tristram Engelhardt (1986), Edmund Pellegrino and David Thomasma (1981) – were all well trained in the traditional ethical disciplines, and were thus able to draw on the rich heritage of ethical theory when formulating their own theories. It is not coincidental that the most prominent works in this domain were advanced by those with extensive theological and philosophical training. However, the early theories also did not simply arise from the top downward, as an application of previously developed philosophical theories. One finds stories such as the following: Beauchamp and Childress, the first a utilitarian and the second a deontologist, tell of how, despite their theoretical differences, they could come to agreement on the concrete problems they confronted in healthcare, and how, from this, they were led to the formulation of their middle level principles. Jonsen and Toulmin similarly recount how their collaboration on casuistry arose from their joint involvement in the President's Commission. While the first team was clearly influenced by the broader debate on the prima facia principles of Ross (1930; 1939), and the latter team saw their task as a recovery of classical casuistical traditions, they all thoroughly rethought their antecedent theories in the light of the realities of the practices of healthcare ethics. These practices constituted the nonphilosophical datum, which in creative ways was brought to thought and recast by the more systematic efforts of formulating bioethical theory.

Within this volume the more prominent bioethical theories are representated, including principlism, casuistry, virtue theory, feminism, narrative ethics, and those theories that see healthcare ethics either as a variant of a more general public ethic or as more specifically wed to the phenomenology of health care practice. Since these all have grown up from the same soil, with similar kinds of experiences, it is not surprising to see common features. Among these are an emphasis upon situated deliberation. In all cases, moral reasoning does not involve a simple application of pure theory or a single principle, but rather requires a subtle balancing which in the end cannot be captured by an algorithm. Put in another way, decisions are not in a straightforward way rule governed. They are instead communally situated and intertwined with a host of complex assessments of situation, rules or general principles, precedent cases, and ends. Some have found the agreement on these themes so great that they even speak of an emergent consensus (Kuczewski, 1997) or, put more negatively, a bioethical orthodoxy (Wear in this volume; Engelhardt, 1996).

Despite the agreement, there are also significant differences, although it is not always clear whether these are differences within a shared theoretical framework or differences between theoretical frameworks. Many of the differences relate to classical disputes within philosophical ethics. These include the following:

- Should one focus on actions or character? The core difference between virtue theory (David Thomasma; Edmund Pellegrino) and other theories such as principlism (Beauchamp and DeGrazia), casuistry (Boyle) and Clouser/Gert's "common morality" rests on this distinction. However, as the essays in this volume will make clear, one should be careful not to formulate these diverse theories in a way that makes the distinction too sharp. Virtue theorists recognize that a focus on character is not enough (Thomasma), and other theorists recognize that virtue is important; for example, Boyle (p. 79) notes how the virtue of practical wisdom is needed for the subtle balancing involved in casuistical analysis (see also Beauchamp/DeGrazia, pp. 66-68; and Clouser/Gert). Aspects of feminist theory (esp. the ethics of care) are related to virtue theory, while others are more act- or system-oriented (Tong); and many narrative ethicists (e.g., Hauerwas and Pinchas, 1997; MacIntrye, 1984; see Nelson in this volume) are directly associated with virtue theory.

- Should one focus on consequences or intrinsic features of an action? Much of the classic dispute between consequentialism (primarily with the variant of utilitarianism) and deontology can be understood in terms of this distinction. However, most bioethicists have eschewed single principle theories in favor of more complex, situated accounts. Where principles or rules are advocated, as in the theories of Beauchamp/DeGrazia and Clouser/Gert, more than one is put forward. Thus the classical dispute between utilitarians and deontologists is reconstructed in terms of balancing concerns associated with these traditions. In particular, principles of beneficence and nonmaleficence (among the principlists) or a rule not to harm (for Clouser/Gert) are put forward to capture the utilitarian concern about consequences, while a principle of autonomy or the rule to advance the norms that uphold society captures core deontological concerns. (Accounts of justice are more complex, with some more utilitarian and others more deontological.) It is worthwhile noting that a similar interest in integrating consequences with deontic constraints can be found in prominent casuistical (Boyle) and virtue oriented accounts (Pellegrino and Thomasma), as well. What used to be regarded as a dispute between theories is, for many bioethicists, a dispute about how best to balance the insights of both theories in the absence of a single, meta-principle or an algorithm for resolving the difference.

- How are moral claims ultimately justified? Problems of justification in bioethical theory take several forms and occur at several levels. For example, in particular cases, a recommended course of action might be justified by appeal to principles, rules, or paradigm cases. There is debate about how effective such principles, rules, or cases are at justifying certain kinds of actions. At another level, a course of action that involves a balance between competing principles might be justified by a fuller account of the situation, which is relevant for the specification of principles, together with a reference to virtues associated with moral judgment (Boyle; Beauchamp and DeGrazia). At the broadest level, justification of the principles and the general theory might involve an appeal to "coherence" or the background "public morality" (Beauchamp and DeGrazia; Clouser and Gert) or, on different accounts, to natural law (Boyle) or a particular meta-narrative (Nelson).

- How should one balance the need for situated deliberation against the threat of relativism? We live in a pluralistic society, with multiple cultures and beliefs informing ethical deliberation. As a result, there are often competing moral norms nested within competing traditions. An appreciation of diversity, and a recognition of the degree to which norms are culturally and historically situated, often comes into tension with the need for unambiguous norms that can mediate conflicting claims among these competing groups. And we often must act together for communal ends. How is such action guided? David Thomasma and Hilde Nelson explore these issues in depth, and they are also addressed by many of the other essays in this volume. Different bioethical theories struggle with these questions in different ways.

- Should one preserve or critically transform the background moral system? The public and health care traditions that inform moral deliberations can have both positive and negative features, and bioethical theories will differ on what these are and the degree to which the background public system needs to be transformed. Some theories will see their primary role as clarifying and bringing to language this presupposed "common morality," while others, such as feminism, will see their role as much more critical, calling into question not just the presupposed systems and norms, but even challenging the forms of ethical deliberation, seeing them as variants of a dominant group's form of reflection. Tong nicely outlines, for example, how care-oriented feminists see traditional modes of moral reasoning, with their abstract, rule-oriented reflection as alienating womens' more particularist, situated, care-oriented reasoning. Nelson similarly considers those who would look to more narrative-based ethical reflection as an alternative to the theoretical-juridical models.

- Do the central norms of healthcare ethics arise out of the specific characteristics of its practice or are they applications of more general public norms? Edmund Pellegrino argues that bioethical norms should arise out of the realities of illness and medical practice; in particular, the healer's response to the suffering patient. Through a phenomenology of the health care encounter, the physician-patient relation is made central, and norms are configured from there. By contrast, others see the norms in terms of a more general ethical theory.

Within bioethics, each of these traditional areas of dispute in ethical theory takes on a new, unique form. The disputes in health care ethics are informed by the broader philosophical debates. However, these broader debates can also be fruitfully informed by current deliberation in bioethics. One of the tasks of this volume is to provide the groundwork for a fruitful exchange between those working in bieothics, informed by the realities of deliberation in a health care context, and those philosophers who have not had this vital link to the ongoing problems of practice.

Picking up on an earlier theme – namely, that the philosopher presupposes a content and then seeks to capture it in thought – one could say that behind all of the classical disputes outlined above, there is a core, interpretive concern: how does one understand the background moral system that informs health care decision-making? Is there a "common morality" apart from the health care context that provides the ultimate justification for any particular bioethical theory? Or are there diverse moral systems, perhaps linked to different religious traditions or background philosophical beliefs, which provide a critical perspective for transforming that public morality? Should bioethics be linked to a more situated kind of ethical reasoning, arising out of the sphere of medicine and a phenomenology of that practice? Are the background moral beliefs and even the forms of reasoning tied to patriarchal structures that have been oppressive rather than liberating? And what roles have broader narratives played within this background system? What roles should they or could they play? As hermeneute or interpreter, the philosopher brings to language and clarifies for thought those features deemed essential to ethical deliberation in healthcare.

V. CORE CONCEPTS IN CLINICAL ETHICS

Bioethics is a compromise between the demands of practice and broader ideals about the ways humans should interact with one another, especially in the context of health care. On one hand, decisions must be made. Health care workers do not have the luxury of endless reflection. They must act. If they linger too long, the uncertainty and delay itself will cause major difficulties. On the other hand, if the patterns of practice are too hastily codified, the deeper ends of a humane medicine will be compromised and more problems will be created. Clinical ethics, as a subset of bioethics, involves a search for the appropriate balance between practical demands and higher ideals in the area of clinical medical practice.

At each stage of development, the norms of practice embody the previous culture, a sense of what is realistic (which includes: what will be acceptable to those practicing), a vision of the ideal clinical encounter, and a sense of the scientific integrity of medicine. A complex accommodation of each of these is present at all times. Each tugs against the other. Sometimes the ideal of a rich, personal encounter between physician and patient comes into tension with the ideal of an evidence-based, scientific practice. At other times these tendencies are complementary. At all times, the ideals of science and ethics are themselves undergoing subtle changes, eventually leading to qualitatively different conceptualizations of the ideal.

When medicine is in the midst of rapid change, or when the diverse tendencies are openly in conflict, then foundational reflection upon the nature and ends of medicine takes place and there is a vital debate about its norms. At other times, when clinicians and patients have grown comfortable with the habits of practice and a reasonable accommodation of diverse interests has been accomplished, there is less reflection. A medical "orthodoxy" emerges. One can draw an analogy here to Thomas Kuhn's distinction between science at times of crisis and "normal science" (1996). During times of conceptual crisis everything is rethought. However, during periods of normal science, most is simply taken for granted, and the majority of effort is directed toward solving a host of little problems that are easily situated within, and understood from, the perspective of the dominant paradigm of thought. Similarly, within the history of the ethics of clinical practice, one can identify periods of radical change and crisis, and one can also identify periods of orthodoxy.

Before the emergence of bioethics in the late 1960s and early 1970s, there was a dominant orthodoxy of clinical practice. To understand this orthodoxy, one must appreciate the accommodation between the science and ethics of that practice; one cannot just look at the background ethical ideals. In particular, one must appreciate how medicine was regarded as a *scientifically grounded practice*. The pre1960s norms of clinical practice are well expressed by Abraham Flexner in his celebrated report, *Medical Education in the United States and Canada* (1910). In this report, the method of a physician's practice is seen as identical to the method of a scientific researcher (pp. 54-57). The history and physical exam are a form of data gathering. From these a disease hypothesis is formulated. Through lab tests and the management of the patient's medical condition, the physician either confirms, disconfirms or modifies the initial hypothesis. When the physician completely fails, the ultimate test is made available: that of the autopsy (pp. 66-67). Since medicine is scientifically grounded, what is "medically indicated" is a function of science, not socioeconomic or individual values.

Presupposed in Flexner's account of medical knowledge is another core assumption; namely, that the physician is the captain of the ship, the central agent of practice, and the focus of this physician's activity is the individual patient. The *clinical domain* is thus associated with the *physician's* realm of discretion. The core ethical relation is between physician and patient. The purpose of the clinical encounter is to cure. This makes disease the focus. And disease is understood as a defect in the body machine (p. 58). Health is restored when this defect is removed.

Until the emergence of bioethics – namely, that novel something that arose in the 1960s and 70s – medicine was very paternalistic. The Flexnerian physician-scientist took it for granted that he knew how the body worked and what the patient needed. When the patient presented, the physician would assess what was needed and simply intervene to promote the patient's best interest. Patients usually accepted that the physician knew what was indeed best, and the patient would generally acquiesce passively to what was proposed. There was little discussion. In this "silent world of physician and patient," the physician, as scientist, would decide (Katz, 1984). An ideal of science directing life was at least partially realized.

Within this context, the ethics of clinical practice focused upon sustaining the conditions of effective, scientifically-based intervention. Two tendencies integral to medicine – the ethics of practice and the ideals of science – were thus viewed as complementary, although the ethical analysis was in many ways subordinated to the science. Confidentiality was to be protected so that patients would not fear disclosure of sensitive information necessary for an effective clinical intervention (one needs all the data). Financial incentives associated with advertizing, fee splitting and other similar business practices were excluded because these nonscientific factors were not supposed to influence how medicine was practiced; they would undermine its scientific integrity. Similarly norms associated with interprofessional collaboration, alternative medical practices ("quackery"), and medical sectarianism (such as homeopathy or osteopathy) were established, all with the intent of assuring that science is upheld (Rodwin, 1993; Khushf, 1998). And while a more cynical, sociological analysis of the first two thirds of the twentieth century might reveal multiple ways in which these "ethical" norms privileged physicians' interests (Starr, 1984), the stated intent of those advancing the norms was to provide a solid foundation for a genuinely modern medical practice. A broad social consensus on the appropriateness of the norms upheld this Flexnerian orthodoxy.

At the heart of modern bioethics was a challenge to the paternalism inherent in the Flexnerian ideal. It was recognized that a patient might conceptualize and give content to "best interest" in a way that was different from the physician. Even further, there were certain biases within medicine associated with the model of the person as body-machine and disease as a defect within that machine. These biases lead physicians to focus on a well functioning body, and to discount the broader lack of ease (the dis-ease) associated with the illness experience. What was "medically indicated" was not necessarily the right thing to do from a patient's perspective. Thus there was a need for dialogue between physician and patient. A new clinical ethic attempted to provide a description of disease and the physician-patient relation, which grasped in a more appropriate way the rich texture of life and provided a basis for the development of moral norms that could guide the clinical encounter. The norms of the new ethic were embodied in doctrines of informed consent and advance directives. Ethics committees were formed to guide clinical decision-making so that the relevant values were then embodied in practice. These are the themes addressed in this volume's section on "core concepts in clinical ethics."

With the Flexnerian ideal, science and the ethics of a humane encounter were regarded as naturally in harmony, since the ideals of the humane encounter were themselves understood in terms of the conditions for the realization of the scientific practice. However, with the new clinical ethic, these two tendencies came into some tension. The split between the cultures of the sciences and humanities, with their sometimes incommensurable languages, now was inherent to clinical ethics. The ethical ideal was often framed in an idiom from the humanities: an ideal of personal, humane interaction. The scientific ideal was still framed in terms of a positivist conception of evidence-based practice. Clinicians, faced with the exigencies of their practice, arrived at ad hoc accommodations. In this case, especially since the more humanistic ideals were largely externally imposed upon medicine, the accommodations gave greater weight to the science. A new orthodoxy emerged where informed consent and the ideals of a humane, personal encounter were largely reduced to legal norms, fulfilled by the provision of a form to be signed by the patient. Clinical ethics came to focus less on the nature of the ideals themselves, and more on pragmatic strategies for realizing them ... and for addressing the tensions between the science and the human dimensions of practice that would now and then emerge. Empirical studies of medicine showed that clinical ethical ideals were rarely realized in the clinical arena (SUPPORT, 1995; see also Wear and Lustig in this volume).

It is here, in considering this new orthodoxy, that the value of philosophical reflection becomes apparent. There are core philosophical themes integral to the new clinical ethics, but which get eclipsed by the practical needs and pragmatic orientation that characterizes much of the field. When faced with practical questions about the what and how of informed consent or advance directives, it is natural for broader questions about the nature of autonomy, human interests, the good life and human flourishing, the character of clinical judgment, or personal identity to be sidelined. Clinicians rarely have the interest or patience to pursue these philosophical issues in great depth. Nevertheless, such questions are integral to the resolutions we give to the practical exigencies; if they are ignored, almost always a naive and often inappropriate resolution will be provided which, in the long run, will work against the ethical fabric and even the clinical success of medicine. The task of the philosopher is to not leave these issues and questions dormant; to probe the orthodoxies of the field, uncover background assumptions, and reorient the norms to the realization of deeper ends. That is what the essays in this section accomplish.

The section opens with an analysis of the nature of health and disease. Arthur Caplan (1992) has spoken of this as a "boundary defining problem" in the philosophy of medicine. Is disease best understood as value-neutral and objectively determined? Is it the primary term, with health simply regarded as the absence of disease? Or is the very nature of health and disease linked to a patient's values? Lennart Nordenfelt addresses these issues, arguing in favor of a holistic concept that sees health in terms of the ability of a person to realize their vital goals, namely, those goals necessary for happiness. Nordenfelt's analysis, and the debate on the nature of health and disease more generally, is integral to a clinical ethic, because the health concepts integrate the science and ethics of clinical practice. If disease is value neutral, then the world of science and that of ethics can be neatly separated, and values can come in as a second, supervening strand. By contrast, a value-laden disease concept means that "ethics" goes all the way to the root of medicine; even the "science" of drawing a line between the normal and the pathological involves ethical/values issues, which, once made explicit, call into question the scientific foundationalism that has informed much of medicine. The criticism of a value neutral disease concept can thus be taken as a challenge to the notion of scientific medicine that has framed the more classical clinical ethic (Khushf, 2001).

To capture the difference between the scientist's and patient's perspective, much of the bioethics literature draws a contrast between disease and illness. Disease is viewed as the atypical diminishment of typical functional ability (Boorse, 1997), a biomedical concept linked to the mechanical model of the body and the ideal of a value neutral science. Illness, by contrast, is seen as a disruption of the patient's life world, entailing an alienation from self, others and the broader frameworks of meaning that orient the life of the well (Cassell, 1976). The physician tends to focus on disease, the patient lives through the illness. In the clinical encounter, these two perspectives mesh. Problems arise when medicine is paternalistic and the physician's understanding of patient interest, guided by the disease concept, trumps the patient's more "subjective" concerns. At times, the very method of medicine can be seen as an attempt to abstract from the patient's subjective experience in order to get at the underlying, "objective" problem (Reiser, 1981; Leder, 1992; Toombs, 1992). This approach can be, and often is, depersonalizing, exacerbating the suffering that attends illness.

Richard Zaner explores these aspects of the physician-patient interaction in his chapter. As the field of bioethics was developing, he utilized the philosophical tools of the phenomenologist to elucidate the clinical encounter, and thereby framed the ethical issues of medicine in terms that captured the dissonance and asymmetry of knowledge and power that characterizes the physician-patient relation (Zaner, 1981; 1988). In his essay for this volume he identifies the challenge a physician has to understand the patient as a person and respond in an appropriate way, thereby framing the ethics of the clinical encounter in terms of the demands for appropriate interpretation of the import and character of that encounter, not just generally, but for each patient.

The demand placed upon a physician to hear, understand, and respond to a patient's own understanding of self-interest is legally and ethically codified in the doctrine of informed consent (Faden and Beauchamp, 1986; Wear, 1998). This doctrine, perhaps more than any other, captures the consensus of the bioethics community about the nature of the clinical encounter and the importance of respecting a patient's autonomy. But is the consensus warranted? In his essay for this volume, Stephen Wear considers bioethical "orthodoxy" on this matter. He argues that there are background notions of autonomy, patient capacity, informational disclosure, and even notions of clinical judgment that are implicit within informed consent doctrine, but needing extensive additional analysis. Wear also explores some of the empirical literature on informed consent, which shows that the ideal is rarely if ever realized in practice. Why is clinical practice so deficient, given the broad consensus on the importance of informed consent?

Is this deficiency a function of physician intransigence or a deeper structural problem in the doctrine itself? Through his philosophical analysis, Wear links this core clinical ethical concept to some of the foundational issues in philosophy more generally.

Cynthia Cohen does for advance directives what Wear does for informed consent: she considers the core assumptions implicit in the clinical ethical doctrine. However, instead of critically challenging the consensus, Cohen provides a defense of advance directives against those who now challenge them. From a philosophical perspective, her analysis is especially interesting because of the way she draws on discussions of personal identity. Within the bioethical literature, advance directives (also referred to as "living wills" or "health care durable powers of attorney") are generally seen as an extension of informed consent. Through an advance directive a patient can continue to exercise control over the process of medical decision-making even after that patient loses the mental capacity to directly participate in those decisions. The directive enables patients to express preferences and values regarding their own care and about the people they think would best represent them as surrogate decision-makers. But does it make sense to hold a future, permanently incompetent patient to wishes expressed by that patient when competent? Is the diminished person sufficiently continuous with his or her former self? And what is involved in extending notions of autonomy to an obviously incapacitated patient? Recently, philosophers and legal theorists have challenged advance directives on just these grounds. Cohen responds by arguing that the notion of personal identity presupposed by these critics is inadequate, and she seeks to place advance directives on more solid philosophical footing.

Issues surrounding patient autonomy, informed consent, and a more symmetrical, dialogical model of the physician-patient relation are central to the new bioethic. When these are emphasized, the literature focuses on shifting power away from the physician and toward the patient. Emphasis is placed upon the mechanisms that assure this transfer. However, this is not the only concern expressed in clinical ethics. Nor does the literature assume a hostile attitude toward physicians. Wear addressed these issues by highlighting some of the legitimate reasons for a more paternalistic, beneficence-oriented approach toward patient care. In the end, many – perhaps even most – of the clinical ethical issues are not a function of clear lines defining appropriate and inappropriate roles. Rather, they are a function of the complexity of modern medicine and the uncertainty of ethical obligations. What is the right thing to do in a complex case? The tools of ethical evaluation are then useful as a way of working through such difficult cases, just as the tools of medicine enable one to work through a complex medical case. Here there is an explicit analogy between physician and ethicist: both have special skills for addressing case-based problems. Ethics committees have been developed as a resource for assisting clinicians in these difficult, situated cases.

In his essay for this volume, Mark Kuczewski considers how ethics committees have been understood as a resource for clinical ethical deliberation in the areas of case consultation, policy review and education. In the end, he challenges the analogy between a specialized medical decision-making and specialized ethical decision-making, suggesting that greater emphasis should be placed on the educational mission of hospital ethics committees, rather than their case consultation function. His arguments rest upon a particular notion of ethics, informed by Aristotelian tradition and aligned with current work on virtue theory, communitarianism, and casuistry. The whole debate on ethical theory, outlined in the first section of this volume, is thus intertwined with the debate on clinical ethics.

VI. THE PUBLIC POLICY CONTEXT

In clinical ethics, the focus is primarily on the relation between two individuals, the clinician and patient. Where values differ between the two, one can arrive at a case-based resolution, which depends on the outcome of dialogue and negotiation between the parties. At least this is the ideal. However, when one moves to the level of policy, the situation is more complex. Policies embody values, and these will inevitably be the values of some and not others. Bioethical policies relate to issues such as human subjects research, euthanasia, reproductive ethics, health policy, and genetics. These are the topics discussed in this volume's section on policy, and they are among the most difficult and divisive issues in our society. Any full resolution of any of these issues depends on a content-full vision of the moral life, involving an anthropology, axiology, politic, etc. And each problem is intertwined with others. Thus, on one hand, the problems addressed by policy require for their resolution a robust moral vision and an appreciation of how one problem is intertwined with others. On the other hand, the fact of moral pluralism and the difficulty of arriving at a resolution demand that these problems be addressed in a partial, fragmentary way. As a result, the actual bioethical policies and policy recommendations will always have an ad hoc character. They will be a result of complex political compromises.

Within much of the bieothical literature, the primary concern is with finding a compromise that sufficiently accommodates the interests and concerns of most people. Since it is recognized that we will not find a universally accepted, content-full resolution on issues such as abortion or euthanasia, the problems are reframed. We now ask what policies are appropriate in a pluralistic society, given the diversity of viewpoints (see Thomasma in this volume; also Engelhardt, 1996). When an ad hoc accommodation is found and the initial furor of a debate dies down (as in earlier debate surrounding in vitro fertilization), then the discussion moves on to another topic. Slowly most (but not necessarily all) people come to accept the compromise as a principled resolution. What is initially seen as ad hoc and penultimate is, after time, taken as the dominant social norm, and the deeper insufficiency of the resolution – both practically and philosophically – is forgotten. And this "forgetting" does not just attend particular topics of debate. It now characterizes much of bioethics generally. Thirty years ago, when the first flowering of bioethics was taking place, the philosophical issues were very much in the air. Now bioethics is so infused with the pragmatic resolutions that it seems difficult or impossible to even make the deeper philosophical issues alive again. One of the central tasks of this volume is to make the forgotten questions, and the penultimate and insufficient character of the answers, explicit again, so that a richer form of discourse can take place. This is something that the philosopher can and should bring to bioethics.

The section opens with a discussion of research on human subjects. As Baruch Brody notes in his essay on the ethics of clinical trials, there is a conflict of interest at the heart of medical research involving human subjects. The interest in conducting successful research, advancing human knowledge, and benefitting future patients can be in tension with the traditional medical obligation to do what is in the best interest of an individual patient. Behind the details of human subjects research protocols one thus finds a classic problem in ethics: the balancing of individual against common good. This problem takes several forms. When interim research results seem to indicate that one arm of a trial is better than another, should one stop the trial and move all subjects into the beneficial arm or should one continue the study to assure more reliable results? Where and how should the line be drawn? When there are limited benefits to a patient and moderate risks, is it legitimate to go ahead with a trial? How should those on an Institutional Review Board (IRB) make those decisions? What kinds of decisions can be made by patients, what information should be provided to them, and what can be done to assure informed consent better approximates the ethical ideal? In all these cases, there are subtle tradeoffs between patient good and the knowledge that advances social good. And there is a balance between what is resolved at the institutionallevel (e.g., those on the IRB say risk outweighs benefit, *even if a patient consents to participate*) versus what is resolved by the patient and clinician (e.g., a patient may choose not to participate, and thus not provide consent). When an IRB decides that the risk/benefit ratio is too high or a study does not have sufficient scientific merit, the decision is taken away from patients, even if they disagree with the institutional assessments. This, by itself, involves a kind of social paternalism, albeit a form generally regarded as legitimate. Brody explores the subtle tradeoffs between individual and common good, and the issues of who determines good and with what information, making explicit the guiding questions for future debate on human subjects research.

The problem of balancing individual and common good, and of balancing the good of a few against that of the many, has its counterpart in other areas; in fact, it can be taken as central to bioethical policy deliberation generally. However, there are areas of policy where the problem has been forgotten. This is the case in health care financing, where recommendations are informed by cost effectiveness analysis (CEA). Dan Brock notes how such analysis embodies a utilitarian standard. However, advocates of CEA are not aware of the problems associated with such a standard. Brock deepens the debate on the prioritization of health care resources by asking questions such as "how should states of health and disability be evaluated?" "what priority should be given to the sickest or worst off?" and "when should small benefits to a large number of persons receive priority over large benefits to a small number of persons?" Through these questions, the classical philosophical debate about a utilitarian standard of justice is now formulated in the idiom of CEA, thus making that debate understandable for those working in the health policy arena.

Some of the most divisive policy issues are associated with reproductive ethics. Christopher Tollefsen's chapter addresses the full range of this debate, including issues of moral status, research on embryos and fetuses, stem cells and cloning, abortion, and more general concerns associated with how to understand reproductive activities. Usually these diverse topics are addressed independently, and they are resolved by a liberal argument: in our pluralistic society such divisive issues, with their competing notions of personhood and the good, should be left to the discretion of individuals. Reproductive autonomy is thus primary. Society should not legislate one vision at the expense of others. Tollefsen challenges this solution, arguing that "a full understanding of the nature and value of reproductive autonomy requires an account of the nature of human good, and its relationship to choice." He then explores the diverse notions of personhood and choice that potentially frame issues of reproductive ethics, and shows how philosophical questions of moral status and the purpose of reproductive activities themselves cannot be sidestepped by more pragmatic attempts at solving these problems.

Modern liberal societies have succeeded in keeping the peace by dividing the world into public and private domains and relegating the resolution of the more divisive issues, with their robust philosophies of life, to the private domain (Rorty, 1989; MacIntyre, 1984). In many ways, Tollefsen's arguments involve a challenge to the usefulness of the public/private distinction as a means of resolving the divisive issues associated with human reproduction. In their chapter on genetics, Kurt Bayertz and Kurt Schmidt further problematize this distinction. They focus on the superindividual character of genes and genomes.

Behind the public/private distinction and the liberal program with which it is associated is the assumption that one can neatly distinguish between issues pertaining primarily to individuals and those pertaining to groups of individuals or society generally. The latter, of public interest, can be addressed in terms of social policy. But the items pertaining to individuals – for example, those related to how a person lives his or her life, what job is chosen, what spouse, etc. – is associated with the realm of individual choice. In fact, notions of autonomy – both ethical and political – assume that individuals are able to make decisions about their own person and property. What happens, however, when an activity is intimately associated with individual notions of the good and choice, but simultaneously has bearing on a group or whole population? Such is the case with activities associated with genetics. A genetic screening test, for example, might show a woman has increased risk of breast cancer, but it simultaneously indicates a higher level of risk for that woman's daughter. How is the normal requirement of confidentiality of medical information preserved when people are intertwined with one another in this way? By considering these kinds of issues, as well as the even more difficult ones associated with genetic enhancements, Bayertz and Schmidt challenge the basic categories of bioethical reflection. In fact, they argue that a revision of our modes of ethical reflection is required. "Once we begin shaping human nature, we shall also be forced to shape the ethical principles which allow or forbid just this." Here the need for a rich dialectic between philosophical reflection and the new technology is especially apparent.

VII. THE FOUNDATIONS OF THE HEALTH PROFESSIONS

A literature search on health care professionals reveals two general kinds of essays. On one hand, one finds the sociologically informed writings. Such writings tend to focus on the power and prestige of the profession, how it manages to establish and sustain self-control, and how external forces might undermine this control and prestige (McNamara, 2001 provides a nice review). Often, there is a distinctly cynical perspective with respect to the ethical ideals. They are more often than not seen as tools by which the professional rationalizes some aspect of their own interest. On the other hand, one finds writings on professionalism by the members of the profession (Bryan, 2000 provides a nice review). Such writing emphasizes the ethical ideals and scientific competency, often seeking to uphold them in the face of some external threat. A good example of such writing can be found in recent attempts to uphold the classical notions of medical professionalism in the face of economic developments which threaten to transform medicine into a business or entrepreneurial activity (American Board of Internal Medicine, 1995).

Largely missing from the literature has been philosophical reflection on the foundations of professionalism as a self-regulating, ethically informed practice (there have been a few exceptional philosophical forays into this domain; see e.g., Cribb, 2001; Schon, 1983). The paucity of philosophical work in this area is unfortunate, not just for the professions who might benefit from a richer discourse on foundations, but also for philosophers, who might find some especially interesting material in this domain. The health professions and the topic of professionalism has a unique status between the micro-ethics of individual-individual relations (as seen in currrent clinical ethics) and the macro-ethics of social interactions (as seen in the debate on policy). Unlike clinical ethics, issues of professionalism cannot be resolved on a case by case basis; some norms for all members are required, at least in so far as the classical ideal of professionalism is approximated. But the resolution differs from normal policy resolutions. Members of a profession ideally share robust values regarding the ends of professional activity and they also share a sense of the norms that govern "good practice." But they do not assume that those outside the profession are governed by those same norms. There is thus a sense of community, with a membership that is only a subset of society generally. Those initiated into the professional body of knowledge are uniquely responsible for forming the standards to which each will be subject. This middle place between micro- and macro-ethics has the potential to bring together issues traditionally divorced from one another by means of the public/private divide.

Another aspect of professionalism that is especially interesting from a philosophical perspective concerns the linkage between professional knowledge and ethical norms of practice. There are three marks of a classical profession. First, the professional is initiated into a specialized body of knowledge. Second, the professional has a unique ethic and serves a social good. Third, those who make up the profession regulate their own behavior. These three are intertwined. The specialized knowledge makes it necessary for self-regulation, since others do not have the expertise to assess whether activities conform to the standards of practice. And the ethical ideals, which include a commitment to serve the good of others or society, are supposed to justify the trust society places in professionals to properly regulate themselves. Much work needs to be done to explore this interface between the epistemological, ethical, and social dimensions of a profession.

The potential for addressing this interface is well exhibited in Andrew Lustig's chapter on end-of-life care. He considers the knowledge base required for healthcare workers involved in care for dying patients, explores the ends that should direct such a practice, and addresses the ethical issues of decision-making. Lustig notes that "[m]uch of the emphasis in palliative medicine is a self-conscious effort to overcome the tendencies toward reductionism at work in models of medicine directed primarily at curing disease." To appreciate the challenge, it is worthwhile to recall the deficiencies: physicians are generally unresponsive to patient preferences about care, they do not sufficiently treat pain, and they don't seem to change their behavior even when a person is hired to directly mediate information on these issues (the deficiencies are well exhibited in the SUPPORT study; see Lustig in this volume). To understand these deficiencies, Lustig considers the assumptions implicit in the knowledge base of medicine; namely, how medicine is conceptualized as a practice, how disease is understood, the nature of care and cure, and the ends of medicine. Palliative care – as advanced, for example, by the Hospice movement – involves a reassessment of each of these, explicitly moving away from medicine's traditional "cure" orientation to embrace a notion of "care." After reviewing these "philosophy of medicine" issues, Lustig considers ethical concerns such as those associated with withholding/withdrawing care and euthanasia. While these ethical issues are addressed in a separate section, Lustig highlights the need for their integration with the foundational, conceptual issues in a single philosophy of palliative care.

Osborne Wiggins and Michael Alan Schwartz consider the foundations of the psychiatric profession. Recently, there have been some major advances in psychiatry, largely associated with the development of the American Psychiatric Association's *Diagnostic and Statistical Manual of Mental Disorders (DSM)*, pharmacotherapy, and advances in neuroscience. These successes prompt some to believe that psychiatry can be effectively modeled after biomedicine. Wiggins and Schwartz challenge this belief and the concomitant reductionism in psychiatry. They argue that a concept of the person is needed to unify the scientific and ethical aspects of psychiatry, and the core questions of personhood require philosophical deliberation. However, this deliberation should not be taken as antithetical to science. Wiggins and Schwartz identify two strands of scientific thought, one abstract and reductionistic, the other descriptive and unifying. They argue that the two must work together, and advocate a descriptive phenomenology of human experience as the means by which the science of psychopathology can be grounded.

A common theme in the chapters on professionalism concerns the interrelationbetween the ethical norms and the knowledge base of the profession.

Sara Fry considers this interrelation in her review of the ethics of nursing. In the early years of the twentieth century, the competence of a nurse was largely understood in terms of her moral conduct; she was to perform a given social role. Cooperation with physicians and the health care institution was integral to that role. "After World War II, the role of the nurse changed from being an obedient, cooperative helper to the physician, to being an independent practitioner." Fry notes how in this new context, the "good nurse" was now "defined in terms of technical competence rather than in terms of personal moral characteristics or even the performance of ethical behaviors." Scientific and ethical reasoning were modeled after clinical judgment in medicine. Fry is critical of both of these periods of nursing history. She argues that nursing should not just be defined in terms of a subservient social role, nor should it be defined in terms of technical competence. The science and ethics of nursing must be integrated in an account of the caring professional. She summarizes the features of this integrated account under the four concepts of cooperation, accountability, caring, and advocacy.

In many of the traditional health professions, the primary focus has been clinical: to address the needs of an individual patient with an identified health problem. The last two essays in this volume shift the perspective, addressing preventive and public health needs and focusing on families and populations in addition to individual patients. In his essay on geroethics, Larry McCullough considers two general issues associated with gerontology and the geriatric speciality of medicine and nursing. First, he looks at preventive ethics. Ethics, like medicine, has tended to focus on acute problems. In the case of the elderly, there are issues surrounding competency for decision-making and ageism, for example, which are often addressed at point of crisis. At that stage it is difficult to appropriately resolve ethical difficulties that arise. McCullough suggests that it is better to anticipate the kinds of changes in mental capacity that the elderly undergo, and then configure procedures for decision-making in a way that accounts for these changes. Such an approach may involve a significant challenge to the orthodoxy in gerontology and geriatrics. It is generally assumed that there are diseases associated with aging, but not diseases of aging. While he does not directly contest this assumption, McCullough holds that recent work on impaired executive control functions may show that autonomy diminishes with age. This has implications for how ethical issues are managed.

The second part of McCullough's chapter considers long term care. In the United States, when the functional status of the elderly declines to the point where special care is needed, there is a general expectation that families will provide that care. The burden usually falls to women, whose job prospects and quality of life may suffer. McCullough questions the justice of such arrangements, and suggests that we need a broader debate about the levels of care that are appropriate. Families may make reasonable decisions about how care for the elderly should be balanced against other family needs, and social policy should be modified to respect these reasonable limits. However, this does not mean that the family considerations are overriding, or that the elderly have no autonomy in making the decisions about the care they need. McCullough argues that a more nuanced notion of autonomy is needed, one that doesn't just focus on rights to make decisions, but rather considers the many subtle ways that the values of the elderly might influence their living environment; for example, how a modest risk might be worthwhile for the elderly, in order to maintain a certain independence. By contrast, many family members and health professionals might place too great an emphasis on safety. Such decisions should not be viewed as objective questions of fact – as if there were some single correct answer to what should be done in a given situation. Rather, these tradeoffs involve a "contested reality," where the elderly are at odds with family or a long term care facility when considering what is in their best interest. McCullough suggests that in the end, such care decisions should be based on clinical ethical research which incorporates philosophical reflection with empirical research. "In such a combination of ethics and health services research is to be found the future of ethics and aging. Ethical analysis generates the research questions and also helps to identify relevant end-points for measurement. Accepted methods of health services research are then used in the design and analysis of such research."

In the final essay of the volume, Douglas Weed explores some of the challenges associated with the integration of philosophical reflection with the scientific tools of epidemiological research. While Weed does not want to draw too sharp a contrast between public health and medicine – there are important areas of overlap in method and goal – they are nevertheless distinguished by "the relationship between the practitioner and the recipient(s) of that practice." Public health intervenes in a community, for preventive purposes, and with the intent of lowering the risks of disease. Medicine intervenes on an individual, who usually comes to the physician for help with an already present health need. Associated with this difference are important philosophical and ethical differences. In public health, a community-oriented intervention may take place without the consent of individuals who are subject to the interventions involved. Issues of autonomy, and of how to balance individual and community interests, are thus framed in a different way than in medicine. There are also important differences in the understanding of "cause," "disease" and "evidence," which have implications for when an intervention is justified.

In many ways Weed's essay is a fitting culmination for the volume. Public health is only now struggling to define its philosophical and ethical foundations. The essay captures the horizon of current debate, as developed by one of the prominent voices in that debate. But even more, Weed's analysis goes beyond public health. In many ways, the issues he identifies are also those of all health care. Whether in the form of managed care or in some other variant, the institutionally-based, population-oriented character of health care is becoming more apparent, and this shift brings with it new philosophical and ethical challenges. To respond appropriately to the ethical issues raised by a rapidly developing health care context, the foundational conceptual and values issues need to be clarified and norms for this new context need to be developed.

Philosophers can provide invaluable guidance in accomplishing this task.

Taken jointly, the essays in this volume provide a review of the most important areas of bioethics, giving a sense of the field as a whole from a philosophical perspective. Integral to bioethics is a vision of a more rational and humane health care. The essays in this volume make clear that much work is still needed to realize that vision. Perhaps it is not too optimistic to say that we are on the way. Some things that are seen as irrational and contrary to the vision (e.g., managed care) are, I am convinced, actually important steps forward, although bioethicists have reacted against the developments.□To the extent bioethics is regressive, its own orthodoxy needs to be challenged. In other areas, however, bioethics clearly points the way ahead, and the task is to put the ideals into practice. The philosopher is needed to provide the critical analysis that enables us to distinguish between what is progressive and what regressive, and to make transparent the current state of health care and of bioethics, so that the next step in formation might be taken. My hope is that many others will follow the examples of the authors in this volume, and that a deeper foundational debate on health care practices might again be made vibrant, just as it was when the discipline was in its formative period. This time, however, the field need not begin de novo. Rather, with its criticism and reconstruction, the philosopher's work can be a celebration of a field that has come of age. It is exciting to take stock of the field at just such a time.

ACKNOWLEDGMENTS

An endeavor such as this one can only be accomplished with extensive guidance and support from many people – too many to sufficiently acknowledge here. Among those who have provided invaluable assistance are H. Tristram Engelhardt, who initially conceived of the Handbook project and gave helpful feedback at several stages in the process; Ruiping Fan, who managed much of the review process; and Adrian Polit and Heather Allen, who provided assistance with the reviews and some of the copy-editing. Many reviewers from the Handbook Editorial Board and among the authors of this volume provided assistance in moving the essays from good to great. Special thanks are due to my wife, Cheryl, and to our children, Abigail, Michael, Hannah, and Joshua, who continually supported me and showed understanding, especially when it became clear that this project would consume much more of our family time than any of us had thought.

For this project, I sought to take stock of the field of bioethics at a time when the torch is passing from the early pioneers to the next generation. Unfortunately, during these transitions, there must come a time when these unquestioned early leaders can no longer be with us to provide their presence, encouragement, wit, and wisdom. This is the case for two of the authors of this volume, who have in the past year passed from this life. A special acknowledgment is due them, not just for their brilliant contributions to this volume, but for their contributions to the field.

K. Danner Clouser was the first philosopher to be employed by a medical school to teach bioethics. He brought a keen analytical mind and clarity of thought, but is perhaps best known for his humble demeanor and exceptional wit. People would attend a session he was moderating just to see how he would introduce the speakers – always a hilarious and insightful experience. Recently his work has been justly celebrated in *Building Bioethics: Conversations with Clouser and Friends on Medical Ethics*, edited by Loretta Kopelman (2000). It was a privilege to know him, work with him on the Editorial Board of *The Journal of Medicine and Philosophy*, and have the benefit of his insights and criticism.

As the field was maturing, it benefitted from the untiring efforts of David Thomasma. Author and editor of far too many books and articles to mention, he also served as Editor-in-Chief of two of the prominent journals of the field, *The Cambridge Quarterly of Healthcare Ethics* and *Theoretical Medicine and Bioethics*. Young scholars knew him as a great mentor, advocate, and kind sage. I have benefitted from his generosity and support more times than I can remember, and his death came as terrible shock and loss.

These great men will be missed by all, and they mark a high standard for all who would follow in their steps.

Department of Philosophy and Center for Bioethics
University of South Carolina
Columbia, South Carolina, USA

NOTE

1. Some may object to this characterization, arguing that ethics is very much a part of philosophy, and bioethics is simply applied ethics. For them, the philosopher relates to bioethics in the same way it would to any of its "own" specialities; for example, as the philosopher relates to John Locke and the debate about empiricism. But even here there is a useful lesson. John Locke and the empirical tradition can be read in different ways. The historian of ideas reads him one way, Jonathan Swift reads him another. The fact that they take up that content does not make them philosophers. And when Locke's empiricism gets codified, becomes a part of the culture, and forms as a new orthodoxy (as Sydenham and Locke's medical Nosography did), then it is no longer philosophy, or at least not just philosophy. To the degree that previous philosophical reflection gets incorporated into the

domain it engages, it becomes something else: early natural philosophy became physics; Locke's empiricism became nosographic medicine; and philosophical reflection on healthcare ethics became bioethics.

REFERENCES

American Board of Internal Medicine: 1995, *Project Professionalism*, American Board of Internal Medicine, Philadelphia.

Beauchamp, T and Childress, J.: 1979, *Principles of Biomedical Ethics*, Oxford University Press, New York (5th edition, 2001).

Boorse, C.: 1977, 'Health as a theoretical concept,' *Philosophy of Science* 44, 542-573.

Bryan, C.: 2000, 'Promoting professionalism: a primer,' *Journal of the South Carolina Medical Association 96, 421-427.*

Caplan, A.: 1992, 'Does the philosophy of medicine exist?,' *Theoretical Medicine* 13(1), 67-77.

Cassell, E.: 1976, *The Healer's Art*, MIT Press, Cambridge, MA.

Cribb, A.: 2001, 'Reconfiguring professional ethics: The rise of managerialism and public health in the UK National Health Service,' *HEC Forum* 13(2), 125-131.

Engelhardt, HT: 1986, *Foundations of Bioethics*, Oxford University Press, New York (2nd edition, 1996).

Faden, R. and Beauchamp, T.: 1986, *A History and Theory of Informed Consent,* Oxford University Press, New York.

Flexner, A.: 1910, *Medical Education in the United States and Canada*, The Carnegie Foundation for the Advancement of Teaching, New York.

Gorovitz, S, Jameton, A.J. and Macklin, R. (eds.): 1976, *Moral Problems in Medicine*, Prentice-Hall, Englewood Cliffs, NJ.

Hauerwas, S and Pinches, C: 1997, *Christians Among the Virtues: Theological Conversations With Ancient and Modern Ethics*, University of Notre Dame Press, Notre Dame, Indiana.

Jonsen, A. and S. Toulmin: 1988, *The Abuse of Casuistry*, University of California Press, Berkeley.

Jonsen, A.: 1998, *The Birth of Bioethics*, Oxford University Press, New York.

Katz, J.: 1984, *The Silent World of the Doctor and Patient*, Free Press, New York.

Khushf, G.: 1998 , 'A radical rupture in the paradigm of modern medicine: conflicts of interest, fiduciary obligations, and the scientific ideal,' *Journal of Medicine and Philosophy* 23(1), 98-122.

Khushf, G.:2001, "What is at issue in the debate about concepts of health and disease? Framing the problem of demarcation for a post-positivist era of medicine," in L. Nordenfelt (ed.), *Health, Science, and Ordinary Language,* Rodopi Press, pp. 123-169; 215-225.

Kopelman, L.: 1999, *Building Bioethics: Conversations With Clouser and Friends*, Kluwer Academic Publishers, Dordrecht.

Kuczewski, M.: 1997, *Fragmentation and Consensus*, Georgetown University Press, Washington, DC.

Kuhn, T.: 1996, *The Structure of Scientific Revolutions*, third edition, University of Chicago Press, Chicago and London.

Leder, D.: 1992, 'The experience of pain and its clinical implications,' in J. Peset and D. Gracia (eds.), *The Ethics of Diagnosis*, Kluwer Academic Publishers, Dordrecht.

MacIntyre, A.: 1984, *After Virtue: A Study in Moral Theory*, second edition, University of Notre Dame Press, Notre Dame, Indiana.

McNamara, K.: 2001, 'The status of physicians in the 20th century,' *Journal of the South Carolina Medical Association* 97, 522-525.

Pellegrino, E. and Thomasma, D.: 1981, *A Philosophical Basis of Medical Practice: Toward a Philosophy and Ethic of the Healing Professions*, Oxford University Press, New York.
Pellegrino, E.: 1993, 'The metamorphosis of medical ethics: A 30-year retrospective,' in *Journal of the American Medical Association* (March 3), 269:9, 1158-1162.
Reiser, S.: 1981, *Medicine and the Reign of Technology*, Cambridge University Press, Cambridge.
Rodwin, M.: 1993, *Medicine, Money and Morals*, Oxford University Press, New York.
Rorty, R.: 1989, *Contingency, Irony, and Solidarity*, Cambridge University Press, Cambridge.
Ross, WD: 1930, *The Right and the Good*, Oxford University Press, Oxford.
Ross, WD: 1939, *The Foundations of Ethics*, Oxford University Press, Oxford.
Schon, D.: 1983, *the Reflective Practitioner: How Professionals Think in Action*, Basic Books, New York.
Starr, P: 1984, *Social Transformations of American Medicine*, Basic Books, New York.
SUPPORT Principal Investigators: 1995, 'A controlled study to improve care for seriously ill hospitalized patients: The Study to Understand Prognoses and Preferences for Outcomes and Risks of Treatment (SUPPORT),' *Journal of the American Medical Association* 274, 1591-1598.
Toombs, SK: 1992, *The Meaning of Illness: A Phenomenological Account of the Different Perspectives of Physician and Patient*, Kluwer Academic Publishers, Dordrecht.
Wear, S.: 1998, *Informed Consent: Patient Autonomy and Clinician Beneficence Within Health Care*, 2nd ed., Georgetown University Press, Washington, DC.
Zaner, R.: 1981, *The Context of Self: A Phenomenological Inquiry Using Medicine as a Clue*, Ohio University Press.
Zaner, R.: 1988, *Ethics and the Clinical Encounter*, Prentice-Hall, Inc., Englewood Cliffs, NJ.

SECTION I

THE EMERGENCE OF BIOETHICS

ALBERT R. JONSEN

THE HISTORY OF BIOETHICS AS A DISCIPLINE

The history of bioethics begins during the decade of the 1960s when savants began to converse with each other about the dangerous and difficult aspects of advances in the biomedical sciences. The history of bioethics as a *discipline* may open with a 1973 article by Dan Callahan, founder and director of the Institute for Society, Ethics and the Life Sciences (now the Hastings Center) entitled "Bioethics as a Discipline." Callahan wrote, "bioethics is not yet a full discipline . . . (it lacks) general acceptance, disciplinary standards, criteria of excellence and clear pedagogical and evaluative norms" (p. 68). This very lack, he suggested, offered bioethics an unprecedented opportunity to define itself. It could move toward "definition of issues, methodological strategies and procedures for decision-making." In defining issues, bioethics would need "the rigor of the unfettered imagination," the ability to envision alternatives, to get into people's ethical agonies (p. 71). In developing methodological strategies, bioethics must use the traditional modes of philosophical analysis -- logic, consistency, careful use of terms, seeking rational justifications – supplemented by sensitivity to feelings and emotions as well as to the political and social influences on behaviors. Finally, the discipline must provide procedures "to reach reasonably specific, clear decisions...in the circumstances of medicine and science." Callahan concluded: "The discipline of bioethics should be so designed, and its practitioners so trained, that it will directly – at whatever cost to disciplinary elegance – serve those physicians and biologists whose positions demand that they make the practical decisions" (p. 72). This sort of discipline, according to Callahan, requires a knowledge of the sociology of the profession and of health care, scientific training, historical knowledge of regnant value theories, facility with methods of ethical analysis common in the philosophical and theological communities, and their limitations when applied to cases – an "impossible list of demands," admitted Callahan, but approachable by "a continuing, tension-ridden dialectic . . . kept alive by a continued exposure to specific cases in all their human dimensions" (p. 73).

This present article will begin with a rough definition of "discipline," and trace the elements of that rough definition through the earliest days of bioethical discourse, that is, through the initial attempts of philosophers,

G. Khushf (ed.), Handbook of Bioethics, 31–51.

theologians and a few other scholars during the 1970s, to define and analyze the moral issues of the biomedical sciences and medical practice. The article will then examine a peculiar feature of this discussion, which will be called "the invention of argument," and suggest how that feature shaped the disciplinary character of bioethics. A final section will speculate on the future directions of bioethics as a discipline.

I. DEFINITION OF A DISCIPLINE

Callahan does not define "discipline" but does delineate the major features that characterize a discipline: "definition of issues, methodological strategies and procedures for decision-making." The noun "discipline," the Oxford English Dictionary informs us, means "a branch of instruction or learning"; the verbal form of the same word means "to bring under control." The Latin word from which both are derived means "the act of teaching or imparting knowledge" and a "discipulus" is a student. An act of teaching requires that the mind of the student be brought under control and that the information also be under control. The OED cites a 1650 text, "Objective disciplines be principally four, Theologie, Jurisprudence, Medicine, Philosophy." So by that date, a "discipline" organized segments of information into compartments that were arranged in an order suited to teaching learners. Those compartments had to be comprised of a certain subject-matter and a method for organizing it. Those "objective disciplines" came into being because of the "rage for order" that inspired the medieval scholastics, the renaissance classicists and enlightenment rationalists: definition, distinction, classification, logical exposition created the compartments into which diffuse information that they had inherited from the minds of the past could be separated and parceled out in the teachable packages of lectures and treatises. Their constructions have existed even to our academic times, with extensive modifications and many additions.

The classical concept of a discipline, however, has become increasingly difficult to maintain. In the modern university, the clear lines of many classical disciplines have diffused into mosaics: even the most definitive ones, such as mathematics and physics, are complex collections of sub-disciplines with quite diverse theories, methods and even definitions of the field. It is not uncommon that professors within the same department, and nominally within the same discipline, have very little common language or few common concepts. Clear subject matters have fragmented into parts that often bear but a nominal relation to the whole: Euclidean geometers and Mobius strip topologists may dwell in the same department but have little to say to each other. New disciplines arise from the merging of old ones: chemistry and biology blend into biochemistry, biology incorporates the physics of crystals to become molecular biology. The lines between the new and the old blur. Methodologies proliferate, so that within the discipline of History or of Literature adherents of classical, modern and post-modern methods may argue about whether the discipline even exists. Thus, to ask whether bioethics is a discipline raises a complex question.

Callahan had been trained in a classical discipline, philosophy. He envisioned the Institute for Society, Ethics and the Life Sciences, which had opened in 1970, as a revolving gathering of persons educated in various disciplines – philosophers like himself, theologians, legal scholars, physicians and biomedical and social scientists – who would discuss the ethical, social and legal questions raised by the new biology and the new medicine. He hoped for a genuine "interdisciplinary" conversation that might illuminate those problems and even manage them through informed policy decisions. During the first few years, the Institute convened many such persons. Their conversations – about topics such as behavior control, genetic engineering, population control, death and dying – were illuminating, as might be expected from intelligent and accomplished persons but they also generated the "impossible lists of demands" that Callahan enumerated in his article. An "interdisciplinary" conversation reveals quickly that the conversants summon fields of knowledge, experience, and sensitivity that have very little communality. A philosopher, conversing with a legal scholar, realizes that he needs to learn more about jurisprudence; as he does, he finds overlaps and gaps between his own discipline's concepts and methods. A sociologist, talking with a theologian, finds that they each mean something quite different by a key word such as "religion," and they attempt to find common ground for understanding. All parties recognize quickly that they do not share any disciplinary method to move investigations toward conclusions. Ironically, any serious interdisciplinary conversation generates a demand for a discipline, a method for defining issues, marshaling data and arguments and moving toward conclusions. It was this demand that Callahan recognized in his "Bioethics as a Discipline."

Early bioethicists became aware of that "impossible list of demands." As soon as they began to discuss the issues with concerned persons, they realized that every question led them into unfamiliar territory. Philosophers found that the levels of abstraction demanded by their discipline flew high above the problems posed by the practitioners of medicine and science. Theologians sensed that the doctrinal commitments from which their discipline flowed were not acceptable to all participants in the conversation. Sociologists realized that the descriptive capacities of their discipline did not close normative gaps. Lawyers learned that the law's minatory face frightened off many to whom they spoke in good faith. Scientists and physicians, often encased in the epistemological positivism and ethical skepticism pervasive at that time, could not easily transcend their highly subjective view of values. It was difficult to imagine a new discipline emerging from any of these traditional disciplines; it was equally difficult to conceive how they could talk to each other with interdisciplinary congeniality.

The early bioethicists, however, were not deterred. Although they all came from disciplinary bases and could not easily abandon disciplinary habits, they tried to shed the more chauvinistic ones. The theologians restrained their reference to scriptural or doctrinal sources and sought to translate the lessons that those sources taught into ecumenical formats. They tempered their use of theological language, although some of that language, such as "sanctity of life," had drifted into secular discourse. The philosophers became less arcane, trying to step down from the rarefied air of deontological and teleological theory into the world of moral discourse more familiar to non-philosophers. The lawyers ventured from the sure paths of procedural justice into unsettled law where little precedent ruled. The physicians and scientists opened their minds to the "soft" data they had previously ignored. These first tentative steps out of familiar disciplinary worlds led the early bioethicists on the paths that Callahan had indicated: a search for theory, an articulation of principles and the formulation of methods for making decisions (Jonsen, 1998).

While all of the early bioethicists came from traditional disciplines and were, almost to a person, academicians, two groups brought disciplinary methods that had been shaped specifically for the analysis of moral issues, the philosophers and the theologians. Moral philosophy had always been one of the integral branches of philosophy; all religious denominations had always attended to the moral dimension of the human relation to God. It was a truism to say that philosophers and theologians, among the academic disciplinarians, were uniquely familiar with moral issues. Closer examination revealed that the philosophers of the time had lost contact with the long tradition of moral philosophy and were struggling to rediscover how to "do" moral philosophy. The theologians took many different approaches to moral issues and many theological scholars were struggling with fundamental questions about the nature of religious morality. It is not an exaggeration to say that, in the 1960s, moral philosophers and moral theologians were excited by the questions raised by the new biology and medicine and, at the same time, poorly equipped to bring disciplinary strengths to the analysis of those questions.

The moral philosophy of that era suffered from lack of long-term memory. Certainly, every philosopher could recall Aristotle's *Nicomachean Ethics* and Kant's *Metaphysics of Morals* but those venerable works and the many others that were catalogued as moral philosophy had lost their allure. For many a modern moral philosopher, G.E. Moore's *Principia Ethica* was just what its title announced, the *beginning* of ethics. Moore, writing in 1903, had challenged the most fundamental assumptions of ethical reasoning, as it had occurred in the Western philosophical tradition: "good" cannot be defined by reference to any natural property, such as happiness, utility, or perfection (Moore, 1903). The subsequent debates about the "non-natural property" that, in Moore's view, had to ground ethical affirmations lead to a denial of any meaning to ethical affirmations. The logical positivists of the Vienna Circle and Cambridge announced that ethical affirmations were nothing but expressions of personal emotion. This opinion spread throughout Anglo-American moral philosophy during the 1930s and 1940s. Moral philosophers, intrigued by these theoretical problems, engaged in "metaethics," the logical analysis of the meaning of ethical terms, and drew away from analysis of any particular moral problem.

Metaethics did not totally absorb normative ethics, which continued to claim that it attended to the analysis of particular moral problems. Yet even those philosophers who remained interested in normative questions did so at a highly speculative level. They had to work under the constant shadow of metaethical critique and thus had to be cautious about the "truth value" of any affirmation. They remained on the relatively safe ground of normative theory, that is, the description of the structure of ethical arguments. One such theory, utilitarianism, had benefited in the late nineteenth century from an exquisite analysis by Henry Sidgwick. His analysis had laid out the moves of utilitarian argument so precisely that subsequent philosophers were lured to further critical analysis and by mid-twentieth-century, rehearsing the strengths and weaknesses of utilitarian or consequentialist ethics was a familiar métier of moral philosophers (Sidgwick, 1874). Utilitarianism had to assault the massive construction of Kant's Categorical Imperative and, although Kant's argumentation had become less persuasive to modern moralists, the persistence of ethical imperatives that contradicted consequentialist reasoning had to be accounted for. C.D. Broad had suggested that all forms of argument about obligation could be distinguished into "deontological" and "teleological" (consequentialist) forms and by mid-century, this distinction become canonical in normative ethics (Broad, 1934). Still, moral philosophers roamed through these constructions of reason and stayed far from moral experience.

Some moral philosophers were aroused from their "dogmatic slumber," as Kant said Hume had awakened him, by the public events of the American 1960s. The civil rights movement swept through the nation. Peaceful demonstration and violence reinforced moral claims to equality among the races and demanded redress of terrible injustices. The doubts of citizens about the wisdom and the rightness of the war in Southeast Asia grew into storm clouds of public protest that pushed a president from office. Both civil rights and war protest, while profoundly political, drew on deep moral convictions. Hence, intense moral conversations engaged citizens everywhere, but nowhere more vigorously than in the universities. Professors of philosophy were lured out of the ivory tower into the maelstrom of debate. Many realized that their discipline did not take them very far on the path toward public moral discourse. In 1971, a cadre of leading philosophers founded a new journal, *Philosophy and Public Affairs*, in which they attempted to move toward the practical dilemmas of moral experience.

The bioethical questions, although not as desperately urgent as racial discrimination and military violence, were at least less entwined with political matters and more susceptible to quiet reflection. So, while a new interest in practical ethics was stimulated by public events, one area in particular formed into something like a discipline. The entry of a few moral philosophers into the world of medicine and science allowed them to bring at least some of their disciplinary virtues to the questions being asked. They had been trained to deconstruct the faulty logic of arguments and to demand clarity of concept and justification of claims. These are the intellectual virtues that build disciplines. They contribute to the definition of issues by seeking for explanatory theories and principles. They also contribute to the development of a method of relating data and theory in ways that can lead to conclusions. Philosophers had always done something like this but the philosophers trained in the latter half of the twentieth century, inspired by the power of logic, were particularly focused on the epistemological and logical requisites for rigorous argument. The turn to practical problems in politics and in bioethics drew them further toward the logic of decision and its intrinsic demand for the integration of fact and opinion into convincing argument. This contact with medical cases that cause moral perplexity inspired philosopher Steven Toulmin to write "medicine had saved the life of ethics" (Toulmin, 1982).

The second group of scholars dedicated to the study of moral questions were the theologians. Roman Catholicism has a distinct discipline of moral theology, which took formal shape in the seventeenth century. It was tightly constructed, with a doctrine of natural law at its theoretical base and a system for addressing particular cases, or "casuistry" as its methodology. At the time bioethics appeared, that venerable system was under internal attack. Many moral theologians found it too rigid, too dependent on Church law, despite its claimed adherence to reason, and too lacking in sensitivity to the human condition. The great church council of the 1960s, Vatican II, had sanctioned a reform in moral theology that enthused many scholars and sent them to search for new ways of formulating their discipline. While a few of these reformers took radical paths, most sought to preserve the strong edifice of the old moral theology while opening it to refreshing analyses of concepts and arguments. Catholic moral theologians who came to bioethics brought a tradition of articulate moral argument and an openness to reexamination of principles in the light of critical logic and of unprecedented factual circumstances in human life and society (McCormick 1981).

Protestant theology had never developed a strong theoretical and practical moral theology (with exception of the Anglican and Lutheran traditions). Although deeply concerned with the Christian moral life, moral questions had been incorporated in the broader fields of biblical and systematic theology. Protestant denominations had abandoned the practice of private confession and thus no longer needed the acute examination of personal behavior that Catholicism required. Moral admonition took broader, more sweeping forms, suited to the pulpit rather than the confessional. In the early years of the twentieth century, a movement called Social Gospel inspired many Protestant theologians and congregations. Its advocates called for a deeper engagement of Christians, as individuals and as churches, in the reform of society according to biblical principles of social justice. The influence of this movement revitalized the study of Christian ethics and, in the era when bioethics appeared, scholars such as the Niebuhr brothers had inspired many theologians to formulate a more articulate and analytic approach to Christian ethics. In doing so, they sought illumination in critical biblical studies, in sociology and political science and even from the distant world of moral philosophy, particularly in its linguistic aspects (Verhay and Lammers 1993; Camenisch, 1994).

The Protestant ethicists who came to bioethics did not bring a discipline. They brought a tradition of concern about human morality at the practical level of personal and social life, and a diverse set of intellectual approaches to the problem of how broad moral injunctions touched the conscience and choices of individuals. Like the Catholics, they were themselves in the midst of a revolution in their own field and so brought as many questions as answers. Jewish scholars brought to the bioethical discussion an ancient tradition of moral argument. As they endeavored to find in Talmudic texts authoritative justification for moral decisions in medicine, they also reflected on the essential features of their method, particularly, the reference to precedential opinion (Rosner and Bleich, 1979). To venture a broad generalization, the philosophers came to bioethics with a sense of critical argumentation about morals but with little sensitivity to moral experience; the theologians came with doctrine, with great sensitivity to moral experience and with a consuming curiosity about the possibility of critical argumentation. At the convergence of bioethics, the philosophers and the theologians were forced to talk about the moral questions with each other (not always a fluent conversation) and with others who knew nothing about philosophical or theological elucabrations concerning morality. The moral philosophers and the theological ethicists could not retire into their own arcane worlds of discourse. They now joined panels at conventions of scientists, participated in committees with other scholars and were appointed to commissions by public officials. They had to ask themselves not merely whether they could communicate intelligibly about the moral issues (most educated persons can do that), but whether their particular disciplinary inheritance could contribute significantly to the "definition of issues, methodological strategies and procedures for decision-making."

Bioethics moved quickly into the classroom. By the early 1970s, a number of medical schools had appointed "ethicists" to teach courses and college philosophy departments found bioethics a popular offering for undergraduates. Teaching stimulates disciplinary organization of material; syllabi and textbooks demand an order with some theoretical and methodological basis. The earliest attempts are crude. One of the first textbooks, *Moral Problems in Medicine* (Gorovitz, *et al.,* 1976), created by a group of philosophers, draws philosophy and the moral problems of medicine together simply by setting side by side articles on the moral problems, usually written by scientists or doctors, and excerpts from the classical treatises of great philosophers. For example, an article by a physician on the difficulty of telling the truth to cancer patients is juxtaposed with the famous discussion of the deontological nature of truth-telling by Immanuel Kant. The editors made little effort to integrate; rather, they hoped that students, after reading both selections, would argue about the problems in the light of the principles. Yet, a first step toward disciplinary organization had been achieved; definition of issues was advanced as ethical reflection and moral experience were moved into approximation.

The test of this technique came in a setting more demanding than the classroom. Congress established the National Commission for the Protection of Human Subjects of Biomedical and Behavioral Research in 1974 and charged it "to identify the ethical principles which should underlie the conduct of biomedical and behavioral research with human subjects and develop guidelines that should be followed in such research." A public body, two of whose eleven members were trained in theological ethics, was given a properly philosophical task. It sought help, asking three moral philosophers and two religious ethicists to explain what it meant to "identify ethical principles" and to suggest what principles might be appropriate for the research enterprise. Several of the essays produced by those scholars are textbook examples of academic ethics, examining what "principle" means and how principles function in various sorts of discourse. Several other essays go further, plunging into the sort of activity called clinical research and seeking a moral structure, that is, a principled way of moving from the ethical question to an ethical answer or, as Callahan had said, "procedures for decision-making." None of the theological authors referred to properly theological sources of morality, such as commands of God or doctrines of the church. Rather they offer the same general considerations as do the philosophers. These essays educated the Commissioners who debated the issues in their light and eventually produced *The Belmont Report.* This report identified as principles "respect for persons," "beneficence," and "justice" and joined to them three "applications," namely, informed consent, risk-benefit assessment and fair selection of subjects. The consulting ethicists and the Commissions had pushed ahead the task of creating a discipline (National Commission, 1979). Their accomplishment was the beginning of a critical discussion about principles and their application to practice that resounded among the scholars for years to come. The principles first enunciated in the Belmont Report were elaborated and refined by two scholars associated with the Commission, theologian James Childress and philosopher Tom Beauchamp, in the successive editions of their influential textbook, *Principles of Biomedical Ethics* (Beauchamp and Childress, 1979).

Among the attributes of a discipline is theory, that is, a set of propositions that explain how the data with which the discipline deals are identified, fitted together and evaluated. It is in light of theory that Callahan's first requisite, "definition of issues" becomes operational. Theory "building," once seen as the preeminent talent of philosophers, had fallen into disfavor with the analytic philosophers, but, at the time bioethics was beginning, theory had achieved renewed respect with the appearance of philosopher John Rawls's *Theory of Justice* (Rawls, 1971). Several of the new bioethicists believed that a theory of bioethics was appropriate, even at the immature stage of the discourse. Robert Veatch based his *Theory of Medical Ethics* on the Rawlsian version of an ethical contract theory that had deep roots in the history of philosophy (Veatch, 1984). H. Tristram Engelhardt reworked Kant's thesis of autonomy as the essence of the moral life into his *Foundations of Bioethics* (Engelhardt, 1986). Both books contributed an essential element to the disciplinary nature of bioethics: they proposed theoretically grounded arguments for the prioritization of principles. Both books, while quite distinct in approach, gave priority to non-consequentialist principles, particularly, respect for the freedom of individuals, over consequentialist and utilitarian considerations. Although other ethicists may disagree with these theoretical formulations and their implications, they were given a clear exposition against which to project their critique. This ability to bring clearly articulated criticism against clearly articulated theory is an important feature of a discipline, since it generates even more clearly articulated theory and counter-theory.

No theory dominates bioethics and no methodology has won universal acceptance. However, during the past thirty years, the original topics of bioethics, namely, human experimentation, genetics, care of the dying, transplantation and reproductive technologies have continued to be debated as new problems appear and new issues, such as the ethics of managed care, have engendered new debates. Many public commissions and professional committees have been charged with reports on these issues. A large scholarly literature grows constantly. Many courses and programs in professional schools, undergraduate colleges and even secondary schools are offered. All of this fits roughly into the form of a discipline: issues are defined by reference to commonly accepted principles based on general, though vague, theory, and a medley of methodologies are used to move through questions, opinions and data toward conclusions.

II. INVENTION OF ARGUMENT

As these activities grew, many persons with no formal training in the ethical disciplines engaged in the discussions. Many other persons were enticed to study the literature of philosophical and religious ethics and became amateur ethicists. Those formally trained in these disciplines, however, continued to play an important role in the formulation of issues and arguments. Yet their contribution was not quite what might have been expected. Trained experts in, say, engineering or economics, usually bring the honed skills and methods of their discipline into broader discussions. The ethicists, as we have shown, did not have such particular skills and methods, although they had a familiarity with general questions about moral argument and, among religious ethicists, a commitment to certain substantive moral positions. The ethicists in bioethics did something unexpected: they began, quite unconsciously, to revive a long neglected function of ethics, namely, the invention of argument.

"Invention" was a central concept of classical rhetoric. Cicero, most eloquent of the classical rhetoricians, wrote, "Invention is the primary and most important part of rhetoric," and defined it as "the thinking out (excogitatio) of valid or seemingly valid arguments to render one's cause plausible (probabilem)" (Cicero, 1976, p. 345, 19). A modern philosophical rhetorician defines invention as "the art of discovering new arguments and uncovering new things by argument ... (it) extends from the construction of formal arguments to all modes of enlarging experience by reason as manifested in awareness, emotion, interest, and appreciation" (McKeon, 1987, p. 58). All classical rhetoricians devoted lengthy treatises to Invention, explaining the ways in which an orator, usually in a forensic or political case, might "discover" the arguments that will "make the case." The classical rhetoricians offer a plethora of rules and methods to guide invention.

These rules and methods were based on the "topics," literally, the "places," where arguments are found. These "topics" of classical rhetoric are resources for invention of argument. The rhetoricians distinguished topics into common and special. Common topics are the constant, intrinsic features of all reasonable discourse, regardless of the subject matter. These were stating definitions, making distinctions, reasoning by induction or deduction, offering examples, proposing analogies, comparisons and contrasts by similarity, difference and degree, stating relationships of cause/effect, of antecedent and consequent, describing circumstances possible and impossible, past and future. These are developed as ways of arguing about any issue: they have no content or matter of their own, but receive all content and matter from the subject under discussion. The special topics applied only to discourse about certain subject matter. Thus, in a judicial case about a crime, causation, occasion and motive would be central topics; in a discourse on foreign policy, the national interest, feasibility, effects on other nations, would be essential elements. These, in contrast to the common topics, have a subject matter that derives from the kind of issue under discussion.

The common and special topics provided frames, already constructed by the nature of reasonable discourse and the subject under debate, within which the particular circumstances of persons, times and places posed by the case at hand could be set. The orator then had the freedom to select, order and emphasize, those particulars within the frame of appropriate topics. In this way, orators invented arguments whereby they moved from the statement of the case to confirmation or proof which, as Cicero wrote, "marshals arguments to lend credibility, authority and support to our case" (Cicero, 1975, p. 69).

The word "invention" or "discovery," even in this technical sense, can be somewhat misleading. It brings to the contemporary mind the image of the products of ingenious creators of mechanical devices or the startling insights of brilliant scientists: Edison invented the incandescent bulb and Watson and Crick discovered the double helix. Perhaps the concept of improvisation, illustrated by musical inventiveness, might better explain to contemporary readers the nature of classical rhetorical and ethical invention. "Improvisation," says a dictionary of music, "is the invention of music at the time it is being performed ... on the spot, without being written down" (Ammer, 1995). This is, of course, the way most music has been made through human history and it is the way much of the best jazz is made today. In the 17th and 18th centuries, as composers perfected the concerto form for orchestra and solo instrument, they often allowed the soloist an opportunity to show technical skill by departing from the composer's notation and playing freely for some time. These "cadenzas" usually came just before the end of the first movement, following the statement of themes and their recapitulation, so that the pianist might pick up the melodies already established in the notated score and modify them in harmony, rhythm, modulation of key and phrasing. The pianist now becomes an improvisationist, allowed to depart from the notation of the composer's score, but still restrained within certain limits as he or she creates music extemporaneously. Melodies, while varied, are still heard; keys are modulated but not forgotten. The sounds of the improvisation must, in some definite way, echo the sounds of the score. Improvisation allows the virtuoso to stray, wander, explore, but not too far from home. It departs from the composition and must return to it and, indeed, even as it flows from the artist's virtuosity, it must remain at least remotely true to the composer's inspiration.

Improvisation is like invention in that both take definite material as the base and frame for creative interpretation. Rhetorical invention finds the common and special topics suited to the case and allows the orator to work creatively with the actual circumstances of persons, times and places. The orator invents with standard materials and with the unique elements that a particular case presents. Musical improvisation provides musicians with melodies and key, then lets them display notes, harmonies and rhythms that expand, emphasize and reorder the original set composition. The simile is not, of course, perfect, but it begins to reveal how ethicists work with their materials, which are the general concepts of morality and the particulars of issues, questions and cases.

III. ETHICS AS IMPROVISATION

Contemporary ethicists do ethics the way a classical pianist improvised a cadenza in a Mozart concerto; the nature of classical improvisation is similar to the task of practical ethics. The classical rhetoricians proposed that rhetoric was, as Aristotle said, "a combination of logic and ethics" (Aristotle, 1941, Rhetoric I, 4, 1359b10). The chief end of rhetoric was to persuade persons to live "the good life" by reason, emotion, example and argument. Thus, its prime and principle part, invention, must also be the prime and principle part of doing ethics. However, the invention of arguments in practical ethics is properly improvisation because, like the classical soloist, the ethicist must improvise by moving from themes already laid down.

Ethicists must improvise. The very nature of ethical discourse and reflection demands it. Improvisation consists in the movement of mind as it seeks to understand concepts and arguments as they appear in particular issues and cases. Aristotle, writing about rhetoric, said, "whether our argument concerns public affairs (or base and noble deeds, or justice and injustice), we must know some, if not all, of the facts about the subject about which we argue. Otherwise we can have no materials out of which to construct arguments" (Aristotle, 1941, Rhetoric II, 22, 1396a5). Note that the material of the argument comes from "the facts about the subject." This might seem peculiar to those who are committed to ethical theory and/or theological premises as the source of ethical arguments. Aristotle certainly does not mean to deny that those constructions are relevant; after all, his *Ethics* and *Politics* are landmark attempts to formulate similar universal truths about life and society and even the *Rhetoric*, in which he makes this statement, contains a mini-treatise on ethics and politics. Still, the "facts of the subject" are, in his view, an indispensable source of the material of argument. The argument is invented and improvised out of these facts; the facts are brought into contact with the broader considerations by the rational ingenuity appropriate to that sort of argument.

Some clarifications are in order. First, the phrase "facts of the subject" is a poor translation of Aristotle's Greek, which comes closer to "the way things are." He is not thinking of facts, as we do today: the supposedly hard edged bits of the material world that one offers in evidence as judicial proof or scientific data. Rather, "the way things are" refers to events in time, institutional structures, cultural conventions, professional practices and many other features of the world of personal and social experience. It is for this reason that Aristotle said, "he who would invent arguments must know some, if not all, the facts about the subject under discussion." This does not mean knowing merely the "data" of the case, but also the nature of the contingent activity, practice or institution under analysis. Thus, one cannot invent arguments about medicine without knowing what medicine is and how it proceeds in its application to the care of patients; one cannot invent arguments about banking without knowing what banks are and how they work. The information contained in these understandings is not only empirical – what is, for example, a surgical operation as distinguished from a medical intervention, or what is rate of interest, but also valuational – under what circumstances of risk would medicine be preferred to surgery, or a line of credit to an equity loan. Within the factual and valued realm of these practices and institutions, arguments are invented. They are invented, in the sense of discovered or found, because they lie within the practices of those institutions (and not of others); but they are invented, in the sense of improvised, by being formulated, justified and criticized by the methods of rhetorical argument. The arguments become regulated reflections or methodic considerations in the context of the subject under scrutiny.

Even the "real facts about justice and goodness," hardly empirical data, must be known and, says Aristotle, "this is the only way in which anyone ever proves anything ... by trying to think out (that is, invent) arguments for special needs as they arise." This invention, however, does not ignore the broader considerations, as the improvisationist does not ignore the themes established by the composer. Aristotle, Cicero and the other rhetoricians give large place to general understandings and appreciations of human nature and the human condition, but never do they pretend that the arguments about the goodness and justice of this or that action, this or that event, can be ascertained by reference only to those general considerations. The activity of invention and improvisation is essentially an exercise of rational ingenuity. Rational ingenuity is the drawing of the broader considerations into the particular issue under scrutiny.

The contribution of philosophers and of theologians to the early bioethics was essentially similar. Scholars from both disciplines "invented and improvised arguments" by exercising rational ingenuity. When they came to the particular issues, they were forced to ask themselves which "broader considerations" were appropriate to the issue and how broadly those considerations should be construed to make a reasonable argument. The first genuine philosophical analysis of a bioethical issue appeared at the very dawn of bioethics: Philosopher Hans Jonas's essay, "Philosophical reflections on experimenting with human subjects" (Jonas, 1967). Jonas was a philosopher trained in the European tradition who had not been frightened by logical positivism's bringing strong theoretical considerations into moral argument. Against a utilitarian defense of the use of human subjects for the benefit of society, he proposed a deontological defense of the free engagement of persons in activities that affect them. He reviewed how a respect for personal freedom should structure the practice of research in various situations and how research so structured should respond to the goal of betterment of society. His invention and improvisation of arguments plays within the established themes of personal liberty and communal welfare. Similar essays appear among the commissioned studies of the National Commission. Philosophers and theologians examine a topic, such as research with children, by drawing general considerations, such as the requirement for consent, into the details of the sort of research that might generate positive benefits for children at large. As these inventions and improvisations are played out, the notions of consent and assent, together with the risks and benefits of various sorts of research interventions, are refined and sharpened. The Commission drew on all these considerations to craft detailed recommendations that would allow certain sorts of research and prohibit others (National Commission, 1973).

The ethicists did not only revive an Aristotelian/Ciceronian view of ethical argument as invention. They also, almost without attribution, resuscitated the ethical thought of two great American philosophers, William James and John Dewey. William James was the dominant figure in American philosophy from the 1870s to his death in 1910. James opened his essay "The Moral Philosopher and the Moral Life" by repudiating "any ethical philosophy dogmatically made up in advance ... there can be no final truth in ethics any more than in physics, until the last man has had his experience and said his say." He then described three questions in ethics: the psychological question, about the origin of moral ideas, the metaphysical question, about the meaning of such words as "good," "evil" and "obligation," and the casuistic question, about the "measure" of the goods and evils which men recognize. He affirmed an experiential and an intuitional origin of moral ideas and gives to moral words a meaning rooted in the experience of the demands which persons make one upon the other. The casuistic question arises when we realize that the demands persons make one upon the other are multiple, various, and often incompatible, and we are forced to ask how they can all be fulfilled. After reviewing the various answers given by philosophers, he proposed that

> [T]he best, on the whole, of these marks and measures of goodness seems to be the capacity to bring happiness. But, in order not to break down fatally, this test must be taken to cover innumerable acts and impulses that never aim at happiness; so that, after all, in seeking for a universal principle we inevitably are carried onward to the most universal principle, namely, that the essence of good is simply to satisfy demand (and) since all demands cannot be satisfied in this poor world, the guiding principle for ethical philosophy must be simply to satisfy at all times as many demands as we can (James, 1891).

James then issued an imperative: "Invent some manner of realizing your own ideals which will also satisfy the alien demands – that and that only is the path of peace!"

Despite this encomium for human moral creativity, James warned that the "laws and usages ... in our civilized society" demonstrate the most satisfying solutions to problems, so that "the presumption in cases of conflict must be in favor of the conventionally recognized good." Those goods are, to use our simile with musical improvisation, the themes and the keys with which the ethicist must play. Still, experience will show again and again that new solutions to new problems must be invented, so as to assure the continued mutual satisfaction of as many demands as possible. So, he concluded, "ethical science is just like physical science, and instead of being deductible all at once from abstract principles, must simply bide its time, and be ready to revise its conclusions from day to day" (James, 1891).

James's view of ethics is preeminently American. It is conservative and religious, though not credal or denominational. At the same time, it is inventive, creative, strenuous. It accepts the worlds of nature and society, yet is ready to change and renew them. It recognizes problems in the facts of situations and resolves them by understanding the facts and exerting the energy of human intelligence and will. Although James, together with almost all the classical American philosophers, ceased to be read in formal academic philosophy for many decades, his genius somehow diffused itself and reemerged, even unnamed, in much of the style and spirit of bioethics.

John Dewey also anticipated the style of bioethics. His philosophical work is enormous and protean and, like that of William James, touches ethics at many points. For over half a century, he incessantly commented on, and engaged in, public affairs. It was said of him, "So faithfully did Dewey live up to his own philosophical creed that he became the guide, the mentor and the conscience of the American people: it is scarcely an exaggeration to say that for a generation no issue was clarified until Dewey had spoken" (Commager, 1950 p. 100). In his *Ethics*, he wrote,

> The moral act is one which sustains a whole complex system of social values; one which keeps vital and progressive the industrial order, science, art, and the State ... It is not the business of moral theory to demonstrate the existence of mathematical equations, in this life or another one, between goodness and virtue. It is the business of men to develop such capacities and desires, such selves as to render them capable of finding their own satisfaction, their invaluable value, in fulfilling the demands which grow out of their associated life ... Such a person has found himself and has solved the problem in the only place and in the only way in which it can be solved: in action" (Dewey and Tufts, 1909, pp. 393, 396).

A major Deweyan thesis appears in the words, "A moral principle is not a command to act or forbear acting in a given way: it is a tool for analyzing a special situation, the right or wrong being determined by the situation in its entirety and not by the rule as such." Writing a decade later, he expanded that notion: "Morals must be a growing science if it is to be a science at all, not merely because all truth has not yet been appropriated by the mind of man, but because life is a moving affair in which old moral truth ceases to apply. Principles are methods of inquiry and forecasts which require verification by the event ... Principles exist as hypotheses with which to experiment." He saw "moral science, not as a separate province, (but as) physical, biological and historical knowledge placed in a human context where it will illuminate and guide the activities of men" (Dewey, 1922, p. 239, 296). Like James, Dewey repudiated a finished system of ethics. The ethical response was always a reflective and creative solution to a problem that presented itself in a unique way. Morality arises with desire and is fulfilled in critical assessment of desire. That critical assessment is the work of intelligence. "That means," says one commentator, "between the problem and its solution, we must always interject a rational method. In the moral situation this means that between casual desire and values approved on reflection there must come a process of knowledge and critical inquiry. Science in this case, knowledge of the potentialities and powers of things together with basic insight into the nature of man and his needs instead of being alien to ethics, turns out to be indispensable" (Smith, 1963, p. 140).

Neither William James nor John Dewey created a system of moral philosophy; rather they created a style of moral philosophy. Moral philosophy was to be an incessant inquiry into the way humans, conceived as bundles of habits vitalized by a curious intelligence and creative freedom, could live in the world of physical nature and cultural creation and could change and challenge that world to meet human purposes. The critical edge of moral philosophy was set against barriers and boundaries in thought, culture, and religion that obstruct human openness and the expansion of mind and mores. There was no critical edge set against progress and novelty, however, and the meaning of these goals remained vague. The Pragmatism that James and Dewey shaped celebrated the dignity of free persons who creatively meet the challenges of their lives and use intelligence and science to direct their evolution. James and Dewey were unquestionably the dominant intellectuals in America for decades. Their thought was not merely speculative, but highly practical and, as recent commentators have convincingly shown, their contributions to "public philosophy" flowed logically from their philosophical convictions. Their spirit and style reemerges, its origins almost unrecognized, in the ethics of bioethics.

IV. THE FUTURE OF BIOETHICS AS A DISCIPLINE.

Bioethics has assumed, in the quarter century since Callahan wrote his essay, all of the trappings of an academic discipline. It has professors. It has textbooks. It has scholarly journals. It has graduate programs. These external trappings cover other more profound elements of a discipline, at least as disciplines now exist in the American university: an interminable argument about the theories and methods which the professors of the discipline should use as they go about their research and teaching. The arguments presuppose that there are alternative theories and methods about which critical claims can be made; they also allow for innovative approaches. These arguments, which go on in every lively disciplinary setting, are symptoms of disciplinary health; if they were to be resolved, disciplinary arteriosclerosis would set in.

During the forty or so years of bioethical discourse, the field has partitioned into somewhat distinct specialties: clinical ethics attends to the moral dilemmas that arise in the care of patients, the ethics of health policy focuses on the ethical aspects of social and political implementation of advances in medicine, science and health care. Finally, philosophical bioethics explores the deeper implications of science and medicine for the human condition. Some bioethicists are fluent in all these specialties; others work and write largely in one or the other. Also new issues have moved to the center of attention. Biomedical experimentation with humans and care of the dying were agenda items for the earliest bioethicists. In recent years, health policy, delivery of health care under various economic and business schemes, such as managed care, remarkable advances in reproductive science, and the growing influence of genomics on medicine and science have summoned the attention of bioethicists. These have fostered the development of "sub-specialties" among bioethicists. The factual complexity of these areas and the elaboration of ethical arguments comprehensible to the participants in the discourse require specialized rather than general knowledge and skills. Each of these specialties and subspecialties calls for invention of different arguments.

Bioethics has accumulated a rich treasure of moral arguments, unprecedented since the casuistic literature of the 16th and 17th centuries. A vast collection of critical studies of practical ethical questions has been produced. The actual arguments that have been constructed about these many issues deserve closer examination. The way in which empirical data is invoked and integrated into the argument should reveal more clearly how "the way things are" originate and influence argument. Similarly, specifically moral notions, such as principles and virtues, and the manner in which they are conceptually sharpened to fit the issue under analysis can advance our understanding of the movements of moral reasoning in practical situations. The manner in which circumstances contribute to the weight and weighing of principles and values is worthy of more precise analysis. A persistent philosophical problem, namely, moral relativity, might be reexamined in the light of the chain of argumentation that has been forged around particular questions; for example, the evolution of arguments about assisted suicide over the last decade shows sophisticated ethicists stepping carefully among arguments and counterarguments to find firm ground. Questions in reproductive technology, such as cloning, reveal similar moves. Positions are changed, but do those changes reflect abandonment of fundamental values or some other influences on ethical judgment?

However, Callahan's words of 1973 are still true: "bioethics is not yet a full discipline." He spoke of the absence of disciplinary standards, criteria of excellence and clear pedagogical and evaluative norms. These features have matured notably over the years, but bioethics is not yet a discipline in another sense. It might rather be called a "demi-discipline." It is half a discipline, concerned with theory and method and standards, and half a public discourse, encouraging vigorous and often disorderly debates in public and political settings. The discipline takes its shape and its vitality from this public discourse. Its issues arise in the world of medicine and science and are defined by the actualities of these enterprises. The people who first experience the issues and who respond to them are the professionals and the patients, the politicians and the public and, only then, do the bioethicists encounter them, already given a certain form by the public discourse. Even as the discourse moves into the discipline, many of the original discussants remain and contribute their views. There are bioethicists who hold tenured professorships and there are "bioethicists" who are articulate collaborators in the discourse.

Bioethics today is less a formal discipline than a "consilience" of many disciplines within a wider public discourse. Today scholars in many disciplines are fascinated by the idea of consilience, "the linking of facts and of fact-based theory across disciplines to create a common groundwork of explanation" (Wilson, 1998, p. 8). Is it possible for the explanatory instruments of one discipline to illuminate the data of another discipline? In recent decades, biologists and chemists have become biochemists, geneticists have become biologists, psychologists have become biochemists, sociologists have become sociobiologists, and so on. Persons trained in a classical discipline have found it useful and necessary to wander from their original home into other territory and to learn methods that they could conciliate into their own. Bioethics, not yet a discipline, does not have thick walls to penetrate. Since its beginnings, it has welcomed those from many disciplines and listened to the discourse of those who had no discipline. Bioethics is a perfect setting for experiments in consilience.

The invention of ethical argument invites the contribution of many voices. The ethicists who are the inventors seek the facts of the case from all sources and preside over the integration of these facts into forms of coherent and criticizable argument. The future of bioethics as a discipline lies in the ability of its practitioners to stay in contact with the public discourse, to shape that discourse into argument that has both relevant empirical references and careful logical construction. Bioethics is rooted in moral experience. The dimensions of moral experience, as manifested in social, political and religious behavior and institutions, must be more deeply explored. The most significant contribution to bioethics in this regard has been the scholarship of feminist bioethics, which brings to the description of moral experience the perspective of women's moral judgment as they move within the social, political and religious institutes of our culture (Wolf, 1996). Connections between cultural ethos and ethical claims should be elucidated. Generalizations about ethical pluralism call for closer reflection. Bioethicists are starting to seek in cross-cultural settings for the intricacies of moral experience within different societies. Serious philosophical and theological work has begun along these lines but these studies can find further breadth and depth. The roots in empirical moral problems and complexity of policy formulation has given bioethics its vitality. The exploration of its deeper foundations will be kept realistic by preserving the practical bent of the discipline. Bioethics must never depart from the advice Dan Callahan gave at its inception: "The discipline of bioethics should be so designed, and its practitioners so trained, that it will directly – at whatever cost to disciplinary elegance – serve those physicians and biologists whose positions demand that they make the practical decisions" (Callahan, 1973, p. 73). Today, almost thirty years later, we would add to his words, "serve those who must make public policy and serve all who must make personal decisions about their own life and health and the life and health of future generations."

School of Medicine
University of Washington
Seattle, Washington, U.S.A.

REFERENCES

Ammer, C. (ed.): 1995, *The HarperCollins Dictionary of Music*, HarperCollins, New York.

Aristotle: 1947, *Rhetoric*, J.H.Freese (ed.), Harvard University Press, Cambridge, Mass.

Beauchamp, T. and Childress, J.: 1979, *Principles of Biomedical Ethics*, Oxford University Press, New York.

Broad, C.D.: 1934, *Five Types of Ethical Theory*, Harcourt Brace, New York.

Callahan, D.: 1973, 'Bioethics as a discipline,' *Hastings Center Studies* 1, 66-73.

Cicero: 1975, *De Inventione*, H.M. Hubbell (trans.), Harvard University Press, Cambridge, Mass.

Comenisch, P.E. (ed.): 1994, *Religious Methods and Resources in Bioethics*, Kluwer Academic Publishers, Dordrecht.

Commager, H.: 1950, *The American Mind*, Yale University Press, New Haven, Conn.

Dewey, J. and Tufts, J.H.: 1909, *Ethics,* H. Holt and Company, New York.

Dewey, J.: 1922, *Human Nature and Conduct*, H. Holt and Company, New York.

Engelhardt, H.T.: 1986, *The Foundations of Bioethics*, Oxford, New York.

Gorovitz, S., Jameton, A.J., and Macklin, R. (eds.): 1976, *Moral Problems in Medicine*, Prentice-Hall, Englewood Cliffs, N.J.
James, W.: 1891, 'The moral philosopher and the moral life,' *International Journal of Ethics*, in 621, 623, 627.
Jonas, H.: 1969, 'Philosophical reflections on experimentation with human subjects,' Daedalus 98, 219-247.
Jonsen, A.R.: 1989, *The Birth of Bioethics*, Oxford University Press, New York.
McDermott J. (ed.), *The Writings of William James*, Random House, New York, pp. 610-628, pp.
McCormick, R.A.: 1981, *How Brave a New World: Dilemmas in Bioethics*, Doubleday, Garden City.
McKeon, R.: 1989, *Rhetoric*, Ox Bow Press, Woodbridge, CN.
Moore, G.E.: 1903, *Principia Ethica*, Cambridge University Press, Cambridge.
National Commission for the Protection of Human Subjects of Biomedical and Behavioral Research: 1979, *The Belmont Report: Ethical Principles and Guidelines for the Protection of Human Subjects of Research.* United States Government Printing Office, Washington, D.C.
Rawls, J.: 1967, *A Theory of Justice*, Harvard University Press, Cambridge, Mass.
Rosner, F. and Bleich, J.D.: 1979: *Jewish Bioethics*, Hebrew Publishing Co., New York.
Sidgwick, H.: 1874, *The Methods of Ethics,* Macmillan, London.
Smith, J.E.: 1963, *The Spirit of American Philosophy*, Oxford University Press, New York.
Toulmin, S.E.: 1982, 'How medicine saved the life of ethics,' *Perspectives in Biology and Medicine* 24, 736-750.
Veatch, R.M.: 1981, *A Theory of Medical Ethics*, Basic Books, New York.
Verhay, A. and Lammers, S.E. (eds.): 1993, *Theological Voices in Medical Ethics*, Eerdmans, Grand Rapids, Mich.
Wilson, E.O.: 1998, *Consilience. The Unity of Knowledge*, Knopf, New York.
Wolf, S.: 1996, *Feminism and Bioethics: Beyond Reproduction*, Oxford University Press, New York.

SECTION II

BIOETHICAL THEORY

TOM L. BEAUCHAMP AND DAVID DeGRAZIA

PRINCIPLES AND PRINCIPLISM

Our objective in this paper is to provide a sketch of how principles developed in bioethics and their philosophical and practical roles. We begin with a brief history of this development and then discuss philosophical problems that have emerged in recent discussions of principles.

I. THE SOURCES OF PRINCIPLES IN BIOETHICS

Prior to the early 1970s, there was no firm ground in which a commitment to general principles or ethical theory could take root in biomedical ethics. This is not to say that there was no prior commitment by physicians and researchers to moral values in their encounters with patients and subjects; of course there were such commitments. Nor is it to say that ethical theory was not well-developed in academic philosophy; it was. However, doctors' dilemmas and value questions in medicine were rarely discussed critically, and the perspective on medical ethics in Europe, America, and most of the world was largely that clinicians are obligated to maximize medical benefits and minimize risks of harm and disease for their patients. This narrow perspective was derived from the tradition-governed perspectives of medicine, public health, and nursing. Particular ethical codes for physicians and nurses had always been written by members of those professions. No bridge had been built to connect work in philosophical ethical theory with the problems of biomedical ethics, and professional ethics was not conceptualized in biomedicine as an interdisciplinary field.

What dawned on many observers in the 1970s was how the once vital Hippocratic tradition had evolved into a minimal and unsatisfactory professional morality. They saw a need for explicit recognition of basic ethical principles that could help identify various clinical practices and human experiments as morally questionable or unacceptable, whether or not they had been so recognized in Hippocratic medicine. The principles of biomedical ethics that are today widely recognized emerged from this felt need for a stable and reflective framework that was lacking in the youthful bioethics of the early 1970s.

G. Khushf (ed.), Handbook of Bioethics, 55–74.

Two publications inspired much of the early interest in principles in biomedical ethics. They were the *Belmont Report* (and related documents) of the National Commission for the Protection of Human Subjects and the book entitled *Principles of Biomedical Ethics*, which one of us jointly authored with James Childress. Both were written and published in the late 1970s. We will outline the views in both works before we move to analysis of principles and so-called "principlism."

The National Commission for the Protection of Human Subjects

After numerous complaints in the U.S. media about abuses of research subjects by investigators and institutions, a National Commission was charged by the U.S. Congress to investigate the ethics of research and to study how research was conducted in United States institutions. It was also charged to determine the basic ethical principles that should govern research with human subjects. In response to its charge, the Commission developed a schema of basic ethical principles and related it to the subject areas of research ethics to which the principles apply (National Commission for the Protection of Human Subjects, 1978, pp. 78-0012):

Principle of	*applies to*	*Guidelines for*
Respect for Persons		Informed Consent
Beneficence		Risk/Benefit Assessment
Justice		Selection of Subjects

This schema functioned as a general framework for handling problems of research ethics. For example, the principle of respect for persons requires that an informed consent must be received from subjects prior to their involvement in research. The purpose of consent provisions is not protection from risk, but the protection of autonomy and personal dignity (including the personal dignity of incompetent persons incapable of acting autonomously, whose involvement requires the consent of a third party). The purpose of the principle of beneficence, by contrast, was to provide protection from undue risk and a proper calculation of benefits in relation to risks. Finally, the principle of justice was concerned with fairness in the selection of subjects of research.

We believe that the Commission did not always delineate these principles in the most appropriate manner. Nonetheless, the *Belmont Report* had an enormous impact on the development of biomedical ethics and contained an early version of the principles that ultimately led to the term "principlism."

Principles of Biomedical Ethics

A second and more philosophically minded grouping of principles was being developed at the time of the work of the National Commission. Programmatic plans were cemented with a publisher in 1975 for *Principles of Biomedical Ethics*, the first systematic, relatively comprehensive work in the fledgling field of bioethics (Beauchamp and Childress, 1979). Virtually every published book in the field at the time was organized by topic. None was organized around principles or philosophical theory.[2]

In this book Beauchamp and Childress advanced a set of moral principles intended to serve as a framework for biomedical ethics that would reach well beyond research ethics and traditional forms of clinical ethics. The principles in this framework were, and still are, grouped under four general categories: (1) respect for autonomy (a principle requiring respect for the decisionmaking capacities of autonomous persons), (2) nonmaleficence (a principle requiring that we not cause harm to others), (3) beneficence (a group of principles requiring that we prevent harm, provide benefits, and balance benefits against risks and costs), and (4) justice (a group of principles requiring fair distribution of benefits, risks, and costs).[3]

The choice of these four moral principles as the framework for bioethics derives in part from professional roles and traditions. Health professionals' obligations and virtues have been framed for centuries, through codes and learned writings, by commitments to provide medical care and to protect patients from disease, injury, and system failure. However, the principles also incorporate parts of morality that traditionally have been neglected in medical ethics, especially through the principles of respect for autonomy and justice. Health care's traditional preoccupation with nonmaleficence- and beneficence-based models of ethics needed augmentation by an autonomy model of patient care and by a conception of social institutions and social justice that would protect vulnerable persons. All four principles, then, are needed to provide an adequately comprehensive framework for biomedical ethics.

Such a conception of principles gave medical ethics in its modern incarnation a shared set of assumptions that could be used to address bioethical problems. This framework also suggested that bioethics has principled foundations such that its considered judgments are not, as some have thought, like so many leaves blowing in the shifting winds of cultural trends, subjective responses, institutional arrangements, and the like.

Connection to the Common Morality

What are the moral foundations of principlism, and why is it an attractive theory? For one thing, its four principles are already embedded in public morality and are presupposed in the formulation of public and institutional policies. In truth, these principles do not deviate from what every morally serious person already knows as a matter of general knowledge. Every moral person believes that a moral way of life requires that we respect persons, take account of their well-being, and treat them fairly. This background in morality is the raw data for theory and is why we can speak of the origins of moral principles as being in the common morality that we all already share.

The common morality should not be regarded as merely one morality that differs from moralities embraced by other individuals or communities. The common morality contains universally valid precepts that bind all persons in all places. In recent years, the favored category in terms of which this universal core of morality has been expressed is human rights, but certain standards of obligation and virtue are also vital parts of the common morality. In our view, there is no more basic source of appeal in ethics than this common morality.

A distinction is needed, however, between "morality" *in the narrow sense* and "morality" *in the broad sense*. The universal principles of the common morality comprise only a skeleton of a well-developed body of moral standards. "Morality" in the narrow sense is comprised of universal principles, whereas "morality" in the broad (full-bodied) sense includes divergent moral norms, obligations, ideals, and attitudes that spring from particular cultures, religions, and institutions. For example, different standards of allocating resources for health care and different standards of giving to charitable causes are parts of morality in the broad sense. A pluralism of judgments and practices is the inevitable outcome of historical developments in cultures, moral disagreement and resolution, and the formulation of complex institutional and public policies. (More is said about this matter below under the subject of "specification.")

It may seem that we are simply packing *our view* of morality into the notion of what morally committed persons believe, while failing to take seriously the claims of the multi-culturalists, post-modernists, and others who reject the idea of the common morality. We believe, however, that it is an institutional fact about morality (not merely our view of it) that it contains certain shared, fundamental precepts. It is exclusively through our grasp of this shared moral substance that we can make justifiable cross-temporal and cross-cultural judgments, such as the judgments that genocide, apartheid, and caste systems are wrong, regardless of the culture in which they occur.

This appeal to the common morality is not intended to suggest that moral reasoning does or should always lead to conclusions that are commonly accepted. Common moral experience provides the starting points of moral discourse, but critical reflection on specific ethical issues may ultimately vindicate moral judgments that are not widely shared.

II. THE NATURE OF PRINCIPLES

A principle is a fundamental standard of conduct from which many other moral standards and judgments draw support for their defense and standing. For example, universal moral rights and basic professional duties can be delineated on the basis of moral principles. Principles and related rules[4] are always binding unless they conflict with other principles or rules.[5] When a conflict of two or more principles occurs, the conflict must be addressed to extract the proper content from each; alternatively, one principle may override the other. This way of handling conflict strikes some commentators as unsatisfactory, because the approach seems relative to individual judgment or to the beliefs of groups, so that principles, conceived in this way, seem unable to resolve hard cases. Other critics, including post-modernists and multi-culturalists, say the reverse: There are *only* principles relative to acceptance by groups.

The Conception of a Supreme Moral Principle

The alternative to this flexibility is to make the principles absolute or inflexible. One version of this approach would feature two or more basic moral principles arranged in a strict hierarchical ordering, so that conflicts between principles are always resolved in the same way. We are not aware of any significant example of a complete ethical theory that has been successfully constructed along these lines.[6]

A more promising and more prominent version of the absolute-principle approach features a single overarching principle that is presented as foundational for the entire moral domain. The most famous theory featuring a supreme moral principle is utilitarianism, which holds that the right action or policy is that which maximizes the balance of beneficial consequences over harmful consequences.[7] By contrast, deontological theories assert (in different ways, depending on the particular theory) that the right action or policy is to be identified by reference to one or more moral principles that cannot be equated with, or fully derived from, the principle of utility. Such theories commonly appeal to respect for persons, individual rights, or both. A deontological theory would have to feature only one supreme principle to fit the present model.[8]

During the 1970s and much of the 1980s, utilitarian and deontological approaches exerted enormous influence on the literature and discourse of biomedical ethics. Although utilitarian and deontological arguments or patterns of reasoning are still common today, the theories themselves hold a much diminished stature in the field. The reasons for the demotion of utilitarian and single-principle deontological theories concern the disadvantages of any approach that attempts to characterize the entire domain of morality with one supreme principle. Three disadvantages are especially worthy of note. First, there is a problem of authority. Despite myriad attempts by philosophers in recent centuries to justify the claim that some principle is morally authoritative, that is, correctly regarded as the supreme moral principle, no such effort at justification has persuaded a majority of philosophers or other thoughtful people that either the principle or the moral system is as authoritative as the common morality that supplies its roots. Thus to attempt to illuminate problems in biomedical ethics with a single-principle theory has struck many as misguided as well as presumptuous or dogmatic.

Second, even if an individual working in this field is convinced that some such theory is correct (authoritative), he or she needs to deal responsibly with the fact that many other morally serious individuals do not share this theory and give it little or no authority. Thus, problems of how to communicate and negotiate in the midst of disagreement do not favor appeals to rigid theories or inflexible principles, which can generate a gridlock of conflicting principled positions, rendering moral discussion hostile and alienating. In our experience – and we believe generally in the experience of teachers of biomedical ethics – even where people disagree at the level of basic theory, they commonly agree at the level of the principles of biomedical ethics enumerated above. These principles, then, seem a far more congenial and fruitful starting point for discussion.

Third, there is the problem that a highly general principle is indeterminate in many contexts in which one might try to apply it. That is, the content of the principle itself does not always identify a unique course of action as right. It has increasingly become apparent that single-principle theories are significantly incomplete, frequently depending on independent moral considerations with the help of which the theories can serve as effective guides to action.

These difficulties support viewing moral principles as fundamental but nonabsolute standards. Moral principles invariably must be specified or balanced when put to use, a claim supported by actual practice. We think through our problems by creatively using principles, rather than finding solutions by applying a principle.

Specification and Balancing

General principles frequently do not fully determine moral judgments because, as just noted, their content is insufficient. Moreover, principles often conflict, pointing in opposite directions. How, then, does one fill the gap between abstract principles and concrete judgments to guide moral decisionmaking sufficiently? The answer is that principles must be *specified* to suit the needs of particular contexts. Specification is the progressive filling in and development of the abstract content of principles, shedding their indeterminateness in particular circumstances and thereby providing action-guiding content (Richardson, 1990).[9]

As Henry Richardson has pointed out, the first line of attack in managing complex or problematic cases involving contingent conflicts should be to specify norms and eradicate conflicts among them. Many already specified norms will need further specification to handle new circumstances of indeterminateness or conflict. Incremental specification will continue to refine one's commitments, gradually reducing the circumstances of contingent conflict to more manageable dimensions. Increase of substance through specification is essential for decision making in clinical and research ethics, as well as for the development of policy (institutional policies as well as public policies).

A simple case will help to illustrate the idea of specification and the related concept of balancing. Consider two rules that come into conflict in a range of cases:

1. It is morally prohibited to let a patient die whose life-threatening condition can be treated.

2. It is morally prohibited to disrespect a parental refusal of treatment.

In some cases, parents refuse treatment for infants – for example, anencephalic infants – in the belief that it is in the infant's best interest not to survive longer than necessary. To handle this conflict, rule 2 can be specified as 2.1:

2.1. It is morally prohibited to disrespect a parental refusal of treatment, unless the refusal constitutes child abuse or violates a right of the child.

Rule 2.1 qualifies as a specification because much of the content of the original rule remains intact, but now we understand that one is only *generally* required to respect parental refusals of treatment. Clearly rule 2.1 is but a start down the road of specification. It will not be adequate to handle all cases and will need further specification regarding the nature of child abuse and the rights of children. Nonetheless, 2.1 does show that there is a path out of the initial conflict between 1 and 2, and it indicates that physicians and hospital administrators are not confronted with an absolute rule requiring respect for parental refusals. With enough additional specification, an entire hospital policy could be constructed that was generally adequate to the range of cases that pediatricians might expect to see.

Consider, as a more extended example of specification, the famous *Tarasoff* case. This case also serves to illustrate the related concept of balancing (Cf. *Tarasoff v. Regents of the University of California*, 1976). The practice of psychotherapy has long honored a principle (or rule) of confidentiality, which states that the information divulged in psychotherapy by a patient to the therapist may not be shared with other individuals without the patient's prior consent. This rule is grounded in the need for the patient to trust the therapist as a condition for fully open discussion of the patient's personal difficulties and in respect for the patient's autonomy. But what if a patient divulges the intention to kill an identified third party? Another commonly accepted rule, even if less often explicitly stated, is that one should take reasonable steps to prevent or warn of major harm to another individual if one is uniquely situated to do so and can do so relatively easily. The strength of this obligation may increase if one occupies a professional role such as that of a psychiatrist, psychologist, or social worker. Clearly we have a conflict, because maintaining confidentiality is inconsistent with the second rule; taking steps to warn the prospective victim would violate the rule of confidentiality.

How should we manage this conflict? Consistent with the legal judgment that was rendered in the case, many therapists and ethicists hold that a serious threat of consequential bodily injury to an identified third party warrants an exception to the principle of confidentiality. Efforts to balance the relevant considerations suggest that the importance of helping the endangered person is weightier than that of confidentiality in such cases. So long as balancing is understood as involving a judgment adequately supported by justifying reasons – and not as a purely intuitive act – the metaphor of balancing fits well with the idea of specification. The reason justifying the resolution can, in effect, be incorporated into a specification of one of the principles or already specified rules (assuming they are regarded as nonabsolute). In the present case the original rule can be *specified* as follows: The information divulged in psychotherapy by a patient to the therapist may not be shared with other individuals without the patient's prior consent, unless the patient expresses an intention to cause severe harm to an identified third party. This specification eliminates the dilemma that originally existed because of a conflict between two rules.

In the light of this account, principles should be understood less as firm directives that are applied and more as general guidelines that are explicated and made suitable for specific tasks, as often occurs in formulating policies and altering practices. This strategy has the advantage of allowing us to spell out our evaluative commitments and to expand them in order to achieve a more workable and coherent body of practical moral guidelines. The field of biomedical ethics has been developed and continues to be developed by tailoring principles to fit the needs of specific circumstances.

However, since it is usually possible to specify a principle in more than one way, and since in conflicts more than one principle can be specified, how we can justify a particular specification is a central issue to which we must return below.

III. ALTERNATIVES TO PRINCIPLES?

With this account of principles before us, we now shift to some criticisms of principles that have been raised by contemporary writers in biomedical ethics. Their approaches are commonly considered competitors to principlism as general frameworks for bioethics

Cases and Principles

Often in contemporary biomedical ethics we concentrate our attention not on principles, but on practical decisionmaking in particular cases and on the implications of those cases for other cases (Jonsen, 1995).[10] Here we proceed by identifying the particular features of and problems present in the case. We may attempt to identify the relevant precedents and prior experiences we have had with related cases, attempting to determine how similar and how different the present case is from other cases. For example, if the case involves a problem of medical confidentiality, analogous cases would be considered in which breaches of confidentiality were justified or unjustified in order to see whether such a breach is justified in the present case.

The leading cases (so-called "paradigm cases") become enduring and authoritative sources for reflection and decisionmaking. Cases such as the Tuskegee Syphilis experiment case (in which a group of men were intentionally not given treatment for syphilis, in order to follow the course of the disease) are constantly invoked in order to illustrate unjustified biomedical experimentation. Decisions reached about moral wrongs in this case serve as a form of authority for decisions in new cases. These cases profoundly influence our standards of fairness, negligence, paternalism, and the like. Just as case law (legal rules) develops incrementally from legal decisions in cases, so the moral law (moral rules) develops incrementally. From this perspective, principles are less important for moral reasoning than cases. Indeed, principles are perhaps even expendable.

Nonetheless, this form of reasoning – now often called casuistry in bioethics – can be misleading. Casuists sometimes write as if cases lead to moral paradigms, analogies, or judgments entirely by their facts alone or perhaps by appeal only to the salient features of the cases. This premise is suspect. No matter how many salient facts are stacked up, we will still need some *value* premises in order to reach a moral conclusion. The properties that we observe to be of moral importance in cases are picked out by the values that we have already accepted as being morally important. In short, the paradigm cases of the casuists are value-laden.

The best way to understand this idea of paradigm cases is as a combination of (1) *facts* that can be generalized to other cases – for example, "the patient refused the recommended treatment" – and (2) *settled values* – for example, "competent patients have a right to refuse treatment." In a principle-based system, these settled values are called principles, rules, rights-claims, and the like; and they are analytically distinguished from the facts of particular cases. In casuistical appeals to cases, rather than keeping values distinct from facts, the two are bound together in the paradigm case; however, the central values are generalizable and must be preserved from one case to the next. For a casuist to reason morally, one or more settled values must connect the cases (hence the necessity of "maxims," or moral generalizations). The more general the connecting norms, the closer they come to the status of principles. At the same time, if the generality is at a very high level, a loss of specific guidance will occur, and some form of specification will be needed to handle complex cases.

Casuists and principlists should be able to agree that when they reflect on cases and policies, they rarely, if ever, have in hand either (1) principles that were formulated without a basis in experience with cases or (2) paradigm cases that have become paradigmatic independently of any prior commitment to general norms. When philosophers now speak, as they often do, about "the top" (principles, theories) and "the bottom" (cases, individual judgments) in moral philosophy, it is doubtful that these poles can be either a starting point or a resting point without some form of cross-fertilization and mutual development.

Rules and Principles

Danner Clouser and Bernard Gert made famous the term "principlism" in an article that alleged certain defects or forms of incompleteness in the use of principles (Clouser and Gert, 1990).[11] They used the term to refer to all theories comprised of a plurality of potentially conflicting prima facie principles. Capturing several widely shared concerns about principles in a sustained critique, they offered an alternative framework centered on impartial rules, inciting an animated discussion in the literature of bioethics.

Clouser and Gert maintain that "principles" function more like chapter headings in a book than as directive rules or normative theories. That is, principles point to important moral themes by providing a general label for those themes, but they do not function as practical action guides. Receiving no helpful or controlling guidance from the principle, a moral agent confronting a problem is free to deal with it in his or her own way and may give the principle whatever weight he or she wishes when it conflicts with another principle.

These deficiencies are alleged to be especially pronounced in the area of justice, because there is no specific guide to action or any theory of justice singled out in the principle (or principles). We know that justice is concerned with distribution and that we should be concerned about it, but the use of "justice" amounts to little more than a checklist of moral concerns. Principlism instructs persons to "be alert to matters of justice," and to "think about justice" – but nothing more (see, especially, Clouser, 1995). Since this lack of normative content deeply underdetermines solutions to problems of justice and has no power to guide actions or to establish policies, the agent is free to decide what is just and unjust, as he or she sees fit. Other moral considerations besides the principle(s) of justice, such as intuitions and theories about the equality of persons, must be called upon for real normative guidance. Clouser and Gert think the same problem afflicts all general principles; principles alert persons to issues, but, lacking an adequate unifying theory, they offer no real guidance on their own.

Is it true that principles lack a specific, directive substance? The claim is most plausible in the case of *unspecified* principles. Any principle – and any rule, for that matter--will have this problem if the norm is underspecified for the task at hand. A basic principle is necessarily general, covering a broad range of circumstances; in this regard, principles contrast with specific propositions. As the territory governed by any norm (principle, rule, paradigm case, etc.) is narrowed, the conditions become more specific – e.g., shifting gradually from "all persons" to "all competent patients" – and along the way it becomes increasingly less likely that the norm can qualify as a principle. For example, the principle of respect for autonomy applies to all autonomous persons and autonomous actions, a norm of respecting informed refusals by competent patients is, due to its narrow scope, more likely to be considered a rule than a principle.

If general principles can be specified and rendered more useful for particular contexts, why continue to think in terms of general principles at all? One practical reason is that principles must be the sort of thing that can be learned by everyone--not just philosophers, but health professionals, ethics committee members, and laypersons. If we thought only in terms of specified principles, their specificity and proliferation would make them very difficult to remember, master, and internalize for practical use.

Any list of general rules must also be specified, or else they will be too general and commonly fail to provide normative guidance. Clouser and Gert's rules, for example, are like our general principles in that they lack specificity in their original general form. Being one tier less abstract than principles, their rules do have a more directive and specific content than abstract principles. However, a set of rules almost identical to the rules embraced by Clouser and Gert is already included in the account of principles and rules that we would defend. We maintain that principles support these more specific and directive moral rules and that more than one principle (for example, respect for autonomy and nonmaleficence) may support a single rule (for example, medical confidentiality). Their rules, then, either do not or need not differ in content from our rules and their rules need not be more specific and directive than ours.

Virtues and Principles

Whether emphasizing principles, rules, or judgments about cases, the approaches we have considered so far all focus on an ethical evaluation of action, that is, on the right action to perform. By contrast, virtue ethics gives virtuous or good character a preeminent place.[12] Moral virtues may be understood as morally praiseworthy character traits, such as courage, compassion, sincerity, reliability, and industry. In virtue ethics, the primary concern is with the agent – with what sort of person he or she is – while action is considered to have secondary importance. Thus, from the perspective of virtue ethics, principlism is likely to be seen as having the wrong moral focus.

There are several reasons for stressing the importance of virtues. One is that we often morally evaluate people's character and motives (which typically flow from and reflect one's character), not simply their actions. Thus, even if we think telling a lie was the wrong choice in a particular case, we are likely to temper our criticism if we believe the agent was motivated by a compassionate desire to spare someone's feelings. In a similar way, we often morally evaluate a person's emotional responses – which tend to reflect one's character – even where no particular action is called for. One might admire a social worker's genuine sorrow at the news that another social worker's patient committed suicide; her expression of sorrow reflects her caring and sympathy. Moreover, in practice, well-established virtues may prove at least as important as mastery of principles, rules, and other action-guides. For example, it may be the case that being truthful, compassionate, perceptive, diligent, and so forth is a more reliable basis for good medical practice than knowledge of the principles and rules of biomedical ethics.

While each of the preceding points about the place of virtue in the moral life is undeniably important, a principlist can accept them and defend the program of incorporating a virtue ethics into a principle-based ethics. Virtue ethicists may insist that this model still inappropriately leaves virtues in a secondary position. They might suggest that even the nature (not just the achievement in practice) of morally appropriate conduct cannot be illuminated without reference to virtues (see Sherman, 1988). It is possible to do the right thing (say, pay off a personal debt) yet not act well (because one pays the debt with inappropriate hostility toward the person who lent the money). This can occur because the manner in which we act – expressing our attitudes and character – has no less moral importance than the type of action we perform. We might say that morally appropriate conduct involves not just right action, but also acting in a manner that expresses appropriate attitudes and motives.

Arguments like this one support the program of integrating action-based ethics and virtue ethics. However, some writers go further. They believe that the advantages of virtue ethics justify making character the focal point of ethics (see Garcia, 1990). The most radical variant of this approach would disregard principles altogether. While we believe that the importance of virtue warrants an integration of principles and virtue considerations (a view taken in *Principles of Biomedical Ethics*, 2001, ch. 2), we are unconvinced that any supporting arguments justify prioritizing virtue over principles. The evaluation of action is as central to the moral life as the evaluation of character. Regarding the former, we have already seen reasons to doubt that action-guides other than principles, such as specific case judgments and rules, merit priority over principles.

Further, it is doubtful that virtue can be adequately conceptualized without some background assumptions about principles and right action. For example, seeing truthfulness as a virtue seems inseparable from seeing truth-telling as a prima facie obligation. If we ask why one should generally be truthful, it seems evasive to say, "because virtuous people are that way." A more adequate response would show how truthfulness displays respect for people's autonomy, tends to promote certain benefits, and ordinarily avoids certain kinds of harm. Principles, we submit, are central to our moral thinking and indispensible to it.

IV. PROBLEMS OF JUSTIFICATION

Having discussed how principles became prominent in biomedical ethics, the nature of principles, and three proposed alternatives to a principle-based ethics, we are prepared to engage two related questions. First, how should we view the relationship between principles and other moral norms, such as rules, particular case judgments, and considerations of virtue? Second, in settling a conflict between principles (or rules) in a particular case – assuming we do not possess a theory that already determines the answer – what justifies one proposed resolution rather than another? Since arriving at an answer to the second question often involves specifying a principle, what justifies a particular specification?

Our responses to the challenges discussed in the previous section suggests an answer to the first question. While principles are indispensible to ethics and deserve a prominent place, particular case judgments, rules, and considerations of virtue are also indispensible. But one might ask, "how is that possible?". A principle-based approach begins with principles and attempts to move 'downward' (sometimes directly, more often through layers of increasingly specific principles or rules) to justify particular case judgments. Yet, to consider just one of the alternative frameworks, casuistry begins with case analysis, taking our judgments at that level to be more reliable and informative than abstract reasoning at a 'higher' level.

Although this image of working "down" from principles to cases grips the imagination of many who work in bioethics, we do not endorse it. It is true that an approach rooted in principles confidently announces several principles that are firmly enmeshed in the fabric of the common morality. However, it does not follow that all, or even most, of the reasoning in this model moves downward. Reflection on cases can itself be a valuable source of moral insight. Often we have more reason to trust our responses to specific cases than a principle or rule to which we had previously not noted any exceptions (or which remained too vague or too unspecified to guide us in the case). In the type of case described earlier, in which a patient in psychotherapy divulges the intention to kill a third party, reflection on what is at stake in the case seems to merit an exception to (a specification of) an entrenched rule of confidentiality. Using a case judgment to motivate changing a rule or principle represents "upward" movement in moral reasoning, but such reasoning presents no problem for our model once the method of specification is properly understood.

Principlism, as we understand it, should take a step beyond *specification* (which is a method of obtaining guidance) and accept what is known as the "reflective equilibrium" or "coherence" model of ethical *justification* (which is a way of showing which form of guidance is best).[13] In this model, no level of moral reasoning – comprehensive theories, principles, rules, or case judgments – is regarded as having priority or as serving as a foundation for all the other levels. On this basis principlism and casuistry are compatible, unless the casuist understands case judgments to have strict priority.[14] Rules also fit comfortably into the overall picture of moral reasoning. Precisely the same may be said for considerations of virtue, as noted earlier.

Thus, principlism, as we understand it, does not treat principles as foundations from which all sound moral reasoning must move "downward." But if a viable principle-based ethics does not simply move "downward," a critic might respond, why promote the concept of specified principles? After all, specification does move from generality to specificity, which is to say "downward." Our answer has two parts.

First, because we are morally confident in certain moral principles, we take them as appropriate starting points. As ethical reasoning progresses, the insights gathered along the way form a developed body of specifications of these principles. One might begin at a different place – say, with cases – but sound reasoning from any starting point should lead eventually to a vindication of fundamental principles.

Second, the idea of specifying general principles (and rules) is consistent with the claim that sound moral reasoning can occur at any level of generality and can motivate revisions of ethical belief either "upward" or "downward." Consider circumstances in which reflection on cases motivates "upward" revisions, since this may seem to challenge the idea of specification. As we saw with the case involving confidentiality in psychotherapy, resolution of troublesome particular cases requires a specification of some more general norm that played a role in the original conflict or indeterminacy. So, even where reasoning moves "upward" from cases, the reasoning that occurs will typically involve some general form of the specification of norms that are already in place.[15]

This takes us, finally, to the question of how particular specifications are to be justified when there are *competing* specifications. The case of a would-be killer in psychotherapy was resolved in our previous analysis by building an exception into the rule of confidentiality. But another and different resolution by specification was possible: Preserve confidentiality as an absolute rule and revise the other norm as follows: "One should take reasonable steps to prevent major harm to another individual . . . unless one's professional duties prohibit the only available means for doing so." So why is the resolution in terms of making a disclosure to a third party more justified than this competing specification of nondisclosure, if it is?

A particular specification, or any revision in moral belief, is held to be justified if it maximizes the coherence of the overall set of beliefs that are accepted upon reflection.[16] This is, admittedly, a very abstract thesis, and employment of the criteria that together constitute "coherence" in the relevant sense is a subtle and somewhat unresolved affair. While we believe at least some of these criteria of coherence are already implicitly accepted by nearly anyone who engages in serious moral reflection and discourse, in the present chapter we cannot do more than list a few of the criteria.

The following, then, are some of the criteria for a coherent (and therefore, according to this model, justified) set of ethical beliefs: logical consistency (the avoidance of outright contradiction among judgments); argumentative support (explicit support for a position with reasons); intuitive plausibility (the feature of a norm or judgment being believable in its own right, whether or not it is also supported by reasons); compatibility or coherence with reasonable non-moral beliefs (in particular, a good fit with available empirical evidence and well-established scientific theories); comprehensiveness (the feature of covering the entire moral domain, or as much of it as possible); and simplicity (reducing the number of moral considerations to the minimum possible without sacrifice in terms of the other criteria).

With such criteria in view, how can we understand the justification for specifying the rule of confidentiality as we did earlier in this chapter? We believe that disclosure is justified because the plausibility of permitting exceptions in this sort of case (while allowing the rule to retain its force in other cases) is higher than the plausibility of remaining silent. Keeping the rule exceptionless requires taking an extreme risk that an innocent person, who could have been warned, will be killed. While maintaining trust and openness in the therapeutic relationship is of vital importance, these conditions can be maintained even if the rule of confidentiality has an exception allowing disclosure (of which patients should perhaps be informed at the beginning of therapy, although we set aside this complication here).[17]

V. FUTURE DIRECTIONS

Despite our statement in the previous section, justification within the present model needs to be explicated in a more satisfying way than it has been to date. We find this model of justification more sensible than its alternatives such as those based on a supreme principle and those that give strict priority to either rules, case judgments, or considerations of virtue, but our ability to use the model fruitfully exceeds our ability at present to explain how justified reasoning occurs within the model. Future research on moral justification should help reduce this gap.

A second matter needing attention is the rich set of relationships between principles and virtues. While some of the major connections between these two types of norms are noted above, further exploration of these connections stand to strengthen both ethical theory and biomedical ethics.

Department of Philosophy and Kennedy Institute of Ethics
Georgetown University
Washington, DC, U.S.A.

Department of Philosophy
George Washington University
Washington, DC, U.S.A.

NOTES

1. On July 12, 1974, P.L. 93-348 authorized the National Commission as an advisory body. Its deliberations on principles came much later.
2. Several books were published in 1976 that influenced the writing of *Principles* before the drafting was completed. These included Brody, 1976; Veatch, 1976; Gorovitz, et al., 1976; Shannon, 1976; Veatch and Branson, 1976. Joseph Fletcher's pioneering *Morals and Medicine* (1954) focussed more on the dying patient's "right to know the truth" than on principles of a system of bioethics. Paul Ramsey's *The Patient as Person* (1970) devoted substantial space to themes such as consent as a canon of loyalty, but the focus was dominantly on guardian consent and on the physician's duties of loyalty, fidelity, and mutuality.
3. As stated above, two of the "principles" – beneficence and justice – are best treated as a group of related principles. However, for convenience, we will simply speak of four principles.
4. Rules are generally thought of as more specific in content than principles, but there is no bright line dividing the two kinds of norm.
5. Ross's term "prima facie duty" indicates that an obligation must be fulfilled unless it conflicts on a particular occasion with an equal or stronger obligation; see Ross, 1930, and 1939.
6. John Rawls, 1971, has developed a theory of justice, not a full ethical theory, that features such a hierarchy of principles.
7. The most widely cited statement of this view is found in Mill, 1863. For significant recent representatives of the utilitarian tradition, see Brandt, 1979; and Hare, 1981.
8. On method, see Jonsen, 1996, pp. 37-49, and 1991. The most classic deontological theory featuring a supreme principle is that of Immanuel Kant; see, e.g., Beck, 1959. For contemporary representatives of this approach, see Donagan, 1977, a book stressing respect for persons; and Gewirth, 1978, which stresses individual rights.

9. For early discussions of this method in bioethics, see DeGrazia, 1992; Beauchamp, 1994, 1994b; and Beauchamp and Childress, 1994. For discussions of what is now sometimes called specified principlism, see, e.g., Davis, 1995, esp. pp. 95-102; Levi, 1996, esp. pp. 13-19, 24-26; and Gert, Culver, and Clouser 1997, ch. 4.
10. For a landmark work in the history of the type of reasoning described here, see Jonsen and Toulmin, 1988.
11. Their views are further developed in later writings. See Green, Gert, and Clouser, 1993; Clouser and Gert, 1994, pp. 251-266; Clouser, 1995; and Gert, Culver, and Clouser, 1997.
12. The most influential representative of the tradition of virtue ethics is Aristotle (see, especially, *Nichomachean Ethics*). Contemporary feminist ethicists often see their views as making common cause with virtue ethics, see, e.g., Carse, 1991; and Little, 1995.
13. For two highly influential works that pioneered this approach, see Rawls, 1971; and Daniels, 1979. Unfortunately, the authors' discussions of this model were as turgid and hard to understand in detail as they were influential. A more accessible and relevant statement of the position for bioethics is found in Daniels, 1996, pp. 96-114.
14. While casuists sometimes seem to take this view (see, e.g., Jonsen and Toulmin, 1988), Baruch Brody apparently does not (see Brody, 1988).
15. Because fundamental principles are so general (and vague), it seems unlikely that reflection on cases will motivate a significant recasting of those principles, as opposed to simply making them more specific. But such a revision is possible. For example, common morality arguably takes nonmaleficence to require not harming other human beings. However, case analysis and other forms of moral reasoning support understanding this principle to apply to anyone who can be harmed (including sentient animals). This example shows the complications in deciding whether a *principle* or some other moral consideration is being modified. Even in this example it seems less likely that the principle is being modified than that one's understanding of the scope or reach of the principle is being modified.
16. For recent efforts to analyze the idea of justification by coherence, see Daniels, 1996, ch. 16; DeGrazia, 1996, ch. 2.
17. One might think it offers greater simplicity, because the rule of confidentiality will remain exceptionless. However, as stated above in enumerating the criteria, simplicity should not be pursued if it entails sacrifice of other criteria.

REFERENCES

Beauchamp, T.L.: 1994a, 'The Four Principles Approach to Medical Ethics,' in R. Gillon (ed.), *Principles of Health Care Ethics*, John Wiley & Sons, London.
Beauchamp, T.L.: 1994b, 'Principles and Other Emerging Paradigms for Bioethics,' *Indiana Law Journal* 69 (3), pp. 1-17.
Beauchamp, T.L. and Childress, J.F.: 1979, *Principles of Biomedical Ethics*, Oxford University Press, New York. The manuscript of this first edition went to the press in late 1977.
Beauchamp T.L. and Childress J.F.: 2001, *Principles of Biomedical Ethics*, 5th ed., Oxford University Press, New York.
Beck, L.W. (trans.): 1959, *Foundations of the Metaphysics of Morals*, Bobbs-Merrill, Indianapolis.
Brandt, R.B.: 1979, *A Theory of the Good and the Right*, Clarendon Press, Oxford.
Brody, B.: 1988, *Life and Death Decision Making*, Oxford University Press, New York.
Brody, H.: 1976, *Ethical Decisions in Medicine*, Little, Brown and Company, Boston.
Carse, A.L.: 1991, 'The 'Voice of Care': Implications for Bioethical Education,' *Journal of Medicine and Philosophy* 16, pp. 5-28.

Clouser, D.K.: 1995, 'Common Morality as an Alternative to Principlism,' *Journal of the Kennedy Institute of Ethics* 5, pp. 219-36.
Clouser, D.K. and Gert, B.: 1990, 'A Critique of Principlism" The Journal of Medicine and Philosophy 15, pp. 219-36.
Clouser, D.K. and Gert, B.: 1994, 'Morality vs. Principlism,' in R. Gillon and A. Lloyd (eds.), *Principles of Health Care Ethics*, John Wiley & Sons, London.
Daniels, N.: 1979, 'Wide Reflective Equilibrium and Theory Acceptance in Ethics,' *Journal of Philosophy* 76, pp. 256-82.
Daniels, N.: 1996, 'Wide Reflective Equilibrium in Practice,' in L. W. Sumner and J. Boyle, *Philosophical Perspectives on Bioethics*, University of Toronto Press, Toronto.
Daniels, N.: 1996, *Justice and Justification*, Cambridge University Press, Cambridge.
Davis, R.B.: 1995, 'The Principlism Debate: A Critical Overview,' *Journal of Medicine and Philosophy* 20.
DeGrazia, D.: 1992, 'Moving Forward in Bioethical Theory: Theories, Cases, and Specified Principlism,' Journal of Medicine and Philosophy 17, pp. 511-39.
DeGrazia, D.: 1996, *Taking Animals Seriously*, Cambridge University Press, Cambridge, ch. 2.
Donagan, A.: 1977, *The Theory of Morality*, University of Chicago Press, Chicago.
Fletcher, J.: 1954, *Morals and Medicine*, Princeton University Press, Princeton.
Garcia, J.: 1990, 'The Primacy of the Virtuous,' *Philosophia* 20, pp. 69-91.
Gert, B., Culver, C.M., and Clouser, K.D.: 1997, *Bioethics: A Return to Fundamentals*, Oxford University Press, New York.
Gewirth, A.: 1978, *Reason and Morality*, University of Chicago Press, Chicago.
Gorovitz, S. et al. (eds.): 1976, *Moral Problems in Medicine*, Prentice-Hall, Englewood Cliffs.
Green, R.M., Gert, B., and Clouser, K.D.: 1993, 'The Method of Public Morality versus the Method of Principlism,' *The Journal of Medicine and Philosophy* 18.
Hare, R.M.: 1981, *Moral Thinking*, Clarendon Press, Oxford.
Jonsen, A.R.: 1991, 'Casuistry as Methodology in Clinical Ethics,' *Theoretical Medicine* 12, pp. 299-302.
Jonsen, A.R.: 1995, 'Casuistry: An Alternative or Complement to Principles?' *Journal of the Kennedy Institute of Ethics* 5, pp. 246-47.
Jonsen, A.R.: 1996, 'Morally Appreciated Circumstances: A Theoretical Problem for Casuistry,' in L. W. Sumner and J. Boyle, *Philosophical Perspectives on Bioethics*, University of Toronto Press, Toronto.
Jonsen A.R. and Toulmin, S.: 1988, *The Abuse of Casuistry*, University of California Press, Berkeley.
Kant, I.: 1959, *Foundations of the Metaphysics of Morals*, Lewis White Beck (trans.), Bobbs-Merrill, Indianapolis.
Levi, B.H.: 1996, 'Four Approaches to Doing Ethics,' *Journal of Medicine and Philosophy* 21.
Little, M.O.: 1995, 'Seeing and Caring: The Role of Affect in Feminist Moral Epistemology,' *Hypatia* 10, pp. 117-37.
Mill, J.S.: 1863, *Utilitarianism*. Many published editions.
National Commission for the Protection of Human Subjects.: 1978, *The Belmont Report: Ethical Guidelines for the Protection of Human Subjects of Research*, Washington: DHEW Publication No. [OS].
Ramsey, P.: 1970, *The Patient as Person*, Yale University Press, New Haven.
Rawls, J.: 1971, *A Theory of Justice*, Harvard University Press, Cambridge.
Richardson, H.S.: 1990, 'Specifying Norms as a Way to Resolve Concrete Ethical Problems,' *Philosophy and Public Affairs* 19, pp. 279-310.
Ross, W.D.: 1930, *The Right and the Good*, Oxford University Press, Oxford.
Ross, W.D.: 1939, *The Foundations of Ethics*, Oxford University Press, Oxford.

Shannon, T.A. (ed.): 1976, *Bioethics: Basic Writings on the Key Ethical Questions that Surround the Major Modern Biological Possibilities and Problems*, Paulist Press, New York.

Sherman, N.: 1988, 'Common Sense and Uncommon Virtue,' *Midwest Studies in Philosophy* 13, pp. 98-101.

Tarasoff v. Regents of the University of California, 17 Cal. 3d 425 (1976).

Veatch, R.M.: 1976, *Death, Dying, and the Biological Revolution*, Yale University Press, New Haven.

Veatch, R.M. and Branson, R. (eds.): 1976, *Ethics and Health Policy*, Ballinger Publishing Co., Cambridge.

JOSEPH BOYLE

CASUISTRY

The term "casuistry" refers descriptively to a method of reasoning for resolving perplexities about difficult cases that arise in moral and legal contexts. The term comes from the Latin *"casus"* which means occasion, event or case. Its focal descriptive meaning in English is conveyed by the sense of the somewhat technical and nowadays quaint expression "cases of conscience." This expression refers to a genre of moral literature, primarily Catholic, but also with some existence within Protestant moral theology. This literature deals with difficult moral decisions, that is, with cases where the morally relevant particularities of a case make its moral character problematic. The purpose of the literature is to resolve the perplexity of the puzzling case by making plain the moral category in which the case properly belongs. Thus, not surprisingly, "casuistry" refers not only to a method of moral reasoning, but also to a genre of literature which employs this method. This literature includes the "cases of conscience" extensively discussed by Western Christians, but also the application of Jewish law to difficult cases, and, nowadays within applied ethics, the non-religious application of moral and legal norms and principles to difficult cases (Jonsen, 1991, pp. 344-348).[1]

The term "casuistry" is also used evaluatively to refer to sophistical reasoning about morally or legally difficult cases in which logical cleverness overwhelms moral seriousness. Used this way, it often also refers to the usually permissive excesses of moral judgment and reasoning that have sometimes been exhibited in casuistical literature. The evaluative use of the term suggests that the method is fundamentally flawed and the literature a debased form of moral teaching (Jonsen and Toulmin, 1988, pp. 11-13). But recent thinking about casuistry and recent casuistical practice suggest strongly that, although casuistry can be abused, it has a legitimate, important and irreducible function in moral thought. Recent apologists for casuistry, most notably Jonsen and Toulmin (1988, pp. 5-15, 279-303), go so far as to suggest that it comprises the center of a more Aristotelian "case based" alternative to moral approaches based on general moral principles.

G. Khushf (ed.), Handbook of Bioethics, 75–88.

I. THE CASUISTICAL METHOD

It is useful to begin by focusing on the sort of difficult case casuistry deals with. Systems of laws, including many moral codes, seek to guide human choices in various ways. They routinely prohibit, enjoin or allow acts of certain kinds. This form of guidance is normative, that is, it directs people's choices. Actions are the things a person can choose to do, or choose not to do, or even simply fail to do. Therefore, the guidance of moral norms and laws must refer to, or have application to, actions insofar as they are projected or described in more or less specific ways that are significant in relation both to the choices to do them and to terms and goals of the legal system or moral code. Thus, for example, acts of killing, or of fulfilling contracts, or of making vocational commitments are among the sorts of human choices many legal systems and moral codes refer to when they provide guidance by either prohibiting, or enjoining, or allowing them. Usually legal systems and moral codes prohibit a significant subset of acts of killing, enjoin many acts insofar as they are necessary to fulfill contracts (or to remedy breaches of contracts), and allow a wide array of vocational commitments.

The reference of norms to actions under some normatively relevant description, as an action of a morally relevant kind, can be spelled out as follows. Some features of an individual act a person could choose to perform provide bases for relating that act to the terms and goals of a normative system; for example, the fact that an action is an act of killing a human being is, for most systems of laws and moral codes, a morally relevant feature of the act. Such features are properties of actions as human undertakings, that is, as performances a person could choose to carry out; specifically they are the properties of human undertakings in virtue of which they are evaluated by legal or moral norms. So, the relevant properties are all those considerations, however consequential or circumstantial, in virtue of which the undertaking is evaluated. Thus, facts such as the following about an action, considered prospectively by one thinking of choosing to do it, are ordinarily normatively relevant: that it harms someone or not, that it has consequences for others that one would not bring upon oneself or upon those one loved, that it violates a promise, and so on.

Although normative guidance applies in the first instance to actions considered as performances a person might choose to carry out, it also applies to past actions which can be evaluated for the sake of repentance or punishment, emulation or praise (all of which can involve further choices, informed by the evaluation of the past action). Indeed, much Catholic casuistry developed in the context of confessional practice, and much legal casuistry deals with the rationale for punishing or excusing wrongdoers.

An action already done is a settled, individual component of a person's unique history. This relative concreteness in comparison to actions projected and more or less vividly imagined as one deliberates about doing them does not alter the fact that retrospective evaluation also considers acts, not in their full concrete detail, but as instantiating morally or legally relevant kinds. Any of the indefinite number of things one might consider about one's past actions might be morally significant, but ordinarily most of these features are irrelevant: for example, in some cases the exact amount of analgesic given to a patient can be morally important as evidence about the intentions of a physician whose action shortens the life of a terminal patient, but in most cases in which analgesics are given, this fact is as morally irrelevant as the lighting in the room or the time of the day generally is.

Even in considering excusing or extenuating factors, which may mitigate responsibility, we understand the concrete, singular action in terms of categories relevant to legal or moral assessment: as compromised by weakness or ignorance, as falling short of a standard no one could be expected to meet, or as just too bad to be excused. Thus, in considering the appropriate punishment for a father who killed his very debilitated and suffering child rather than continuing to struggle with her continued difficult life, we focus not on all the tragic details but on the father's capacity to understand the law and his situation, his ability to cope with it, to find hope and support from others and so on.

Plainly, legal systems, or the moral codes they are meant to provide, require that what people consider choosing to do be identifiable as an act of a legally or morally relevant kind. In other words, a person seeking guidance from the legal system or moral code must be able to determine whether what he or she is thinking of doing is an act, for example, of prohibited killing or of fulfilling a contract, or of some other morally relevant kind. That determination involves identifying the properties of the action which make it prohibited killing (or something else) or fulfilling a contract (or something else).

In many cases the identification of the features in virtue of which an action is prohibited, enjoined or allowed, or, after the fact, is excused even if prohibited, is so straightforward that the reference of the norm to the action is unproblematic. Sometimes the normatively relevant feature is so salient among the features of the concrete act that it can serve as a paradigm of the kind the legal or moral norm identifies for its purposes. For example, the famous Tuskegee experiments for tracking syphilis are virtually defined for us by the deception of the subjects and by their later, discriminatory neglect; these experiments epitomize our moral and legal repugnance for those kinds of acts.

However, it happens frequently enough to make normative guidance problematic, that it is not immediately clear whether the choice one seeks normatively to guide has features that relate it to the terms of moral norms. For example, it is frequently unclear whether withholding this life sustaining treatment is a choice to end a patient's life, or rather a case of declining futile, harmful or otherwise inappropriate treatment. These are the difficult cases casuistry addresses, and reasoning is directed to clarifying the morally relevant features of the perplexing choices we sometimes face.

This unclarity about the proper identification of an act can emerge when people disagree about how an action should be categorized, but can also arise when perplexity on this matter exists within a person's individual moral reflection. One source of such disagreement and perplexity is the simple opacity of some actions. These actions appear initially to lack any features that relate them to the terms or goals of normative systems. At the same time, these actions seem also to have social or human significance which brings them under the direction of legal or moral norms. The new forms of human reproduction made possible by developing biotechnology are opaque to many people in this way: they believe that law and morality should direct such actions, but their morally relevant features are difficult to perceive and articulate. These technologies promote fertility and help people solve very distressing personal problems, yet they seem to manipulate human procreation and treat new humans as objects. The moral perplexities surrounding mutual nuclear deterrence between nations capable of mutual destruction also arise because of this kind of opacity: how can one seriously threaten to kill innocents when the whole point is never to execute the threat?

One reason for this kind of opacity is that some actions appear to fall, or really do fall, along the fuzzy boundaries of existing norms. For many people the telling of the harmless but necessary or helpful white lie falls into this category: lies have several wrong making features which may seem to exist only minimally, if at all, in harmless white lies, communicative acts that nevertheless remain deliberately deceptive.[2]

Another source of the disagreement and perplexity that trigger casuistical reasoning is that some actions have, at the same time, features which make acts prohibited, and, in addition, features which render acts permissible or required. This can happen because actions appear to fall, or do fall, on the fuzzy boundaries of several norms, some of which allow or enjoin what others prohibit. Thus, for example, some acts of killing, such as those involved in separating Siamese twins, seem more like deflecting harm from an innocent party than intentionally imposing it on the party harmed. The latter is a paradigm of prohibited killing and, according to many moral codes, is absolutely prohibited. Yet the former can be good and even obligatory. Similarly, some acts of abortion seem more like evicting an unwelcome visitor, which can be permissible, than killing an innocent and dependent person, which is prohibited.

Controversial or perplexing characterizations of actions can trigger an inquiry in which one seeks grounds for determining whether an opaque action really has the suspected morally relevant feature, or to which category an action seeming to have conflicting moral features belongs. That inquiry necessarily involves reasoning, but the reasoning involved is not drawing conclusions from premises, either deductively or inductively. For the logic involved here must be that of rational categorization, classification or definition, and this, though connected in various ways with well known logical procedures, is neither deductive inference nor inductive generalization. Efforts to classify proceed by comparisons and contrasts with similar and dissimilar things. They are tested by appeals to ordinary language, intuitions, counter-examples, considerations of logical implications, and controlled by a steady focus on the purpose of the classification or taxonomy; the inquiry is broadly dialectical (Miller, 1996, pp. 237-241).

The devices that implement this logic of classification are varied. Analogies, stories, comparisons to simple paradigm cases, are standard fare in casuistical writing. Even more than the devices, the styles of casuistry are varied. Western Christian casuistry is scholastic, worked out in a set of technical categories with close relatives in canon law and Church teaching; moreover, Catholic casuistry is dominated normatively by a conception of morality as unified in the principles of the natural law. Although this normative framework is usually present, it is sometimes not more than a horizon for the application and development of a moral code whose structuring by moral principle is not obvious. The secular descendants of Western Christian casuistry, in areas such as bioethics and international ethics, largely reject the natural law horizon of Catholic casuistry in favor of a more pluralistic normative theory, and abandon or recast many of the scholastic categories. Jewish casuistry is neither governed by a general moral theory nor scholastic in its analysis. Traditional texts and authorities are creatively used to illuminate the moral character of actions; the ordering of moral concerns that a unified theory can provide seems deliberately eschewed.

Plainly, there is a creative moment in casuistical reasoning. The right analogy or contrast, the relevant story or precedent that clinches one's understanding of an action's normative character, cannot be programmed. One must come to understand the normative meaning of one's and others' choices, and that requires insight into both one's action and the point of the norm. This creative moment is most likely to arise in the thinking of those who care deeply for the purposes of moral and legal norms and seek seriously to organize life around these normative aims. So, the view that virtue, and in particular the virtue of practical wisdom, may be closely connected to the practice of casuistry, and, furthermore, may be needed to distinguish its use from its abuse (Jonsen and Toulmin, 1988, pp.58-74; Miller, 1997, pp. 8-10).

Nevertheless, this creative moment occurs in a process of reasoning which results in propositions about one's actions and their normative relevance. These propositions can be true or false and more or less well supported. To determine generally the conditions of truth and epistemic warrant for these propositions, we need a broadly logical study of the process of casuistical reasoning; this is hardly begun in the current literature on casuistry. The ongoing, rigorous practice of casuistry provides an extensive and formidable subject matter for logical and methodological reflection which can surely improve practical moral thinking. Yet the current writing about casuistry hardly broaches the logical and methodological issues.[3] The needed logical and methodological work on casuistical reasoning – not general ethical observations about the importance of practical wisdom – will contribute to removing the cloud of sophistry that continues to hang over this necessary area of moral thought.

When successful, casuistical reasoning removes a block to evaluation by making clear the normatively morally relevant features of the action being evaluated. Thus, although casuistry normally ends with a practical verdict, that does not come from casuistical reasoning alone, but from the normative conviction whose relevance the casuistry clarifies. In the simplest case, a norm, justified within a legal code or moral system, is seen to apply unambiguously to the action clarified by casuistical reasoning. For example, a decision to stop feeding a person in PVS can in some cases be revealed by casuistical reasoning to be morally identical to a choice to abandon that person, the norms generally forbidding the latter and then seen as applying to the PVS decision.

In many cases, however, things are more complicated. In particular, the casuistical clarification of an action may starkly reveal that there are several aspects of the act which point in different normative directions. Here, plainly some kind of ranking of normative considerations is needed if a normative verdict is to be reached. The ranking of normative considerations in ways sensitive to the results of casuistical analysis is commonly thought to be constitutive of casuistry (Miller, 1996, pp. 25-32). But normative ranking involves a different kind of judgment than that to which the classificatory techniques of casuistry are directed: "morally relevant feature P is more important than (or trumps) morally relevant feature Q" is a different sort of judgment than "action A has morally relevant features P and Q." The reasoning to establish the truth of the latter hardly affects the warrants for accepting the former. The rhetorical power of the most effective forms of casuistry may cause people to believe that the analysis itself settles normative and not simply classificatory issues. If there is a genuine connection here, it has yet to be developed, as I hope to reveal in the immediately subsequent paragraphs.

The effort to classify acts so as to reveal their morally relevant features is part of people's larger project of organizing and directing their lives by moral principles. Honest casuistry, therefore, presupposes that we have moral convictions and a will to virtue. Still, the justification of moral convictions is logically distinct from their practical application and development in casuistry; the acceptance of norms as authoritative is a propositional attitude distinct from judging that some feature of an action – "a morally appreciated circumstance" (Jonsen, 1996, pp. 44-45) – relates the action to the norm. And this difference obtains even when one's normative convictions emerge articulately only in reflection on particular cases.

The analogies and comparisons used in casuistical analysis frequently embody normative convictions and rankings. Therefore, a person can come to see the moral relevance of some feature of an action and in that very understanding come to accept the norm or normative principle which makes the feature relevant. This is not logically different from accepting an already articulated norm from a legal or moral code or from some other paradigm, such as the example of a hero or a saint (Jonsen, 1996, pp. 44-45).

Nevertheless, when normative acceptance emerges only from the result of casuistical reasoning, critical normative questions can be overlooked. In particular, when questions concerning the justification of the norm are overlooked, the generality and priority of that norm in relationship to other norms are likely to be settled tacitly and uncritically. If there are concerns about such matters, more straightforward normative analysis and critical reflection seem required (Arras, 1991, pp. 39-41). Thus, although a good deal of practical ethical thinking can be conducted by considering cases alone, not all the questions raised by casuistical clarification can be settled without reference to the more abstract arguments of moral theory.

The particularity-sensitive virtue of practical wisdom is often invoked at this point to provide an alternative to introducing abstract moral arguments. However, when the issue is the ranking of normative considerations, shown by casuistry to have bearing on an action, this alternative seems to be either intuitionist or authoritarian, and lacks an account of how the intuition works or on what the authority is founded. In the light of what considerations does the practically wise person rank norms? Questions such as this are raised by the practice of casuistry and are especially important for those who wish to conduct moral reasoning wholly or primarily by reasoning from cases.

The fact that casuistry is embedded in discourse shaped by normative convictions implies that casuistical techniques can be used rhetorically, not simply to clarify action, but to express preferences and to persuade. This persuasive capacity obtains even when there is no need to rank the norms applying to the action. For the sharp perception that an obscure action does have a morally decisive feature or the recognition that an overlooked norm is saliently embodied in an action can motivate the choice to perform that action. Consequently, there is reason for the frequently asserted connection between casuistry and classical rhetoric. Cicero's *De Officiis* (1961) placed classical antiquity's most influential casuistry in the context of moral exhortation; St. Augustine's casuistry on such things as lying, suicide and marital ethics were in a similarly persuasive style and context. The contexts of preaching, repentance and penance have surrounded much of the good casuistical work ever since (Mormando, 1995, pp. 55-84). This link cannot be accidental or simply historical, but its significance for moral and legal inquiry remains unexplored. Is there something in the rhetorical character of casuistry that throws light on the unresolved logical and methodological concerns already noted? Does this rhetorical context provide some of the grounds for distinguishing the use from the abuse of casuistry? Such questions have not yet been carefully pursued.

II. THE IMPORTANCE OF CASUISTRY

Since casuistry deals with perplexing or controversial cases, and is most commonly used to deal with them in a practical context of guiding choices by legal or moral norms, its importance as a part of practical moral thinking, and as part of any adequate account of moral life, is plain. Casuistry becomes a necessary part of the rational direction of actions by norms whenever the fit between the norm and an action to be directed is obscure. So, casuistry is an essential ingredient in ethics, as even such allegedly anti-casuistical and surely non-casuistical moralists as Sidgwick and Moore acknowledged (Sidgwick, 1966, p. 99; Moore, 1971, pp. 4-5). Sidgwick and Moore understood casuistry as the application of general moral norms to particular actions, and so as part of what they quaintly referred to as scientific ethics.

Jonsen and Toulmin introduced the current discussion of casuistry by characterizing it as a way of avoiding what they call the "tyranny of principles." They present a reading of the academic moral philosophy formed by philosophers such as Sidgwick and Moore, and dominated by metaethical abstractions and moral theory preoccupied with general principles. The suggestion is that casuistry is most at home in an open textured moral theory that is neither dominated by general principles or preoccupied with ethical abstractions.

Casuistry is plainly needed to develop and apply a legal system structured like the common law; it is also needed to develop and apply moral codes like those of Judaism and Christianity in which there are a plurality of precepts without a rigorous ordering. The currently popular form of intuitionism, in which commonly accepted moral judgements are taken as basic, makes extensive use of casuistry to extend moral conviction from clear cases to obscure cases (Brody, 1979, pp. 446-451; Keenan, 1993, pp. 294-315) . Only a very extreme form of intuitionism, in which every concrete moral judgment is taken to be based on a discrete and unargued moral intuition escapes the need for casuistical clarification. But casuistry also has a role within general moral theories, as Sidgwick's and Moore's appreciation of it indicate. To change the example, it is not clear why the medieval Christian effort to systematize the precepts of morality around basic principles such as the Love Commandments or the Golden Rule should make the natural law theory thus constructed any less friendly to or dependent upon casuistical reasoning than less systematic approaches to morality. Even a tightly structured moral theory based on a single foundational principle will need a clarification of human action to allow the application of its central moral predicate, for example, loving one's neighbour as oneself, or respecting rational nature as an end in itself, to actions beyond those to which it paradigmatically applies. This clarification will consider actions under more or less specific descriptions; to be complete, it will take into account all the morally relevant features of any action to be evaluated. Alan Donagan's account of the structure of a moral theory based on a Kantian first principle makes plain that the deductive structure of the derivation of precepts requires premises that link action descriptions to the basic moral predicates. His account shows that these premises, even those concerning actions very generally described, are vindicated by the same sort of informal, classificatory reasoning as is found in casuistry (Donagan 1977, pp. 66-74; cf Miller, 1996, pp. 236-240).

Only the strictest form of intuitionism about particular moral judgements, mentioned above, and the most direct forms of consequentialism seem able to dispense with casuistry. Direct conseqentialism applies its good promoting norm to acts by considering only one factor about them, their contribution to the net goodness of the undertaking in comparison with the other available alternatives. The inquiry into which alternative is on balance best leads to the only morally relevant classification of actions such consequentialists recognize, so perhaps even this kind of projection and comparison of consequences might be labelled casuistry. However, this is hardly the inquiry into the clarification of action which casuistry has ordinarily undertaken: analogies and taxonomies are very different than projecting consequences and weighing values. Indirect consequentialisms, by contrast, can have an interest in the plurality of morally relevant features of actions which makes moral taxonomy valuable, even if the final account of that relevance is consequentialist.

In short, the connection between casuistry and various moral approaches is far from clear. Its methods appear to be needed by virtually all approaches to morality. They certainly have proved vital for the conduct of bioethical inquiry as it has actually developed over the last quarter century.

In addition to its apparent necessity as an element within virtually all normative approaches that hope to guide choices in a precise way, casuistry has held out the promise of delivering practical consensus in the face of deep moral controversy. Jonsen and Toulmin report their own experience of the capacity of casuistical reasoning in the pluralistic context of a National Commission to reach practical agreement about a set of ethical guidelines for research on children. The moral consensus that emerged about the core practical issues was based on casuistical reasoning and not ethical theorizing. That consensus would have been impossible if the focus were on the divergent theoretical rationales for the consensus (Jonsen and Toulmin, 1988, pp. 16-19).

In a practical context as vexed as bioethics, that of international affairs, Michael Walzer has provided an account of why such consensus should be possible. He thinks that, for all our differences generated by allegiances to diverging moral theories and by membership in particular moral communities, modern humans inhabit a common moral world. That social world is structured by moral convictions and arguments that have passed muster to the point that they are widely accepted as having some interpersonal objectivity. This objective moral world is structured by norms that are propositionally formulated and whose implications can be debated as part of the effort to confirm and develop the existing consensus (Walzer 1977, pp. xv, 44; Boyle 1997, pp. 83-89). Casuistry is the main form of reasoning for developing the consensus, as Walzer's careful development of the categories of just war doctrine reveals.

Health care plainly falls within the common moral world. The common expectations people have in dealing with its institutions and professions presuppose significant common understandings of its purposes and constraints. The work of ethics committees, the practices of informed consent and so on all assume that, whatever the deep disagreements among those who participate in health care, some agreements and some bases for discussing moral issues exist. The legal and moral norm requiring the consent of competent patients to the health care offered them is a striking example of a norm with a firm place in the common moral world, one with casuistical potential to clarify other matters, such as the consent to treatment for minors and living wills, and possibly even issues related to euthanasia and assisted suicide

Plainly, the use of casuistry to develop and expand areas of moral agreement will be as limited as the areas of agreement people begin with (Arras, 1991, pp. 42-44). But Jonsen and Toulmin's example, Walzer's just war casuistry, and the primacy of patient choice in health care decision making, suggest that there are more areas of agreement than we suppose, and that there may be some agreements where we think they cannot exist. So, the effort to engage in the casuistry of the common moral world in the context of recognizing deep differences of moral outlook and judgment does not seem pointless. There are bound to be urgent practical issues whose resolution requires agreement among people deeply in disagreement, but where casuistry that seeks to expand or reveal existing consensus cannot generate the needed agreement. This seems true of many of the great issues that now divide people: abortion, suicide, nuclear deterrence, and so on. But there is no a priori way to tell, prior to doing the casuistry, whether the resolution of any given case lies outside the boundaries of potential moral consensus. One need not accept a common awareness of the natural law to suppose that a person's recognition through casuistical argumentation of the moral relevance of a feature of an action can expand that person's moral vision or even change his or her moral views.

Casuistry's potential for expanding a moral consensus is also important within moral communities. The practical implications of common moral allegiances are often obscure. The results of compelling casuistry can, over time, clarify these implications so as to remove group bias and overcome disagreements that can fracture communities. The working of Roman Catholic casuistry in recent years has not overcome all the moral conflicts dividing Catholics, but on some important issues, for example limiting capital punishment and resisting the idea that patients in PVS may be abandoned without food and water, seems to be moving towards consensus.

III. THE LIMITS OF CASUISTRY

It is clear that casuistry is an essential element in moral reasoning, however, it is not the whole of moral reasoning if that includes the entire effort to bring moral life and discourse under rational scrutiny. In particular, casuistry's method aims to identify morally relevant features of actions, but casuistry itself is not a method capable of rationally assessing the normative factors – precepts, principles, character traits or other paradigms – in virtue of which these features become "morally appreciated circumstances." A person's awareness of such normative factors may emerge in casuistry, but that recognition does nothing to justify the normative factors (Arras, 1991, pp. 42-44).

Unless the norms one comes to recognize in casuistry are capable of critical scrutiny, the casuistry as clarification is as likely used to rationalize social bias or personal immorality or prejudice as it is to bring one's choices and morally justified standards into contact. The point here is not the obvious one that casuistical reasoning can break down like any other kind of reasoning; it is that casuistry is abused if put to morally unworthy purposes. This cannot be avoided unless the norms whose relationship to actions casuistry reveals are understood to be sound. In short, casuistry must be carefully done, but it must also be driven by moral seriousness, and moral seriousness requires critical reflection and sound moral conviction. Virtue alone, without such reflection, is either blind or not real moral seriousness.

Consequently, the laxist abuse of casuistry that has given the method a bad name is not inherent in the enterprise, but an abuse based on a desire to excuse, a desire itself unjustified if it includes a refusal to apply the moral truth to the case at hand. For example, in the laxist abuse of the idea of mental reservation, the fact that an expression could in some context have a different meaning than the meaning it is used to communicate has no moral significance, but is taken as sufficient for that communication's not being a lie.

A preoccupation of casuists since the time of Cicero – the balance between expedience and duty – seems to mark another aspect of the normative limitation of casuistry. Cicero devoted book III of *De Officiis* (1961, pp. 270-403) to this question. He showed by casuistical analysis that some conflicts between casuistry and duty disappeared when all the circumstances relevant to determining concretely one's duty were considered (p. 287). However, his more general view that duty and true expedience can never conflict is based on specifically normative convictions that most people now find incredible. Without the harmony guaranteed by this general normative position, Cicero would have nothing to say about when expedience should prevail over morality, and vice versa. His Stoicism, not his casuistry, settles the matter by preventing its being a real question.

Michael Walzer's treatment of nuclear deterrence and what he calls "supreme emergency" also illustrates this limitation. He argues that there are, in political contexts, circumstances of supreme emergency which allow considerations of necessity to override common moral considerations. The moral significance of these circumstances does not emerge from the broad consensus on just war, but from a set of normative considerations, more "realistic" or consequentialist in form and content, which claim to overturn the application of the consensus (Walzer 1977, pp. 251-283; Boyle, 1996, 92-97). Here, a normative conviction, perhaps made more acceptable by Walzer's casuistical narrative, is used to block the direction the casuistry seems to go. Whether or not the normative commitment is justified, it is beyond the limits of casuistry, as Cicero's contrary judgment suggests (1961, p. 317).

The attempt to extend the implications of the bioethical consensus on the right of competent patients to refuse medical treatment so as to justify euthanasia and physician assisted suicide also reaches beyond the boundaries of the common moral world, not by the compelling logic of casuistic analysis, but by imposing on this right an interpretation based on a controversial conception of autonomy or an extension based on consequentialist reasoning. But the reality of these normative limits does not devalue the essential role of casuistry in moral life and reasoning, or the importance for bioethics and other areas of applied ethics of investigating its logical and methodological structure, its role in the history of ethics, and its relationships to the sources of normative judgment, to the dynamics of virtuous living, and to the logic of persuasion.

St. Michael's College and
The Joint Centre for Bioethics
University of Toronto

NOTES

1. Jonsen and Toulmin (1988) provide a useful introduction to Western Christian casuistry; R.M. Wenley (1910) and N. Biggar (1989) provide information about Protestant practice of and views about casuistry; D. Feldman (1968) and B. Brody (1989)provide examples of Jewish casuistry; S. Freehof (1973), provides some history and many examples; M. Walzer (1977) provides a justly famous casuistical treatment of just war doctrine, making use of religious casuistry and international law.
2. St. Augustine rejected such lies as immoral. But many people do not see how the wrong making features of lies that are plainly wrong obtain in such white lies. Casuistical reasoning about cases of lying comprise an important part of Roman Catholic casuistry beginning with St. Augustine, who argued on a case by case basis that no type of lying could be justified. Augustine's absolutism on lying incorporates a conception of lying as intentional deception by affirming what one believes false. This was widely accepted in the Western Church, but the development after Aquinas of the idea of broad mental observation – the idea that uncommunicated ambiguities in one's language are sufficient to render a deceptive act something distinct from a lie – seemed to reduce truth telling to cleverness. History's verdict on this particular form of casuistry shows that the fuzzy edges of normative concepts can be made more precise. Jonsen and Toulmin provide an entre into this rich discussion (1988, pp.195-215).
3. Miller (1996, pp. 17-38) provides a rich account of the logical workings of casuistical reasoning, with ample illustration. On his account, casuistry includes specifically normative elements such as ranking norms. On the contrary, I believe that what is specific to the taxonomic activity of casuistical reasoning is the clarification of action.
4. The literature on casuistry contains much history indispensable for the needed logical work (Keenan and Shannon, 1995). The focus in the literature is on the normative significance of casuistry, not on the details of its logical workings. Grisez (1997, pp. 849-897)provides a careful, analytical treatment of categories pertinent to Catholic casuistry, that is, the ideas such as intention, side effect, and cooperation; continuing disputes about the application of these notions suggest how difficult it is to have confidence that casuistical clarification has succeeded.

REFERENCES

Arras, J.: 1991, 'Getting Down to Cases: The Revival of Casuistry in Bioethics,' *The Journal of Medicine and Philosophy* 16, 29-51.

Biggar, N.: 1989, 'A Case for Casuistry in the Church,' *Modern Theology* 6, 29-51.

Boyle, J.: 1997, '*Just and Unjust Wars:* Casuistry and the Boundaries of the Moral World,' *Ethics and International Affairs* 11, 83-98.

Brody, B.: 1979, 'Intuitions and Objective Moral Knowledge,' *The Monist* 62, 446-455.

Brody, B.: 1989, 'A Historical Introduction to Jewish Casuistry on Suicide and Euthanasia,' in B. Brody, (ed),*Suicide and Euthanasia: Historical and Contemporary Perspectives*, Kluwer Academic Publishers, Dordrecht, The Netherlands, pp. 39-76.

Cicero, M.: 1961, *De Officiis* with an English translation by W. Miller, Loeb Classic Library, William Heineman, Ltd. and Harvard University Press, London and Cambridge, MA.

Donagan, A.: 1977, *The Theory of Morality*, University of Chicago Press, Chicago and London.

Feldman, D.: 1968, *Birth Control in Jewish Law: Marital Relations, Contraception and Abortion as set forth in the classic texts of Jewish Law*, New York University Press and University of London Press Limited, New York and London.

Freehof, S.: 1974, *The Responsa Literature and A Treasury of Responsa*, KTAV Publishing House, Inc., New York.

Grisez, G.: 1997, *The Way of The Lord Jesus: Volume Three; Difficult Moral Questions*, Franciscan Press, Quincy, Illinois.

Jonsen, A.: 1991, 'Casuistry,' in W. Reich (ed.), *Encyclopedia of Bioethics: Volume 1*, Simon and Schuster, New York and London, pp. 344-350.

Jonsen, A.: 1996, 'Morally Appreciated Circumstances: A Theoretical Problem for Casuistry,' in W. Sumner and J. Boyle (eds.), *Philosophical Perspectives on Bioethics*, University of Toronto Press, Toronto, pp. 37-49.

Jonsen A. and S. Toulmin: 1988, *The Abuse of Casuistry: A History of Moral Reasoning*, University of California Press, Berkeley and Los Angles.

Keenan, J.:1993, 'The Function of Double Effect,' *Theological Studies* 54, 294-315.

Keenan, J. and T. Shannon, (eds.): 1995, *The Context of Casuistry*, Georgetown University Press, Washington, D.C.

Miller, R.: 1996, *Casuistry and Modern Ethics: A Poetics of Practical Reasoning*, University of Chicago Press, Chicago and London.

Moore, G.E.: 1971, *Principia Ethica*, Cambridge University Press, Cambridge.

Mormando, F.: 1995, 'To Persuade Is a Victory: Rhetoric and Moral Reasoning in the Sermons of Bernardino of Sienna,' in J. Keenan and T. Shannon (eds.), *The Context of Casuistry*, Georgetown University Press, Washington, D.C., pp. 55-84.

Sidgwick, H.: 1966, *The Methods of Ethics*, Dover Publications Inc., New York.

Walzer, M.: 1977, *Just and Unjust Wars: A Moral Argument with Historical Illustrations*, Basic Books Inc. Publishers, New York.

Wenley, R.: 1910, 'Casuistry,' in J. Hastings (ed.), Encyclopedia of Religion and Ethics: Volumee III, T&T Clark Charles Scribner's and Sons, Edinburgh and New York, pp. 239-247.

DAVID C. THOMASMA[†]

VIRTUE THEORY IN PHILOSOPHY OF MEDICINE

Let us dedicate ourselves to what the Greeks wrote so many years ago: to tame the savageness of man and make gentle the life of this world (Kennedy, 1999).

I. INTRODUCTION

Can virtue theory be helpful for philosophy of medicine and health care today? I have proposed with Pellegrino that it can (1993a; 1996). However, in those arguments, we cautioned that virtue theory cannot stand alone, as it might have been able to do in earlier, more homogenous times. The postmodern world emphasizes differences, built on cultural fractures, relativism, and almost spectacular individualism – the opposite of homogeneity. The postmodern crisis in culture also affects our thinking about bioethics and professional responsibility in medicine. We have to take seriously the challenge of pluralism and its impact on assumptions about the variability of goals and values in health care (Engelhardt, 1996). These changes affect the very foundations of medicine and medical practice. Without common agreement about goals and values, virtue theory cannot survive in health care.

Virtues have long been associated with a mean between two extremes of personality, emotion, or character. The theory itself today is also at the mean between extremes in philosophy of medicine and health care, and those found in bioethics. On the one hand virtue theory corrects for over-reliance on abstraction and a search for absolutes by emphasizing the developing and concrete interactions in being sick and in healing. On the other hand, virtue theory avoids over-emphasizing the concrete and particular to the point of relativism. Thus it is positioned between the abstract and universal on one hand and the pluralistic and relative on the other. The reason the theory avoids the extremes is that it is based on common human structures of existence and on personal and social developmental standards.

This essay, therefore, examines the role of virtue theory in medicine today. Can it help "tame the savages of man," and aim human lives engaged in the practice of medicine at the good?

G. Khushf (ed.), Handbook of Bioethics, 89–120.

II. BACKGROUND

In general, the history of virtue theory is well-documented (Sherman, 1997; O'Neill, 1996). Its relationship to medicine is also recorded in our work and in that of others (Pellegrino and Thomasma, 1993b; 1996; Drane, 1994; Ellos, 1990). General publications stress the importance of training the young in virtuous practices. Still, the popularity of education in virtue is widely viewed as part of a conservative backlash to modern liberal society. Given the authorship of some of these works by professional conservatives like William Bennett (1993; 1995), this concern is authentic.

One might correspondingly fear that greater adoption of virtue theory in medicine will be accompanied by a corresponding backward-looking social agenda. Worse yet, does reaffirmation of virtue theory lacquer over the many challenges of the postmodern world view as if these were not serious concerns? After all, recreating the past is the "retro" temptation of our times. Searching for greater certitude than we can now obtain preoccupies most thinkers today. One wishes for the old clarity and certitudes (Engelhardt, 1991). On the other hand, the same thinkers who yearn for the past, like Engelhardt sometimes seems to do, might stress the unyielding gulf between past and present that creates the postmodern reaction to all systems of Enlightenment thought (1996).

Still, it is simplistic to think that the breach between past and present is either ignorable or equally, irrevocably unbridgeable. MacIntyre notes that practices give rise to intrinsic goods that might lead to specific values and norms from which an ethic may be derived and grounded (MacIntyre, 1990). This approach is followed in this essay, with an effort at grounding the goal of healing in medicine in a foundational reality that is cross-cultural and trans-historical.

III. VIRTUE THEORY IN MEDICINE

Virtue theory in medicine has been associated with other movements against principlism in modern medical ethics and might seem to contribute to the blurring of norms and standards of conduct (Veatch, 1985, pp. 338-340). In fact, the increase in interest in the virtues is prompted by the realization that theory is too abstract to contribute much to the discussion. Regarding abstraction Rorty notes:

> Somewhere we all know that philosophically sophisticated debate about whether human nature is innately benevolent or innately sadistic, or about the internal dialectic of European history, or about human history, or objective truth, or the representational function of marriage, is pretty harmless stuff (1989, p. 182).

The rich phenomenological and existential context of illness, finitude, disease, and death provides a personal biography for each individual that requires a moral theory of balance – a balance between an ethic that is too rationalistic and abstract and one that is too relativistic and pluralistic. Thus, Rorty argues that instead of being formed through theory one's identity and self-consciousness as a moral agent is formed from very specific and unique associations occurring in one's life (1989, p. 153-154). This viewpoint corresponds very well with the thesis that moral growth and development is guided by the virtues rather than by abstract principles. The virtues, and virtue theory, permit one to examine a moral life over time and, by learning from mistakes and good work, to choose more wisely in the future.

At the very least, notwithstanding the obvious benefits, one might be left with a suspicion that if increasingly we adopt virtue theory in medicine, we would simultaneously deemphasize objective standards associated with rights and duties. These are hard-won concepts, achievements in a pluralistic society that we would not want to lose (Veatch, 1985: 336-337). Of course their loss is not necessary for the addition of virtue theory to theory about medicine (Dunne, 1993). A good part of this essay will demonstrate why.

At this point it may suffice to note that the standards on which the virtues must rest in order to point the way for developing a moral and professional life within medicine arise from common structures of human existence. These existential structures concern finitude, becoming sick, having diseases, and facing death. The common structures are not "strange" to any human being. Thus those who profess to care for these occurrences, for the outbreaks of our finitude as it were, are not operating as totally moral strangers either. They can be expected to meet certain standards raised by the expectations and promises to treat.

The fundamental philosophical position for this view is a relational one. Unlike other communities such as family, civic unit, religion, or state, medicine rests on the existential fact of illness in the deepest structures of human life (Pellegrino and Thomasma, 1981). At its root medicine is a devotion to sick persons by the community. Health care is more than a commodity; it is a commitment to one another in the community. As Stanley Hauerwas says of medicine:

> Medicine involves the needs and interests that we all share. All of us wish to avoid untimely death. All wish to avoid unnecessary suffering. All wish to be cared for when we are hurt...Medicine provides a powerful reminder...of our "nature" as bodily beings beset by illness and destined for death. Yet medicine also reminds us it is our "nature" to be a community that refuses to let suffering alienate us from one another (1986).

Glaser underlines the importance of community when considering beneficence, arguing strongly that a "community of concern" devolves around the very notion that individuals are persons-in-relationships. The normal state of such persons is to be "reciprocal, responsive, and engaged." By this he means that beneficence is required for a well-functioning community; it is linked to our very nature as social beings (Glaser, 1994).

Alastair B. Campbell has taken an argument Pellegrino and I made in this context forward to a more controversial step, agreeing with us that autonomy and self-determination is a moral presumption in health care that tends to emphasize individualism and, therefore, an inadequate account of the therapeutic [and I would add, research] relationship (Pellegrino and Thomasma, 1988). His more controversial step comes in the claim that a fundamental characteristic of human life itself is dependency, a kind of "creatureliness" that necessarily entails dependency on others for the fulfillment of many of our needs. He says of this state:

> To be a creature is to be born of others, to know ourselves through them, to depend upon them and create dependency, to know the pain of losing them and finally to be the instance of that pain to others (Campbell, 1994).

This fundamental existential characteristic of human life means that dependency on one another is a common structure of human existence. Many times there are induced dependencies that are temporary, e.g., nursing at the breast, or anesthesia for surgery. But there are also necessary dependencies that arise from choices we make in life. The obvious example is one created by our love of others, or the nurturing of our parents. But it is important to consider health care itself as one even beyond a chosen environment of dependency, not so much chosen or imposed on us as accepted as part of our creatureliness. I will return to this point in the concluding section of the essay.

Patients, as citizens, are responsible to one another for the kind and manner of care they will receive in the future. Yet their power to shape that care is in jeopardy by the current vulnerability of some of the potential subjects to whom they are related through the existence of illness in the community, among other things.

As a result of these and like reflections, it can be reasonably asserted that the norm that arises from the practice of medicine upon which to build a moral basis of medicine is beneficence, acting in the best interests of others (Pellegrino and Thomasma, 1988). The goal of medicine is healing and the requirements to heal inform the virtues that must be practiced (Pellegrino and Thomasma, 1981, pp. 170-191).

IV. THE RECOGNITION OF ETHICAL COMPLEXITY

Current moral discourse includes an honest recognition of the complexity of life and the practices about which we reflect and argue. This complexity defines our efforts to adequately describe the situation of modern medicine, its ethics, and public policy. It characterizes the "critical turn" of our times, not so much a turn to metaphysics as to questions about the interrelatedness of human concourse and practices. Due to increasing recognition of this complexity, particularly the awareness of different cultural viewpoints and their apparent lack of relation to one another, an avenue is created for ethical and philosophical theories that are more complex than standard accounts.

So, for example, the virtue of professional integrity (Pellegrino and Thomasma, 1993b, pp. 127-143) is one that formerly might have been analyzed in singular terms, underscoring the heroism of a person of principle. Today, though, we recognize that the virtue may impinge, not only on the physician's self-interest, but also on the moral center of professional life itself. Putting the patient first sometimes must even mean acting against one's convictions. That very statement, however, demonstrates how variable can be the context in which the virtue of integrity, or any of the virtues, is practiced. This is why the standards provided by principles are so important to virtue theory as well as to other theories of medical ethics.

Ruth Macklin's reaction to a South American physician who allowed a young woman to bleed to death after a botched abortion is a case in point. The physician's view was that her abortion was immoral and that he could not add to the evil by saving her from the consequences of her action. Despite a conviction that cultural pluralism should be honored, Macklin was horrified (1999). In her book she also recounts other cultural practices and beliefs that belie fundamental "ethical universals" (distinct from moral absolutes in that they admit of a variety of interpretative practices). These are seen to pertain to human rights in health care (Macklin, 1999, 220).

Like most of us, Macklin apparently assumed that physicians should be devoted to acting in the best interests of patients, and in this case, saving the woman's life. We could even postulate a rule derived from the goals of medicine to heal that, without exception, physicians must treat an illness or accident regardless of its cause. This experience, among others, prompted Macklin to write about moral standards in a pluralistic age. She argues that among the ethical universals are respect for persons, affirmations of privacy, support of liberty, and the duty to avoid pain and suffering (Veatch, 2000).

This and similar reactions lead to a conviction that, despite the pluralism of our times, some standard must be instilled in every physician according to which the source of any physical difficulty is irrelevant to the duty to save life or to honor the wishes of patients. Camus' dictum about the purpose of human concourse could readily by adopted for medicine: "The Other Person in Need, rather than my moral convictions, sets the norm" (Camus, 1948).

The problematic assumption in such cross-cultural perceptions, however, is that medicine is a social entity in its own right which can be superimposed upon the various cultures in which it is practiced without much adaptation. In fact, however, medicine exists enculturated, as all human activities do.

Concern about the relation of medical practice to contemporary culture and the inbuilt values and changing expectations accompanying it prompt the exploration of virtue theory in medicine today. Was that South American doctor being virtuous in answering to a higher norm of ethical responsibility or was he neglectful, criminally neglectful, by summoning his own beliefs about moral action and consequences to override his dedication to save life in the emergency room? This is not a trivial question. It appears in many contexts today.

Bioethics Critical Turn

After almost thirty years of successful growth, modern secular medical ethics has now turned to critical self-reflection. Indeed this reflexive turn would occur in the natural development of any discipline. It is particularly timely at this period in history. Modern communication technologies not only permit us, but actually force us, to consider the linkages of all persons and cultures.

Instantaneous communication with other patients also enable individuals to take greater and greater responsibility for their choices in health care. They can by-pass professionals altogether (Koch, 1995). Simultaneously, managed care in the United States and reorganization and privatization of health care in national health care plans elsewhere raise questions of standards of care, the common good, and personal responsibility for one's health, even to the point of tracking one's test results (Landers, 1999).

Among the responses to all of these changes in health delivery and health systems is a fundamental questioning of the validity of ethical recommendations, policies, norms, and judgments themselves. This is especially true of ethical recommendations based on a monochromatic cultural assumption about the importance of autonomy (Holm, 1995). Actually by "respecting autonomy," patients seem to understand the desire to be informed, to know something of their physician's values, to be assured the physician is acting for their best interests, to retain veto power over suggested treatments, and enjoy a variable degree of freedom depending on their personal values (Schneider, 1998).

Similarly doctors also must have respect for their own autonomy. The resulting interaction is a far cry from the usual connotation of unrestrained freedom and self-governance. In fact, autonomy, like all other values in health care, arises from the relationship of healing itself. It is shared (Bergsma and Thomasma, 2000).

The same critique of the primacy of any cultural assumptions could apply to any set of culturally-embedded values as applied to bioethics. A common mistake is to criticize Western value assumptions, and then canonize those of other, sometimes more primitive cultures as alternatives to Western thought. A good example is the excellent work done on Navajo culture regarding the institution of life support systems, or the exploration of Gypsy abhorrence of IV lines (Marshall et al., 1998). If these explorations were to be interpreted as providing a "better" model of health care intervention than those in Western culture, such an interpretation would constitute cultural hegemony in reverse, albeit a kind of trumping of the strong by the weak.

Inadequacy of Principlism

To properly place the role of virtues and virtue theory in medicine today, we must examine first the growing self-reflection about ethics, its critical turn, by looking primarily at the criticism of principlism. If at least some of the criticism of principlism is valid, then we are justified in turning away from purely objective standards to more intersubjective ones, among them the virtues themselves. At first this seems odd in itself, since a partial critique of objective standards, namely that they do not contain sufficient information about the particularities of cases, the richness of the emotional and relational conflicts that arise in each instance of a moral dilemma, does not in itself preclude at least some validity for setting objective standards. Indeed this is the reason Pellegrino and I argued earlier that virtue theory in medicine required at least some reference to the principles in the role of setting objective standards. As an Aristotelian moderate realist I most often find that the truth lies somewhere in the middle where most of the debate occurs.

In the last thirty years, the philosophical underpinnings of medical ethics have undergone a profound metamorphosis. Physicians and other health workers must be familiar with shifts in contemporary moral philosophy if they are to maintain a hand in the restructuring of the ethics of their profession. They all need to provide a reality check on the nihilism and skepticism of contemporary philosophy. Medical ethics is too ancient and too essential a reality for physicians, patients, and society to be left entirely to the fortuitous currents of philosophical fashion.

Ethical principles appear abstract – or better, speculative – because they do not possess the same degree of social legitimacy as the values of everyday life. Moral abstractions frequently are seen by non-philosophers as empty of the normal ingredients of moral concerns people have in their day-to-day life. No doubt they can and do seep into that daily life, but the process of connecting theory to practice is a long and subtle one in most cases (Graber and Thomasma, 1989). How often do we encounter physicians and patients who become impatient with "thinking" that has no practical consequence (Thomasma, 1988)?

In bioethics the four-principle tradition is now so widely accepted that some of its more whimsical critics have labeled it a "mantra" applied automatically and without sound moral grounding (Clouser and Gert, 1990). W. D. Ross' theory of prima facie principles (1988, p. 19) had a particular appeal to physicians and soon became the dominant way of "doing ethics." This approach was adapted to medical ethics in the text that has most influenced clinicians, Beauchamp and Childress' *Principles of Biomedical Ethics* (1989). They followed the direction taken by W. D. Ross and opted for prima facie principles, i.e., principles that should always be respected unless some strong countervailing reason exists which would justify overruling them. Four principles in this prima facie category were especially appropriate for medical ethics – non-maleficence, beneficence, autonomy, and justice.

This set of principles had the advantage of compatibility with deontological and consequentialist theories, and even with some aspects of virtue theory. It was quickly applied to the resolution of ethical dilemmas by medical ethicists, and especially by health professionals. For clinicians, this four-principle schema has several appeals. First, it promised to reduce some of the looseness and subjectivity that had characterized so many ethical debates. Second, it provided fairly specific action guidelines. Third, it offered an orderly way to "work-up" an ethical problem in a way analogous to the clinical work-up of a diagnostic or therapeutic problem. Fourth, such ethical workups could enhance educational programs in medical school (Thomasma and Marshall, 1995). Finally, it avoided direct confrontation with the intractably divisive issues of abortion, euthanasia, and the use of reproductive technologies.

The authors of the four principle approach were, of course, well aware of the limitations of Ross' system of prima facie obligations – e.g., the difficulties in putting any set of abstract principles into practice in particular cases, the difficulty of reducing conflicts between prima facie principles without some hierarchical or lexical ordering of the principles. Ross' rather vague formula of taking the action that gives the best balance of right over wrong really begs those questions. We are left with the need to employ some interpretative principle by which to measure the appropriateness of the balance. Also, Beauchamp and

Childress recognized the difficulties of attaining agreement on the most fundamental foundations of ethics, on the nature of the good, on the ultimate sources of morality, or on the epistemological status of moral knowledge.

To accommodate those shortcomings, Beauchamp and Childress proposed four requirements that must be met to justify "infringements" of a prima facie principle or obligation: (1) the moral objective sought is realistic; (2) no morally preferable alternative is available; (3) the least infringement possible must be sought; and (4) the agent must act to minimize the effects of infringement (1989). These authors hope in this way to steer a course between the absolutism of principles and the relativism of situation ethics. Their guidelines are helpful but do not eradicate the inherent limitations of any set of prima facie principles which is not lexically ordered.

Many remedies are offered to replace, prioritize, complement, or supplement prima facie principles. Engelhardt, for example, puts autonomy in the first order of priority (1996), placing it ahead of beneficence (Engelhardt and Rie, 1988); we favor beneficence-in-trust for that position (Pellegrino and Thomasma, 1988); others choose non-maleficence or justice. Additional alternatives to principle-based theories include an ethic based in virtue, caring, "experience," casuistry, or a return to theological and biblical sources as the only reliable grounding for medical morals.

The Virtue Theory Alternative

These limitations of principlism were the subject of serious criticism in an issue of the *Journal of Medicine and Philosophy* (1990). In that issue, Brody called the four principles "mid-level" principles, meaning that they are, themselves, in need of rational justification and of a firmer grounding in one of the great moral traditions (1990). Clouser and Gert decried the lack of a unifying moral theory which would tie the principles together and give them the conceptual grounding they need (1990). Were such a theory available, it would make the principles unnecessary. Holmes, like MacIntyre earlier, contended that philosophical ethics, itself, is of limited value (Holmes, 1990; MacIntyre, 1980). He advocated "moral wisdom" for which philosophy does not prepare us. Gustafson argued that philosophy is an insufficient tool for confronting the broad agenda of biomedical ethics. He calls for the inclusion of prophetic, narrative, and public policy elements in the discourse. He considered these more suited than principles to resolution of key ethical issues in health care (Gustafson, 1990).

There are other criticisms of the four principles coming largely from outside the philosophical community. Principles, it is said, are too abstract, too rationalistic, and too removed from the moral and psychological milieu in which moral choices are actually made; principles ignore a person's character, life story, cultural background, and gender. They imply a technical perfection in moral decisions which is frustrated by the psychological uniqueness of each moral agent or act.

It is clear that "principlism" in its present form is unlikely to survive unscathed through the next decade. But, its limitations notwithstanding, (for several reasons) I do not believe principles will disappear.

First, "principles" – that is to say, fundamental sources from which specific action guides like duties or rules derive and are justified – are implicit in any ethical system. The Hippocratic ethic, for example, was virtue-based, but its action guides were rules and principles. Second, there are equally serious limitations to any alternative theory to principlism. Third, the necessity and utility of principles become increasingly evident when we try to apply the alternative theories to actual cases; and, finally, principles are not inherently incompatible with other theories. The real question, as old as moral philosophy itself, is how to go from universal principle to individual moral decisions and back again.

Surely the character of the agent is crucial to medical ethics since the health professional interprets and applies whatever theory is used. Yet MacIntyre has shown brilliantly how irretrievable is the metaphysical consensus virtue theories require (1981). Virtue ethics by itself does not provide sufficiently clear action guides; it is too private, too prone to individual definitions of virtue, or the virtuous person, as we saw in the case of the South American physician Macklin identified.

Virtue theory must be anchored in some prior theory of the right and the good and of human nature in terms of which the virtues can be defined. It also requires a community of values to sustain its practice (the matter for the final section of this essay). In an integrated medical ethics virtue and character will be folded into any future version of biomedical ethics. This will require a conceptual link with duties, rules, consequences, and moral psychology, in which the virtue of prudence plays a special role (MacIntyre, 1990).

V. THE VIRTUES IN MEDICINE

For many centuries, virtue ethics was the dominant ethical theory in general ethics, as well as in medical ethics. If anything, as Pellegrino notes, virtue ethics in medicine resisted for a longer period of time the erosion found in general ethics (1995). This ethic, joined with the Hippocratic corpus and major religions, became a world-wide basis for a community of moral values and common dedication to the sick. Only in recent times has the four-principle approach supplanted or complemented it (Gillon, 1994). Today, then, the basis of professional duties as well as clinical and public policy bioethics, is sought elsewhere than in the virtues or virtue theory. This is because those traditional medical duties are subject to deconstruction and challenges about the conceptual foundations.

Particularly important in this questioning environment are the cultural expectations of the role of care givers and patients, since many are strangers to one another. One must be able to trust that a person in a white coat, in the role of nurse or physician, has certain precast standards that can be relied upon within the variabilities of social standing, culture, cities and rural areas, even countries. At the very least, principlism accounts for such standards. Questioning the validity of reasoning on the *sole* basis of such principles does not preclude some relevance for the standards in a different moral scenario. Thus, even if principlism itself is flawed, it does not necessarily follow that objective standards cannot be formulated from within the medical practice itself, as I shall argue.

Within medicine, virtue theory stands upon the goals of the practice itself, as MacIntyre has suggested (MacIntyre, 1988). I will amplify upon this in my final section. At this stage in the argument, however, I reflect on how virtues relate to specific practices.

Virtue practices go as far back as the earliest moral shaping of a child by a community. Virtue theories can be traced to Socrates, who, through Plato's eyes, discussed the merits of virtue and its importance in living a human life. Aristotle found the discussion of the virtues in Plato inadequate, largely because they were compared in humans to ideal norms in the realm of ideas. Instead, Aristotle formulated virtue theory in his ethics as a branch of politics, or the study of the larger virtues of public life. The virtues were to be grounded in both human psychology, the potentialities, proclivities, personalities, and emotions of persons, and in human affairs, the real relations of persons to one another in friendship and community.

For many centuries then, virtue theory was largely identified with an Aristotelian view of human nature and human social life, but later, during and after the Enlightenment, the basis of virtue theory was also expanded towards instinct, common sense, and gentlemanliness. Gradually as the cultural basis of morality became more and more apparent, efforts were made to preserve virtue theory and its correlate in natural law, by appealing to an ideal observer. Aristotelian virtue theory in its broad strokes could survive in a multicultural environment without such an appeal, of course, if one could acknowledge that different societies produced different social expectations and different emphases of moral character in individuals. However, this view would rapidly become relativistic as well, since what one culture considered worthy, say absolute dedication to Hitler among Nazi physicians, other cultures like the Allies would find morally abhorrent. There would be no way to adjudicate the disparate viewpoints.

By appealing to the natural law or to human psychology one avoids this kind of relativism in favor of a cross-cultural, trans-historical "nature" that provides the basis for judgments of moral rightness and wrongness that still is not the same as appealing to objective moral principles and mandates. In essence Aristotelian and Thomistic virtue theory argues that all human beings have an inborn nature that tends to the good in moral actions, but needs molding and direction, and most especially repeated habitual action, to refine that nature away from vices and towards the good. Virtues, in fact, are defined as good operative habits that intensify the potentialities of human nature ranging from one's emotions to one's intellect and will. These habits are aimed at good human actions.

Clearly anyone who grew up in a strong community will have been shaped this way, trained by parents and the community, secular and religious, in what sort of person one should be. Our language and arts are filled with stories and pictures of moral virtues essential for a decent human society: courage, love, friendship, responsibility, truth-telling (for example, Pinocchio), faithfulness, and wisdom. The point of these stories and artistic expression is to emphasize the individual's responsibility for choosing the good in every situation (Bennett, 1993). Additionally, as Gregory Pence notes, "Certain core virtues are always necessary for any decent society ... physicians need additional virtues, such as humility (the opposite of arrogance), compassion, and respect for good science (integrity)" (Pence, 1980; See also Pellegrino and Thomasma, 1993a, 1996).

It is helpful to consider that ethics has traditionally been concerned with the agent, the motive of the agent, the action itself, and the goal or end of the action. Ethical theories tend to stress one or another of these concerns. How does virtue theory relate to these concerns and the other theories?

Virtue theory shares with deontological theory an emphasis on the moral agent. It adds to the moral rightness of an action the requirement to analyze the motives of the agent as well. However, it shares with teleological theory an analysis of the goodness of actions too, since, as Aristotle argued, "all agents act for an end." This means that, independent of a good motive and a good human being, an action can be wrong in itself. Thus, virtue theorists might argue that euthanasia, although performed out of compassion, is morally wrong since it involves killing, itself an evil act. Alternatively, a virtue theorist might argue that providing uncompensated care for the poor is a good human act, even if done for illicit motives like personal pride, since the act has a quality of goodness independent from the agent.

Virtue theory thus can combine the strengths of both of the other traditional theories in medical ethics. Its basic principle was articulated by St. Thomas Aquinas as "Do good, and avoid evil." Aquinas saw prudence and the other cardinal virtues as working together to help persons make moral judgments. The natural law itself functions, in his view, only as a guide in this process (Nelson, 1992). Obviously, our social training helps us decide what is good or evil. Avoiding evil and doing good as a principle is derived from a natural law theory, a notion that in human existence itself, in nature, is a "law" that impels people to do good as they perceive it. Part of the Enlightenment Project already mentioned was also based on an appeal to this law, in effect when Enlightenment figures intoned, somewhat solemnly, the basis of morality to be "Nature and Nature's God..." Nonetheless, the dominant trend of moral thought subsequently has been towards increased formalism that abstracts from the rich texture of everyday life (Schmitz, 1999).

An additional strength of virtue theory is its explicit grounding in the community. The individual is not perceived separate from his or her own community. This can also be a weakness, as when a young member of a gang "learns" loyalty to the gang by killing in a drive-by shooting when ordered to do so. Further, virtue theory is less apodictic than either teleological or deontological theory. There is room for moral judgment that is "generally, for the most part, true." As such it occupies the middle position between absolute moral certitude and a view that there are no moral answers at all. Because the virtues are located in the moral character of the individual, they are much more "at hand" than more abstract theories and principles, thus contributing to ethical decision skills of moral imagination, attentive listening, and critical self-reflection (Gauthier, 1997).

We may conclude this section by following the suggestion that virtue theory might be resurrected on the basis of a moral community such as medicine, and that this effort may bear fruit despite the moral pluralism and relativism that is a characteristic of our society. There are obstacles of course. Virtue theory must be anchored in some prior theory of the right and the good and of human nature in terms of which the virtues can be defined. Erich Loewy asks, for example, how we would know what habits (virtues) to develop without a prior normative conception (1997). Virtue theory thus requires a community of values to sustain its practice (Pellegrino, 1995). This requires a conceptual link with duties, rules, consequences, and moral psychology, in which the virtue of prudence plays a special role (MacIntyre, 1990). But what of a link to the practice of medicine itself? Could we turn to a practice rather than to moral theory? Could not the moral community of medicine itself provide some of the needed context for virtue theory?

VI. VIRTUES AND PRACTICE

One can readily imagine that even in primitive groups, surely by the time hunting and gathering began to appear, roles important for group survival and solidarity appeared. At the very least, one or two were responsible for hunting, others became specialists in gathering food. Some of the latter sub-specialized in gathering healing herbs and techniques of repairing wounds, while some of the former sub-specialized in stalking different prey, each type presenting different challenges to the hunters. The elderly acquired the role of passing on wisdom for the next generation.

If this reconstructed scenario is authentic, then we can imagine that specific virtues began to appear that would help the community and the individuals themselves choose what role to play; although every person has all the virtues to some extent, those with a predominance of courage might be chosen as fierce hunters or warriors, those with wisdom as adjudicators, those with patience for fishing, those with prudence with locating the group's habitat, and the like. Among those specialties grew experts in healing practices, and later, the profession of medicine.

Practice

By a "practice" is meant an activity that combines structured direction for actions and special practical reasoning space. MacIntyre defines a practice as a coherent, socially organized activity with notions of good practice within the practitioner's understanding and skillful comportment (1981). In other words, a practice combines theory and practice in a unique way, since theoretical guidelines direct actions, but do not predetermine them. The goals of the practice provide judgments about the good actions within the practice. Clinical ethics itself is a practice in this regard (Crigger, 1995). As Patricia Benner notes, "a practice...is not a mere carrying out of an interiorized theory; it is a dynamic dialogue in which theories and new understandings may be created. The expert [in a practice] is called on to think in novel, puzzling, or breakdown situations"(Benner, 1997).

This unique form of human activity requires a different sort of reasoning from scientific reasoning. In scientific reasoning, as Taylor argues, we require formal objective characteristics for rational justification (1993). Scientific reasoning rests on the ability to spell out in snapshot-like fashion, all the relevant criteria and essential features of the situation. It requires a logic of breaking down a problem into component parts and objectifying them as far as possible.

By contrast, clinical and ethical reasoning is a form of practical reason Taylor calls "reasoning in transition." By that is meant that an expert in a practice is required to reason about gains and losses in understanding over time, as in a moving picture, while simultaneously thinking ahead about future possibilities:

> Practical reasoning ... is a reasoning in transitions. It aims to establish, not that some position is correct absolutely, but rather that some position is superior to some other. It is concerned, covertly or openly, implicitly or explicitly, with comparative propositions ... The argument fixes on the nature of the transition from A to B. The nerve of the rational proof consists in showing this transition is an error-reducing one. The argument turns on rival interpretations of possible transitions from A to B, or B to A. The form of the argument has its source in biographical narrative. We are convinced that a certain view is superior because we have lived a transition which we understand as error-reducing and hence an epistemic gain (Taylor, 1989).

This skill, for skill it is, requires more of the thinker than a scientific model does. It requires trust both in an expertise in theory along with an expertise in adjusting and practical reasoning in the very absence of strict objective guidelines. Good practice, then, requires more than scientific knowledge and/or its application. It also requires trustworthy, moral practitioners who have a skill in adjusting their thinking to the patient's condition and the patient's best interest (Pellegrino and Thomasma, 1988). Aristotle linked such instances of practice with skillful thinking this way:

> But let us take as agreed in advance that every account of the actions we must do has to be stated in outline, not exactly ... the type of accounts we demand should reflect the subject-matter; and questions about actions and expediency, like questions about health, have no fixed (and invariable) answers. And when our general account is so inexact, the account of particular cases is all the more inexact ... and the agents themselves must consider in each case what the opportune action is, as doctors and navigators do (Aristotle, 1985).

Viewed in this way, a practice, like medical care, requires virtue in the care giver. Furthermore, this virtue is tied directly to the goals of the practice. In other words, the moral basis for the practice is derived not from moral theory itself, but from the predetermined moral force of the goals of the practice.

Now it seems that some practices may have morally neutral goals, practices such as farming, or gardening, architecture, navigating, sculpturing, or military science perhaps. The ultimate morality of such practices depends to a great extent on the usages of the product of the practice, on the social order itself. Good navigation is moral if it aims at preserving the lives of passengers. Defense of one's country through good military skills arguably may be virtuous for practitioners in some nations, but not in other nations with dictators at their head. The practices are intrinsically morally neutral, but when carrying them out, performing them, one becomes immersed in moral values. We may call these practices of external morality. A particularly difficult problem for these practices is to judge the morality of the ends while being immersed in and engaged with the particularities of the practice itself, e.g., a second-rank Nazi physician working on experiments in a camp. Nonetheless, it is clear that world citizenry requires this moral assessment in practitioners and holds them accountable for it. Regarding navigation, we hold accountable a drunken captain who grounds an oil tanker and poisons the environment. Regarding military science, witness the Nuremberg Trials or the new trials about atrocities in Bosnia. Practitioners are accountable not only to the internal goals of the practice, but to broader moral claims that should make them pause about accepting those internal goals uncritically when performing the practice (Pellegrino and Thomasma, 2000).

Medicine as Practice

Unlike practices of extrinsic morality, which may arguably be morally neutral and derive their moral nature from external sources either with respect to the practice itself or its actual performance, other practices, while responsible to broader morality too, also have intrinsic moral ends. These include nursing, medicine, law, teaching, and ethics itself. There is some debate about such moral ends in a pluralistic society. Yet that debate is focused on determining what the proper moral ends are, not whether such ends are, indeed, moral ones. Engelhardt, in earlier works, argues that in a pluralistic society one cannot assume agreement on the internal moral ends of medicine, and that these are negotiated through "peaceable dialogue" (1986). Yet it seems to me that the goals of medicine, as held by both practitioners and patients alike, are non-negotiable human goods. Perhaps particular goals of therapy are negotiable, as are their content and actions, but the healing aim itself should not be. This means that the healing aim is a normative principle for medicine. Without adherence to and focus on it, the practice of medicine loses its bearings. Thus the "non-negotiabillity" of the healing aim of medicine stems from practical experience of medicine being thwarted by other principal goals and aims, most especially by the aims of serving the state, insurance companies, or other third parties.

Two points arise from this reflection. The first is that the virtues are based on roles in the community that contribute to its survival. The second is that once the role is chosen, role-specific duties emerge that can be viewed as expectations or even standards arising from the practices of the community as a whole. This is why, again, objective expectations or standards are important for virtue theory in any practice, especially so in medicine.

Thus far I have argued that the goals of a practice can be extrinsically or intrinsically moral, and that the pertinent moral relationship of practitioners of the latter, like those in medicine, is not so much to moral theory as it is to the practice itself, in this case, to the profound act of healing within the context of the doctor-patient relationship (Thomasma and Pellegrino, 1994). Since that relationship is a transactional one, as well as one in transition, as the illness develops and plays out in response to therapies, practical reasoning skills, along with scientific ones, are required. These skills absolutely require an internal morality in the practitioner that cannot depend on saluting abstract principles, since these do not always apply with the same weight to individual situations, as has been long recognized, but also may not offer the degree of "systematic grasp, some way of ordering the subject matter," as Nussbaum describes a *techne* (1986), that is needed to resolve both the disease and the needs of the patient.

Instead another path is necessary, a path towards an ontology of the body. Practical reason must be involved in the processes of interpreting the patient's body as an object, as a living organism, and as the living being the patient is him- or herself, with all the contexts of values and relationships that entails (Davis, 1996). It is crucial to keep in mind that illnesses befall us, sometimes without obvious reasons, sometimes even after much observation has taken place, such that the causes remain obscure. Scientific reasoning remains insufficient to adjust the outlines of treatment when there is no template for what has happened. The expert is required to make those adjustments anyway, holding in trust the values of the patient along with his or her dedication to the patient's good. The question must now be asked, whether reliance on the virtues of such a practitioner can be enough to sustain a good outcome, a good outcome in the sense of respect for the patient's well being and values.

The answer is yes, if social expectations are made explicit. In the past, these expectations were included within the Hippocratic ethos around the world. Today, however, it is insufficient to make assumptions when physicians and other care givers are not necessarily "Hippocratic" in the sense that they aim at the goal of healing. They may have conflicts of interest. Two things need to be said about this.

Requiring negotiation about specific goals of therapy did not entail requiring negotiation about the goals of medicine itself. Does the recognition that physicians have many social roles today counteract my own claim? At first it seems to. If the doctor is to aim not at healing but, let us say, at entrepreneurship (in owning laboratories to which he or she sends patient specimens) (Thomasma, 1994), or at a social good such as the health of the community as a whole (Veatch, 1981), or at the good of a managed care corporation, then this new role must be negotiated with patients (Thomasma, 1996). At the very least, new social expectations that will change the goals of a particular transaction require mutual consent of both doctor and patient. This does not entail an abrogation of primary social virtues by the physician, as my second point will argue.

The very fact that physicians and other care givers sometimes assume other roles than that of a traditional healer, i.e., a person who "puts the patient's needs first," to paraphrase Camus, requires before one lays hands on the patient, informed consent to do so under some new rubric. For the most part, this recognition has arisen from the goals of medical research, which are not always conjoined to the aim of the best interests of the patient, especially when that research is intended not for benefit but for the discovery of new knowledge. In effect, then, the requirement for informed consent is a form of telling patients that the normal social assumptions about the role of the physician are to be suspended for this or that experiment.

This practice, however, underlines the very point I am making. The social duty of a physician to act in the best interest of the patient, the virtue of benevolence, is so much to be assumed that if it is to be suspended, patients must be fully informed and must freely consent to this suspension. The goal of medicine has not changed, the objective standard of beneficence has not changed, only the negotiated circumstances of its suspension have changed the character of the actions of the physician. The same form of argument can be made about other specific practices, such as limiting access to health care in managed care or government health care systems, and the like.

In all particular forms of health care transactions, great variability takes place. This has always been the case. The duties of physicians, to aim at healing the patient, to obtain their permission to act on their bodies towards that end, to persuade the patient to empower them to heal (Brody, 1992), to garner the resources necessary by referring them to other specialists or hospitals, to enlist the support of the family, all the detailed acts, are moral acts precisely because healing is a moral act. Healing sets off medicine as a practice in all its richness. These moral actions are all forms of virtues that refer to the goals of medicine. They would be so with or without moral theory in which to anchor them since they are anchored, instead, in the relationship that healing requires.

There are many practices as defined by MacIntyre. Some of these are not moral, but rather morally neutral, since their aims or ends are not moral values. Medicine, however, has a moral value as an end, the healing of a person who is vulnerable and in need of help. The imbalances this creates between such persons and their doctors have been long recognized as requiring specific duties of physicians in order to protect the vulnerabilities of patients. These duties are virtues with respect to the practice that is medicine, since without them the goal of healing is not attainable (Pellegrino and Thomasma, 1981).

Curing a person may in fact take place even if healing does not, as might occur, for example, when a person might be treated for severe diverticulitis by an obnoxious and imperious physician who has little regard for patient wishes or values. This cannot be described as a legitimate practice of medicine, however, since the objective standards, not so much moral theories or principles, but standards of care in medicine, have been violated even in this act of curing.

I have argued that, despite our growing awareness of cultural pluralism and the development of human thought and value systems, objective standards exist in medicine as a practice. These standards are derived, not solely from moral theory, but also from the goal of healing, and from long experience in categorizing and treating illnesses (Crigger, 1995, 91-92). The virtues of good physicians and good patients alike are derived from this goal, such that to violate the goal is to violate the relationship itself. This is not to say that moral theory is not important, and that the four principles, among others, are not critical. I do suggest, however, that these principles are derived inductively from the moral character of the practice of medicine just as much as they might be applied to it externally, from ethical theory itself (Graber and Thomasma, 1989).

If this possible view can be defended, then virtue theory could stand without explicit reference to other ethical theories, especially that of principlism. Virtue theory also could provide a basis for a response to anti-foundationalism by grounding medical and professional responsibilities in a culturally-sensitive practice rather than in abstract, conceptual theories of ethics. Professional responsibilities in this model would therefore be forms of social virtues, the society for this being the practice of medicine seen as a social entity. Within this practice, one might exercise the virtue of restraint (as Pellegrino and I analyzed the cardinal virtue of temperance, usually applied to food and drink) regarding available technologies when the patient is dying or in futility cases (1993a, p. 119).

Interpretative Differences

I already mentioned the South American physician's apparent interpretation of the meaning of integrity: he felt he could not participate in saving the life of a woman who aborted her fetus, since to do so would be cooperating with evil. Equally one could imagine a view of integrity that would lead a physician to treat anyone in need of help, regardless of the triggers by which they became ill, sometimes in fact, against the very conscience of the physician. Above, I stressed the importance of role-specific duties in defining the virtues within a practice. These duties establish a social expectation that, in turn, prompts the exercise of the virtues. In this case, one can readily see that even if one agrees that integrity is an essential virtue of a health professional, what that means in practice will vary among individuals, communities, civil societies, and cultures.

Some thinkers are concerned with profound arguments about traditional commitments to the value of human life within the patient care relationship as contrasted to respect for autonomy alone (Gaylin et al., 1988). Thus Leon Kass presents a thoughtful articulation of what is owed a dying patient by health professionals. He argues that humanity is owed humanity, not just "humaneness," (i.e., being merciful by killing the patient). Kass argues that the very reason we are compelled to put animals out of their misery is that they are *not* human and thus demand from us some measure of humaneness. By contrast human beings demand from us our humanity itself. This thesis, in turn, rests on the relationship "between the healer and the ill" as constituted, essentially, "even if only tacitly, around the desire of both to promote the wholeness of the one who is ailing" (Kass, 1989).

The temptation to employ technology rather than one's personhood in the process of healing we might call "the technological fix." The technological fix is not only easier to conceptualize and implement than the more difficult processes of human engagement, but is also "suggested" by technology itself. The training and skills of modern health professionals are overwhelmingly nurtured within an environment of technological fixes. By instinct and proclivity, all persons in a modern civilization are tempted by technical rather than personal solutions to problems.

The most dramatic examples of taking such responsibility for the particularities of a case are culled from the problems of euthanasia and withholding and withdrawing care from the dying. But integrity is also required to assess properly the interventions to be given to the weak and debilitated elderly, to the demented, to individuals with metastases, and other vulnerable persons, and to individuals who wish to exercise their autonomy in ways that are easily judged to be self-destructive.

Since the virtues are intimately linked with social mores and goals, interpreting the meaning of one's acts and reactions itself becomes the higher virtue of practical reasoning or prudence, or the still-more comprehensive virtue of wisdom itself. Even appeal to the certitudes of the golden rule, or universalizability, cannot disguise the variability of the virtues, despite their obvious importance. Hence, while there might be widespread agreement that integrity involves not abandoning the dying patient (as the South American physician seemed to do), differences with Kass's argument about integrity would focus on different interpretations of what the "humanity" of the professional might owe the dying patient. That is not all, of course.

Operational Differences

A second point about virtue theory also requires elaboration. Should agreement be reached about the contextual interpretation of a specific virtue like integrity, the actual operations may be conceptualized differently. Subtle nuances in thinking might separate otherwise unified points of view. Using the example of physician aid-in-dying, we may detect those subtle differences regarding integrity-in-action with respect to a duty to assist dying patients who request euthanasia. This is one of many operational sequella to an agreement that integrity equals non-abandonment during dying.

In their essay on what doctors owe a dying patient, Miller and Brody argue that integrity demands caring for the patient to the extent that one would be able to bend the normal standards of conduct in order to relieve suffering. Essential to their argument is the notion of integrity as continuous adjusting to suit the values of the case and the challenges of the moment. Although they call this notion of integrity a substantial principle and not a virtue, what they describe in terms of adjusting is precisely what is meant by the virtue of integrity explored earlier. Too little adaptation leads to formal rigidity; too much leads to accusations of a loss of moral grounding (Miller and Brody, 1995). In their view there is no need to reference integrity back to a formal principle for its moral grounding.

Battin follows a similar line of argument in her examination of moral duties to the dying in the presence of legal opting-out clauses. In legalizing assisted death or euthanasia, such laws provide for the health care professional's conscience. If she does not wish to participate then she can refer the patient to another physician. This is a commonly-practiced privilege professionals enjoy in daily interactions and in the law. Battin wants to argue a more difficult point however; namely, that moral integrity might require the physician not to refer the patient, since at the stage of dying, the values they have shared precludes such transfer. It would be equivalent to abandonment (Battin, 1998).

My focus is on the way the virtue of integrity is invoked in both essays. Apparent agreement exists about how integrity involves addressing the patient's suffering while not abandoning the patient. There may be a duty of physicians to respond to requests by patients for assistance in dying, especially if the patients are unable to act on their own behalf due to debilitation. Thus, there seems to be interpretive agreement. Nonetheless, operational distinctions still remain.

For Miller and Brody, the physician may still morally opt out of the situation based on professional integrity. For Battin this may not be possible, also for reasons of personal and professional integrity.

This one comparison of the literature in bioethics and professional responsibility, from countless others that are possible, demonstrates how the virtues, and therefore, virtue theory, must be related to interpretation and contexts (operational expectations). This need for grounding in something other than more formal principles leads to the final section.

VII. RELATIONAL FOUNDATIONALISM

As so far argued, virtue theory in medicine is one of a number of efforts to recapture a less formal acknowledgment of complexity and character in both the theory and the practices of medicine. Yet those who argue that virtue theory cannot survive without some grounding that was once provided by a more homogeneous culture are correct. One way to interpret the meaning of the virtues is to set standards and principles that everyone would salute. Even so, the interpretation of the agent and the acts, as shown, would also involve the changing practices of medicine, and historical and cultural development.

Finding the balance between absolutes and a radical relativism today is difficult. It is reinforced by the worldwide surge of cultural hegemony in morality. On this view, the medical ethic which has supplanted the Hippocratic ethic is a Western product and is incompatible on various grounds with other cultures, particularly with reference to autonomy. As the Western version comes into contact with other cultures, we can expect sharper definitions of points of conflict and agreement (Flack and Pellegrino, 1992). The most ancient ethical theory, then, is the theory of virtues whereby the community praised its heroes through sagas and stories, establishing its expectations. In medicine, virtue theory relates to expectations and standards as well. But there is a problem – in a pluralistic society, one person's definition of virtue is another's definition of vice. Indeed, many thinkers have tired of the Enlightenment Project, as it is called. Among them are MacIntyre and Rorty, already mentioned. According to MacIntyre, the effort to ground ethical certitude in a set of trans-cultural principles is doomed to failure, since ethics is so culturally-embedded and traditional (in the sense of a set of privileged starting points that are accepted by a thinker, who at once becomes thereby part of a tradition of "doing" ethics). Persons in different spheres of moral enquiry cannot properly engage in a dialogue with one another, but only with those with whom they share the values of that tradition of enquiry (MacIntyre, 1981).

In this regard, it is important to realize that even formal systems, rules, principles, and the like, are the result of interpretation. There can be no positivist, objective "fact" without a corresponding interpreter. In order to avoid the variability of relativism, then, it is insufficient to appeal to principles, as these too have cultural determinants that might make them vulnerable. This problem reveals the lack of a social and philosophical fundament. I now turn to a proposal for a theory of relationals that might support objective goals and standards for the virtues account by relating the virtues to both objective and subjective structures. Let us revisit the contemporary debate about ethics in this regard.

Reconstituting Medical Ethics

MacIntyre's analysis of the breakdown of ethics, beginning with the Enlightenment, targets the loss of a *telos*. In earlier times, he argues, moral terms such as "good," "just," or "fair" were attached to specific roles in society and to social goals that everyone understood. The goal of humankind gave them meaning. Once detached from a teleological scheme that gave them meaning, moral terms became intrinsically debatable and variable. The Enlightenment tried to anchor the meaning of morality in pure reason, but reason alone (principles and axioms) does not govern morality, practice does. For MacIntyre it is like being told to travel, but not being told where. He suggests that such moral terms can survive within a specific practice (MacIntyre, 1988).

We saw that Rorty's objection to ethics is even more radical. He agrees that ethics requires commonly accepted foundations that are simply rejected in our pluralistic society. Philosophical efforts to acknowledge and accept this pluralism and lack of foundations are essentially self-defeating, he thinks. Philosophy itself is not possible. So he is consistent with his convictions, leaving his post as a professor of philosophy, and taking up instead, a position as University Professor of the Humanities at the University of Virginia. Discourse about values, rather than the discovery of truth, is the essential feature of his argument. This discourse replaces the Enlightenment Project, the quest for universal truth discoverable by reason, with what we might call the "Conversation Project." Rather than a search for truth, a quest for honoring freedom becomes the goal of respectful discourse (Rorty, 1989).

Engelhardt has also participated in a philosophical *metanoia* from a wholesale libertarianism that directly derived from the Enlightenment to a rejection, too, of the Enlightenment Project. In its place, though, Engelhardt moves toward religious belief, declaiming the Enlightenment Project as in principle unachievable (Engelhardt, 1995). At first blush his move from a libertarian emphasis on autonomy to a religious emphasis on faith seems to be an extreme reversal. Yet the two are, in fact, searches for a type of moral certitude that is unobtainable. Note Engelhardt's ruing the loss of certitude in the Church in his early footnotes to *Bioethics and Secular Humanism* (Engelhardt, 1991). Indeed, the argument of that book seemed to be, "I wish it were not so, but without a moral consensus in society based on faith that was present in earlier times, we must appeal to secular humanism instead, its poor cousin." Since then, Engelhardt has banded together with Andrew Lustig and Fr. Kevin Wildes to bring out a journal devoted to "non-ecumenical" positions in religion and bioethics: *Christian Bioethics: Non-Ecumenical Studies in Medical Morality*. By "non-ecumenical" is meant a no-holds barred approach to defending religious insights in bioethics, such that there is no facile reduction of moral insights to the lowest common denominator (Wildes, 1995; Engelhardt, 1995). I suggest, however, that the project of searching for moral certitude continues in this new guise. Yet by moving dialectically through extreme positions, one misses the chance to see the truth that may lie in the middle.

Is it necessarily true that a respect for pluralism leads even the reluctant bioethicist either to religious foundationalism or to a commitment to relativism? I once witnessed a strong reaction to a Spanish thinker's conviction that pluralism and relativism do not necessarily entail one another. At an international bioethics conference, he kept insisting that one could respect pluralism without succumbing to relativism (Marshall et al., 1994). Afterwards he was roundly criticized by colleagues from other Western nations, who found the assertion "odd." After all, from the perspective of analytic medical ethics, either one accepts certain norms that govern behavior and are adjusted based on the context, or one must abandon a sense of right and wrong altogether and instead argue that morality is culturally embedded. Does not a respect for pluralism, by that very fact, entail a requisite judgment that no single set of norms can govern human conduct? Conduct is, alternatively, judged by cultural verities, some of which vanish once one considers another cultural context. Being subject to a single culture, raised in a single place, no matter how cosmopolitan, means that one is still "provincial." And so, it seems, is one's ethics.

Yet for several reasons this jump to relativism need not be true. First, universally consistent moral rules have been postulated by turning from cultural diversity to common structures that are postulated to be trans-historical and cross-cultural. For some, these moral rules could be gleaned from common cultural structures, like kinship piety and group solidarity, a kind of modern version of Aristotle's *jus gentium,* the law of all peoples. For others, the structure was even more certain since it rested on biological "laws" built into all human beings as a *jus naturale*. This full-blown natural law theory still finds a place today among those who argue that much, if not all, of human behavior rests on or is determined by genetic codes. Once these two ancient positions, the law of peoples or nations and the natural law, were seriously challenged, other objective trans-cultural structures were posited through social stratagems. Good examples are the theory of social contract as a tool of thought. Through thinking about the goods all would want, by consent one would abrogate natural rights in favor of social security. Another example is the theory of the ideal observer who could determine right from wrong behavior. Kant's effort to make ethics a science, immune from cultural criticism or relativity, rests especially on an assumption of objective reason, practical reason, by which judgments about the good can be made, *as if* what one claims would become a universal law. We are familiar with the many and sundry arguments for and against these many objectivist stances. They are so diverse and so potentially damaging to objectivist claims that Ortega Y Gassett suggested that reading the history of Western philosophical thought is like taking a trip through a pleasant insane asylum (1973).

However, there are second and third approaches to escape relativism. One might quite possibly be the one our Spanish ethicist held, and was not considered by his critics. The first of these two is a form of "soft" objectivism. In this approach one finds the conduct of others reprehensible, but acknowledges that the condemnation one issues is based on a moral *mien* rather than objective standards *per se*. Such an approach avoids both the arrogance of objectivist imposition on the conduct, but also avoids the pusillanimity of relativism, in which everything might have some justification. One is able to hold firm opinions of right and wrong without ultimately justifying these on objective truths alone.

The second of two alternative approaches to relativism is that of virtue theory. According to this approach, ethics is not a system of laws or objective norms, but a lived reality among highly complex human beings in many different familial, social, political, and cultural settings. It is important to note that this position does not require the same level of objective certitude as other theories about judging good and evil acts, persons, or consequences. We now turn to a brief description of the relational realms that would require much more thorough analysis.

Relational Reality

The theory of relationals involves three levels because human beings are interdependent with one another and with their environment, social and natural, on three levels: the existential, the historical/cultural, and the foundational. I can only briefly note them here.

(1) Existential: There is in medical ethics more hope for a better grounding of principles, rules, virtues, and moral psychology than in any other field of ethics. That hope rests on the universality of the phenomena of the experiences

of illness and healing and on the proximate and long-term aims of medicine, if they are grounded in the realities of the physician-patient relationship. All human beings experience the dependency on others which their own finitude informs and establishes. That finitude requires that they seek help from one another. At the existential level they are not strangers, but fellow-travelers. On this basis expectations about compassion and healing can be developed (Loewy, 1991).

Traditionally, the community emphasized compassionate care of individuals by providing the structures needed for individuals who were sick to be surrounded by those who both loved them the most and also knew their values. Decisions about health care were made within a context of compassion and respect for the values of the patient. The care of such individuals was impervious to marketplace economics. It was an act of mercy, not a commodity to be traded or delivered. By contrast, today, the community seems more concerned about the resources the sick divert from our other projects. Rationing care that is to be given appears to be more important in a scale of values than providing it. As a result of concern about rationing, there is a danger of a concomitant shrinking from sacrifice – of time, emotions, energies, and money – that the care of the sick so much requires. These traditionally constituted vital service.

But none of the changes in society or the technology of medical care in recent times can alter the call the sick themselves press upon care givers so insistently. They call them as fellow creatures in need (Pope John Paul II, 1983). Recognizing their need as persons can only be done if the care giver's own self-perception is one of being an agent of mercy and compassion. In the effort to be compassionate, all healthcare providers must try to permeate society as a whole with their vision of mercy. What is the meaning of this compassion within the context of biomedical decisions?

Compassion is more than pity or sympathy. It transcends social work, philanthropy, and government programs. It is the capacity to feel, and suffer with, the sick person – to experience something of the predicament of illness, its fears, anxieties, temptations, its assault on the whole person, the loss of freedom and dignity, the utter vulnerability, and the alienation every illness produces or portends. True compassion is more than feeling. It flows over in a willingness to help, to make some sacrifice, to go out of one's way. "No one can help anyone without entering with her whole person into the painful situation; without taking the risk of becoming hurt, wounded, or even destroyed in the process" (Nouwen 1972, p. 2).

Existential finitude, then, engenders compassion. Compassion in turn entails a comprehension of the suffering experienced by another. When individuals themselves have suffered they are sometimes better able to understand it in others. As De Unamuno says, "Suffering is the substance of life and the root of personality, for only suffering makes us persons" (1972, p. 224). Compassion for the suffering of others thus enriches self-understanding, especially of what we too must some day pass through. Compassion helps us, therefore, to realize that sick brothers and sisters are not aliens. They are still very much part of the human family. They are vital to one's existence and moral growth. The healthy need the sick to "humanize" them as much as the sick need the healthy to humanize their sickness.

(2) Historical/Cultural: We often think that the moral life is captured by doing good and avoiding evil. This is the ancient injunction of natural law ethics, based as it is on the insight (now contested) that human beings have an inbuilt capacity for happiness that is partially based on their nature and partially on their choices. The best moral life is achieved by meshing and harmonizing the structures of our nature with our personal and social choices. This meshing occurs within relationships to one another and to our social and physical environment. No where does this meshing appear more readily than in illness and dying, when our embodiment becomes at risk, as are the many relationships it has created.

The sick and the care-givers alike cannot escape the confrontation with mortality which even a minor illness may entail. Human illness is always illness of the whole person – body, mind, and spirit. Hence, the illness and/or dying process is more than some aberration in an organ system. The illness transcends the biological. It encompasses the whole person and his or her value system. Illness fractures self-image, upsets the balance the patient has struck between aspirations and limitations. Illness is nothing less than a deconstruction of the self. Compassion enables the decision makers to assist in healing, if by healing we can mean the reconstruction of the person. Involved here is an effort to put back together a ruptured self that has separated into an ego and a body that has betrayed that individual (Bergsma and Thomasma, 1983). The particularities of culture, ethnicity, and language are what makes illness a unique experience for each person. True healing can only take place when all of these particulars are taken into account (Pellegrino and Thomasma, 1997). The decisions to be made, then, entail a rich awareness of particulars and values. These decisions, although guided by general, abstract principles, involve the emotions and values of all the parties involved in the process of caring for the sick person.

(3) Foundational: In a pluralistic society, those radical personal and social choices arising in illness and dying are or seem to be isolated from any moral certitudes or foundations. Without a moral consensus, however, people get caught up in what John Paris and Richard McCormick call the "casuistry of means"(1987). Discussion of appropriate goals in medicine suffers from an abundance of means and a poverty of ends. The reason for this is not hard to find.

There is precious little agreement on the fundamentals. So instead, the debate shifts to the means rather than the ends, the points and counterpoints of the debate leading to interminable and often unresolvable disputes.

Recall the earlier discussion of medicine as a moral enterprise, and its conclusion that healing is a requirement of interpersonal life, the survival of the community, and that this relationship is not so much chosen or imposed, but a structure of human existence. All have a primeval and foundational dependency on others and the environment in order to survive and prosper. Bergsma and I call this, in Dutch, the *begane grond* (literally, the beginning ground, or ground floor). By it is meant that certain relationships are foundational, giving rise to all others. Among these are the husband-wife, the parent-child, and among others, the healing relationship (Bergsma and Thomasma, 2000). Similarly in his work in philosophy of medicine, Welie has argued that a noncontractual foundation in the doctor-patient relationship provides an immediacy for medical ethics (Welie, 1999).

Compassionate care also means that the patient who cannot be cured by medical sciences – especially the dying cancer patient – may still be "healed." Even the dying patient can be healed if we help him or her to express the meaning of a life in the final days of that life by respecting, insofar as possible, the patient's values and commitments.

VIII. CONCLUSION

Since the beginning of modern clinical ethics, it has been no secret that the reasoning patterns of clinical judgment in medical care parallel those of ethical judgment. This realization is important for many reasons, not the least being the relation of the virtue of prudence or practical reasoning to reasoning about healing the body and the person. Articles and books on the philosophy of health care have underscored the relation of the ethics of healthcare to clinical judgment (Graber and Thomasma, 1989). More pointedly for our purpose, the nexus between clinical judgment and clinical ethics can help reveal structures of good decision making in patient care that are not simple products of contractual models of the provider-patient relationship. More is going on in that relationship than initially meets the eye: the humanity of care giving is dependent on the character of the participants, their common existential, historical and cultural, and foundational experience that has given rise over time to more formal, objective expectations and role-specific duties in the form of moral principles.

We are too challenged by the dilemmas of our time to stress some aspects of medical ethics to the detriment of others, as has been done with principle-based ethics in the past. Radical incorporation of the virtues in medical theory itself is not enough, unless we also pay attention to the personal, social, existential, and foundational structures of the healing relationship.

Many challenges remain for virtue theory in medicine:

(1) Continued thinking about the relation of the virtues to more objective standards is needed.

(2) These standards and the growth in character of physicians and patients alike during an episode of illness are articulated by narrative ethics. The relation of virtue theory to narrative ethics is only intuitively recognized as yet, and has not been thoroughly explored.

(3) The rise of alternative standards in different cultures, and the multitude of cultures in modern society, can create confusion about expectations, which in turn leads to confusion about the virtuous life. Clarifying expectations is essential to a postmodern theory of virtues in medicine.

Stritch School of Medicine
Loyola University Chicago
Maywood, Illinois

REFERENCES

Aristotle:1985, *Nichomachean Ethics* (tr. Irwin, T.), Hackett Publishing Co., Indianapolis, IN.

Battin, M.: 1998, 'Is a physician ever obligated to help a patient die?,' In L.L. Emanual (ed), *Regulating How We Die. The Ethical, Medical and Legal Issues Surrounding Physician-Assisted Suicide*, Harvard University Press, Cambridge, MA.

Beauchamp, T.L., Childress, J.F.: 1989, *Principles of Biomedical Ethics, 3rd Edition*, Oxford University Press, New York.

Benner, P.: 1997, 'A dialogue between virtue ethics and care ethics,' *Theoretical Medicine* 18, 47-61.

Bennett, W.J.: 1993, *The Book of Virtues: A Treasury of Great Moral Stories,* Simon & Schuster, New York.

Bennet, W.J.: 1995, *The Moral Compass: Stories for a Life's Journey*, Simon & Schuster, New York.

Bergsma, J., Thomasma, D.C.: 2000, *Autonomy and Clinical Medicine: Renewing the Health Professional Relation with the Patient*, Kluwer Academic Publishers, Dordrecht/Boston.

Bergsma, J., Thomasma, D.C.: 1983, *Health Care: Its Psychosocial Dimensions,*.Duquesne University Press, Pittsburgh, PA.

Brody, B.A.: 1990, 'Quality of scholarship in bioethics,' *Journal of Medicine and Philosophy* 15(2), 161-178.

Brody, H.: 1992, *The Healer's Power*, Yale University Press, New Haven, CT.

Campbell, A.V.: 1994, 'Dependency: The Foundational Value in Medical Ethics,' in K.W.M Fulford, G. Gillett, and J.M. Soskice (eds), *Medicine and Moral Reasoning*, Cambridge University Press, Cambridge, pp 184-192.

Camus, A.: 1948, *The Plague*, Modern Library, New York.

Clouser, K.D. and Gert, B.: 1990, A Critique of Principlism, *Journal of Medicine and Philosophy* 15(2), 219-236.

Crigger, B.J.: 1995, 'Negotiating the moral order: Paradoxes of ethics consultation,' *Kennedy Institute of Ethics Journal* 5, 89-112.

Davis, F.D.: 1996, *Clinical Reasoning and the Hermeneutics of Embodiment.* Ph.D. Dissertation; Georgetown University, Washington, D.C.

De Unamuno, M.: 1972, *The Tragic Sense of Life, Bollingen Series, LXXXV, 4* (tr. Anthony Kerrigan), Princeton, New Jersey.

Drane, J.F.: 1994, *Clinical Bioethics: Theory and Practice in Medical Ethical Decision-Making,* Sheed and Ward, Kansas City, MO.

Dunne, J.: 1993, *Back to the Rough Ground: Practical Judgment and the Lure of Technique,* University of Notre Dame Press, Notre Dame, IN.

Ellos, W.J.: 1990, *Ethical Practice in Clinical Medicine,* Routledge, New York.

Engelhardt, H.T., Jr.: 1991, *Bioethics and Secular Humanism*, Trinity International Press, Philadelphia.

Engelhardt, H.T., Jr.: 1995, 'Towards a Christian bioethics,' *Christian Bioethics* 1, 1-10.

Engelhardt, H.T., Jr.: 1996, *The Foundations of Bioethics, 2nd Edition*, Oxford University Press, New York.

Engelhardt, HT., Jr., Rie, M.A.: 1988, 'Morality for the medical-industrial complex – a code of ethics for the mass marketing of health care,' *New England Journal of Medicine* 319(16), 1086-89.

Flack, H.E.,Pellegrino, E.D. (eds): 1992, *African-American Perspectives in Biomedical Ethics,* Georgetown University Press, Washington, D.C.

Gauthier, C.C.: 1997, 'Teaching the virtues: Justification and recommendations,' *Cambridge Quarterly of Healthcare Ethics* 6, 339-46.

Gaylin, W., Kass, L., Pellegrino, E.D., Siegler, M.: 1988, 'Commentaries: Doctors must not kill,' *JAMA* 259, 2139-40.

Gillon, R.: 1994, *Principles of Health Care Ethics*, John Wiley & Sons, England, New York.

Glaser, J.W.: 1994, *Three Realms of Ethics,* Sheed & Ward, Kansas City, MO.

Graber, G.C., Thomasma, D. C.: 1989, *Theory and Practice in Medical Ethics*, Continuum Books, New York.

Gustafson, J.M.: 1990, 'Moral discourse about medicine: A variety of forms,' *Journal of Medicine and Philosophy* 15(2): 125-42.

Hauerwas, S.: 1986, *Suffering Presence,* T & T Clark, Edinburgh.

Holm, S.: 1995, ' Not just autonomy – the principles of American biomedical ethics,' *Journal of Medical Ethics* 21, 332-8.

Holmes, R.L.: 1990, 'The limited relevance of analytical ethics to the problems of bioethics,' *Journal of Medicine and Philosophy* 15 (2), 143-60.

Kass, L.: 1989, 'Arguments against active euthanasia by doctors found at medicine's core,' *Kennedy Institute of Ethics Newsletter* 3, 1-3, 6.

Kennedy, R.F.: 1999, *Make Gentle the Life of This World: The Vision of Robert F. Kennedy,* M.T. Kennedy (ed), Broadway Books, New York.

Koch, T.: 1995, 'The gulf between: Surrogate choices, physician instructions, and informal network responses,' *Cambridge Quarterly of Healthcare Ethics* 4 (3), 185-92.

Landers, A.: 1999, 'Some advice never gets out of date,' *The Chicago Tribune,* July 12.

Loewy, E.H.: 1991, *The Suffering and the Beneficent Community: Beyond Libertarianism*, State University of New York Press, Albany.

Loewy, E.H.: 1997, 'Developing habits and knowing what habits to develop. A Look at Virtue in Ethics,' *Cambridge Quarterly of Healthcare Ethics* 6, 347- 55.
MacIntyre, A.: 1980, 'A crisis in moral philosophy: Why is the search for foundations so frustrating?,' In H.T. Engelhardt, D. Callahan (eds), *Knowing and Valuing,* Hastings Center, Hastings-on-Hudson, NY, pp. 18-43..
MacIntyre, A.: 1981, *After Virtue: A Study in Moral Theory*, University of Notre Dame Press, Notre Dame, IN.
MacIntyre, A.: 1988, *Whose Justice? Which Rationality?,* University of Notre Dame Press, Notre Dame, IN.
MacIntyre, A.: 1990, *Three Rival Visions of Moral Enquiry,* University of Notre Dame Press, Notre Dame, IN.
Macklin, R.: 1999, Against Relativism: Cultural Diversity and the Search for Ethical Universals in Medicine., Oxford University Press, New York.
Marshall, P.A., Thomasma, D.C., Bergsma, J.: 1994, 'Intercultural reasoning: The challenge of international bioethics,' *Cambridge Quarterly of Healthcare Ethics* 3, 321-328.
Marshall, P.A., Koenig, B.A., Barnes, D.M., Davis, A. J.: 1998, 'Multiculturalism, bioethics, and end-of-life care: case narratives of Latino cancer patients,' In J.F. Monagle, D.C .Thomasma (eds), *Health Care Ethics: Critical Issues for the 21st Century,* Aspen Publishers, Gaithersburg, MD.
Miller, F.G., Brody, H.: 1995, 'Professional integrity and physician-assisted death,' *Hastings Center Report* 23(8), 8-17.
Nelson D.M.: 1992, *The Priority of Prudence: Virtue and Natural Law in Thomas Aquinas and the Implications for Modern Ethics*, The Pennsylvania University Press, University Park, PA.
Nouwen, H.: 1972, *The Wounded Healer*, Doubleday, New York.
Nussbaum, M.C.: 1986, *The Fragility of Goodness: Luck and Ethics in Greek Tragedy and Philosophy*, Cambridge University Press, Cambridge.
O'Neill, O.: (1996). *Toward Justice and Virtue: A Constructive Account of Practical Reasoning,* Cambridge University Press, Cambridge.
Ortega Y Gassett, J.: 1973, *An Interpretation of Universal History* (tr., Adams, M.), Norton, New York.
Paris, J.J., McCormick, R. A. : 1987, 'The Catholic tradition on the use of nutrition and fluids,' *America*, 356-61.
Pellegrino, E.D.: 1995, 'Toward a virtue-based normative ethics for the health professions,' *Kennedy Institute of Ethics Journal* 5(3), 253-77.
Pellegrino, E.D., Thomasma, D.C.: 1981, *A Philosophical Basis of Medical Practice: Toward a Philosophy and Ethic of the Healing Profession*, Oxford University Press, New York.
Pellegrino, E.D., Thomasma, D.C.: 1988, *For the Patient's Good: The Restoration of Beneficence in Health Care*, Oxford University Press, New York.
Pellegrino, E.D., Thomasma, D.C.: 1993a, *The Virtues in Medical Practice*, Oxford University Press, New York.
Pellegrino, E.D., Thomasma, D.C.: 1993b, ' Integrity,' *The Virtues in Medical Practice*, Oxford University Press, New York: 127-43.
Pellegrino, E.D., Thomasma, D.C.: 1996, *The Christian Virtues in Medicine*, Georgetown University Press, Washington, D.C.
Pellegrino, E.D., Thomasma, D.C.: 1997, *Helping and Healing*, Georgetown University Press, Washington, D.C.
Pellegrino, E.D., Thomasma, D.C.: 2000, 'Dubious premises – evil conclusions: Moral reasoning and the Nuremberg trials,' *Cambridge Quarterly of Healthcare Ethics* 9, 261-74.
Pence, G.E.: 1980, *Ethical Options in Medicine*, Medical Economics Company, Oradell, NJ.

Pope John Paul II: 1983, *Humanize Hospital Work, Address to the Sixty-First General Chapter of the Hospital Order of St. John of God*, L'Osservatore Romano.

Redman, B.K.: 1995, 'Drug misadventuring as an issue in the ethics of leadership,' *American Journal of Health-System Pharmacy* 52(4), 404-5.

Rorty, R.: 1979, *Philosophy and the Mirror of Nature,* Princeton University Press, New Jersey.

Rorty, R.: 1989, *Contingency, Irony, and Solidarity*, Cambridge University Press, Cambridge.

Ross, W.D.: 1988, *The Right and The Good,* Hacket, Indianapolis.

Schmitz, K.: 1999, 'Postmodernism and the Catholic tradition,' *American Catholic Philosophical Quarterly* 73(2), 233-52.

Schneider, C.: 1998, *The Practice of Autonomy: Patients, Doctors, and Medical Decisions,* Oxford University Press, New York.

Sherman, N. : 1997, *Making a Necessity of Virtue: Aristotle and Kant on Virtue,* Cambridge University Press, Cambridge.

Taylor, C.: 1989, *Sources of the Self: The Making of the Modern Identity*, Harvard University Press, Cambridge.

Taylor, C.: 1993, *The Quality of Life,* M.C. Nussbaum, A. Sen (eds.), The Clarendon Press, Oxford: 208-31.

Thomasma, D.C.: 1988, 'Applying general medical knowledge to individuals: A philosophical analysis,' *Theoretical Medicine* 9, 187-200.

Thomasma, D.C.: 1994, 'The ethics of medical entrepreneurship,' in J.F. Monagle, D.C. Thomasma (eds.), *Health Care Ethics: Critical Issues*, Aspen Publishers, Inc., Rockville, MD: 342-50.

Thomasma, D.C.: 1996, 'The ethics of managed care: Challenges to the principles of relationship-centered care,' *Journal of Allied Health* 25(3), 233-46.

Thomasma, D.C., Pellegrino, E.D.: 1994, 'Fidelity to trust,' In G.P. McKenny, J.R. Sande (eds.), *Theological Analyses of the Clinical Encounter, Volume 3*, Kluwer Academic Publishers, Dordrecht, The Netherlands: 69-84.

Thomasma, D.C., Marshall, P.A.: 1995, *Clinical Medical Ethics: Cases and Readings,* University Press of America, Lanham, MD/New York/London.

Veatch, R.M.: 1981, *A Theory of Medical Ethics*, Basic Books, New York.

Veatch, R.M.: 1985, 'Against virtue: A deontological critique of virtue theory in medical ethics,' In E.E. Shelp (ed.) *Virtues in Medicine. The Philosophy and Medicine Series, Volume 17,* Reidel, Dordrecht/ Boston.

Veatch, R.M.: 2001, 'Ruth Macklin: *Against Relativism: Cultural Diversity and the Search for Ethical Universals in Medicine*,' *Theoretical Medicine* 21, 385-92.

Welie, J.V.M.: 1999, 'Towards an ethics of immediacy: A defense of a noncontractual foundation of the caregiver-patient relationship,' *Medicine, Health Care and Philosophy* 2 (1), 11-19.

Wildes, K.W.: 1995, 'The ecumenical and non-ecumenical dialectic of Christian bioethics,' *Christian Bioethics* 1(2), 121-7.

K. DANNER CLOUSER[†] AND BERNARD GERT

COMMON MORALITY

I. INTRODUCTION

The mission of this Handbook is to focus on the philosophical aspects of bioethics: to sort out what they are, to show what they contribute, and to suggest what needs further attention. This mission is to serve two goals: to keep the field of bioethics from becoming so subdivided by issues that each subspecialist is no longer in touch with nor profiting from what the others are doing, and to keep the field of bioethics from being so full of itself as a new discipline that it loses awareness of the distinctive contributions of each of its ingredient disciplines.

The focus of this chapter is not on any of the content issues within bioethics, but rather on the ethical framework within which all those issues are considered. That in itself is a unifying move for the field since it is arguably something that all aspects of bioethics have in common, and in itself is also a unique contribution of philosophy to the field of bioethics. In the earlier days of the bioethics revival there was not much explicit attention paid to ethics as such. Everyone was very much caught up in dealing with first order concerns, those immediate matters that called out for solutions. General intuitions were mostly sufficient for carrying out this sort of moral first aid. Tribute was paid to deontological theories and consequentialist theories in passing, but that was mostly to give a theoretical gloss to these matters in teaching or discussion. After all, all could agree on most solutions to the problems; most were primarily engaged in highlighting the problems and analyzing them. That was often sufficient. And everyone had a contribution to make – physicians, journalists, scientists, lawyers, novelists, etc. But one knew that sooner or later these theoretical matters were bound to float to the top (or sink to the bottom) and demand explicit attention. It is more or less expected that any new enterprise after two or three decades of enthusiastic work and involvement will slow down and become more reflective about its foundation. However, this theoretical turn might have been accelerated in this instance because more and more philosophers (who, by and large, avoided the field early on) were beginning to enter the field of bioethics.

G. Khushf (ed.), Handbook of Bioethics, 121–141.

II. PRELUDE TO THEORY

Many people believe there is no substantial agreement on moral matters and that there is even less agreement on the adequacy of any account of morality. These views are prompted and perpetuated by concentration on only the controversial moral issues without realizing that such controversial matters form only a small part of those matters about which people make moral decisions and judgments. Indeed, most moral matters are so uncontroversial that people do not even make conscious decisions concerning them. For example, no one hesitates to make negative moral judgments about those who harm others simply for their own personal interests. Nor do they hesitate to make moral judgments condemning unjustified deception, breaking promises, cheating, disobeying the law, and not doing one's duty. The universal moral praise that is given to those who help the needy or who devote themselves to relieving pain and suffering also illustrates the uncontroversial nature of most moral matters. Although the prevalence of hypocrisy shows that people do not always behave in the way that morality requires or encourages, it most notably also shows that even hypocrites know the general kind of behavior that morality requires, prohibits, discourages, encourages and allows.

We believe that an explicit, clear, precise, and comprehensive account of morality is possible and that it would help to make clear the uncontroversial nature of many moral decisions. We also believe that such an account would help in understanding, and sometimes even in resolving, some of the controversial moral problems that arise in the field of medicine. Our account provides a common framework on which all of the disputing parties can agree, for this account does not *settle* most controversial issues. By clarifying the causes of disagreement and allowing for more than one alternative to be morally acceptable, it provides a better chance to manage that disagreement in a productive way. That is, an adequate account of morality will provide a detailed blueprint of the dispute showing precisely the source of moral disagreement as well as exposing what non-moral features of the disagreement are posing as moral, thereby clearing the way for achieving agreement or, at least, a morally acceptable compromise. Making the common moral system explicit actually has practical benefits.[1]

Those who deny the possibility of a comprehensive account of morality may intend to deny that any systematic account of morality can provide an answer to every moral problem. But we have already acknowledged that the common moral system does not provide a unique solution to every moral problem. Common morality does provide a unique answer in many cases, though most of these cases are not interesting because they are not controversial. Only in a very few situations does an explicit account of morality settle what initially seemed to be a controversial matter, e.g., whether dying patients should be allowed to refuse food and fluids as well as life-sustaining medical procedures. Although most controversial cases do not have unique answers, even in these cases an explicit account of morality is often quite useful because it shows that there are significant limits to legitimate moral disagreement. That there is no agreement on the right answer does not mean that there is no agreement on the boundaries of what is morally acceptable, that is, there is agreement on which of the alternatives are wrong.

The fact that moral disagreement on some issues is compatible with complete agreement on many other issues seems to be almost universally overlooked. Most philosophers seem to hold that if equally informed impartial rational persons can disagree on some moral matters, they can disagree on all of them. Thus most philosophers hold either that there is a unique right answer to every moral question or that there is no unique right answer to any moral question. The unexciting but correct view is that some moral questions have unique right answers and some do not. There is far more moral agreement than there is moral disagreement, but the areas of moral disagreement are much more interesting to discuss, and so are discussed far more often, thus skewing the overall account of morality.

III. COMMON MORALITY

Common morality is a subtle and complex system. It is far more subtle and complex than the systems of conduct that philosophers such as Kant and Mill generate from their moral theories, and then offer as improvements on common morality. The moral theory presented in this chapter is not a creation intended to improve on common morality; rather, it is an attempt to clarify, describe, explain, and justify the common moral system. We try to provide a description of common morality that does justice to its subtlety and complexity; we explain it nature by relating it to universal features of human nature such as fallibility and vulnerability; and we justify it by relating it to impartiality and rationality.

Common morality is the system people use, often unconsciously, when they are trying to make morally acceptable choices among several alternative actions or when they make moral judgments about their own actions or those of others. That everyone uses exactly the same moral system does not mean that everyone, even all fully informed impartial rational persons, agree in all of their moral decisions and judgments. Some of these disagreements are quite significant, e.g., on the moral acceptability of abortion, nonetheless, there is overwhelming agreement on most moral decisions and judgments.

It is an essential feature of common morality that everyone who is judged by it knows what it prohibits, requires, discourages, encourages, and allows. That is why moral judgments are not made about the actions of lions and tigers, or the actions of infants, small children, and severely mentally retarded persons. Since moral judgments are made about all rational persons, it follows that morality is universal. This has the surprising consequence that common morality cannot be based on any findings of modern science, for most rational persons do not know about these findings. It also follows that common morality cannot be based on any particular religion, for no particular religion is known, let alone accepted, by all rational persons.

Although most moral decisions are not made by explicitly employing any account of morality, agreement about the framework of common morality accomplishes three tasks: (1) makes clear that there is agreement on the overwhelming majority of cases; (2) makes clear the sources of moral disagreement and explains why there may sometimes be no unique best solution; and (3) makes clear that new and perplexing problems can be described in a way that shows their similarity to more familiar moral matters.

Our account of common morality makes clear to those entering the health care professions that the moral framework that is used in medicine is the same moral framework that they have always used. Although doctors, nurses, and other health care professionals have specific duties that cannot be deduced from common morality, these duties cannot be incompatible with it either. The difficult moral problems that health care professionals commonly face require knowledge of common morality, of the workings and relationships within the health care world, and of the specific duties and ideals of health care professionals. These specific duties and ideals may sometimes seem to conflict with the requirements of common morality, but a proper understanding of both will show that the moral system provides a method for dealing with those conflicts in the same way that it deals with conflicts within common morality.

The following definition of morality captures what most people mean by that term. *Morality is an informal public system applying to all rational persons, governing behavior that affects others, and includes what are commonly known as the moral rules, ideals, and virtues and has the lessening of evil or harm as its goal.* Since the phrase "public system" forms such an important part of our definition of morality and it is not a commonly used phrase, it is important to explain what we mean by it.

A public system has the following two characteristics. (1) All persons to whom it applies (i.e., those whose behavior is to be guided and judged by that system) understand it (i.e., know what behavior the system prohibits, requires, discourages, encourages, and allows). (2) It is not irrational for any of these persons to accept being guided and judged by that system. The clearest example of a public system is a game. A game has an inherent goal and a set of rules that form a system that is understood by all of the players; and it is not irrational for all players to use the goal and the rules of the game to guide their own behavior and to judge the behavior of other players by them. Although a game is a public system, it applies only to those playing the game. Morality, on the other hand, is a public system that applies to all moral agents; it applies to all rational persons simply in virtue of their being responsible for their actions. This may explain the force of Kant's claim that the demands of morality are categorical.

In order to ensure that all rational persons know what common morality requires, prohibits, discourages, encourages, and allows, knowledge of the moral system cannot involve any beliefs that are not held by all rational persons. Only those beliefs which it would be irrational for any moral agent not to hold are essential for knowledge of morality. We call such beliefs *rationally required* beliefs. On the other hand, beliefs that it would be irrational for any moral agent to hold, we call *rationally prohibited* beliefs. Beliefs that are neither rationally required nor rationally prohibited, we call *rationally allowed* beliefs.

Rationally required beliefs include general factual beliefs such as: people are vulnerable, that is, they can be harmed by other people; and people are fallible, that is, they have limited knowledge. Having these beliefs is necessary in order to be a moral agent. On the other hand, *rationally allowed* beliefs such as scientific or religious beliefs cannot be necessary to be a moral agent, even though some rationally allowed beliefs, e.g., beliefs about the facts of the particular case, are necessary for making particular moral decisions or judgments. However, personal beliefs about one's own race, gender, religion, abilities, etc., are not rationally required, and so these beliefs cannot be used to develop the moral system.

Although all those who are held responsible for their actions know what morality requires, prohibits, discourages, encourages, and allows, morality is not a simple system. A useful analogy is the system of grammar used by all competent speakers of a language. Almost no competent speaker can explicitly describe this system, yet they all know it in the sense that they use it when speaking and when interpreting the speech of others. If presented with an explicit account of the grammatical system, competent speakers have the final word on its accuracy. It would be a mistake to accept any description of a grammatical system that rules out speaking in a way that they know is commonly regarded as acceptable or allows speaking in a way that they know is commonly regarded as completely unacceptable. Similarly, our account of morality should be tested by seeing if it yields any judgments that are incompatible with those that are commonly accepted.

Morality is an informal public system, i.e., a public system that has no authoritative judges or procedures for determining a unique correct answer. A formal system such as law, or a formal public system, such as a game of a professional sport, has procedures for arriving at a unique correct answer within that system. But when people get together to play a game of cards or backyard basketball, they are involved in an informal public system. For the game even to get started, there must be overwhelming agreement on how the game is to be played, but disagreements can arise which have no established method of resolution. These disagreements are either resolved in an ad hoc fashion, e.g., flipping a coin or asking a passerby, or are not resolved at all, e.g., a replay is initiated or the game is disbanded.

Morality, like all informal public systems, presupposes overwhelming agreement on most matters that are likely to arise. However, like all informal public systems, it has no established procedures or authorities that can resolve every moral disagreement. There is no equivalent in morality to the United States Supreme court in deciding legal disputes, or the Pope in deciding some religious matters for Roman Catholics. The claim that the Pope is a moral authority is the result of failing to distinguish between morality and religion. There can be religious authorities, but there are no moral authorities. No one has some special knowledge of morality not available to others, for everyone who is subject to moral judgment must know what it requires, prohibits, discourages, encourages, and allows. When morality does not provide a unique right answer and a decision has to be made, the decision is often made in an ad hoc fashion, e.g., asking a friend for advice. If the moral disagreement is about some important social issue, e.g., abortion, and a decision has to be made, the problem is transferred from the moral system to the political or legal system.

Abortion is an unresolvable moral question because it involves the scope of morality, that is, it has to do with what beings are impartially protected by the moral rules. Although everyone agrees that all moral agents are protected by the moral rules, there is no agreement on what other beings are protected, e.g., potential moral agents. Since it has to be decided which, if any, abortions are to be allowed, the question is transferred to the legal and political system. The question is resolved on a practical level, but the moral question is not resolved, as is shown by the continuing intense moral debate on the matter. But almost everyone, including those who are opposed to abortion, agrees that it is not morally justified to cause harm, that is, to kill or injure doctors and others involved in providing abortions. Thus, despite what anti-abortion groups claim about the immorality of abortion, most of them seem to recognize that it is not justified for individuals to harm others in attempting to prevent abortions.

In addition to differences concerning the scope of morality, another source of unresolvable disagreement in moral judgment is due to differences in the rankings of harms, e.g., whether death is worse than a painful disability, including differences in weighing the probabilities of harms. Disagreements on the proper speed limit is a disagreement about whether the certain deprivation of some freedom to millions, namely, the freedom to drive between 55 miles an hour and 70 miles an hour, is justified by the high probability that some lives will be saved. Everyone agrees that there should be some speed limits on most roads, but there is disagreement as to what that speed limit should be. Most of this disagreement, like most moral disagreement, is due to a disagreement about the facts, e.g., how many less accidents, injuries, deaths, would there be if the speed limit were ten miles per hour less? But even if people agreed on the facts, they still might disagree, e. g., about whether the saving of ten lives justifies reducing the speed limit by ten miles per hour.

Similarly, everyone agrees that competent patients should be allowed to refuse medical treatment in order to hasten their deaths, but there is disagreement about whether their doctors should be allowed to provide them with the medical means to commit suicide. Some of this disagreement is due to factual disagreements about the effects of allowing physician-assisted suicide, but there may also be unresolvable disagreements concerning the rankings of the harms and benefits involved, e.g., whether the decrease in suffering by dying patients justifies the increase in the number of unwanted earlier deaths. Hence even with agreement on the facts there may be disagreement on whether physician-assisted suicide should be allowed.

IV. THE MORAL RULES

In this chapter we focus on the moral rules and their place in the moral system, but it is important to make clear that morality includes ideals as well as rules. The existence of a common morality is shown by the widespread agreement on what counts as a moral rule, i.e., a rule whose violation is immoral unless one has an adequate justification for violating it. Everyone agrees that harming people, i.e., killing them, causing them pain or disability, or depriving them of freedom or pleasure, is immoral unless one has an adequate justification. Similarly, everyone agrees that violating those norms that are essential to the successful functioning of all societies, i.e., deceiving, breaking a promise, cheating, breaking the law, and neglecting one's duties, also needs justification in order not to be immoral. No one has any real doubts about any of this. Nor is there any doubt that helping the needy and working to alleviate pain and suffering are morally good actions, that is, they count as following moral ideals.

For cultural and aesthetic reasons, as well as for clarity and precision, we prefer to formulate the moral rules in a way that results in there being ten rules, but as indicated above, one could formulate the rules so that there are only two rules, "Do not cause harm." and "Do not violate the norms necessary for society." But if one chooses to have only these two rules, it would be necessary to list the different kinds of harms, in order to make clear that rational persons can disagree in their ranking of these harms. It would also be necessary to explicitly identify the universally justified norms, i.e., those that are necessary for any society to function, and distinguish them from norms that are not so justified. Nonetheless, there are contexts in which it may be appropriate to list only two moral rules, or perhaps only one, e.g., "Do not cause harm or increase the probability of harm being suffered." (Similarly all of the moral ideals could be summarized as "Prevent or alleviate the suffering of harms.") Although having only one or two rules can be advantageous in certain contexts, we shall list ten rules which would be supported by all impartial rational persons who used only rationally required beliefs. Because these rules are so supported we call them the justified moral rules.[2]

The justified moral rules:

1. Do not kill.
2. Do not cause pain.
3. Do not disable.
4. Do not deprive of freedom.
5. Do not deprive of pleasure.
6. Do not deceive.
7. Keep your promises.
8. Do not cheat.
9. Obey the law.
10. Do your duty.[3]

As indicated above, these rules are not absolute; they all have justified exceptions, and most moral problems involve determining which exceptions are justified. Although there is some disagreement about the scope of these rules, e.g., whether animals or embryos are protected by them, that should not lead to any doubt that harming moral agents, i.e., adults like us, needs justification. People also disagree about what counts as an adequate moral justification for some particular violation of a rule, e.g., deceiving or breaking the law, and on some features of an adequate justification, but everyone agrees that what counts as an adequate justification for one person must be an adequate justification for anyone else in the same situation, i.e., when all of the morally relevant features of the two situations are the same. This is part of what is meant by saying that morality requires impartiality.

In order to capture both the fact that the moral rules must be obeyed impartially and the possibility of disagreement concerning what actions count as justified violations, it is necessary to make explicit the attitude that all impartial rational persons take toward the rules. We call this the moral attitude, for it expresses the morally correct attitude toward the moral rules: *Everyone is always to obey a moral rule except when a fully informed rational person can publicly allow violating it.* If no fully informed rational persons would publicly allow the violation, it is unjustified; if all fully informed rational persons publicly allow the violation, it is strongly justified; if fully informed rational persons disagree about whether to publicly allow the violation, it is weakly justified.

To "publicly allow" a violation is to claim that one would allow that kind of violation to be part of the "informal public system" that is common morality. In less technical terms, to publicly allow a violation is to claim that one would be willing for everyone to know that kind of violation was allowed. What counts as the same kind of violation is determined by the morally relevant features of the violation. If two violations have all of the same morally relevant features, then they are the same kind of violation. Determining the same kind of violation is the first step of what we call the two-step procedure for justifying violations of moral rules. The second step of this two-step procedure is estimating and comparing the consequences of publicly allowing this kind of violation versus the consequences of not publicly allowing it.

When deciding whether or not an impartial rational person can advocate that a violation of a moral rule be publicly allowed, the kind of violation must be described using only morally relevant features. Since the morally relevant features are part of the moral system, they must be understood by all moral agents. This means that any description of the violation must be such that it can be reformulated in a way that all moral agents can understand it. Limiting the way in which a violation must be described makes it easier for people to discover whether their decision or judgment is biased by some consideration which is not morally relevant. All of the morally relevant features that we have discovered so far are answers to the following questions. It is quite likely that other morally relevant features will be discovered, but we think that we have discovered the major features. Of course, in any actual situation, it is the particular facts of the situation that determine the answers to these questions, but all of these answers can be given in a way that can be understood by all moral agents.

1. What moral rules would be violated?
2. What harms would be (a) avoided (not caused), prevented, and (b) caused? (This means foreseeable harms and includes probabilities as well as kind and extent.)

3. What are the relevant beliefs and desires of the people toward whom the rule is being violated?
4. Does one have a relationship with the person(s) toward whom the rule is being violated such that one sometimes has a duty to violate moral rules with regard to the person(s) without their consent?
5. What benefits would be caused? (This means foreseeable benefits and also includes probabilities, as well as kind and extent).
6. Is an unjustified or weakly justified violation of a moral rule being prevented?
7. Is an unjustified or weakly justified violation of a moral rule being punished?
8. Are there any alternative actions that would be preferable?
9. Is the violation being done intentionally or only knowingly?
10. Is it an emergency situation that no person is likely to plan to be in?

When considering the harms being avoided (not caused), prevented, and caused, and the benefits being promoted, one must consider not only the kind of benefits or harms involved, one must also consider their seriousness, duration, and probability. If more than one person is affected, one must consider not only how many people will be affected, but also the distribution of the harms and benefits. Anyone who claims to be acting or judging as an impartial rational person who holds that one of the two violations be publicly allowed must hold that the other also be publicly allowed. This simply follows from morality requiring impartiality when considering a violation of a moral rule.

V. ALTERNATIVE MORAL THEORIES[4]

Casuistry[5]

Casuistry involves concentrating on a particular case and comparing it to other cases so as to determine what rules are most applicable to it, and how these rules should be interpreted when dealing with this case. Casuistry, when it is not explicitly a part of the kind of moral system that we are describing, may help to resolve problems, but it can do nothing to help avoid them, because casuistry, considered by itself, has no concept comparable to that of a moral rule. Of course, casuistry, properly understood, is part of the kind of moral system that we present; it is a useful method for interpreting and applying the moral system. It is not always clear what kind of behavior counts as killing, deceiving, etc. and casuistry can help with such interpretation. Casuistry is also helpful in determining whether the case under consideration should be viewed as a justified exception to the rule. Concentrating on the particular case and comparing it to other cases, may make more salient the morally relevant features of the case. Further, this comparison of cases can also help one to see whether one would want everyone to know that this kind of violation is allowed. Divorced from the moral system, however, casuistry is of little value and simply promotes ad hoc solutions to problems.

Of course, to use casuistry successfully one need not explicitly accept the moral system; one need only employ it implicitly, as most of us do, in interpreting common morality. Casuistry, for example, can help in deciding whether a particular case of not telling a patient some information counts as withholding that information and thus as deception, or whether, on the other hand, there is no moral requirement to provide that information. Indeed, it is not even clear what it would be to use casuistry without using the moral system, at least implicitly. Without the moral system, casuistry seems to be nothing more than ad hoc reasoning about moral matters. It contains no way to resolve disputes if people choose different cases as models that they claim should be used to make a decision about the case under consideration. More importantly, casuistry does not make clear what is even causing the dispute and, independent of a moral system, it does not even identify what counts as a moral matter.

Casuistry makes clear that the moral rules need interpretation and that such interpretation is often essential before one can apply a rule to a particular case. Casuistry also emphasizes the need to look for all of the morally relevant features of the case, although it does not provide a list of such features. Casuistry realizes that a particular detail, e.g., the relationship between the parties involved, may change the act from one that is morally unacceptable to one that is morally acceptable, but it provides no explanation of when or why this is so. Since casuistry requires a moral system, it is not an alternative to our account of morality. Rather it has the subsidiary but important role of helping to apply common morality to particular cases.

Virtue Theory

Virtue theory emphasizes a dimension of moral behavior that is not part of the basic description of the moral system, namely, that morality is usually concerned with a consistent pattern of behavior and does not call for special decisions in every particular case. Thus it is perfectly appropriate for parents to present morality to children as the acquiring of the moral virtues and to teach them by example. However, in order to be a useful guide to behavior, the virtues must be understood in light of the moral system. Parents can teach morality by the virtues only if they appreciate the connection of the virtues to the moral system. It may be valuable to select a role model, but some way of selecting the right role model is needed, and also some way of determining if the role model is acting in the right way in a particular situation. All of this requires a clear understanding of the moral system and of how to apply it. Virtues are not fundamental features of morality, hence they are not an alternative to our account of the moral system, but rather they are based on the moral system and are an important and practical supplement to it.

We call those who take virtues to be fundamental to morality, virtue theorists. A significant problem with many virtue theorists is that they usually do not distinguish the moral virtues, e.g., honesty and kindness, from the personal virtues, e.g., courage and temperance. The moral virtues are those virtues which all rational persons want other people to have, whether or not they want them themselves. The personal virtues are those virtues that all rational persons want to have themselves. The moral virtues are directly related to the interests of others, and only indirectly related to one's own interests, while the opposite is true for the personal virtues. Understanding the personal virtues does not require understanding morality, whereas understanding the moral virtues does. We are concerned with virtue theory only insofar as it purports to provide an account of morality that makes moral virtues more fundamental to morality than the moral system, i.e., the moral rules, ideals, the morally relevant features, and the two-step procedure for justifying violations.

Virtue theory, like casuistry, is closely related to the moral system, however, unlike casuistry, it is not a method of applying morality to a particular case, but an alternative and incomplete way of formulating the moral system. A complete account of morality must, of course, include an account of the moral virtues and vices.[6] Such an account lists the moral virtues and vices and relates them to particular moral rules and ideals, but it is clear that having a particular virtue, e.g., truthfulness, does not mean that one never lies; a truthful person lies when lying is justified. The moral system enables one not only to identify those who have the virtues, but also to describe the ways virtuous persons could act in a particular situation. One's intentions and motives are crucial for determining if one is a virtuous person, but what intentions and motives are the appropriate ones is determined by the moral system.

Some virtue theorists claim to provide a useful guide to conduct by enabling virtuous persons to be identified and then used as role models. This account of virtue raises several critical questions, e.g., how does one pick out a virtuous person and how does the virtuous person decide how to act? Further, it is not only possible, but common, for a person to have some of the virtues, but not others. Even those who have the virtue in question do not always exemplify it.

For virtue theorists who advocate the use of role models as basic guides, these are serious problems. If no particular person can serve as a role model for all situations – and there is general agreement that few, if any, can – there needs to be some way to determine what virtue is called for in a particular situation so that a role model with the relevant virtue can be selected. Even more serious, no virtuous person can be depended upon to act virtuously 100% of the time, so there has to be some independent way of determining when he is acting virtuously and when not. These are not new problems that we have just discovered; indeed, in a different context, Kant explicitly raised these same points. (Kant, 1993, Ak. 408-409, pp. 20-21)

For virtue theory to be of practical use there must be some way to determine what counts as acting virtuously in a particular situation. Common morality does provide a way to make that determination. Having a moral virtue is being disposed to obey the moral rules, or to follow the moral ideals, in a given situation in the way that at least some fully informed impartial rational persons would do. Although fully informed impartial rational persons do not always agree, there are always limits to their disagreement, that is, they all do agree on the immorality of some alternatives. Providing a clear account of common morality, including the moral rules and ideals, the morally relevant features, and the two-step procedure for determining whether a violation of a moral rule is justified, does provide a way of determining what counts as a virtuous way of acting in any particular situation.

If the moral virtues are understood primarily as possessing the appropriate motivation for one's actions, then serious problems arise, particularly with regard to paternalistic behavior. All cases of genuine paternalism with regard to patients involve the health care professional being motivated to act for the benefit of the patient. On the motive reading of virtue theory, the doctor who acts paternalistically is necessarily a virtuous person. By acting benevolently, trying to help others, he demonstrates the virtue of beneficence. Unfortunately, his beneficence may obscure the fact that he is also violating a moral rule, e.g., the rule prohibiting deceiving or the rule prohibiting depriving of freedom, and so he may not even recognize that there is a moral problem. Virtue theory not only makes one less likely to consider whether a particular paternalistic act is justified or not, it also provides no method for determining whether or not it is justified. A proper account of the moral virtues must explain not only when but why an impartial rational person should or should not violate a moral rule or follow a moral ideal in specific kinds of circumstances.

VI. USING PATERNALISM AS A TEST OF THEORIES

Two of the most important philosophical accounts of morality, which also have great prominence in bioethics, give incompatible answers to the question of how paternalistic behavior is justified: act consequentialism says that only the consequences of one's actions are morally relevant, and strict deontology says that, without consent, only the conformity or nonconformity of one's action with a moral rule is morally relevant. Paternalism is arguably the most pervasive moral issue throughout health care practice. Paternalistic behavior requires justification because it involves violating a moral rule. Since the violation is done in order to benefit the person toward whom the rule is violated, the consent of that person would make it a justified violation but, of course, with consent the violation is no longer paternalistic. All impartial rational persons would publicly allow violating a moral rule with regard to a person if that person gives valid consent to the violation and the violation benefits her. Paternalism is interesting because it involves breaking a moral rule without consent toward a person who believes she is able make her own decision, but it is done to benefit that very person.

For the purpose of using paternalism as a test of various moral theories, it is convenient that all of the positions that can be maintained with respect to the justifiability of a common kind of paternalistic behavior, that in which the action provides a net benefit to the person and no other person is harmed, can be put into three categories: 1) it is always justified, 2) it is never justified, and 3) it is sometimes justified. It is also convenient that with regard to this kind of paternalism, act consequentialism says that paternalism is always justified, strict deontology says that it is never justified, while common morality holds that is sometimes justified. This is what makes the discussion of the justification of paternalism so important philosophically; it provides a real test of the various accounts of morality. Since the opportunity and temptation to act paternalistically is ubiquitous in the field of health care, it is of great practical value to show which account of morality best determines when acting paternalistically is justified.

Act Consequentialism - Paternalism is Always Justified

Act consequentialism is a very simple guide to conduct. It claims that an act should be done if it will result in the best overall consequences. There are many sophisticated variations of this view, but in its simple form it is held by many who do not regard themselves as holding any philosophical view at all. People who hold this view often state that all that really matters is that things turn out well; if they do, you can do anything you want. It is interesting to note that, ignoring alternative actions, the ethical theory of act consequentialism implies that all paternalistic behavior which provides a net benefit to the person toward whom one acted paternalistically (and no other person is harmed) is justified. Of course, in many cases of paternalism it is not clear that the patient gains a net benefit, for the physician may rank the harms differently than the patient does, and both rankings may be rational. Further, because there is so much self-deception, as well as so many mistakes about the outcomes of paternalistic intervention, even accepting the physician's rankings, paternalism often does not result in any net benefit for the person who is being treated paternalistically.

On the most plausible interpretation, act consequentialism holds that in any situation, a person does what is morally right by choosing that action which, given the foreseeable consequences, will produce at least as favorable a balance of benefits over harms as any other. This is the type of ethical theory that underlies what is sometimes called "situation ethics." Since this theory denies that there are any kinds of acts which need justification, i.e., denies the significance of moral rules, it denies that violations of moral rules need justification. According to act consequentialism, if the foreseeable consequences of the particular paternalistic act provide at least as favorable a balance of benefits over harms as any other act, then the act is morally right, and if the foreseeable consequences are not as favorable, then the act is not only morally wrong, there is no justification for it.

That act consequentialism is false is seen most clearly when considering a case of cheating on an exam in a course taken on an honor system that is not graded on a curve. If the foreseeable consequences are that no one will be hurt by the cheating and the cheater will benefit by passing, act consequentialism not only says that cheating is acceptable, but that it is morally unacceptable not to cheat. Common morality, however, correctly judges cheating in this kind of situation as morally unacceptable, for no impartial rational person would publicly allow such a violation. Act consequentialists, however, are unconcerned with the consequences of *this kind of act* being publicly allowed, and consider only the foreseeable consequences of *this particular act*. Thus they must make up facts about human nature, e.g., given human nature, the foreseeable consequences of cheating in this kind of situation will never result in as favorable a balance of benefits over harms as not cheating. But this is fudging the facts in order to prevent the theory from conflicting with the moral judgments that everyone would actually make.

Strict Deontology - Paternalism is Never Justified

According to the strict deontological view, it is never justified to break a moral rule without the valid consent of the person with regard to whom you are breaking it. Some hold that even valid consent does not justify violating some moral rules, e.g., against disabling, so that it is even immoral to violate these moral rules with regard to oneself. This extreme position usually has a religious foundation, e.g., that the moral rules were ordained by God to govern the behavior of human beings. However, this position can also have a metaphysical basis, e.g., as in Kant, where reason takes the place of God as the author of the moral rules. Without such a religious or metaphysical foundation, the view that it is never justifiable to violate a moral rule with regard to a competent person who has rationally consented to your violating the rule toward her, has no support.

Further, almost everyone who holds a deontological view holds that it is justified to violate a moral rule, sometimes even the rule prohibiting killing, with regard to someone who has himself violated a moral rule. Punishment, even capital punishment, is accepted as justified by most deontological thinkers including Kant and most religious philosophers. It is only with regard to the innocent that they hold that it is never justified to violate a moral rule without consent. It is quite common for strict deontologists to hold that only consent can justify violating a moral rule with regard to an innocent person and even on this less radical account, no paternalistic behavior is justified.

Some strict deontologists, like some act consequentialists, claim that they are presenting a description of the common moral system. However, like act consequentialism, strict deontology does not provide an accurate account of common morality. Common morality sometimes justifies deception if necessary to save an innocent person's life. If the only way to prevent very serious harms is by breaking a moral rule and the violation prevents so much greater harm than it causes that one could publicly allow such a violation, common morality holds that such a violation is at least weakly justified. Common morality also differs from Kant, who holds that in order for a violation to be justified an impartial rational person must will that everyone act in that way; common morality correctly claims that all that is necessary for the justification of a violation of a moral rule is that an impartial rational person can publicly allow such a violation.

Similarly, common morality holds that paternalistic behavior is sometimes justified, e.g., paternalistic deception, if it is the only way one can prevent very serious harm to the patient. Paternalistic deprivation of freedom, in the form of involuntary commitment, is even sanctioned by law if there is a high enough probability of the person seriously harming herself. Common morality sometimes even allows harming an innocent person who has not given consent for such a violation, if the harm to be prevented is sufficiently great to publicly allow such a violation. Some strict deontologists, like some act consequentialists, have put forward implausible views of human nature which make their views seem more acceptable. They have claimed that any violation of a moral rule with regard to an innocent person without his consent inevitably results in wholesale violations of moral rules with disastrous consequences. It is interesting that this defense of strict deontology seems to depend upon consequences, but closer inspection shows that it presupposes the view that we have put forward, i.e., that the decisive factor in determining the morality of an act is the consequences of everyone knowing that the violation is allowed.

To avoid obvious problems, some strict deontologists characterize paternalism as limited to violating a moral rule toward a *competent* person without her consent, and thus hold that paternalism is never justified. A serious problem with limiting paternalism in this way is that it may sometimes allow morally unacceptable behavior toward those who are not competent to make a rational decision. Just because people are not competent to make a rational decision does not mean that it is justified to violate any moral rule with regard to them as long as they benefit from that violation. If the benefit is small, it generally is not justified to deceive or deprive of freedom. By making competence of the patient a necessary feature of paternalism, it seems to justify treating large numbers of incompetent patients in a way that would be paternalistic if they were regarded as competent. Act consequentialist and strict deontological views seem to justify what would be viewed by common morality as unjustified paternalistic behavior. The act consequentialist would claim that his act is justified because there is a net benefit to the patient, while the strict deontologist would claim that his behavior is not even paternalistic because the patient is not really competent. Someone who simply concentrated on the relevant behavior, however, might not be able to distinguish the strict deontologist from the act consequentialist when dealing with those patients toward whom physicians are most tempted to act paternalistically.

Defining paternalism in a narrow way so that one can only act paternalistically toward those who are competent to make a rational decision, may lead physicians not to be concerned with justifying their violations of moral rules with regard to someone they view as incompetent. Since many cases of medical paternalism are with regard to patients whom the physician regards as not competent, this is a serious problem. The strict deontologist shifts the emphasis from the genuine moral problem of justifying violating a moral rule for a patient's benefit without his consent, to the problem of determining if the person is competent. If he is not, the interference is not paternalistic and need not be justified.

According to common morality, even if the patient is incompetent, as long as he believes he can make his own decision, it still may not be justified to intervene. The harm prevented by interfering may not be great enough to justify the violation of a moral rule with regard to the patient. On the strict deontological proposal there is an absolute dichotomy between the way the two classes of patients may be treated. No matter how great the harm prevented and how minor the violation of the moral rule, competent patients can never be interfered with for their own benefit without their consent. With regard to incompetent patients, interference is justified by determining that even a little more harm is prevented than caused. But there is not such a sharp line separating competence from incompetence and, even if there were, it has not yet been reliably enough determined to allow it to play such an important role in determining whether it is justified to break a moral rule toward someone without consent.

The strict deontologist, who holds that genuine paternalism (paternalism toward the competent) is never justified, like the act consequentialist, who holds that genuine paternalism (paternalism that has a net benefit for the patient) is always justified, has a serious problem. Neither presents us with a practical way of deciding whether or not particular patients should be deceived or deprived of freedom when the foreseeable consequences are that they would benefit from this violation of moral rules with regard to them. Both of these views present overly simple accounts of how physicians do and should go about determining whether or not it is justified to act paternalistically. Of course it is just their simplicity which makes them so attractive, for if physicians accept either of these views, they have a simple way of dealing with troublesome cases. However, those physicians who are serious about the matter have to be prepared to look at the morally relevant features of each case and only then decide whether they would publicly allow acting paternalistically. Even when they have done all of this correctly, they must accept that sometimes not everyone will agree with them.

Common Morality - Paternalism is Sometimes justified

Showing that common morality applies to cases of medical paternalism provides strong reasons for thinking that there is no need for a special ethics for medicine. This is very important, for holding that common morality does not apply in medical situations may lead some physicians to think that they are not subject to the same moral constraints that all other people are. This may be one explanation for the many instances of unjustified paternalism in medicine. Like most moral theories, most philosophical discussions of the justification of paternalism oversimplify. The act consequentialist considers morally relevant only the consequences of the particular act, and the strict deontologist considers morally relevant only whether a moral rule is being broken with regard to an innocent person who has not given consent for the violation. But there are also many other morally relevant features. Failure to take into account all of these features often leads to a failure to distinguish between cases that might differ in only one crucial respect, e.g., whether that situation is an emergency. Common morality does not make this mistake.

Perhaps the most overlooked feature, but one which is often the most important, is the feature concerning alternative actions. If there is a non-paternalistic alternative that does not involve any unconsented to violation of a moral rule and does not differ significantly in the harms and benefits to the patient, then paternalistic behavior cannot be justified. This is a very significant matter, for often there is an alternative to paternalistic behavior, viz., long conversations with the patient trying to explain the benefits of accepting a treatment. Often it is lack of time to spend with the patient rather than lack of alternatives that leads to paternalistic behavior.

Act consequentialists claim that the answers to questions about harms, benefits and alternatives, are the only morally relevant features that need to be considered in determining what morally ought to be done. This claim is incorrect. Not only are there other morally relevant features that are necessary to determine the kind of moral rule violation, but determining the kind of moral rule violation is only the first step. The next step requires considering whether or not one would advocate publicly allowing that kind of violation, that is, all violations of that rule in those kinds of circumstances. Failure to move to the next step puts one back into a kind of act consequentialism, considering only the consequences of the particular act. The function of the morally relevant features, including the foreseeable consequences, is to determine the kind of violation which, although absolutely crucial, is only the first step of a two step procedure.

The second step of the two step procedure is answering the morally decisive question, which is: would the foreseeable consequences of that kind of violation being publicly allowed, i.e., of everyone knowing that they are allowed to violate the moral rule in these circumstances, be better or worse than the foreseeable consequences of that kind of violation not being publicly allowed? Consequences are crucial, but it is not the consequences of the particular act, rather it is the consequences of that *kind* of act being publicly allowed that are decisive. This account of common moral reasoning incorporates the insights of both Kant and Mill. Previous philosophers have oversimplified common morality. We not only recognize the diverse nature of the morally relevant features, we realize that moral reasoning involves a two step procedure: (1) using the morally relevant features to determine the kind of violation and (2) estimating the foreseeable consequences of that kind of violation being publicly allowed.

Disagreement about moral decisions and judgments can occur in either step. People can disagree about the kind of act, or they can disagree about whether or not they favor that kind of act being publicly allowed. Of course, one cannot even begin to decide whether one favors a kind of act being publicly allowed until the kind of act has been determined. This explains why discovering all the relevant facts is so important. It also explains why it is important for all of the morally relevant features to be recognized, for they tell one what facts to look for. Only after determining the kind of violation, by finding all the facts indicated by the morally relevant features, can the morally decisive question be asked: Does the harm avoided or prevented by this kind of violation being publicly allowed outweigh the harm that would be caused by it not being publicly allowed? If there is disagreement, we call it a weakly justified violation, and whether it should be allowed is a matter for decision. Since some disagreement is unavoidable in many cases, having a public policy about who makes the decision is essential. Further, whenever possible, we think that in cases of disagreement, there should be a public policy that involves consulting some experienced advisory body, e.g., an ethics committee.

Our goal is not to provide a solution to every case, but rather to provide a framework that enables fruitful discussion of moral problems. We think that this framework always provides a range of morally acceptable answers. Further, we believe that using this explicit account of moral reasoning makes it less likely for mistakes to be made, e.g., failing to consider a patient's religious beliefs. Perhaps, most important, it enables people to disagree without any party to the dispute concluding that the other party must be ill informed, partial, or acting irrationally or immorally. Providing limits to legitimate moral disagreement and at the same time leading people to acknowledge that, within these limits, moral disagreements are legitimate and to be expected, provides the kind of atmosphere which is most conducive to fruitful moral discussion.

Department of Humanities
Penn State University College of Medicine
Hershey, Pennsylvania, U.S.A.

Department of Philosophy
Dartmouth College
Hanover, New Hampshire, U.S.A.

NOTES

1. See Gert (1998) for a more extended account of morality, and of the moral theory that justifies it. For a shorter account of this theory and an account of its application to bioethics, see Gert, Culver, and Clouser (1997).
2. For an account of the applications of these general moral rules to the field of healthcare, see Gert, Culver, and Clouser (1997, ch. 3).
3. In this formulation, the term "duty" is being used in its everyday sense to refer to what is required by special circumstances or by one's role in society, primarily one's job, not as philosophers customarily use it, namely, simply as a synonym for "what one morally ought to do."
4.We have discussed principlism in so many places that we do not think it is worthwhile to discuss it again in this chapter. For one such discussion see Gert, Culver, and Clouser (1997, ch. 4).
5. We are primarily concerned with the general approach of casuistry, not any particular version of it. However, the most prominent defense of casuistry is *The Abuse of Casuistry* (Jonsen and Toulmin, 1988.)
6. We do not discuss the virtues in this article because we do not consider them to be useful in helping to resolve the moral problems that arise in medicine, but an account of virtue is provided in Gert (1998, ch. 9).

REFERENCES

Gert, B.: 1998, *Morality*: Its Nature and Justification, Oxford University Press, New York and Oxford.

Gert, B., Culver, C. M., and Clouser, K. D.: 1997, *Bioethics: A Return to Fundamentals*, Oxford University Press, New York and Oxford.

Jonsen, A., and Toulmin, S.: 1988), *The Abuse of Casuistry*, University of California Press, Berkeley and Los Angeles.

Kant, I. 1993, *Grounding for the Metaphysics of Morals*, translated by James W. Ellington, Third Edition, Hackett Publishing Company, Inc., Indianapolis and Cambridge.

ROSEMARIE TONG

FEMINIST APPROACHES TO BIOETHICS

I. INTRODUCTION

Like non-feminist approaches to bioethics, feminist approaches to bioethics are many and diverse. Nevertheless, despite their rich variability, most feminist approaches to bioethics fall into one of two clusters: care-focused approaches and power-focused approaches. To understand these two types of approaches to feminist bioethics is to understand a great deal about the ways in which feminist thought has influenced and improved not only traditional bioethical theory but also health care practice and policy over the last quarter of a century. Therefore, my aims in this essay are relatively simple. First, I seek to describe these two types of feminist approaches to bioethics in considerable detail. Second, and more ambitiously, I consider ways in which these two approaches to feminist bioethics are merging to constitute fully-developed feminist theories of bioethics.

II. CARE-FOCUSED FEMINIST APPROACHES TO BIOETHICS

Feminist care-focused approaches to bioethics have several features in common with non-feminist ones. Specifically, they share what Alisa L. Carse (1995) has termed a two-fold commitment to "qualified particularism" and "affiliative nurture." In their commitment to "qualified particularism," all care-focused approaches to bioethics "highlight concrete and nuanced perception and understanding – including an atunement to the reality of other people and to the actual relational contexts we find ourselves in" (Carse, 1995, p. 10). Relatedly, in their commitment to all "affiliative virtue," care-focused approaches to bioethics "assert the importance of an active concern for the good of others and of community with them, of a capacity for sympathetic and imaginative projection into the position of others – and of situated-attuned responses to others' needs" (Carse, 1995, p. 10). The factor that differentiates feminist care-focused approaches from non-feminist ones is the emphasis the former place on the role *gender* plays in a person's gravitation towards or away from a bioethics of care. Care-focused feminist bioethicists are, as a rule, so-called difference

G. Khushf (ed.), Handbook of Bioethics, 143–161.

feminists or cultural feminists. They believe that one of their central tasks is to challenge as inadequate, lacking, or even "bad" those traits and values that many cultures have labeled "masculine." The list includes: "independence, autonomy, intellect, will, wariness, hierarchy, domination, culture, transcendence, product, asceticism, war and death" (Jaggar, 1992, p. 364). Closely related to, and even more important than this task, is the effort of difference or cultural feminists to rehabilitate, valorize, or present as "good" those traits and values that many cultures have labeled "feminine." Here the list includes: "interdependence, community, connection, sharing, emotion, body, trust, absence of hierarchy, nature, immanence, process, joy, peace and life" (Jaggar, 1992 p. 364). Typically, feminists who stress care-focused approaches to bioethics rely on the writings of Carol Gilligan and Nel Noddings, both of whom correlate "being female" with a tendency to cultivate such culturally-associated "feminine" virtues as care and "being male" with a tendency to cultivate such culturally-associated "masculine" virtues as justice.

In her book, *In a Different Voice* (1982), Gilligan offers an account of *women's* moral development that challenges Lawrence Kohlberg's account of what he describes as *human beings'* moral development. According to Kohlberg, moral development is a six-stage, progressive process. Children's moral development, he says, begins with Stage One, "the punishment and obedience orientation"; this is a time when, to avoid the "stick" of punishment and/or to receive the "carrot" of reward, children do as they are told. Stage Two is "the instrumental relativist orientation." Based on a limited principle of reciprocity – "You scratch my back and I'll scratch yours" – children in this stage do what satisfies their own needs and occasionally the needs of others. Stage Three is "the interpersonal concordance or 'good boy-nice girl' orientation," when immature adolescents conform to prevailing mores simply because they seek the approval of other people. Stage Four is "the law and order orientation," when mature adolescents do their duty in order to be recognized as honorable as well as pleasant people. Stage Five is "the social contract legalistic orientation," when adults adopt an essentially utilitarian perspective according to which they are permitted considerable liberty provided that they refrain from harming other people. Finally, Stage Six is "the universal ethical principle orientation," when some, though by no means all, adults adopt an essentially Kantian perspective universal enough to serve as a critique of any conventional morality, including that of their own society. These individuals are no longer governed by self-interest, the opinions of others, or the rule of law but, rather, are motivated by self-legislated and self-imposed universal principles such as those of justice, reciprocity, and respect for the dignity of human beings as intrinsically valuable persons (Kohlberg, 1971).

Puzzled by the fact that, according to Kohlberg and his scale, women rarely climbed past the good boy-nice girl stage whereas men routinely ascended at least to the social contract stage, Gilligan hypothesized that the deficiency lay not in women but in Kohlberg's scale. She alleged his scale provided a way to measure not *human* moral development as he claimed, but only *male* moral development; she further explained that if women were measured on a scale sensitive to women's style of moral reasoning, women would prove just as morally developed as men. For example, while studying how each of twenty-nine women decided whether it was right or wrong to have an abortion, Gilligan noted that, as a group, these women referred little to their rights and much to their relationships. She also noted that in the process of developing (or failing to develop) as moral agents, these women moved in and out of three understandings of the self-other relationship: (1) an overemphasis on self; (2) an overemphasis on the other; and (3) a proper emphasis on self in relation to the other. On the basis of this and several other empirical studies of women's moral reasoning patterns, Gilligan concluded that for a variety of cultural reasons, women typically utilize an ethics of care which stresses relationships and responsibilities, whereas men typically employ an ethics of justice which stresses rules and rights (Gilligan, 1982).

In a similar vein, Nel Noddings writes in her book *Caring: A Feminine Approach to Ethics and Moral Education* (1984) that traditional ethics has favored theoretical and "masculine," as opposed to practical and "feminine" modes of thinking. Eschewing the interpretive style of reasoning which is characteristic of the humanities and social sciences, most traditional ethicists have instead embraced the deductive-homological style of reasoning which is characteristic of mathematics and the natural sciences. Therefore, says Noddings, most traditional ethicists have been so focused on "principles and propositions" and on "terms such as justification, fairness, and justice," that "human caring and the memory of caring and being cared for . . . have not received attention except as outcomes of ethical behavior" (Noddings, 1984, p. 1).

Convinced that in its nearly exclusive emphasis on justice non-feminist ethics has gone awry, Noddings proposes to privilege what she terms "eros, the feminine spirit" over what she terms "logos, the masculine spirit" (Noddings,

1984, p. 1). She does not, however, argue that logos understood as logic or reasoning has no role to play in ethics. Rather, she argues that eros – understood as an attitude "rooted in receptivity, relatedness, and responsiveness" – is a more basic approach to ethics than logos, since ethics is about human relations. In ethics, logos is the handmaiden of eros rather than vice versa (Noddings, 1984, p. 2).

Noddings's ethics of care borrows heavily from the moral sentiments theory which frames David Hume's ethics. Like Hume, Noddings believes both that the sentiments of sympathy are *innate* and that these sentiments must be *cultivated* lest they fail to guide one's everyday moral decisions and actions. In explaining the complex relationship between what she terms "natural" caring on the one hand and "ethical" caring on the other, Noddings notes that most people's initial experiences of care come easily, even unconsciously. Among the examples of *natural* caring which Noddings provides is that of a little boy who helps his mother fold the laundry because she does so many things for him. He wants to be connected to her and have her recognize him as her helper. Later, when he is an adolescent and he doesn't feel like helping his mother because he'd much rather be out with his friends, his childhood memories flood over him as the "feeling I must" (Noddings, 1984, p. 79). This new feeling prompts the adolescent to be late for a party so he can help his mother in "remembrance" of his old little-boy feelings. Through this kind of decision and action rooted in feelings, says Noddings, *ethical* caring comes into existence, a form of caring which is more deliberate and less spontaneous than *natural* caring.

Significantly, Noddings does not describe moral development as the process of replacing natural caring with ethical caring. As she sees it, our "oughts" build on our "wants." Noddings comments: "An ethic built on caring strives to maintain the caring attitude and is thus dependent upon, and not superior to, natural caring" (Noddings, 1984, p. 80). Moreover, morality is not about serving others' interests through the process of disserving one's own interests. Rather, morality is about serving one's own and other's interests simultaneously. When we engage in ethical caring, we are not denying, negating, or renouncing ourselves in order to affirm, posit, or accept others. Rather, we are acting to fulfill our "fundamental and natural desire to be and to remain related" (Noddings, 1984, p. 83).

As might be expected, many health care practitioners – particularly females – are very attracted to Gilligan's and Noddings' ethics of care. Julien Murphy, for example, observes that "those who work with the human vulnerabilities of sick and dying patients may find their work rewarding in part because of the integration of feelings, reason, and emotions that they are able to effect with their patients" (Murphy, 1995, p. 150). An ethics of care enables health care practitioners to transform mundane and even disgusting tasks into opportunities for human connection and moral growth. Yet, as Murphy also observes, the ethics of care is an underdeveloped ethics. Feminists need to distinguish more clearly between genuine and distorted understandings and acts of care.

Care may be distorted in two ways. First, care may disserve the caregiver's interests. It can become a trap for women, particularly women who work in the service industry. Society has viewed women far more than men as being responsible for the care of the young, the old, and the infirm. It has expected women to be the ones to sacrifice their own lives on behalf of others. Continuing to associate women with caring, as Gilligan and Noddings do, might have the effect of reinforcing the idea that because women can care and have cared so well for others, they should always care – no matter the cost to themselves. Thus, critics such as Sara Lucia Hoagland fault Noddings for imposing unfair moral obligations upon caregivers. Hoagland insists there is more to the moral life than being responsive to other persons' needs and wants. She states: "I must be able to assess any relationship for abuse/oppression and withdraw if I find it to be so. I feel no guilt, I have grown, I have learned something. I understand my part in the relationship. I separate. I will not be there again. Far from diminishing my ethical self, I am enhancing it" (Hoagland, 1991, p. 256). Ethics is about knowing when *not to care* for others as well as when *to care* for them. It is also about knowing when to care for one's self as well as when *not to care* for one's self.

Care may also be distorted in a second way, one which disserves the care receiver's interests. For example, in a discussion of disability and feminist ethics, Susan Wendell notes that in American culture, characteristics of autonomy and independence, being in control of one's own destiny, and doing things on one's own terms are treasured. Because persons with socially-recognized disabilities have trouble performing certain tasks "on their own," persons without socially-recognized disabilities view them as dependent. However, as Wendell sees it, persons without "official" disabilities are also dependent. She comments that,

... 'independence,' like 'disability,' is defined according to a society's expectations about what people "normally" do for themselves and how they do it. Few people in my city would consider me a "dependent" person because I rely on others to provide me with ... food and clothing in markets where

> I can buy them instead of producing them myself. Perhaps some would consider me "dependent" because I rely on a gardener to do the heavy work in my yard and a cleaner to clean my house. ... Yet if I needed to rely on someone else to help me out of a bed, help me use the toilet, bathe me, dress me, feed me, and brush my teeth, most people would consider me very "dependent" indeed (Wendell, 1993, p. 145-146).

Stressing that our ideas about which of us are dependent/disabled and which of us are independent/non-disabled are culturally constructed, Wendell laments the fact that all too often it is assumed that persons with socially recognized disabilities who receive care have nothing to give to their caregivers. Wendell argues against the view that the care-giver/care-receiver relationship is unidirectional, pointing to the emotional care care-receivers often give care-givers (Wendell, 1993, p. 150). Hoagland adds that the value of emotional care cannot be overemphasized. She relates a conversation she once had with Karen Thompson, a lesbian who fought against tremendous odds to obtain legal guardianship of her brain-stem injured partner, Sharon Kowalski. Impressed by all the difficulties Karen encountered and overcame, Hoagland asked Karen whether she viewed all the money, time, and energy she had spent on Sharon as a "sacrifice." Karen responded, "No, I did this to make my life more meaningful" (Hoagland, 1992). Sharon, the care-receiver, had provided Karen, the care-giver, with the opportunity to act courageously within oppressive circumstances (a legal and moral system that discriminates against lesbians) and to create meaning for herself and Sharon within these circumstances.

In an attempt to help both care-givers and care-receivers form morally-good relationships with each other, Sheila Mullett has articulated several criteria for genuine or authentic care. She stresses that women in particular should continually ask themselves whether the kind of caring in which they are engaged:

1. fulfills the one caring;
2. calls upon the unique and particular individuality of the one caring;
3. is not produced by a person in a role because of gender, with one gender engaging in nurturing behavior and the other engaging in instrumental behavior;
4. is reciprocated with caring, and not merely the satisfaction of seeing the ones cared-for flourishing and pursuing other projects; and
5. takes place within the framework of consciousness-raising practice and conversation (Mullett, 1989, pp. 119-120).

Although Mullett's criteria for genuine care are, as she admits, partial and provisional in nature, they are certainly useful for anyone interested in using care-focused feminist approaches to bioethics in the realm of health care. One can easily imagine a nurse, for example, asking herself questions such as the following: Do I feel that nursing is my vocation *only* or mostly because I am a woman? Would I feel the same about nursing if I were a man? Have I been "brainwashed" into being a caring person? Why does nursing remain a female-dominated profession despite the fact that doctoring is no longer a male-dominated profession? Why isn't nursing as socially esteemed as doctoring?

Clearly, developing care-focused feminist approaches to bioethics is a pioneering task. Not only must concerns about gender be addressed; concerns about making care the foundational value or virtue of moral life must also be addressed. Care-focused feminist bioethicists must ask themselves, for example, whether care, understood as having certain kinds of sentiments, feelings, or emotions, is *essential* to an action's goodness or whether it is simply an *enhancement* of an action's goodness. Lawrence Blum argues that doing one's duty solely because it is one's duty, and not because one cares about the object of one's action threatens the "goodness" of one's action. He comments that "Emotion itself is often part of what makes the act morally right in a given situation" (Blum, 1980, p. 42). But if Blum's view is correct, then caregivers must actually *feel* something in the way of empathy or sympathy for the people entrusted to them in order to be said to truly care for them. The problem with this view, should care-focused feminist bioethicists embrace it, is that although it may be possible for an individual to recognize at the cognitive level that persons, simply because they are persons, are worthy of equal respect and consideration, and therefore entitled to *just* treatment, it may not be possible for that same individual to truly *care* for anyone at the affective level. The world is full of uncaring people, only some of them responsible for their "cold" hearts. Can we expect a person who has never felt he matters to anyone, or who has never been the object of truly caring words and deeds, to care for others, when he himself is so desperately in need of care? In similar vein, what should our moral assessment be of a person who discovers that no matter how hard she *tries* to feel the emotions that caring persons feel, she, like the skeptic who cannot force himself to believe in God, cannot get herself to feel much of anything at all? Do steps exist for such a person to overcome her moral limitations?

To this probing set of questions, I suppose that care-focused feminist bioethicists may respond, as Aristotle did, that by imitating the behavior of caring persons, uncaring persons may be able either to experience for the first time in their lives the feelings and desires associated with the virtue of care, or to rekindle the natural feelings and desires of care they once had (Aristotle, 1962). Of course, caring role models are not always available. Should this be the case, uncaring persons might have an obligation to seek professional help. If the ground of ethical caring is natural caring, as seems plausible, then psychologists who teach self-awareness, personal decision-making, stress and feeling management, empathy, communication, self-disclosure, insight, self-acceptance, personal responsibility, assertiveness, group dynamics, and conflict resolution may have a vital role to play in cultivating uncaring persons' dormant emotions (Goleman, 1995).

III. POWER-FOCUSED FEMINIST APPROACHES TO BIOETHICS

Although feminists developing power-focused feminist approaches to bioethics generally applaud developments in care-focused feminist approaches to bioethics, they are reluctant to make care the quintessential moral concept. For example, Claudia Card is bothered by Noddings' characterization of a "feminine approach" to morality as that of "one attached" and a "masculine approach" as that of "one detached" (Card, 1990, p. 104). As Card sees it, we are closely attached only to a tiny fraction of the world's people, and yet, given the way in which the economies and technologies of nations are intertwined, we invariably affect through our actions not only our families, friends, neighbors, and colleagues, but also a multiplicity of strangers from whom we are profoundly "detached." Therefore, says Card, we require "an ethic that applies to our relations with people with whom we are connected *only by relations of cause and effect* as well as to our relation with those with whom we are connected by personal and potential encounters" (Card, 1990, p. 105). Such an ethic need not view universal principles as "masculine" impediments to deeply personal relationships.

As valuable a virtue as caring is, Card insists it is not the only valuable moral virtue. Justice is also a valuable moral virtue. Indeed, properly interpreted, justice is necessary for our defense against the sexism, racism, ethnocentrism, homophobia, and xenophobia which plague our "poorly integrated, multicultural society" (Card, 1990, p. 105). Given the fact that so many social groups knowingly or negligently, willfully or unintentionally, fail to care about those whose sex, race, ethnicity, religion, or even size and shape differ from their own, justice must be treasured. We cannot have a *caring* society, suggests Card, until we have a *just* society. However, in our current society, all persons are not treated equally well. Justice must not be dismissed simply as the abstract, alien tool of the "fathers," for it can be used to protect the weak as well as the strong. Justice often is correctly blind to particulars in order to prevent details of sex, race, and creed from determining whether we care for someone or not.

In addition to questioning whether it is wise to underemphasize justice in the world in which we live, feminists who are developing power-focused feminist approaches to bioethics question the wisdom of singling out certain personal relationships – for example, the mother-child relationship – as the ideal or paradigm for any and all good human relationships. Jean Grimshaw has found particular fault with those care-focused feminist approaches to bioethics that seek to model all relationships on the mother-child relationship, given the inequalities inherent in this relationship. First, Grimshaw notes that the parent-child relationship lacks the kind of reciprocity that typically characterizes good human relationships. Whereas parents have a duty to maintain and promote their childrens' physical and psychological well-being, children have no such duty with respect to their parents. Second, parents are permitted – indeed sometimes required – to tell their children what to do, but not vice versa. Third, parents are expected to "tolerate, accept, and try not to be hurt by behavior" from their children that would be intolerable, unacceptable, or hurtful in most adult relationships (Grimshaw, 1986, p. 25).

Given these three asymmetries, modeling all relationships on the mother-child relationship seems like a prescription for disaster. The features which tend to make a mother-child relationship work are precisely the ones that are likely to damage or destroy a relationship between a husband and wife, for example. For an adult relationship to work, both parties must be responsible for each other; neither must presume to know the other's good better than s/he herself/himself knows it; and both must behave equally well, since the small manipulations, name callings, and temper tantrums parents should accept from children are not ones that adults should accept from other adults.

Grimshaw's reservations about using the mother-child relationship as a paradigm for all relationships are small compared to philosopher Jeffner Allen's reservations about this relationship, however. Allen rejects the mother-child relationship model because she rejects motherhood. According to Allen, biological motherhood causes women to be viewed as "breeders," confining them to the kind of genital sexuality that is oriented toward procreation only (Allen,

1984, p. 310). Allen insists that motherhood, and therefore the mother-child relationship, has little to do with female virtue and much to do with female oppression, and that women will never be able to develop a true ethics of care until they "evacuate" motherhood (Allen, 1984, p. 310). Therefore, she recommends that women stop having children so that "women's repetitive *reproduction* of patriarchy" may at last be replaced by the "genuine, creative, *production*" of women (Allen, 1984, p. 326). Were women to decide not to have any children for a period of twenty years or so, claims Allen, "the possibilities for developing new modes of thought and existence would be almost unimaginable" (Allen, 1984, p. 326). At last freed from the reproductive roles and responsibilities which have historically held women back, women might be able to develop an ethics of care based on the model of an ideal female friendship model. Such an ethics of care, implies Allen, would be egalitarian enough to withstand the distortive degenerations of servility on the one hand and authoritarianism on the other.

Although most power-focused feminist bioethicists are not ready to "evacuate" motherhood, they all aim to achieve justice and equality for women. They agree with philosopher Alison Jaggar that their most important tasks are (1) to provide moral critiques of actions, practices, systems, structures, and ideologies that perpetuate women's subordination; (2) to devise morally justifiable ways (e.g., public policy initiatives, peaceful protests, boycotts) to resist the economic, social, and cultural causes of women's subordination; and (3) to envision morally desirable alternatives to the world as we know it: sexist, racist, abelist, heterosexist, ethnocentric, and colonialist (Jaggar, 1992). The practice of health care is, in the estimation of power-focused feminist bioethicists, full of systems, structures, and institutions which neglect, downplay, trivialize, or ignore women's moral interests, issues, insights, and identities in ways that reinforce women's subordination to men (Jaggar, 1992). The puzzling thing for power-focused feminist bioethicists, then, is why so many people in the health care system, including so many women, fail to see just how much inequality, particularly gender inequality is there. For this reason, power-focused feminist bioethicists maintain that little progress can be made towards the goal of reforming the practice of health care until at least women develop what Mullett has termed a "feminist consciousness" (Mullett, 1989, p. 109).

Mullett, who straddles the divide between care-focused approaches to ethics and power-focused ones, describes feminist consciousness as beginning with moral sensitivity to one's own and others' pain, proceeding to a moment of profound ontological shock, and concluding in praxis (Mullett, 1989, p. 114). According to Mullett, when a woman becomes morally sensitive, she focuses on the harm done to women. She begins to feel the pain that many women feel when, for example, they find themselves old, poor, disabled, and very much alone in a nursing home. The more she feels other women's pain as her own, the more disturbed she becomes. So sensitized, the woman undergoes something akin to a religious experience. In the same way that the world metamorphoses for the unbeliever turned believer, the world changes dramatically for the ontologically shocked woman. Suddenly, her eyes are open to realities to which she was previously blind, and she sees the damage done to herself and other women as something she must stop. No longer is she able to view, for example, the thousands, perhaps millions of women with inadequate long-term care as "just one of those things." On the contrary, she now views this state of affairs as manifestly *unjust*. It is this new vision that impels a woman to action or praxis. Together with other women who see what she sees, a feminist determines to *do* something to ameliorate women's lot.

Something like feminist consciousness was probably at work in 1990 when Congresswomen became aware of gender bias in medical research at the National Institutes of Health (NIH). Shocked that important clinical studies had been conducted only on males, Congresswomen decided to take the matter into their own hands. In particular, they encouraged the National Institutes of Health (NIH) to create an Office of Research on Women's Health (ORWH). The goal of the office, established in September 1990, remains threefold: (1) to increase women's participation in health research studies, (2) to insure that NIH-supported research pays due attention to women's health issues, and (3) to promote the number of women in biomedical and biobehavioral careers. In April 1991, Dr. Bernadine Healy, the first woman to head the NIH, launched a $625 million study of over 160,000 women ages 50 to 79 at forty-five clinical centers across the country to investigate the causes and potential prevention of major diseases of women – particularly heart disease, cancers of the breast, colon and rectum, and osteoporosis (Angier, 1991, p. 88). Now well underway, this study and others like it have been so successful that some critics fear that the health interests of men are being slighted. They note, for example, that although 78.3 percent of 1991 NIH funds went to study diseases afflicting both women and men, 16 percent went to diseases exclusive to women and only 5.7 exclusive to men (Rubin, 1991, p. A3).

Despite such strides in women's health, power-focused feminist bioethicists have not rested content. In an effort to make certain that recent gains in serving women's health interests are not only maintained but increased, they have proposed a women's health specialty that would focus exclusively on women's health concerns, both non-reproductive and reproductive. Interestingly, this recommendation has not met with feminist bioethicists' universal approval. In the same way that U.S. feminist academics in the 1970s debated whether to establish separate women's studies programs, or instead to "mainstream" materials related to women into the traditional academic disciplines, present-day U.S. feminist bioethicists debate whether to establish a separate women's health specialty, or instead to "mainstream" women's health concerns into traditional medical specialties. So far the emerging consensus seems to be a "both-and" approach that supports a women's health specialty until women's health concerns are fully integrated into the relevant existing specialties (Clancy and Massion, 1992, p. 1920). No feminist wants to create a low-paying, unprestigious "ghetto" for women's health concerns.

Power-focused feminist bioethicists do not stop at urging the health care establishment to pay equal attention to women's and men's health care concerns. They also consider the consequences of specific medical treatments, scientific studies, and health care policies on women's overall well-being and freedom. They ask, for example, whether the reproduction-controlling technologies of contraception, sterilization, and abortion on the one hand, and the reproduction-assisting technologies of in vitro fertilization and embryo donation on the other are necessarily "women-liberating." Although these technologies have benefited many women in the U.S. and enabled them to take charge of their destiny, they have not benefited all U.S. women (let alone all women in other nations) equally well. Poor women and women of color are far more likely to be discouraged from procreating than middle and upper class women and white women are. Moreover, they are less likely to have access to expensive reproduction-assisting technologies, even though their desire to have a child genetically and gestationally related to them may be just as great as that of women who can pay out of pocket the considerable costs of several cycles of in vitro fertilization.

Clearly, feminist consciousness about gender inequities has enabled power-focused feminist bioethicists to identify ways of thinking that typically lead to the creation of oppressive relationships between people, particularly between men and women. They have noted that ontologies separating the self from others, epistemologies privileging the insights of supposedly "Impartial Spectators," and methodologies encouraging people to debate each other in adversarial fashion tend to produce the kind of distorted ethics and bioethics which favor those people who, according to Marilyn Frye, have a vested interest "in being good and/or in others' being good" (Frye, 1991, p. 53). Frye claims that in our society all privileged people – especially privileged men – tend to conceive of themselves as the "leaders" and those who are less privileged than themselves as the "followers." This being the case, the hospital president, chief of medical staff, or bioethicist at a prestigious institution bases his (far less often her) conception of himself as a "judge, teacher/preacher, director, administrator, manager, and in this mode, as a decision maker, planner, policymaker, organizer" on his conviction that he is in the right, that he knows what is the best course of thought and action for all people (Frye, 1991, p. 54). As long as the people who are forced, more or less coercively, to accept this kind of moral authority do not rebel against it, they have, says Frye, but one of two choices: (1) to become themselves moral "authorities" (i.e., oppressors, "know it alls"); or (2) to become moral "lackies" (i.e., persons with no moral sentiments and convictions of their own). Viewing both of these choices as undesirable, Frye urges feminists to reject traditional ethics and bioethics, and to create in their stead *feminist* ethics and *feminist* bioethics.

IV. CONCLUSION

Looking into the new millennium, I see care-focused and power-focused feminist bioethicists coalescing their forces. Specifically, I predict the development of several full-fledged feminist *theories* of bioethics, the application of which will help to further erode human insensitivities and invidious power relations that make our health care system less than just and caring. In this connection, Susan Sherwin and Mary Mahowald have already taken the lead. Sherwin develops a power-focused theory of feminist bioethics in her book *No Longer Patient (1992)*, applying it to a variety of pressing issues in health care. She stresses the importance of "empowering those who have been traditionally disempowered" (Sherwin, 1992, p. 39) and creating a health care system that "not only would be fairer in its provision of health services but would also help to undermine the ideological assumptions on which many of our oppressive practices rest" (Sherwin, 1992, p. 240). Mahowald also stresses issues of power in her book, *Women and Children in Health Care: An Unequal Majority (1993)*. However, unlike Sherwin who has enough serious concerns about the ethics of care to de-emphasize it throughout her work, Mahowald is more favorably disposed to an ethics of care properly interpreted. In fact, Mahowald states that she is trying "to combine an emphasis on equality with a 'feminine' approach to issues in health care" (Mahowald, 1993, p. 4). Moreover, she concludes her book with a chapter entitled "Just Caring: Power for Empowerment," insisting that "If the *equality* [my stress] of women, men, and children is a worthwhile social goal, it relies rudimentarily on the nurturance of *caring* [my stress] that is essential to individual as well as social health" (Mahowald, 1993, p. 269).

More recently than Sherwin and Mahowald, Eva Feder Kittay has developed a moral, political, and social theory which feminist bioethicists, who wish to further merge care-focused and power-focused perspectives, might wish to adapt. In her book, *Love's Labor (1999)*, Kittay looks around the world as it is rather than as some have imagined it to be. She sees dependent people everywhere: children, people with mental and physical disabilities, infirm people, people with major addictions, the frail elderly, homeless people, indigent people, and people who have simply lost their way. She also sees the people who care for these dependent individuals; formal and informal caregivers, a disproportionate number of whom are women who work for nothing or relatively little. Sometimes these dependency workers – Kittay's term for those who care for dependent people – choose their work; but, more often than not, their work is "chosen" for them by a society that seems too willing to exploit them. In order to eliminate this very large inequality, Kittay offers a moral, political, and social theory (1) based on the assumption that we are, more often than not, unfree and dependent; and (2) dedicated to the cause of empowering dependency workers (caregivers). The way to measure equality in a society, implies Kittay, is to check how well or how badly its most dependent members and their caregivers are doing (Kittay, 1999).

Because she views her theory of equality as feminist in methodology, Kittay begins her analysis with a survey of other feminist critiques of equality. Kittay rejects the popularly-accepted view that equality for women consists in women being given the same opportunities to enjoy "the resources and privileges now concentrated in the hands of men" (Kittay, 1999, p. 8). She reasons that the "sameness" approach to equality will not work for women in general and for dependency workers in particular because the reference class for women's aspirations (that is, men as they have been socially constructed) is a class whose self-conception, behaviors, values, and virtues have caused a great amount of injustice, with dependent people and their caregivers the least of their concerns.

Thus, for women to want to be like men is to want to create a world in which very few people do dependency work. Whether life in such a world is at all desirable for either men or women is, of course, another matter. Absent its dependency workers, a world is likely to be little more than a jungle where life is indeed short, nasty, and brutish.

Having provided her main reason for rejecting a sameness interpretation of equality, Kittay proceeds to discuss in some detail the three standard feminist alternatives to it: (1) the difference approach favored by care-focused feminist ethicists and (2) the dominance and (3) diversity approaches favored by power-focused feminist ethicists. Although Kittay borrows ideas from each of these three feminist approaches to equality, she ultimately offers her own "dependency" approach to equality. Her view is predicated on the conviction that enlightenment liberals were wrong to conceive of society as an association of independent and equally-empowered individuals. Not only does this view of society obscure the inevitable asymmetries and dependencies present in most human relationships, says Kittay, it also results in the systematic relegation of dependents and their caregivers to the underbelly of society: the domain where the so-called Other resides. We would, says Kittay, have less inequality in our society, had society instead been defined as an association of unequals, who are dependent on each other throughout life to a greater or lesser extent.

Kittay supports her view of society with some childhood memories. When Kittay's mother used to sit down to dinner after serving her and her father, she would justify her action with the statement "after all, I'm also a mother's child" (Kittay, 1999, p. 25). Kittay claims that this statement reveals the fundamental source of human equality. All human beings are both radically dependent for an extended period of time and the product of the work of one (or more) mothers. Were it not for the kind of work mothers and other dependency workers do, human society would not be possible. For Kittay, the paradigm dependency worker is a close relative or friend who assumes daily responsibility for a dependent's survival for no financial reward, and whose labor is characterized by care, concern, and connection to the dependent person. Closely related to the paradigm case of a dependency worker is the worker who is paid more or less well to care for an unrelated person, but who views her job as much more than a mere job. Kittay provides an example of such a dependency worker; namely, Peggy, the woman who has cared for her severely developmentally-disabled daughter, Sesha, for over a quarter of a century, and to whom she has distributed many of her motherly tasks. Without Peggy's help, Sesha would not be able to do the range of things she is currently able to do, and Kittay and her husband would not be able to pursue, as they do, successful careers as academic professionals. On the contrary, most of their energies, and particularly Kittay's, would be devoted to caring for Sesha.

Unlike the subject of traditional liberal theory, a dependency worker is anything but an independent, self-interested, and fully autonomous agent. Rather, she is, in Kittay's estimation, a transparent self; that is, "a self through whom the needs of another are discerned, a self that, when it looks to its own needs, it first sees the needs of another" (Kittay, 1999, p. 51). As Kittay sees it, to the degree that the dependent relies on the dependency worker, to that degree is the dependency worker obligated to put the dependent's needs and interests before her own. Because of the dependency worker's weighty obligations of care, and because true dependents are able to reciprocate with little more than love and affection, if that, others have an obligation to care for the dependency worker. According to Kittay, dependency workers and dependents exist in a "nested set of reciprocal relations and obligations" (Kittay, 1999, p. 68). Ordering this relational network is a principle Kittay calls "*doulia*," from the Spanish term for a postpartum caregiver (a *doula*) who assists the new mother so that the new mother can care for her child (Kittay, 1999, p. 60). Kittay reasons that since everyone is some mother's child, it is only fair that third parties or general society take care of the dependency worker. For third parties or general society to do anything less than this for the dependency worker is to treat the dependency worker unequally – as not some mother's child – and, therefore, unjustly.

Kittay's theory is a fully-developed hybrid theory of ethics that combines considerations about care and justice in approximately equal measure. Kittay spends as much time discussing possible ways to increase the power and well-being of dependency workers as she spends articulating the necessary and sufficient conditions for the constitution of a caring relationship. Kittay stresses, for example, that in an ideal society all dependency workers would have the material and personal resources to preserve the lives of their children, to foster their growth, and to help them become contributing members of a society to which they feel connected (Kittay, 1999, p. 154). Unlike Card who argued that unless a society is just, it cannot be caring, Kittay argues that unless a society is caring, it cannot be just.

Although Kittay did not specifically shape her dependency theory as a bioethical or health care ethics theory, her theory is easily applied to the type of cases and controversies that emerge in the health care and bioethics realms. In fact, as I see it, Kittay's dependency theory better fits these realms than Rawls's justice theory, which many bioethicists and health care ethicists sought to apply in their work. But as good as Rawls's liberty and difference principles are – and as much as they have been able to narrow the gap between the kind of health care the "least advantaged" and most-advantaged members of society receive – the gaps remain ominously large. Specifically, the gap remains large between those who do most of society's dependency work – caring for the young, the old, the infirm, and the disabled – and those who do little of this work. Although society extols those who care for others, it does not make it easy for caregivers to do "love's labor." In fact, it structures itself in ways that make it difficult for dependency workers to care for dependents, punishing rather than rewarding dependency workers for "choosing" to do that work without which society would devolve into a jungle-like state of nature. Our nursing homes, for example, are full of mostly old women who find themselves in dire financial straights largely because they took care of others throughout their lives and/or because they worked in one of society's poorly-compensated "caring" professions (Abel, 1991). No wonder, then, that Kittay adds to Rawls's two principles of liberty and difference a third principle she terms "the principle of social responsibility for care." This reads: "To each according to his or her need for care, from each according to his or her capacity for care, and such support from social institutions as to make available resources and opportunities to those providing care, so that all will be adequately attended in relations that are sustaining" (Kittay, 1999, p. 113).

Equipped with both Rawls's two principles and Kittay's one principle, feminist bioethicists will have a full set of conceptual tools with which to work. If feminist bioethicists wish to eliminate not only gender-based but also all inequities in the health care system, they will need to draw upon the perspectives of both power (justice) and care. Feminist bioethicists have within their reach the opportunity to transform the entire field of bioethics so that it becomes a safe haven for vulnerable people – a place where the right kind of theory springs into many good policies, practices and actions.

Center for Professional and Applied Ethics and
Department of Philosophy
The University of North Carolina at Charlotte
Charlotte, North Carolina, U.S.A.

REFERENCES

Abel, E. K.: 1991, *Who Cares for the Elderly? Public Policy and the Experiences of Adult Daughters,* Temple University Press, Philadelphia.

Allen, J.: 1984, 'Motherhood: The Annihilation of Women,' in J. Trebilcot, (ed.)., *Mothering Essay in Feminist Theory*, Rowman and Allanheld, Totowa, NJ.

Angier, N.: 1991, 'Women Join the Ranks of Science but Remain Invisible at the Top,' *The New York Times*, Tuesday, May 21, p. 88.

Aristotle: 1962, *Nicomachean Ethics*, M. Ostwald (trans.), Bobbs-Merrill, Indianapolis.

Blum, L. A.: 1980, *Friendship, Altruism and Morality*, Routledge and Kegan Paul, London.

Card, C.: 1990, 'Care and Evil,' *Hypatia* 5, 31, pp. 104-105.

Carse, A. L.: 1995, 'Qualified Particularism and Affiliative Virtue: Emphasis on a Recent Turn in Ethics,' in *Revista Medica de Chile* 19, pp. 7-15.

Clancy, C. and C. Massion: 1992, 'American Women's Healthcare: A Patchwork Quilt with Gaps,' *JAMA* 260:14, 1920.

Frye, M.: 1991, 'A Response to Lesbian Ethics: Why Ethics?' in C. Card (ed.), *Feminist Ethics*, University of Kansas Press, Lawrence, Kansas, pp. 52-59.

Gilligan, C.: 1982, *In a Different Voice*, Harvard University Press, Cambridge, Mass.

Goleman, D.: 1995, *Emotional Intelligence*, Bantam Books, New York.

Grimshaw, J.: 1986, *Philosophy and Feminist Thinking*, University of Minnesota Press, Minneapolis, p. 25.

Healy, B.: 1995, *A New Prescription for Women's Health*, Viking Press, New York.

Hoagland, S. L.: 1990, 'Some Concern About Nel Noddings' Caring,' *Hypatia*, 5:1, p. 114.

Hoagland, S. L.: 1991, 'Some Thoughts About *Caring*' in C. Card (ed.), *Feminist Ethics*,University Press of Kansas, Lawrence, pp. 246-263.

Hoagland, S. L.: 1992, 'Why Lesbian Ethics?,' *Hypatia*, 7:4, p. 198.

Jagger, A. M.: 1992, 'Feminist Ethics,' in C. Becker and L. Becker (eds.), *Encyclopedia of Ethics*, Garland, New York, pp. 363-364.

Kittay, E.F.: 1999, *Love's Labor: Essays in Women, Equality, and Dependency*, Routledge, New York and London.

Kohlberg, L.: 1971, 'From is to Ought: How to Commit the Naturalistic Fallacy and Get Away With It in the Study of Moral Development,' in T. Mschel (ed.), *Cognitive Development and Epistemology*,Academic Press, New York, pp. 151-232.

Lugones, M. and E. Spelman: 1992, 'Have We Got a Theory for You! Feminist Theory, Cultural Imperialism, and the Demand for the Woman's Voice,' in J. Kourney, J. P. Sterba, and R. Tong (eds.), *Feminist Philosophies*, Prentice-Hall, Englewood Cliffs, NJ, pp. 382-383.

Mahowald, M.: 1993, *Women and Children in Health Care: An Unequal Majority*, Oxford University Press, New York.

Mullett, S.: 1989, 'Shifting Perspectives: A New Approach to Ethics,' in L. Code, S. Mullett and C. Overall (eds.), *Feminist Perspectives*, University of Toronto Press, Toronto, pp. 109, 114, 119 -120.

Murphy, J.: 1995, *The Constituted Body: AIDS, Reproductive Technology and Ethics*, State University of New York Press, New York.

Noddings, N.: 1984, *Caring: A Feminine Approach to Ethics and Moral Education*, University of California Press, Berkeley, California.

Rawls, J.: 1971, *A Theory of Justice*, Harvard University Press, Cambridge, Mass.
Rubin, J.H.: 1993, 'Neglected Women's Health Research Wins Funds,' *The Philadelphia Inquiry*, (Wednesday, March 31), p. A3.
Sherwin, S.: 1992, *No Longer Patient: Feminist Ethics and Healthcare*, Temple University Press, Philidelphia.
Tong, R.: 1993, *Feminine and Feminist Ethics*, Wadsworth, Belmont, California
Wendell, S.: 1993, *The Rejected Body: Feminist Philosophical Reflections on Disability*, Routledge, New York..

HILDE LINDEMANN NELSON

FOUR NARRATIVE APPROACHES TO BIOETHICS

The "theoretical-juridical" model of morality, as Margaret Urban Walker (1998) has usefully styled it, has been under siege for some time now. A number of moral theorists have rejected the idea that morality consists of a solitary judge applying lawlike principles logically deduced from a comprehensive, impartialist, universalist moral theory to the specifics of an instant case. Bernard Williams (1981), for example, famously observed that the theoretically unlimited demands of impartialist systems of morality elbow out much that gives meaning to life, including anything that could inspire us to take any moral goal seriously. Michael Stocker (1987) criticized such systems on the grounds that their impersonal stance prevented them from capturing what is morally significant about such interpersonal relationships as friendship, love, or community. Feminist ethicists argued that the model's transcendental "view from nowhere" was actually a view from male privilege that failed to take women's interests and experiences seriously. Postmodernists challenged the model's representation of the moral subject as a unified self with a private, disembodied consciousness.

Within bioethics, the classic challenge to the theoretical-juridical model was issued by David Burrell and Stanley Hauerwas (1977). They argued that because the model separated moral agents from their interests, provided no account of how moral selves are formed, and reasoned from principles stripped of all cultural content, it presented a distorted view of the moral life. Reason divorced from a specific historical community with its own particular standards of rationality, they claimed, was too abstract to be action-guiding for that community. Moral reasoning, in other words, must always be linked to a culture's story. Accordingly, Burrell and Hauerwas urged a form of rationality based on *narrative*. The capacity of narrative to connect contingencies – this because of that, then the other – made it possible to understand the relationships among them and so, they claimed, made them morally intelligible. Moreover, narratives were morally normative in that they shaped our perceptions and molded our moral sensibilities. While other communitarians within bioethics were later to pick up many of these themes, narrative has not played a specially prominent role in their work (although H. Tristram Engelhardt, who might be dubbed a communo-libertarian, opens his

G. Khushf (ed.), Handbook of Bioethics, 163–181.

monumental *Foundations of Bioethics* with a foundation myth depicting the "collapse of the hegemony of Christian thought in the West" and an Enlightenment unable to fill the void, as a means of motivating his own version of ethics and public policy). Outside of bioethics, however, Alasdair MacIntyre (1984) is perhaps the best-known proponent of the approach to ethics advocated here.

At about the same time as Burrell and Hauerwas threw down the narrative gauntlet, courses in the humanities began to be introduced into the curricula of U.S. medical schools. Drawing on literature, philosophy, anthropology, sociology, religious studies, history, and law, such courses offered first- and second-year medical students new perspectives from which to reflect on the ethical issues surrounding the profession for which they were fitting themselves. The people who taught these courses from the "home" discipline of literary criticism were, naturally enough, interested in the contributions that reading great works of literature might make to these reflections. At the same time, however, they applied the tools of literary criticism to various practices in medicine that they treated as literary texts. Because much of the work in what might be called narrative bioethics has emanated from people trained in literary criticism, most of the claims for the moral work that narratives can do in a medical context center on literary narratives and the tools of textual criticism. Joanne Trautmann Banks (1982), Tod Chambers (1999), Rita Charon (1994), Robert Coles (1989), Anne Hunsaker Hawkins (1993), Anne Hudson Jones (1987), Martha Montello (1997), Kathryn Montgomery (formerly Hunter) (1991), Lois LaCivita Nixon (1997), and Suzanne Poirier (1999), among others, have done important work in this area.

Another narrative approach to bioethics has centered less on literary narratives or textual criticism than on the stories that patients tell about their experience of being ill. Illness narratives have long been a popular literary genre, but in the late 1980s, bioethicists began to argue for the *moral* importance of telling one's story of sickness. Patients needed to tell these stories, it was claimed, as a way of responding to what was happening to their bodies, and in particular, to the impact that their illness and its treatment was having on their self-understanding. Further, if patients needed to tell these stories, doctors also needed to hear them, for only by understanding and responding to the patient's story could the physician hope to heal. Those who have written about the importance of telling and hearing stories of sickness include Howard Brody (1987), Arthur Frank (1995), Arthur Kleinman (1988), David Hilfiker (1994), Kathryn Montgomery (Hunter) (1991), Oliver Sacks (1985), Richard Selzer (1994), and Abraham Verghese (1994).

A fourth locus of narrative activity within bioethics has been the revival of casuistry as a method for reasoning morally about problematic clinical cases. As Albert Jonsen and Stephen Toulmin (1988) have described the method, a clinical case is analyzed in terms of such formal topics as, for example, Medical Indications, Patient Preferences, Quality of Life, and Context of Care. The analysis permits the deliberator to note the case's similarities to a paradigm case—one that displays most visibly the moral maxims and principles that guide action in "cases of this sort." Analogous cases are those in which particular circumstances justify qualifications of the principles or exceptions to the general maxim. By clustering cases around a paradigm case in which one already knows what to do, one can reason by analogy about what to do in the case at hand. The method has a handful of adherents, among them Carson Strong (1997) and Mark Kuczewski (1997), but it has not acquired the popularity that Jonsen and Toulmin predicted for it.

In what follows, I examine a number of claims that have been made concerning the moral work that narratives are capable of – in particular, those that propose to unseat the theoretical-juridical model in favor of a unique narrative ethics. As each of the four moments in the history of bioethics sketched out above works with narratives in a distinct way, I use these four moments as rubrics for organizing the claims under examination. I then offer my own account of the differences between theoretical-juridical and narrative approaches to ethics.

I. THE STORIES WE INVOKE

The first of the four narrative approaches to ethics works with stories by *invoking* them. In a flat-out rejection of the theoretical-juridical model, Burrell and Hauerwas (1977) and MacIntyre (1984) claim that reason and rationality are always characteristic of a specific historical tradition. It is the story of one's community – whether it be ancient Greece, medieval Paris, or eighteenth-century Edinburgh – that develops one's capacity to see things as reasonable, appropriate, valuable, and so on. Because the narrative tradition of the community subtly shapes all its members' knowing and valuing, there is no one model of moral reasoning that can be used as a vantage point from which to pass judgment on all the many cultures within which human beings live.

The theoretical-juridical model, taking itself to be precisely such a vantage point, is thus dismissed by the proponents of historical narrative as a philosophical cul-de-sac, a dead hope. In its place they call for an ethics based on what might be called the "just-so" stories of a culture: like Rudyard

Kipling's stories of how the elephant got his trunk or the camel his hump, the high mythical and historical narratives of our communities explain who we are as a people and how we came to be this particular "we." These foundational narratives serve as the community's source of moral normativity, and we invoke them as a means of ethical justification. When we have run out of reasons for acting as we do, we appeal to the norms that are internal to the social practices of the community – norms which are in turn justified by the community's "just-so" stories. So, to take an obvious example, physicians might look to the history of medicine and the physician's traditional role in society as justification for refusing to participate in physician-assisted suicide: the norms internal to the practice of medicine forbid physicians to kill their patients, and the rightness of these norms is explained by invoking the foundational story of the Hippocratic tradition.

The advantages of a historical narrative approach are evident: it generates an ethics that is much more useful for guiding action than are the transcendental principles of the theoretical-juridical model. A complaint commonly lodged against Kantian ethics, for example, is that any maxim to be universalized in accordance with the Categorical Imperative must be couched at such a high level of abstraction that it cannot tell us what we ought to do in a given set of circumstances, but only (sometimes) what we mustn't do (Nussbaum, 1992). The corresponding complaint against utilitarian ethics is that its prescriptions, when they are not platitudes, are often counterintuitive or hopelessly vague. They depend heavily on uncertain predictions and interpersonal comparisons of utility that are difficult to make (Williams, 1981). On the historical approach, by contrast, one need merely invoke one's story to find the rationale for one's action.

However, the drawbacks to the approach are perhaps as obvious as its merits. For one thing, in many communities, including the United States, the "just-so" stories forcibly include and oppress certain people at the same time as they exclude and silence certain others. "Our" culture's traditions are not apt to yield a picture of the good life that could ever be adopted by those who find themselves on the margins of the culture, or those who are subordinated within the culture. This use of narrative unjustly excludes certain people from the community altogether, while at the same time it problematically includes certain other people by characterizing their lives within the community as unsuccessful or fit only for others' purposes. And what of people within the culture whose positions are liminal? What, for example, of the Capulet in love with the Montague, the person of mixed race, the transgendered person? How are *they* to understand themselves with respect to the community's traditions?

If the historical narrative approach is fraught with problems of inclusion and exclusion *within* a community's traditions, it is also at a disadvantage when it comes to reconciling the traditions *between* communities. As John Arras has pointed out, "foundational stories not only tell us who we are; they also tell us who we are not" (Arras, 1997, p. 75). As a community, we define ourselves against other communities, positioning ourselves as not-Protestant if we are Catholic, not-Palestinian if we are Israeli. Often, these "not" relationships involve the domination of the one over the other, and usually this Other figures in the community's foundation myths as an objectified element rather than a subject community with myths of its own. So, as Arras points out, "for contemporary Palestinians the relevant story is the history of their oppression at the hands of the Jewish state; conversely, for contemporary Israelis the relevant foundational story is the history of Palestinian aggression and terrorism. The subjects of these historical narratives are thus locked in a perpetual struggle, not only over land, but also over the meaning of their common history" (p. 75).

When cultures clash and traditions are at odds, the exhortation to invoke one's own tradition offers no political or moral relief. If the impasse is to be resolved peacefully, each side will likely have to make an effort to hear and understand the other's story. Often, such efforts fail or never get off the ground, but supposing the parties did somehow manage to hear one another, it is possible that they might come to question their own stories as a result, and then they would need some standard for judging which story is better – or indeed, whether either story is morally adequate. Invoking their own tradition to settle the dispute will not help, for that is to use as a yardstick the very thing that one is trying to measure.

Burrell and Hauerwas's (1977) solution to this problem is to set out criteria for evaluating competing stories. "The test of each story," they claim, "is the sort of person it shapes" (p. 136). Moreover, a good story has: (a) the power to release us from destructive alternatives; (b) ways of seeing through current distortions; (c) room to keep us from having to resort to violence; (d) a sense for the tragic: how meaning transcends power (p. 137). The trouble with these criteria, however, isn't merely that some are redundant or that the list is incomplete, but that they derive their morally normative force from something outside a particular culture's traditions. Arras thinks these criteria are translatable into something like transcendental, lawlike principles (Arras, 1997, p. 77), and if he is right, then we have floated off into precisely the context-free, theoretical-juridical dead end that the historical approach purported to show us a way out of. My own view is that the criteria need not be understood in this way, but in using them at all, Burrell and Hauerwas seem at least to have forfeited the supremacy of narrative over other forms of rationality.

MacIntyre (1988) refuses to give up on narrative in his attempt to solve the problem of competing stories. He claims that some new stories do a better job than the old ones of solving the problems that prompted people to call the old stories into question in the first place. When members of a narrative tradition come to see the tradition as ultimately unable to resolve its problems, the tradition experiences an "epistemological crisis" (p. 361) for which its resources are no longer adequate. At that point, says MacIntyre, the members may look to outside traditions to show them not only a new set of social roles and new narratives to justify them, but also a way out of the crisis. MacIntyre insists that the adoption of new stories constitutes epistemological and moral progress. The new story is *better* than the old at solving the problems set for it – "better" according to the evaluative standards inherent in the old narrative tradition – so we have not settled for a mere succession of one story after another. In this way, he claims, the historical narrative approach to ethics can remain critical without ultimately abandoning narrative for transcendental principles.

MacIntyre's solution may well be the right one, especially when one considers that the evaluative standards inherent in any tradition are often at odds with one another, such that a new story can appeal to norms that were present but not taken seriously enough in the old ways of understanding the tradition. The sticking-point, however, is MacIntyre's failure to recognize that disenfranchised or subordinated members of the community were in epistemological crisis *all along*. They cannot rely on the modes of thought and evaluation made available by the tradition, because they are either alienated from those modes or connected to them in morally troublesome ways. This is not to say that they cannot at all draw on the resources implicit in the standing narratives. But it does mean that they have to approach those narratives with suspicion and distrust, and that they must continually challenge their authority. Once it is acknowledged that the normativity of the "just-so" stories must be called into question by some of the community's members all of the time, and by all of the community's members in times of crisis, we have a more complicated picture than the one MacIntyre offers us of how stories perform the work of moral justification.

II. THE STORIES WE READ

The second approach is to *read* stories (or listen to them, or view them), where the story is a work of literary or theatrical or cinematic art. Here the idea is to attend carefully to the nuances and complexities of great literature, films, or plays as a means of broadening one's morally formative experiences and so sharpening one's moral sensibilities. Martha Nussbaum (1990) is perhaps best known for her exploration of the role of artistic literature in developing the moral emotions. By reading serious fiction, she argues, one can make of oneself a person "on whom nothing is lost." One does this by allowing the author of the work to direct one's attention to the rich and subtle particulars of the narrative – the moral, intellectual, emotional, and social nuances. When the author has set these out with skill and imagination, overlooking no meaningful detail, the educated reader can see what is morally at issue in the narrative: she becomes, in the words Nussbaum borrows from Henry James, "finely aware and richly responsible." "Moral knowledge," Nussbaum writes, "is not simply intellectual grasp of propositions; it is not even simply intellectual grasp of particular facts; it is perception. It is seeing a complex, concrete reality in a highly lucid and richly responsive way; it is taking in what is there, with imagination and feeling" (Nussbaum, 1990, p. 152). Having broadened her field of vision and refined her moral perception, Nussbaum argues, the reader is in a better position to respond excellently to actual people in the world.

Within bioethics, the "stories we read" approach has been espoused by many writers. Rita Charon (1997), for example, urges attention not only to the complexity of the narrative's content, but also to the means the author uses to achieve the desired effects. In offering an interpretation of Henry James's *The Wings of the Dove*, for example, she claims that "the methods James uses to tell his story instruct his readers in the fundamental and profound skills of apprehending, amid great conflict and human pain, the good and the right – whether in a fictional world or in the ordinary world of one person with another" (p. 92). And again: "Literary methods help doctors and patients to achieve contextual understandings of singular human experiences, supporting the recognition of multiple contradictory meanings of complex events" (Charon, 1996, p. 244). Like Nussbaum, Charon sees the literary narrative approach to ethics as a matter of acquiring morally formative experiences that refine one's moral perception, arguing in particular that physicians ought to read great works of literary art so that they may better grasp "the texture, the conflicts, the regrets, and the hopes" of their patients' lives (Charon, 1997, p. 109). As Martha Montello (1997) puts it: "The same literary skills that critical readers use to interpret the meaning of events in a story allow clinicians to see the way ethical issues are embedded in the individual and contingent nature of people's beliefs, culture, and biography" (p. 186).

Kathryn Montgomery (Hunter, 1991) concurs. Believing that physicians can broaden their knowledge of human beings, cultivate the power of observation, and equip themselves to confront the pain of a patient's illness or death by reading serious fiction, she claims that "physicians who read more than the bodies of their patients and are acquainted with more life stories than the ones in which they are asked to intervene ... see more clearly the mix of pain, pleasure, and loss in most people's lives and know what, if anything, suffering may be good for" (159). Because literature "cultivates moral sensibility and provides models for behavior," it is "central to practical reason both in its substance and in the process of interpretation that it exemplifies and requires" (Hunter, 1996, pp. 311, 316).

As Montello (1997) has quite rightly pointed out, the literary narrative approach requires the cultivation of narrative competence: the inquirer must learn the same literary skills that critical readers use to interpret the meaning of events in a story. Montello identifies three sorts of narrative competence, claiming that they "enhance the ability of physicians to find meaning in the complex lives of unique individuals" (p. 191). While Montello has worked out the connection between narrative competence and moral competence in greater detail than other proponents of the approach, they all agree that the one enhances the other. There is reason, though, to doubt that the relationship between the two kinds of competence is as straightforward as these writers have taken it to be.

Narrative competence may be defined, first, as the ability to *choose* good literature, and second, as the ability to *read* good literature with care, skill, and critical judgment. It is the capacity for selecting works of fiction that are worth one's time and attention, and then reading these works shrewdly, with an awareness of the literary techniques the author employs and a sense of how the text relates to other literary texts. It involves knowing the standards for what makes a work of literary art good or great and either using those standards to evaluate what one reads or calling them into question.

Note, however, that the relevant standards are primarily *aesthetic* ones – the ones we use for judging literature as a work of art. But if I am trying to refine my *moral* perception, why should those be the right standards? Might it not be more important for me to learn how to choose books – even books of poor literary quality – that can improve my moral position, and then learn how to read these books in ways that make me a morally better person? If choosing and reading books for these purposes involve standards other than aesthetic ones, then attaining *narrative* competence will not help me become "richly responsible."

Suppose, to use an example of Michael DePaul's (1993), a young man who is naive about women decides to read novels for the purpose of cultivating a fine-grained moral response to some future soulmate. If he were to ask what kind of novel to read, and were told to pick one that people with highly developed levels of narrative competence have found worthy, the young man might reasonably wonder what these people know about love that he does not. He has probably met literary scholars who are just as naive about relationships with women as he is – and if he has not, I have. The English professor who is a renowned Shakespeare scholar but an inept human being does not read great authors with a view to her own moral improvement. She does not approach the texts in the right spirit, or use them for the purpose of cultivating her moral perception. Why, then, should her standards for good literature guide the naif's choice of what to read? Nussbaum, Montello, Montgomery, and Charon all appreciate the importance of reading great literature in the proper frame of mind if one is to improve one's moral position. But the proper frame of mind has less to do with one's *narrative* competence than with one's *moral* competence, and about this the literary scholar, *qua* literary scholar, has no special expertise. If I am right about the difference between the ability to read good literature and the ability to read literature in a way that makes you good, it turns out that narrative competence required for this approach is at best insufficient to the task of improving one's moral perception.

There is a deeper problem. Even if the inquirer were to combine narrative competence with a high degree of moral competence, it is not clear that the fine awareness and rich response one could then display toward fictional characters is as readily transferable to actual human beings as the proponents of this approach suppose. As the poet Alan Shapiro observes, "To enrich perception isn't necessarily to make perception more amenable to virtue, or to a particular moral code" (Shapiro, 1996, p. 3). For one thing, even the most artfully drawn, complex, and fully realized fictional character does not much resemble, either in her subjectivity or in the range of her experiences, any actual, flesh-and-blood person, because the author carefully selects only certain incidents and actions as a means of bringing a character to life. For another, we can see and know things about characters in a novel or play that we cannot possibly hope to know about actual people, as the author or playwright carefully directs our attention toward desires, emotions, and thoughts that ordinarily remain hidden from view. René Magritte put the point nicely: "Ceci n'est pas une pipe." It is not a pipe. The image on the canvas is only a painted representation. You can't fill it with tobacco and you cannot smoke it, either.

Proponents of the literary narrative approach are typically not so much interested in overthrowing the theoretical-juridical model as in refining the sensibilities of the inquirer who uses it. Nussbaum insists that "perceptions 'perch on the heads' of the standing terms; they do not displace them" (Nussbaum, 1990, p. 155). Charon too is content to augment rather than supplant the understanding of morality that has dominated the last few centuries: "The principlist methods of ethical inquiry remain as the structure for clarifying and adjudicating conflicts among patients, health providers, and family members at the juncture of a quandary. The principles upon which bioethics decisions have been based . . . continue to guide ethical action within health care" (Charon, 1994, p. 277). There would be no conceptual incoherence in embracing the literary narrative approach while at the same time repudiating the theoretical-juridical model, but I am not aware of any attempt to work out this position.

III. THE STORIES WE TELL

The third approach is to *tell* stories – stories of sickness. Patient's stories of their illness are deemed morally valuable either because they help the physician "both to know and treat the patient" (Hunter, 1991, p. 46), or because they "enhance the self-consciousness of the ill and aid them in developing their distinctive community" (Frank, 1997, p. 43).

Anne Hunsaker Hawkins introduces her study of stories of sickness by asking, "What impact, if any, can such a study have on the rapidly changing patterns of medical practice today and – even more important – tomorrow?" (Hawkins, 1993, p. x). Like Hawkins, Howard Brody (1992) is interested in how attention to the stories patients tell can inform clinical practice. Brody argues that if he, a family practice physician, is to enter into a therapeutic relationship with his patients, both he and his patient must understand what the illness or debility means to that patient. "As a general rule," he writes, "if physicians are most effectively to understand the meaning an illness has for the patient so as to be able to alter it positively, they must be attuned to the role that the illness plays in the unfolding story of the patient's life" (p. 133). In the face of patient suffering, Brody calls for the physician to join the sufferer in a project of constructing a new narrative of the patient's life – one that appears relevant to the patient himself and that also reconnects the sufferer to the broader community (Brody, 1992, pp. 256-57).

If Arthur Frank (1995) were in the business of offering advice to clinicians he would probably echo Brody's sentiments. His focus, however, is not on the clinical encounter but on encouraging the sick to tell their own stories rather than submit to the reductionistic and objectifying categories that the practice of medicine imposes on them. He argues that patients have a duty to tell the story of their sickness so that they do not give up their own experience of it to the medical chart and in that way lose a part of themselves. By this he does not mean that only the patient has a valid perspective on the illness and its treatment. In telling the story, the patient simultaneously claims membership in and constructs a community of those who share her story.

"Storytelling," Frank writes, "is *for* an other just as much as it is for oneself. In the reciprocity that is storytelling, the teller offers herself as a guide to the other's self-formation. The other's receipt of that guidance not only recognizes but *values* the teller. The moral genius of storytelling is that each, teller and listener, enters the space of the story *for* the other" (Frank, 1995, pp. 17-18). In this way Frank underscores the significance of an ethics of listening.

By making narrative sense of a serious or prolonged illness, Frank claims, patients redefine themselves, for such illness and its treatment has an impact on the patient's identity. Indeed, the story is seen as an act of resistance whereby the patient attempts to stave off assaults on how others perceive her and on her own self-understanding, inflicted by either the illness itself or by a medical ideology that sees her as a passive object of care rather than a moral agent. The moral value of the story, then, lies in its ability to reveal both to the patient and to those in her storytelling community who the patient is and should be seen to be.

The personal narrative approach need not reject the theoretical-juridical model of morality, but Frank clearly does. Against the falsely totalizing discourses of modern medicine or Enlightenment moral philosophy, he sets a quest for personal self-development that offers its own subjective account of what is true. For Frank, the stories of the patient's quest for meaning "are their own truth," and he confesses himself unsure of "what a 'false' personal account would be" (Frank, 1995, p. 22). Even a wildly distorted account, he claims, tells the truth about the patient's desire to have lived out a different story. The danger here, as John Arras has pointed out, lies in "mistaking the authenticity of the narrator for ethical truth" (Arras, 1997, p. 85). If the narrator is self-deceived, the story she tells about herself does not give her any critical perspective from which to assess the stories of who she actually is. Identity is not simply a matter of one's beliefs about oneself, no matter how sincerely held (Nelson, 2001). Authenticity isn't enough. A true story also requires *accuracy*.

If this failing were an ineradicable feature of the personal narrative approach, we would have to think very seriously about whether this is the best we can do. I believe, however, that the approach contains resources that allow its proponents to avoid having to fall back on the theoretical-juridical model, on the one hand, or sinking into a quagmire of subjectivity that precludes moral judgment, on the other. If, in telling a story of sickness, the patient simultaneously claims membership in and constructs a community of those who share her story, the others in the community serve as a check on the story, challenging faulty accounts of what happened or was done and making their own assessments of the story's accuracy. Within this storytelling community, the patient's self-understanding and how others perceive the patient are both open to reconstruction, in a narrative process of negotiation among all the parties involved.

IV. THE STORIES WE COMPARE

The fourth approach is to *compare* a story that is morally puzzling to a paradigm story or stories whose moral meaning is clear. The paradigm story, according to proponents of casuistry, displays summaries of shared moral understandings that are already embedded in our actual practices, and these summaries serve as action-guiding principles. Thus, when we come across a new case that leaves us uncertain how to proceed, we can look to the paradigm case for guidance.

The principles displayed by the paradigm case are always subject to revision in light of new cases. This is so not only because casuistical principles are derived from practice (they emerge from the 'bottom up' rather than being applied from the 'top down,' as on the theoretical-juridical model), but because discerning just which particulars of a story are morally relevant is a messy and uncertain business, and may require a number of passes before we get it right. As subsequent cases reveal new implications of principles embedded in the paradigm case, our view of the paradigm case may shift, causing us to modify our understanding of the original principles or perhaps even to replace them with new principles altogether. As Arras puts it, "Casuistical analysis might be summarized as a form of reasoning by means of examples that always point beyond themselves. Both the examples and the principles derived from them are always subject to reinterpretation and gradual modification in light of subsequent examples" (Arras, 1991, pp. 35-36).

One of the advantages of the casuistical approach to bioethics is that our society is familiar with this form of reasoning: it is how judges, lawyers, and juries in the U.S. and other English-speaking countries build the body of common law. A second advantage is that clinicians are accustomed to thinking in terms of cases. Indeed, the clinical ethics case study dominates the bioethics literature, although the discussion of a morally troublesome case does not often make use of casuistical methodology. The third advantage is that case-based reasoning, like the historical narrative approach, does a much better job than the theoretical-juridical model of telling the inquirer just exactly what she ought to do. But unlike the historical narrative approach, which relies on a community's morally problematic "just-so" stories, casuistry makes use of more local and more easily revisable stories that need not exclude or subordinate anybody.

Still, the method has its disadvantages. For one thing, some of its proponents claim that the work of comparing cases for moral guidance requires no theoretical apparatus, which seems to be a claim that the stories speak for themselves. They do not, of course. If Thomas Kuhn and other philosophers are right to argue that all perception is theory-laden, then so is the perception of events and circumstances that constitute a case. The particulars that are selected out of the manifold of experience for inclusion in the case are chosen according to some theory about their relevance to the story. Moreover, the instant case is compared to the paradigm case on the basis of some theory about the relationship between the two. This means that theoretical disagreements can arise not only about what facts are morally salient in the instant case, but also about whether an analogy actually exists between an instant case and any particular paradigm case.

Moreover, the shared moral understandings that are supposedly displayed by the paradigm case will be action-guiding only if they are indeed shared. So, for example, the treatment team on a burn unit might well know what to do for a horribly burned patient who, after three months' treatment, begs to be allowed to die: they can look to Dax's case. Because there is widespread agreement now among clinicians and bioethicists that patients' right to self-determination must be respected, the story of Dax Cowart *can* be taken as a paradigm instance of what physicians must not do. Where there are no shared understandings, however, there can be no agreement about what constitutes a paradigm case. Should we give antibiotics to a badly damaged neonate with a Do Not Attempt to Resuscitate order, or does that count as futile treatment? Are we killing non-heart-beating organ donors when we use the Pittsburgh Protocol? Should we treat a pleasantly demented Alzheimer's patient according to her current best interests or according to an advance directive that was predicated on the assumption that life with dementia would not be worth living? The case-based approach works less well in these instances because there is no story that the deliberator can invoke as paradigmatic.

And once again the problem of moral justification rears its ugly head. For many casuists, the source of normativity lies within the paradigm stories, much as it lay within the "just-so" stories of a community's traditions on the historical narrative approach. So when no paradigm story is available, or there is no agreement about whether an instant case is relevantly similar to a particular paradigm story, the proponents of the case comparison method are left without the resources for judging one course of action to be better than another. Because the absence of or disputes over paradigm cases do not rock the inquirers' very foundations, they are in better epistemic shape than the historical narrativist who can no longer rely on his narrative tradition, but they will have to look to something other than narrative as the basis for making moral judgments.

Putting that problem to one side, there remains the problem of where the normativity goes when subsequent cases show that something is wrong with the paradigm case. When the paradigm is called into question, its principles are no longer taken to be binding. If the casuistic approach is to be nonarbitrary, however, some principle must be used that tells the inquirer that an adjustment in the equilibrium between the paradigm and the subsequent case is necessary. The standard for deciding this obviously no longer resides in the paradigm case. Is it, then, in the subsequent case? Does it keep moving from case to case as the inquirer just *sees* that the old understandings won't do? If so, the case comparison approach comes down to an elaborate refinement of the inquirer's intuitions regarding cases. And if her intuitions are shaped by racism, sexism, or other forms of oppression, then casuistry will merely refine her prejudices. While the particulars of successive stories can help her to get clearer about our shared social meanings, these stories are not apt to display the imbalances of power that distort social group relations, so they provide little standpoint from which to criticize those relations.

Mark Kuczewski's solution to these problems of normativity is to point out that even in the face of moral disagreement, the members of a given community can look to the moral understandings they *do* hold in common (Kuczewski, 1997). Rejecting the image of the unpositioned, solitary inquirer with discretionary power to decide which of "our" judgments are well considered, his model offers instead the picture of actual moral communities, whose members express themselves and influence others by appealing to mutually recognized values, and use those same values to refine understanding, extend consensus, and eliminate conflict. On Kuczewski's communitarian view of casuistry, the authority for a moral intuition rests on its embeddedness in a shared form of moral life, while the basis for moral criticism lies in the tensions between, and the fissures within, the stories that circulate widely in the community. Anyone who does not share enough of the important intuitions is either a morally incompetent member of the moral community or not of the community at all.

V. THE MORAL WORK THAT STORIES DO

Any judgment as to whether stories can do the various kinds of moral work that are claimed for them depends on what, precisely, one takes morality to be. The theoretical-juridical model represents morality as a codifiable set of moral formulas, typically conceptualized as rules or principles at a high level of generality, which any agent can apply to a situation to guide action or assess wrongdoing. On this model the (solitary) moral philosopher's task is to construct, test, and refine covering laws that exhibit moral knowledge. The task of moral justification is carried out by the covering principles or procedures that make up the moral theory.

On the theoretical-juridical model a story might instantiate a principle, but it is the principle, not the story, that justifies action. Stories could serve other purposes, however. First, even the most ardent proponents of the theoretical-juridical model will acknowledge that most people learn most of what they know about morality from stories of one sort or another, beginning with nursery tales told to children just learning to talk. Second, on the theoretical-juridical model, stories might also serve as heuristic devices for the abstractly impaired – the people who lack the talent to grasp, or to learn how to use, the propositions that are the proper medium for moral reasoning. Third, even if, as I argued earlier, the link between reading literary narratives skillfully and applying moral principles skillfully is not as straightforward as Nussbaum and others have supposed, there is surely more that could be said about this and the role that moral perception plays here.

For some of us who have lost faith in the theoretical-juridical model, however, a uniquely narrative ethic has seemed a promising alternative. To be that, it would have to be shown that the stories do moral work that theoretical-juridical propositions either do not do as well or cannot do at all. I think there are at least four such tasks, but here I can only suggest what they are – a detailed account is sorely needed. The first is that stories are better than propositions at displaying relationships among people, between people and institutions, and among responsibilities that are in some tension with one another. The second is that *only* stories are capable of representing a temporal sequence. The explanation of how a particular situation came to be a moral problem is precisely a story: if you arranged a set of propositions that indicated what happened first and what followed because of it, the propositional set would *be* a story. The third is that a story does a much better job than propositions can at working with the difference between a first-person and a third-person point of view. It can therefore accommodate the distinction between what I know, or take responsibility for, and what I believe someone else knows or is bound by, where these differ. (The distinction is epistemologically crucial. Think of G. E. Moore's classic [1942] "I went to the pictures last Tuesday but I don't believe that I did," vs. "I went to the pictures last Tuesday but she doesn't believe that I did.") The fourth is a difference noted thirty years ago by William Gass (1970). In "The Case of the Obliging Stranger, " Gass tells the story of a man who approaches a stranger and asks if he will help him with an experiment. The stranger is willing to oblige. The man hits him on the head with the broad of an axe, takes him home, and roasts him in the oven. Unfortunately, he leaves him in the oven until he is burnt to a crisp, so the experiment is ruined. "Something has been done wrong," Gass concludes, "or, something wrong has been done." Gass goes on to point out that no set of propositions derived from a moral theory is required to show us that this man has done something worse than cook his specimen incorrectly. The story is morally more fundamental than any theory that might be developed to justify the point. Nothing is gained by reformulating what is at issue as a moral proposition (how would the proposition go? Roasting obliging strangers is wrong? How does that tell us anything the story did not tell better?).

A uniquely narrative ethic need not eschew moral norms. Indeed, in the Gass example, the norm against roasting people lies right at the heart of the story. But the norms that are put into play could be thought of as moral markers that show us something we need to take into account rather than as inflexible, lawlike standards. And if in addition our narrative ethic lays stress on the social and interpersonal nature of morality, seeing moral life as something lived in common, rather than as the activity of solitary judges, then moral deliberation stops being a matter of whether *I* can make the proper distinctions and nuanced judgments that produce moral knowledge, and instead becomes a collaborative activity among some group of "us" who need to know how we can best go on in a shared life.

To my mind, a collaborative view of morality requires two different types of stories. First, it requires stories that *represent* the moral problem, displaying who the relevant parties are, how they got into this mess, what relationships hold among them, perhaps what social or institutional constraints shape their options. Second, the *resolution* of the problem also takes a narrative form—it is a set of stories of how best to go on from here, and what this going on will mean to the various affected parties. The stories that display the problem are backward-looking stories. They are the narratives that explain what has happened to bring everybody to this current, problematic point. The backward-looking stories also explain to each of the inquirers who the others are, morally speaking: what they care about, on what basis they assign responsibility and blame to the various parties, and how all this informs their understanding of the problem at hand. The stories of what will now be done are forward-looking stories. They are the narratives that set the field for future courses of action, displaying the possibilities that are now open to the inquirers and what might have to be given up so that the parties can all stay in relationship with one another. Sometimes the forward-looking stories show the impossibility of maintaining these relationships. Moral resolutions arrived at through this process of collaborative narrative construction are justified ultimately by the habitability and goodness of the common life to which they lead (Nelson, 1999).

The work on narrative approaches to bioethics has barely begun, so the false starts, unsubstantiated claims, and gaps in various accounts are scarcely to be wondered at. Despite these failings, there have also been a number of promising developments, and work in this area is gradually gaining intellectual respectability. Because stories can represent both the complexity and the subtlety of the moral life, they are an invaluable medium for moral deliberation – a medium whose strengths and weaknesses we are only just beginning to understand. For those of us who have become increasingly disenchanted with the theoretical-juridical model of morality, the narrative turn in bioethics holds out the hope of genuine progress.

Department of Philosophy
Michigan State University
East Lansing, Michigan, U.S.A.

NOTE

An ancestral version of this essay appeared in the *New Zealand Bioethics Journal* 1, no. 2 (October 2000): 10-21. I am grateful to James Lindemann Nelson, John Arras, and Margaret Urban Walker for helping me refine my thinking about the moral work that stories can – and can't – do.

REFERENCES

Arras, J.: 1991, 'Getting down to cases: The revival of casuistry in bioethics,' *Journal of Medicine and Philosophy* 16, 29-51.

Arras, J.: 1997, 'Nice story, but so what? narrative and justification in ethics,' in H. Nelson (ed.), *Stories and Their Limits*, Routledge, New York, pp. 65-88.

Banks, J. T.: 1982, 'The wonders of literature in medical education,' *Mobius* 2B, 23-31.

Banks, J. T.: 1990, 'Literature as a clinical capacity: Commentary on the "Quasimodo" complex,' *Journal of Clinical Ethics* 1, no. 3, 227-31.

Brody, H.: 1987, *Stories of Sickness*, Yale University Press, New Haven, Conn.

Brody, H.: 1992, *The Healer's Power*, Yale University Press, New Haven, Conn.

Brody, H.: 1997, 'Who gets to tell the story? Narrative in postmodern bioethics,' in H. Nelson (ed.), *Stories and Their Limits*, Routledge, New York, pp. 18-30.

Burrell, D., and S. Hauerwas: 1977, 'From system to story: An alternative pattern for rationality in ethics,' in H.T. Engelhardt, Jr., and D. Callahan (eds.), *The Foundations of Ethics and Its Relationship to Science: Knowledge, Value, and Belief*, vol. 2, The Hastings Center, Hastings-on-Hudson, New York, pp. 111-152.

Chambers, T.: 1996, 'From the ethicist's point of view: The literary nature of ethical inquiry,' *Hastings Center Report* 26, 25-32.

Chambers, T.: 1997, 'What to expect from an ethics case (and what it expects from you),' in H. Nelson (ed.), *Stories and Their Limits*, Routledge, New York, pp. 171-184.

Chambers, T.: 1999, *The Fiction of Bioethics*, Routledge, New York.

Charon, R.: 1993, 'The narrative road to empathy,' in H. Spiro et al. (eds.), *Empathy and the Practice of Medicine*, Yale University Press, New Haven, Conn, pp. 147-159.

Charon, R.: 1994, 'Narrative contributions to medical ethics: Recognition, formulation, interpretation, and validation in the practice of the ethicist,' in E.R. DuBose, R. Hamel, and L.J. O'Connell (eds.), *A Matter of Principle? Ferment in U.S. Bioethics*, Trinity Press International, Valley Forge, Pa, pp. 260-283.

Charon, R.: 1997, 'The ethical dimensions of literature: Henry James's *The Wings of the Dove*,' in H. Nelson (ed.), *Stories and Their Limits*, Routledge, New York, pp. 91-112.

Charon, R., et al.: 1996, 'Literature and ethical medicine: Five cases from common practice,' in K.D. Clouser and A.H. Hawkins (eds.), *Journal of Medicine and Philosophy*, pp. 243-265.

Childress, J.F.: 1997, 'Narrative(s) versus norm(s): A misplaced debate in bioethics,' in H. Nelson (ed.), *Stories and Their Limits*, Routledge, New York, pp. 252-271.

Clouser, K.D. and A.H. Hawkins (eds.): 1996, special issue, literature and medical ethics, *Journal of Medicine and Philosophy* 21B, no. 3.

Coles, R.: 1989, *The Call of Stories: Teaching and the Moral Imagination*, Houghton Mifflin, Boston.

Davis, D.: 1991, 'Rich cases: The ethics of thick description,' *Hastings Center Report* 21B, no. 4,12-16.

DePaul, M.: 1993, *Balance and Refinement: Beyond Coherence Methods of Moral Inquiry*, Routledge, London.

Engelhardt, H.T.: 1986, *The Foundations of Bioethics*, Oxford University Press, New York.

Epstein, J.: 1995, *Altered Conditions: Disease, Medicine, and Storytelling*, Routledge London.

Frank, A.: 1991, *At the Will of the Body Reflections on Illness*, Houghton Mifflin, Boston.

Frank, A.: 1995, *The Wounded Storyteller: Body, Illness, and Ethics*, University of Chicago Press, Chicago.

Frank, A.: 1997, 'Enacting illness stories: When, what, and why,' in H. Nelson (ed.), *Stories and Their Limits*, Routledge, New York, pp. 31-49.

Gass, W.: 1970, 'The case of the obliging stranger,' in *Fiction and the Figures of Life*, Knopf, New York.

Hawkins, A.H.: 1993, *Reconstructing Illness: Studies in Pathography*, Purdue University Press, West Lafayette, Ind..

Hawkins, A.H.: 1996, 'Literature, philosophy, and medical ethics: Let the dialogue go on,' in K.D. Clouser and A.H. Hawkins (eds.), special issue, literature and medical ethics, *Journal of Medicine and Philosophy*, 21, no. 3, pp. 153-170.

Hawkins, A.H.: 1997, 'Medical ethics and the epiphanic dimension of narrative,' in H. Nelson, (ed.), *Stories and Their Limits*, Routledge, New York.

Hilfiker, D.: 1994, *Not All of Us Are Saints: A Doctor's Journey with the Poor*, Hill and Wang, New York.

Hunter, K. (now K. Montgomery): 1991, *Doctor's Stories: The Narrative Structure of Medical Knowledge*, Princeton University Press, Princeton.

Hunter, K.: 1996, 'Narrative, literature, and the clinical exercise of practical reason,' in K.D. Clouser and A.H. Hawkins (eds.), special issue, literature and medical ethics, *Journal of Medicine and Philosophy*, 21, no. 3, pp. 267-286.

Johnson, A., and S. Toulmin: 1988, *The Abuse of Casuistry*, University of California Press, Berkeley and Los Angeles.

Jones, A.H.: 1987, 'Literary value: The lesson of medical ethics,' *Neohelicon* 14, no. 2, 383-92.

Jones, A.H.: 1996, 'Darren's case: Narrative ethics in Perri Klass's *Other Women's Children*,' in K.D. Clouser and A.H. Hawkins (eds.), special issue, literature and medical ethics, *Journal of Medicine and Philosophy*, 21, no. 3.

Kleinman, A.: 1988, *The Illness Narratives: Suffering, Healing, and the Human Condition*, Houghton Mifflin, Boston.

Kuczewski, M.: 1997, *Fragmentation and Consensus: Communitarian and Casuist Bioethics*, Georgetown University Press, Washington, D.C.

MacIntyre, A.: 1984, *After Virtue*, second edition, University of Notre Dame Press, Notre Dame.

MacIntyre, A.: 1988, *Whose Justice? Which Rationality?* University of Notre Dame Press, Notre Dame.

Montello, M.: 1997, 'Narrative competence,' in H. Nelson (ed.), *Stories and Their Limits*, Routledge, New York, pp. 185-197.

Moore, G.E.: 1942 (2d ed. 1952), 'A Reply to My Critics,' in P. Schlipp (ed.), *The Philosophy of G. E. Moore*, Tudor, Evanston, Ill, pp. 533-677.

Murray, T.H.: 1997, 'What do we mean by "narrative ethics"?'in H. Nelson (ed.), *Stories and Their Limits*, Routledge, New York, pp. 3-17.

Nelson, H.L.: 1999, 'Context: Backward, sideways, and forward." *Healthcare Ethics Committee Forum* 11, n. (1), 16-26.

Nelson, H.L.: 2001. *Damaged Identities, Narrative Repair,* Cornell University Press, Ithaca, N.Y.

Nelson, H.L. (ed.): 1997, *Stories and Their Limits: Narrative Approaches to Bioethics*, Routledge, New York.

Newton, A.Z.: 1995, *Narrative Ethics*, Harvard University Press, Cambridge, Mass.

Nussbaum, M.C.: 1992, *Love's Knowledge*, Oxford University Press, New York.

Poirier, S.: 1999, 'Voice, structure, politics, and values in the medical narrative," *Healthcare Ethics Committee Forum* 11, no. 1, 27-37.

Poirier, S., and D.J. Brauner: 1990, 'The voices of the medical record," *Theoretical Medicine* 11, 29-39.

Rorty, R.: 1989, *Contingency, Irony, and Solidarity*, Cambridge University Press, Cambridge, Mass.

Sacks, O.W.: 1985, *The Man Who Mistook His Wife for a Hat and Other Clinical Tales*, Summit, New York.

Selzer, R.: 1994, *Raising the Dead: A Doctor's Encounter with His Own Mortality,* Viking, New York.

Shapiro, A.: 1996, *The Last Happy Occasion*, University of Chicago Press, Chicago.

Stocker, M: 1987, 'The schizophrenia of modern ethical theories,' in R.B. Kruschwitz and R.C. Roberts (eds.), *The Virtues: Contemporary Essays on Moral Character,* Wadsworth, Belmont, Cal.

Strong, C.: 1997, *Perinatal Medicine: A New Framework,*Yale University Press, New Haven, Conn.

Verghese, A.: 1994, *My Own Country: A Doctor's Story of a Town and Its People in the Age of AIDS*, Simon and Schuster, New York.

Walker, M.U.: 1998, *Moral Understandings*, Routledge, New York:.

Williams, B.: 1981, 'Persons, character, and morality' in *Moral Luck*, Cambridge University Press, New York.

EDMUND D. PELLEGRINO

PHILOSOPHY OF MEDICINE AND MEDICAL ETHICS: A PHENOMENOLOGICAL PERSPECTIVE

I. INTRODUCTION

Since the beginnings of the Bioethics movement, a plurality of ethical theories have been used as a foundation for medical ethics. One approach just beginning to be examined is the grounding of medical ethics in a philosophy of the physician-patient encounter (Pellegrino, 1979). On this view the phenomena of being ill, being healed and promising to heal are taken as the staring points for ethical reflection. An ethics based in the clinical encounter promises to be more closely related to the concrete experiences of doctor and patient than the application of pre-existing ethical theories (Pellegrino and Thomasma, 1981). This approach is "phenomenological" in a loose sense in that it begins with a reflection on the shared experience of physician and patient and attempts to uncover the deeper structures and meanings of their experiences in the clinical encounter. A formal phenomenological methodology is not intended. Rather, the approach is more analogous to what Natanson might call a "phenomenological orientation" (Natanson, 1972, pp. 14).

A phenomenological approach makes use of four concepts drawn more directly from phenomenology, i.e., the concepts of epoché, reduction, life-world, and inter-subjectivity. These concepts are used heuristically to illuminate the foundations of medical ethics and in this way to respond both to current divergent accounts of those foundations and to the post-modernist assertion that foundations for ethics are not even tenable. The focus is on the life-world of doctor and patient, their shared experience in the clinical encounter, and the ethical meanings and imperatives embedded in that encounter.

The inquiry proceeds in three steps: first, by a brief delineation of the present state of philosophical ethics; second, by a brief account of the phenomenological way of approaching philosophy and the way concepts drawn from one type of phenomenology can be used heuristically in elucidating the ethics of medicine; and, finally, by showing specifically how the life-world of doctor and patient might serve to ground the ethics of the healing professions.

The overall aim of this essay is to ground medical ethics in the concrete realities of being ill and being healed. The methodology employed is a form of

G. Khushf (ed.), Handbook of Bioethics, 183–202.

empirical analysis but not that of positive science. Rather, it has similarities to the "realist phenomenological position" developed by Dietrich Von Hildebrand, a "pre-philosophical contact with how things are." (Hildebrand, 1991; Seifert, 1973, pp. xlvi-xlvii). This inquiry is undertaken in that spirit rather than a strict application of Hildebrand's methodology.

The concrete reality which is the focus of attention is the "life-world," where "all projections of more specialized realms – the law, the government, the professions – takes place and it is the life-world, itself, which becomes philosophy's theme" (Natanson, 1972, pp. 17, 127; Husserl, 1960, pp. 50). In this case, the life-world of doctor and patient is the object of philosophical inquiry.

II. PRESENT STATE OF MEDICAL ETHICS

Medical ethics, the ethics of the physician as physician, has been the subject of systematic philosophical inquiry for only a quarter of a century (Pellegrino, 1993). Before that, medical ethics was grounded in a set of moral precepts freely and unilaterally asserted and derived from the Oath and Ethos of the ancient Hippocratic school. In recent years, philosophers have begun intensively to question these moral groundings and each of the precepts drawn from there (Veatch and Mason, 1987). As a result, many today question whether any enduring moral foundation for medical ethics beyond societal or professional consensus is tenable.

To be sure, there have been proposals aplenty to substitute for, or replace, the traditional groundings. They vary with the particular philosophical stance one takes as the analytical tool – deontology; consequentialism; *prima facie* principlism; Aristotelian or Thomistic virtue theories; feminist, caring, narrative, or casuistic philosophies; etc. These theories and others have been applied intensively to medicine, clinical dilemmas, and professional conduct. The resulting moral diversity is simultaneously salubrious and confusing, but also philosophically interesting.

On the salubrious side is the opening of the previously protected sanctuary of physician-patient relationships to moral scrutiny. This was inevitable in an era of self-determination when medical ethics has become everybody's concern. How physicians conduct themselves in the face of the universal human experiences of illness and healing is of universal interest. Medical ethics cannot responsibly ignore the unprecedented scientific, societal, and political challenges posed in our times to the traditional accounts of physician-patient inter-relationships.

On the confusing side are the conflicting moral precepts, divergent answers to moral dilemmas and variant justifications so many different theories of ethics can generate. The facts of moral diversity and philosophical pluralism notwithstanding, can so many opposing views all be true? Or, must we, as the post-modernists insist, give up any notion of a generalizable foundation for medical ethics and settle for what eventuates from the practices or social constructions of the moment? (Toulmin, 1997; Wartofsky, 1997)

Few would argue for the traditional method of free moral custom found in professional codes. But not all would agree that there is no need for, or deny the possibility of, a durable philosophical foundation for medical ethics or the philosophy of medicine (Pellegrino, 1997). For many the question remains: Now that the moral assertions of the past have been challenged, how do we deal with the fractured foundation? The questions posed by today's moral philosophers cannot be ignored. But there are too many perils in an easy acquiescence to coherence, social constructivist, or dialogical ethics to justify them as the only or the best answers.

In the long term, foundations cannot be avoided. However we designate it, some philosophical theory will be used to justify particular moral choices – even if that theory holds that no foundation is conceptually tenable. The central problematic is how to deal with the fragments of the fractured foundation. Can the insights of a quarter of a century of philosophical inquiry be assimilated without following that inquiry to its current deconstructionist conclusion?

Beneath and beyond the skepticism of contemporary moral philosophy so far as medical ethics is concerned, there is an undeniable reality. That reality is the encounter between one human person who is ill and seeks assistance and another person who freely professes to be able to heal. The patient's predicament and the professional's response to that predicament center in another reality – the intersection of the life-worlds of doctor and patient within which the acts of medicine take place. If we can understand something of this intersection, we can grasp more firmly the origins and essence of the moral encounter which makes medicine the special kind of human activity it is or should be.

III. THE PHENOMENOLOGICAL PERSPECTIVE

In the broad array of philosophical perspectives to which medicine and medical ethics has been recently exposed, phenomenology has had little prominence in Europe or the United States (Dell'Oro and Viafora, 1996). To be sure, there have been several excellent commentaries from a phenomenological perspective, e.g. Spicker (1970) and Zaner's (1964) work on embodiment, Straus, Natanson amd Ey (1969) on the philosophy of psychiatry, Sokolowski (1989) on the art and science of medicine, Zaner (1988) and Ricoeur (1996) on the clinical encounter and Straus' approach to clinical psychiatry (1976). These studies offer valuable insights into the philosophy of medicine and lead ultimately to questions of medical ethics. These studies use the methodology of phenomenology specifically to examine some of the realities of the clinical encounter in the inter-subjectivity and life-world of patient and physician. They do not link the ethics of the health professions as closely as they might to the three specific phenomena of being ill, being healed, and professing to heal as the present essay attempts to do.

Phenomenology is notoriously difficult to define precisely because it is not a philosophical system in the usual sense (Honderich, 1995). It is often likened to a movement or, more loosely, as a "standpoint for viewing the world" (Molina, 1962, pp. 38-39). It has been used as a method by a variety of philosophers whose works are very different to examine the complex structures of imagination (Sarte, 1956), embodiment (Merleau-Ponty, 1962; Marcel, 1960), the emotive structure of abstract knowledge (Heidegger, 1962, 1970; Zaner and Ihde, 1973), and human personhood (Wojtyla, 1979).

The word "phenomenology" itself has undergone sharp transformations of meaning since its introduction in the 18th century (Edward, 1967). Its contemporary usage derives from a group of German philosophers earlier in this century. Unquestionably, the most influential member of this coterie has been Edmund Husserl with whose name phenomenology is now most often associated. Attempts have been made to define the features of philosophizing characteristic of phenomenologists, but even these are open to disagreement (Audi, 1995, pp. 578-579).

What all seem to agree upon, however, is that the phenomenological perspective hinges on two methodological moves, both deriving from the work of Husserl. These two moves are the epoché and the eidetic reduction. Even these two ideas are interpreted and practiced differently by different phenomenologists. It is essential, therefore, at the outset to clarify how I intend to use these as heuristic devices in exploring the ground of medical ethics.

Natanson, who is as clear and reliable a guide as can be found to the work of Edmund Husserl, warns us repeatedly that phenomenology is not easy (Natanson, 1972). He says it takes years to comprehend, much less apply effectively, the phenomenological perspective as expressed in the notions of epoché and reduction. This essay does not, therefore, pretend to orthodox or sophisticated use of the phenomenological perspective. It does, however, take that perspective in a very limited sense, and attempts to show its potential relevance to the question of a grounding for medical ethics.

Epoché and Reduction

The definitions of epoché and reduction are deceptively simple. Epoché refers to the act of abstention, of holding in suspension the natural interpretation of an experience in order to get at its essence, its "eidos." This is not to deny the "world" of fact. Rather, it is to disconnect, abstain from, or "bracket," what the natural sciences tell us about the world. As Husserl puts it:

> I disconnect them all (the sciences which relate to the natural world). I make absolutely no use of their standards, I do not appropriate a single one of their propositions that enter into their systems, even though their evidential value is perfect, I take none of them, no one of them serves me for a foundation – so long that it is understood in the way these sciences, themselves, understand it as a truth concerning the realities of the world. I may accept it only after I have placed it in the bracket (Husserl, 1962).

Natanson takes this idea of epoché to be the "clue" to the larger method of phenomenology, which is reduction: "Most simply, reduction refers to a shift in attention from factuality and particularity to essential and universal qualities" (Natanson, 1972, pp. 65). Natanson discerns two "species of reduction" – one is the turn from fact to essence (i.e., the eidetic reduction), the second is the movement from believingness to transcendental subjectivity (i.e., the phenomenological reduction)(Natanson, 1972).

I will limit my use of the phenomenological perspective to the first of these species, to the eidetic reduction, the attempt to characterize the essence of what is given in experience, to comprehend more clearly what it is to be ill, to be healed, and to promise to heal. These are the three phenomena of the clinical encounter to which I have alluded elsewhere as a grounding for the ethics of medicine (Pellegrino, 1979). I will not follow Husserl all the way to the second species of reduction, i.e., to transcendental subjectivity. Husserl's transcendentalism implies a conceptual idealism beyond methodology per se.

More specifically, I propose to make limited use of the device of the epoché to bracket existing philosophical systems, particularly moral philosophy as it pertains to medical ethics. Without denying the assertions of any philosophical system for the moment, I wish to bracket all their ethical opinions about the physician-patient relationship in an attempt to uncover the essence or "eidos" of that relationship from the phenomena in which it is embedded. In short, I will use the epoché of philosophical systems in a way analogous to the epoché of the physical sciences in Husserl's account. This is similar to the epoché of philosophy employed by Straus and Natanson in their study of the philosophy of psychiatry (Straus, Natanson and Ey, 1969).

Life-World and Intersubjectivity

In addition to the notions of epoché and reduction, there are two other inter-related ideas taken from phenomenology that seem applicable to the question of a clinical foundation for medical ethics. These are the ideas of the life-world and inter-subjectivity, both of which were best developed by Husserl's pupil Alfred Schutz (Shutz, 1967; Shutz and Luckman, 1973). Schutz defines the life-world as "that province of reality which the wide awake and normal adult simply takes for granted in attitude of common sense. This is the 'natural attitude.'" (Schutz, 1973, pp. 3.) This attitude does not suspend beliefs in the existence of the outer world or its objects. What it brackets "is the doubt whether the world and its objects could be otherwise than just as they appear" (Schutz, 1973, pp. 27).

Natanson labels Schutz' point of view as a "phenomenological orientation" – "a style of thought, a way of attending to phenomena as they are initially entertained by the mind" (Natanson, 1973, p. 114). This he distinguishes from the phenomenological attitude as a more formal feature of Husserl's method (Natanson, 1973, p. 114). This orientation is characterized "by an insistence on a return to what is basic to science, its grounding in a reality taken for granted by the learned and the vulgar" (Natanson, 1973, p. 115).

At the outset, Schutz also takes it for granted that this life-world is not private, that others exist in this life-world, are endowed with similar consciousness and share the same fundamental realities more or less (Schutz, 1973, p. 4). The Life-world is therefore inter-subjective from the outset; it comprises not only the subject's experience but also the social world in which he or she finds herself and which she shares with others. Different persons with different biographies can share a life-world because of the structure of common sense, which allows for reciprocity of perspective and a sharing of the same world (Natanson, 1973, p. 110).

The Life-world to which Schutz' phenomenological orientation can be applied is the Life-world of doctor and patient – their shared experience of illness, healing, being healed and promising to heal – the phenomena of the clinical encounter. The ground for the ethics of medicine can be found, I will argue, in an exploration of this Life-world through a limited epoché which implies acceptance neither of Schutz' total construction of social reality nor Husserl's transcendental Ego.

Attractions of a Phenomenological Orientation

In the current matrix of conflicting visions of bioethics, a phenomenological perspective has several attractions (Pellegrino, 1997, pp. 23). Its emphases on the everyday world, on things in themselves and essences, as revealed in the mundane world of medicine, are particularly apposite. First, phenomenology is primarily a method of philosophizing and not the whole content of a philosophical system applied from without, as is the case with the standard ethical theories today, like Kantian deontology, Millsian utilitarianism, or Ross' *prima facie* principles as elaborated by Beauchamp and Childress (Beauchamp and Childress, 1996). There are many ways of doing phenomenology and one may employ it with a certain "bias," as Erwin Straus did without accepting the whole of Husserl's transcendental reduction or Heidegger's hermeneutic. We can include the anti-foundationalist argument in our epoché of philosophy. An epoché of philosophy would bracket anti-foundationalism just as it brackets other philosophical constructs. Phenomenology is a way in which one may begin to contemplate the life-world physicians and patients occupy unfettered by the apparatus of a preconceived philosophical system.

There is, however, nothing in the method of phenomenology, which is inconsistent with standard philosophical systems, except in so far as those systems oppose the method of phenomenology, for example, scientific positivism which would deny epistemic status to anything but the results of scientific method. In the long run, a phenomenological analysis based in the realities of the clinical encounter should be consistent with the essentialist, teleological, virtue-oriented philosophy and ethics of medicine derivable from more classical sources of philosophical inquiry (Pellegrino and Thomasma, 1981).

IV. STRAUS AND NATANSON: PHILOSOPHY AND PSYCHIATRY

Straus and Natanson have envisioned a way in which phenomenology and psychiatry can illuminate each other (Straus, Natanson and Ey, 1969; Straus, 1969). Though both focus, in two seminal essays presented in 1963, on psychiatry and philosophy, their perspective is relevant to every branch of clinical medicine. This essay takes an approach to the ethics of medicine analogous to Straus' approach to the philosophy of psychiatry.

Straus started with the phenomena of clinical psychiatry, i.e. the psychopathology of mental illness, and sought through the phenomenological analysis to see what questions psychiatry would pose to philosophy. The leitmotiv of his work he defined as discovery of "the unwritten constitution of everyday life" and "the problems hidden in the familiar." In this way he hoped to respond to Husserl's appeal to go "back to the things themselves" (Straus, 1966, pp. xi). He saw psychoses as experiments arranged by nature giving insight not only into the genesis of psychosis but in the structure of the life-world itself (Straus, 1966, p. 257). Psychoses could be defined as disorders of the life-world and this, in turn, told us something about the life-world of the psychotic and the psychiatrist as well.

In his essay entitled "Psychiatry and Philosophy," Straus invoked a philosophical epoché, bracketing all philosophies. Thus starting with the psychiatric situation he let psychiatry "lead us to philosophical questions" (Straus, 1969, pp. 18) and perhaps even some solutions. In a companion piece entitled "Philosophy and Psychiatry," Natanson reverses the order and has philosophy direct its questions to psychiatry, questions like normalcy, communication, etiology and therapy. Natanson depicts the major challenge of phenomenology to psychiatry as the focus on man as a being in the world to be respected on his own terms and not in the terms of a Galilean model of scientific explanation (Natanson, 1969, p 110).

Like Straus, Natanson underscores the utility and the necessity of the phenomenologic epoché, which suspends belief in the reality of the world and thus counters the "natural attitude" of scientific positivism, which would otherwise dominate psychiatry and obscure the "miracle of the ordinary" (Natanson, 1969, pp. 109). In this way, Natanson proposes to "gain access to a knowledge of what is given in experience apart from pre-judgment and epistemic prejudices of naive realism"(Natanson, 1969, pp. 82). With this outlook, Natanson proceeds to define the underlying philosophic problems in psychiatry.

Straus and Natanson see a reciprocal relationship between psychiatry and philosophy, each with its own methodology contributing to and challenging the other. The adherence of each to Husserl's terminology and especially acceptance of his transcendental idealism, ontology, and anthropology differs in degree. This is true of most of the practitioners of phenomenological method. There are also differences between them in the way the phenomenological method is used. It is necessary, therefore, to specify what element of the more general notion of the phenomenology I wish to employ in this examination of the realities of the clinical encounter in order to uncover the essence of a philosophy and ethics of medicine.

For purposes of this essay, it is important to re-emphasize that I will eschew Husserl's transcendental reduction and the spiritual anthropology and ontology that accompany it (Bidney, 1973, pp. 109-140.). The first three steps of the practice of phenomenology as outlined by Spiegelberg are relevant to an analysis of the philosophy and ethics of medicine in terms of clinical reality. These three statements are: investigating particular phenomena, investigating general essences, and apprehending essential relationships between essences (Spiegelberg, 1960, pp. 659). These three aims can be approached, so far as clinical medicine is concerned, by use of the phenomenological technique of the epoché and by focusing on intersubjectivity and the Life-world of everyday clinical practice. One must, in this, remain cognizant of the difficulties of the phenomenological mode of philosophizing (Natanson, 1973, pp. 11-14)

The aim is "to practice epoché, to attempt descriptions of presentations without prejudicing the results by taking for granted the history, causality, intersubjectivity, and value we ordinarily associate with our experience and to examine with absolute care the fabric of the world of daily life so that we may grasp its source and direction" (Natanson and Pehno, p.8). The aim is to gain insight into essences through the experiences of their exemplifying particulars. Such essences are akin to, but not identical with, the notion of essence in Scholastic philosophy (Bidney, pp, 117-118). They bear a relationship to essence as seen by Zubiri and Scheler. Their notions of essence are different, but both bear the marks of objectivity, in opposition to more subjective transcendental concepts of truth (Scheler, 1973, pp. 394-395; Zubiri, 1980, pp. 283-285).

In its broadest sense, this epoché involves suspension of belief about existence of the natural world, or of any other presupposition (philosophical and moral) about the essence and meaning of things in the everyday life of doctor and patient. By the act of prescinding from preconceived notions of essence, it is hoped the true essence of things can be apprehended. The crucial epoché for such a project would be an epoché of philosophy such as Straus performed for psychiatry (Straus, 1969, pp. 18). Specifically in examining the clinical encounter, such an epoché would involve prescinding from all theories of medicine and theories of medical ethics.

A phenomenological ethics should be an ethics based in experience. It is not an ethics justified, however, by practice per se. It begins with experiences within the perimeter of reality of the clinical encounter, i.e., the Life-world of doctor and patient. Only when we have the "meaning" of that encounter can we proceed to the ethics embedded in that meaning. The meaning is not a raw datum but our human perception of that datum, its meaning in our Life-world. However, it is not merely a subjectivist meaning but our interpretation of a shared reality of the world which doctor and patient experience in the clinical encounter.

A phenomenological ethics in general searches for the "oughtness" of human intentional actions in the meanings of the experiences peculiar to the activity in question – law, medicine, ministry, citizenship, etc. At the outset, one prescinds even from the realm of things that ought to be done and things that ought never be done. But ethics itself remains the science of oughtness. Without oughtness, there is no ethic. If the realm of oughtness is denied as a universal human experience then ethics collapses into solipsism. This surely is the case for dialogue with a moral person, whose psychopathology is characterized by absolute moral solipsism. The concept of oughtness in relation to others is simply not part of the life-world of the sociopath. For the sociopath, oughtness means only what others owe him; there is no mutuality of obligation for the amoral person.

For those who admit the existence of moral obligations as part of the life-world, however, the search for their basis in reality, in things and events in themselves is valid. In the case of medical ethics, these events are the events of medicine, those activities that characterize medicine, and those activities that set it apart from other activities like law, ministry or teaching. Each profession is set apart from the other by a specific arena of human activity with its own telos within which it properly functions. Each has its own shared life-world, e.g. lawyer-client, minister-parishioner, teacher-student, psychologist-person counseled. These, like the physician-patient encounter, are intersubjectively experienced life-worlds – each with its own unique phenomenon. In a phenomenology of medical ethics the method of phenomenology – especially the methods of epoché and reduction – are applied to the everyday experiences of medicine. The arena of experience most characteristic of medicine is the healing encounter, the clinical reality of physician-patient, and the nurse-patient relationship. Medical knowledge may, of course, be used beyond the clinical dyad, for example, in preventive and social medicine. These uses can be set-aside for the time being. On another occasion, possible parallels can be examined in preventive medicine and public health by the same phenomenological method.

Erwin Straus, in his incisive essay on the philosophy of psychiatry, set out to find the questions for philosophy in psychiatry. Similarly, we can examine the questions of philosophy as they pertain to medicine in the thing-in-itself, i.e. in the existential and experiential phenomena of the everyday world of the clinical encounter. A reflection on the phenomena of medicine should help to discern what medicine itself is, what its proper task (telos) should be, and what this entails for the ethics of physicians or nurses. If there is to be a philosophical medical ethics there must first be a philosophy of medicine which itself might be derived from a phenomenological analysis of the events of medicine.

This approach then is neither anti-metaphysical nor anti-epistemological nor anti-empirical. Rather, it grounds the ontology of the clinical encounter in the shared experience of that encounter. Its epistemology is an epistemology of the real world apprehended through reflection on the mundane world of the clinical encounter.

V. THE LIFE-WORLD OF THE CLINICAL ENCOUNTER

In the Life-world of doctor and patient, two persons encounter each other in a specific way distinct in part, though not wholly, from other types of personal relationship. The patient comes to the physician because he feels "sick." He has detected some aberration in his body or psychic functioning that he considers "abnormal", i.e., a deviation from his usual or expected functioning. Whatever that aberration may be – some symptom, some sign, some affective state – the person's perception of health, well-being or normality is put into question. The patient may for a time minimize, deny, or try to cope with, or treat, this new predicament but if it persists, worsens or causes anxiety, that predicament effects a change in the existential and experiential state of the sick person.

From a state of well-being, the person enters a state of illness whether demonstrable disease is present or not. To be "sick" is literally a statement of *dis-ease* – a loss of well-being characterized by a constellation of changes in Life-world and lived body. Anxiety about the meaning of the sign or symptom takes the place of "ease." What does this encounter mean? Is it serious? Will it mean death, disability or inability to do what I want to do? Immediate and future plans are put on hold, or if pursued, they are approached warily and fearfully, with uncertainty that they will be fulfilled as anticipated. Even minor aberrations in function will threaten some degree of loss of freedom to do what one wants to do on one's own terms. Even minor illness, as experienced clinicians know, can for the patient be an encounter with finitude. Sickness puts the whole fabric of the sick person's life-world at risk. How the patient's life-world will be reassembled is an open question neither patient nor physician can fully answer in the first stages of their encounter.

As the perceptions of sickness expand and strengthen there arises a loss of existential freedom with respect to the lived body. No longer is it the "quiet handmaiden of the soul." It becomes a potential or actual enemy of the self, shattering the unity of the awareness of self. Sickness verges on an ontological assault on the self-body unity. We speak of the "war" against cancer, the "struggle" against a disease or pain or suffering. The self wages this war. Alienation of the body from the self, a true ontological disassociation occurs. The person altered by illness asks if he is the same person who became ill. He does not know if he will ever be again that person.

As sickness grows, it begins to occupy the center of the stage of the patient's life-world, to make demands, and to discolor all of life's plans and actions. The range of future choices and activities is limited until the direction of the sickness becomes manifest. At some point, the sick person is forced to seek out the doctor, to submit his or her "complaint" to the doctor's consideration and judgment. At this juncture the sick person becomes a "patient" – someone bearing a burden, someone suffering (*patior*) as the Latin etymology of the word suggests. At this point, freedom is further compromised. The desire to cope on one's own terms is eroded or even seen as impossible. If one wants to be healed, freedom is necessarily curtailed to some degree. By "going to the doctor" the patient is now dependent on the doctor's authoritative knowledge and his perception of the patient's experience of illness.

It is precisely when the sick person seeks out the physician that his life-world as afflicted person encounters the life-world of the un-afflicted doctor. Indeed, the patient invites the physician to enter his life-world, just as the physician offers to enter that world by his offer to heal. The life-worlds of doctor and patient inter-penetrate each other. Their relationship is conditioned henceforth by intersubjectivity. This is the moment of clinical encounter, the confrontation of one human person bearing the burden of illness seeking to be healed by another human person possessed of the knowledge and skill needed by the sick person in search of healing. This is an encounter between the life-worlds of two persons who up to the moment of confrontation were strangers independent of each other. Now, they are intimately enmeshed with each other in the most intimate and special sorts of ways, some of which may make for a healing relationship and some for a harming one.

The life-world of the patient is sharply altered by virtue of this transition from being well, to being sick and then being sick enough to seek healing from a professed healer. A new state of dependency is introduced, i.e., dependency on the knowledge and skill of the professed healer, the doctor or the nurse. The patient is at the mercy of the characters who profess being healers. Patients are dependent too on the institutions, the personnel, and the facilities doctors marshal to relieve the sick person of his burdens. This unavoidable dependency generates a state of serious vulnerability and exploitability, which the responsive healer can recognize and respect, and the callous healer can use to further his own interests at the expense of the patient's.

For brevity's sake, I have chosen just a few of the more common perceptions of the experience of sickness in a person undergoing the existential transformation of his life-world from that of wellness to sickness to patient-hood. This is an obvious over-simplification. How each of these perceptions is manifest in any particular person is a reflection of the uniqueness of that person. The task of the phenomenological method is to locate this person and this doctor in the full texture of the complexity of their respective and shared life-worlds. It is within this intersection of life-worlds that the whole specialized activity of medicine is located.

Yet this locating can never be complete. The patient's life-world is lived in intersection with the life-worlds of all who live in relationship with him. Family, friends, other clinicians also become enmeshed in the patient's world via the reality of sickness and disease. The meaning of illness to each person therefore will differ as it is interpreted in the infinity of variations possible in individual life-worlds. No two persons experience the predicament of illness exactly the same way. No life-world is exactly congruent with any other.

Nevertheless, there is a sense in which the alterations in life-worlds can be generalized, as a universal human experience. For even as there is uniqueness there is also commonality, since the life-world is a human creation and certain responses in that world are typified as "human" responses. As Straus pointed out, Canaletto and Guardi both painted the same Venice (Straus, 1969, pp. 29). Though they saw it with different eyes, one can still identify the palaces and churches of Venice. Of course, with an abstractionist we might not be able to do that. But even then the abstract painting is a human creation sharing some things with the other human creations we call paintings and with abstractions as experienced by others. Uniqueness lies in the way particulars are construed but not in the ontological reality of those particulars or the typology of their impact on our life-worlds.

When we turn to the life-world of the physician, we begin with a very different perspective and horizon. The physician is well and therefore occupies a radically different world from the patient. Even if he is ill himself, he prescinds, at least in part, from that illness when he acts a healer. He retains his freedom to act, limited by the science and ethics of medicine but not by his own disability. The disabled physician remains able to heal – sometimes even more sensitively than the non-handicapped. If he were disabled or ill to the point that he could not exercise his healing functions the physician would be disqualified as a healer. The physician's conduct is held within bounds by an ethical commitment that requires a certain suppression even of his legitimate self-interest.

The physicians' life-world is constituted by all those things that derive from his private and personal life as they are transmuted by being a physician, as one committed to healing the sick. For every physician there is a unique mix of personal and cultural realities with the ethical, social and cultural meaning and requirements of being a physician. How each physician balances these personal, professional and psycho-social modalities is just as unique as the patient's balance. The physician differs from the patient in that the balance he has struck is not at issue as it is with the patient. In their shared world, he is not in the vulnerable, exploitable, dependent, relatively powerless state of the patient. He is the professional freely offering to "help" the patient. It is true that he may be a "wounded healer," but he is a healer nonetheless – even when his own vulnerability is part of his life-world.

When the physician encounters the patient and asks: "What can I do for you?," "How are you?" and "What is wrong?," he invites himself into and enters the life-world of the patient and begins to share his own life-world with the patient. The doctor, in asking about the patient's problems, is offering himself as a healer. Otherwise, he would have no legitimate access to the intimate knowledge he requires to effect his healing. In asking he is saying: "I am competent. I have the knowledge and skill you require." He is also promising to act in such a way as to benefit the patient and not to harm him. This is his existential act of "*pro-fession*," his declaration of his willingness to heal and to do so in the patient's interests, not his own, not the family's, not society's, or the managed care organization's.

In making this profession, the physician also invites *trust* – in his knowledge, his competence and his character. The physician invites this trust and makes his promise of competence in the presence of another human person who is in the altered existential state we call sickness, with all the alternations of life-world I listed above. The patient is not just playing a "sick role," he is now fashioning that role, creating it as he re-creates his whole life-world. That person is vulnerable, dependent, anxious and eminently exploitable should the physician be a vicious and not a virtuous person. The patient, for his part, is forced to trust the physician even if he wishes not to do so. At least he must submit even if he does not trust – if he wishes to be healed by *this* doctor. Yet, he may not have chosen *this* doctor. Or, having chosen him, he still does not know exactly how he will act in *this* illness now, or in its uncertain and threatening future.

The doctor and patient's life-worlds intersect in countless, complex ways, each of which may affect the success of the healing relationship. In an effective healing relationship these two different life-worlds must somehow interact positively around the common intention of healing which is the end or telos of their mutual clinical encounter. I cannot in the space of this essay examine all those ways of intersection, nor explore the full implications of the inter-subjectivity binding or separating these two life-worlds. I can only briefly note the shared intentions and expectations inherent in the clinical encounter.

Both physician and patient expect healing to be the intentional focus of their encounter. Each share intentionality with reference to healing – the telos of the encounter. For both this means making the patient "whole" again, repairing the damage to bodily or mental integrity, restoring the state of well-being or, if this is impossible, ameliorating the impact of sickness and disease. This healing is the "act" of medicine in which patient and physician come together. In a sense they heal together since the patient's cooperation with a technically right and morally good clinical decision is essential. What is technically right derives from whatever there is of science and manual skill in medicine. What is morally good is a far more complex aim. It is the intersubjective apprehension of what is healing and what is harming to *this* patient that is crucial (Pellegrino and Thomasma, 1987).

In all of this, several things emerge as common to the clinical encounter – the existential state of the sick patient, the act of promise and invitation of trust by the physician, and the expectations generated by the physician when he makes his promise of help. All of this is transacted in the presence of the patient's vulnerability. These insights emerge as "essential" in our philosophical epoché and our look at the intersection of life-worlds of doctor and patient as realities internal to medicine and healing. They are also the entry points to a philosophy and ethics of medicine.

If this analysis of the clinical encounter from a phenomenological orientation has uncovered the essence of medicine, that which sets it apart as a human activity and experience, then it should help to structure a philosophy of medicine. Then, knowing what medicine is, some statement can be made about the moral obligations of those who profess to practice it. It is here that the phenomenological orientation converges with a more classical mode of reasoning which defines medicine by its telos (Pellegrino, 1998). By using the device of the philosophical epoché – bracketing ethical theories at the outset – we arrive at an essence not too far from a more classically derived essence.

The telos of medicine is healing, and medicine qua medicine is that set of human activities that has as its end and purpose – both for doctor and patient – the act of healing, of "making whole again." In the clinical encounter, the telos is a right and good healing action for *this* patient. As we have shown elsewhere the virtues, duties and axioms of medicinal ethics are derived from this telos (Pellegrino and Thomasma, 1993). The virtues of medicine are those traits of character that predispose the physician habitually to act in a way that effects healing. But these virtues cannot be defined without some genuine grasp of the life-world of the patient by the doctor, and of the doctor by the patient.

Thus, fidelity to trust is an indispensable virtue of the physician because trust is an ineradicable reality in the life-world of both patient and physician. Trust is crucial to the clinical encounter, since the patient is forced to trust the doctor who promises to be trusted. The ends of medicine cannot be attained without trust. The patient entrusts the intimate details of his life-world and an essential part of its reassembly to the doctor. For the same reasons trust is a virtue of the patient, entailed by the phenomena of the clinical encounter. Without being able to trust the patient, the doctor cannot fulfill his promise to help. The doctor needs privileged access to the life-world of the patient, a truthful rendition of the impact of sickness, if he is to heal.

In the same way, intellectual honesty, truthfulness, courage, suppression of self-interest, and compassion can be shown to be virtues entailed by the telos of medicine as well as by the interplay of life-worlds of doctor and patient. These virtues are always understood implicitly and intersubjectively. Just how they are understood in a particular physician or patient encounter lies in the intersubjective world – patient and doctor enter in the intimacy of their unique clinical encounter.

Thus, through the telos of medicine – the defining point of a philosophy of medicine – and the virtues entailed by that telos, a philosophy of medicine and an ethic of medicine are joined. The task of a phenomenological inquiry is to perform the philosophical and ethical epoché, to suspend judgment in pre-existing theories of medicine and pre-existing theories of ethics and medical ethics so that the "essence" of the central phenomenon of the clinical encounter can be revealed and apprehended. Once apprehended, a philosophy and ethic of medicine can be grounded "internally" – i.e., in recognition of the nature of medicine as it is revealed in its essential form in the phenomena of the clinical encounter. As suggested elsewhere, we can then also ground the *prima facie* four principles of beneficence, justice, non-maleficence and autonomy in clinical realities, as well as the duties and obligations of both doctor and patient (Pellegrino, 1994, pp. 353-366).

VI. MEDICINES' CONTRIBUTION TO APPLIED PHENOMENOLOGY

The main focus of this essay has been to illustrate how a phenomenological perspective, or more precisely, how the method of phenomenology can illuminate the complex structures of the physician patient relationship. Medicine, for its part, however, can also contribute to phenomenology because medicine is deeply concerned with such complex problems as health and illness, embodiment, anxiety, personhood, the self, healing, suffering, finitude and dying, hope, etc. They are all parts of the life world of patients. They are subjects that have engaged the existentialist philosophers, who use the phenomenological method, and the phenomenologists who take the transcendental turn with the later Husserl as well.

Medicine per force seeks, as did the early Husserl, to get at the things themselves as they are presented to us in our experiences. One may think of a phenomenological realism (Seifert, 1987; Von Hildebrand, 1991) in which medicine and phenomenology study the world and man from a mutually productive stance.

Perhaps the clearest exposition of the possibility is the description the phenomenologist Richard Zaner has provided from his personal experience. Zaner became engaged with medicine when he joined a medical school faculty in the late sixties of the last century. Teaching medical students, making rounds, conversing daily with medical faculty enriched his own project as philosopher and phenomenologists (Zaner, 1993).

Erwin Straus, a physician and psychiatrist, was one of the leaders of phenomenology in the United States. His work with Natanson and Ey is an excellent example of the way medicine can help to illuminate the life world of the psychologically disturbed patient (Straus, Natanson, and Ey, 1969). It is through their mutual interest in different aspects of the "life-world" that medicine can facilitate the cogitations of the phenomenologists.

Natanson defines the life world according to Husserl, as "the realm of experiential immediacy from which all interpretation derives..." (Natanson, 1973, pp. 125). Or, "...the sum of man's involvement in everyday affairs: his knowledge, interpretation, response and organization of his experience" (Natanson, 1973, pp.127). Even more directly, Straus establishes the point of intersection of medicine and phenomenology. To the psychiatrist he says, "the wards are the natural laboratories where we begin to wonder about the structure of the Lebenswelt" (Straus, 1966, pp. 256).

The phenomena of illness, health, vulnerability, healing, etc., as they exist in the life world of the patient, become the natural laboratory for reflecting on the life world of those who are experientially patients. If phenomenology is to be applied in the sense Alfred Schutz used it (Schutz, 1967, 1973), it would seek to understand the structures of the everyday realities in terms of the phenomenological attitude. Schutz did for the social sciences what Straus did for psychiatry. One can envision an applied phenomenology of medicine in which medicine provides the point of departure for the phenomenological reduction.

Perhaps the relationship between medicine and phenomenology was best epitomized by Erwin Straus when he said, "the quest for a foundation for psychiatry remains unfinished. I believe that it will ultimately be accomplished with the help of phenomenology, applying it in spirit rather than in letter" (Straus, 1964, p.8). The same may be said of the foundation for a philosophy of medicine and the ethics of medicine more generally.

The interchange between medicine and philosophy is most fruitful when each retains its own perspective in the clinical encounter. This should not eradicate the distinction between the natural and the phenomenological perspectives. But neither should they be hermetically sealed off from each other. Medicine cannot solve the problems of philosophy. Philosophy cannot solve the problems of medicine. Medicine can contribute to philosophy by presenting new problems, questions and dilemmas for phenomenological analysis (Hildenbrand, 1991, pp. 211-214). This is a promising region for independent and cooperative research between the disciplines.

VII. SUMMARY

The phenomenological orientation I have taken differs from the usual derivations of the ethics of medicine from the applications of existing moral philosophies to medical practice. It departs from the many current alternative ethical theories like narrative, casuistry, caring and principlism. Each of these, however, grasps some aspect of the life-world of doctor and patient. None is by itself sufficient as a philosophy or ethic of the clinical encounter. By bracketing these theories at the outset, the phenomenological orientation offers an alternative which may come closer to the thing in itself; namely, what medicine is as a human experience and a moral enterprise.

Center for Clinical Bioethics
Georgetown University Medical Center
Washington, DC, U.S.A.

REFERENCES

Audi, R. (ed..): 1995, *Cambridge Dictionary of Philosophy*, Cambridge University Press, Cambridge.

Beauchamp, T. and J. Childress: 1996, *Principles of Bioethics*, 4th Edition, Oxford University Press, New York.

Bidney, D.: 1973, 'Phenomenological Method and the Anthropological Sciences,' in M. Natanson (ed.), *Phenomenology and the Social Sciences*, Northwestern University Press, Evanstan, IL, pp. 117-118.

Edward, P. (ed.): 1967, *The Encyclopedia of Philosophy*, McMillan, New York, pp. 135.

Dell'Oro, R. and C. Viafora: 1996, *Bioethics, A History*, International Scholars Publications, San Francisco.

Heidegger, M.: 1962, *Being and Time*, J. Macquarrie and E. Robinson (trans.), Harper and Row, New York.

Heidegger, M.: 1970, *The Idea of Phenomenology*, J.N. Deely, J.A. Novak and E.D. Leo (trans.), New Scholasticism, XLIV, 325-344.

Honderich, T. (ed.): 1995, *The Oxford Companion to Philosophy*, Oxford University Press, New York.

Husserl, E.: 1960, *Cartesian Meditations: An Introduction to Phenomenology*, D. Cairns (trans.), The Hague, M. Nijhoff.

Husserl, E.: 1962, *Ideas: General Introduction to Pure Phenomenology*, Collier Books, New York. pp. 100.

Husserl, E.: 1965, *Phenomenology and the Crisis of Philosophy*, Q. Laver (trans. and intro.), Harper Torch Books, Harper and Row, New York.

Marcel, G.: 1960, *The Mystery of Being*, Gateway Edition, Henry Regenery Co., Chicago.

Merleau-Ponty, M.: 1962, *Phenomenology of Perception*, Colin C. Smith, Humanities Press, New York.

Molina, F.: 1962, *Existentialism as Philosophy*, Prentice-Hall, Englewood Cliffs.

Natanson, M.: 1969, 'Philosophy and Psychiatry', in E. Straus, M. Natanson and H. Ey (eds.), *Psychiatry and Philosophy*, Springer Verlag, New York, pp. 85-110.

Natanson, M.: 1972, *Edmund Husserl, Philosopher of Infinite Tasks*, Northwestern University Press, Evanston.

Natanson, M.: 1973 ' Phenomenology and the Social Sciences,' in M. Natanson (ed), *Phenomenology and the Social Sciences*, Vol. **1**, Northwestern University Press, Evanston, IL, pp. 11-14.

Pellegrino, E.D.: 1979, 'Toward a Reconstruction of Medical Morality: The Primacy of the Act of Profession and the Fact of Illness,' *Journal of Medicine and Philosophy* 4(1), 32-56.
Pellegrino, E.D. and D. C. Thomasma: 1981, *A Philosophical Basis of Medical Practice: Toward a Philosophy and Ethic of the Healing Professions*, Oxford University Press, New York.
Pellegrino, E.D. and D. C. Thomasma: 1987, *For the Patient's Good: The Restoration of Beneficence in Health Care*, Oxford University Press, Oxford.
Pellegrino, E.D.: 1993, 'The Metamorphosis of Medical Ethics: A 30-Year Retrospective,' *Journal of the American Medical Associationi*, 269(9), 1158-1163.
Pellegrino, E.D. and D. C. Thomasma: 1993, *The Virtues in Medical Practice*, Oxford University Press, Oxford.
Pellegrino, E.D.: 1994, 'The Four Principles and the Doctor-Patient Relationship: The Need for a Better Linkage,' in Raanon Gillon (ed.), *Health Care Ethics*, John Wiley, New York, pp. 353-366.
Pellegrino, E.D.: 1997, 'Praxis as a Keystone for the Philosophy and Professional Ethics of Medicine: The Need for an Archsupport: Commentary on Toulmin and Wartofsky,' in R. Carson and C. Burns (eds.), *A Twenty Year Perspective and Critical Appraisal*, Kluwer Academic Publishers, Dordrecht, pp. 69-83.
Pellegrino, E.D.: 1998, 'What the Philosophy of Medicine Is', *Theoretical Medicine and Bioethics*, 19, (6), 315-336.
Ricoeur, P.: 1996, 'Le Trois Niveaux du Judgement Medicale,' *Esprit* 12, 21-33.
Sartre, J.P.: 1956, *Being and Nothingness: An Essay of Phenomenological Ontology*, Philosophical Library, New York.
Scheler, M.: 1973, *Formalism in Ethics and Non-Formal Ethic of Values: A New Attempt Toward the Foundation of Formalism,* M.S. Frings and R. Frank (trans.), Northwestern University Press, Evanston, pp. 394-395.
Schutz, A.: 1967, *The Phenomenology of the Social World*, with an Introduction by , G. Walsh and F. Lehnert (trans.), Northwestern University Press, Evanston.
Schutz, A. and T. Luckman: 1973, *The Structures of the Life-world*, R.M. Zaner and H. T. Engelhardt, Jr. (eds.), Northwestern University Press, Evanston.
Seifert, J.: 1973, 'Introductory Essay,' in D.Von Hildebrand, *What is Philosophy?*, Routledge, London and New York, pp. xlvi-xlvii.
Sokolowski, R.: 1989, 'The Art and Science of Medicine,' in E.D. Pellegrino, J.P. Langan, and J.C. Harvey (eds.), *Catholic Perspectives on Medical Morals: Foundational Issues*, Kluwer Academic Publishers, Dordrecht, pp.263-275.
Spiegelberg, H.: 1960, *The Phenomenological Movement: A Historical Introduction*, Martinus Nijhof, The Hague, Vol. I, pp. 659.
Spicker, S. (ed.): 1970, *The Philosophy of the Body, Rejection of Cartesian Dualism,* Quadrangle Books, Chicago.
Straus, E.: 1964, 'Opening Remarks,' in E. Straus (ed.), *Phenomenology: Pure and Applied, The First Lexington Conference*, Duquesne University Press, Pittsburgh, pp. 8.
Straus, E.: 1966, *Phenomenological Psychology, Selected Papers of Erwin Straus*, Basic Books, New York.
Straus, E.: 1969, 'Psychiatry and Philosophy,' in E. Straus, M. Natanson and H. Ey (eds.), *Psychiatry and Philosophy*, Springer Verlag, New York, pp. 1-83.
Straus, E., M. Natanson, and H. Ey (eds.): 1969, *Psychiatry and Philosophy*, Springer Verlag, New York.
Straus, E.: 1976, 'The Existential Approach to Psychiatry,' in J. Smith (ed.), *Psychiatry and the Humanities*, Yale University Press, New Haven.
Toulmin, S.: 1997, 'The Primacy of Practice: Medicine and Post-Modernism in Philosophy of Medicine and Bioethics,' in R. Carson and C. Burns (eds.), *A Twenty Year Perspective and Critical Appraisal*, Kluwer Academic Publishers, Dordrecht, pp. 41-53.

Veatch, R. and C. Mason: 1987, 'Hippocratic Vs. Judaeo-Christian Ethics: Principles in Conflict,' *Journal of Religious Ethics*, 15:86-105.

Wartofsky, M.: 1997, 'What Can the Epistemologists Learn from the Endocrinologists? Or, Is the Philosophy of Medicine Based on a Mistake?' in R. Carson and C. Burns (eds.), *A Twenty Year Perspective and Critical Appraisal*, Kluwer Academic Publishers, Dordrecht, pp. 55-68.

Wojtyla, K.: 1979, *The Acting Person: Translated from the Polish by Andrze J. Potocki*, Reidel, The Netherlands.

Zaner, R. M.: 1964, *The Problem of Embodiment: Some Contribution to a Phenomenology of the Body*, The Hague, Martinus Nijhoff Phenomenologica 17.

Zaner, R.M. and D. Ihde: 1973, *Pehonemology and Existentialism*, Capricorn Books and G.P. Putnam and Son ress, New York.

Zaner, R.M.: 1988, *Ethics and the Clinical Encounter*, Prentice-Hall, Englewood Cliffs.

Zaner, R.: 1993, *Troubled Voices*, Pilgrim Press, Cleveland.

Zubiri, X.: 1980, *On Essence,* A.R. Caponigri (trans. and intro.) Catholic University Of America Press, Washington, D.C., pp. 283-285.

SECTION III

CORE CONCEPTS IN CLINICAL ETHICS

LENNART NORDENFELT

THE LOGIC OF HEALTH CONCEPTS

I. INTRODUCTION

It is often maintained that health is one of the major goals of medicine or even *the* goal of medicine. This idea has been eloquently formulated by the American philosophers of medicine Edmund Pellegrino and David Thomasma in their book *A Philosophical Basis of Medical Practice* (1981, p. 26):

> Medicine is an activity whose essence lies in the clinical event, which demands that scientific and other knowledge be particularised in the lived reality of a particular human for the purpose of attaining health or curing illness through the direct manipulation of the body and in a value-laden decision matrix.

Although some other goals of medicine exist, such as saving lives and advancing quality of life, health is still taken to be the central goal of medicine and health care in general. However, the formidable task of interpreting the nature of health remains. What more specifically is health? To what more precise goal shall we direct our efforts in medicine and health care?

These questions are not simply academic. They are of great practical and thereby ethical concern. The consequences for health care diverge considerably, not least in economic but also in social and educational terms, if health is understood as people's happiness with life, or their fitness and ability to work, or just the absence of obvious pathology in their bodies and minds. There are adherents of all these ideas in the modern theoretical discussion on health.

One of the major problems in this discussion is to establish the relation between the notion of disease and that of health. Are the two notions directly linked, so that health is the total absence of disease, or is there a much looser connection? Is health something over and above the absence of disease? Is health even compatible with the existence of disease?

We seem to have varying intuitions in this regard. We seem also inclined to interpret health slightly differently in different contexts. In this paper I will attempt to disentangle such issues by presenting, in some detail, two prominent theories of health (a biostatistical theory of health, *BST*, and a holistic theory of health, *HTH*) and try to assess these using two criteria for

G. Khushf (ed.), Handbook of Bioethics, 205–222.

assessment, viz. their usefulness in medical practice and in public health contexts. My general conclusion will be that the holistic theory, *HTH,* is the more plausible theory of health.

II. TWO FUNDAMENTAL APPROACHES TO HEALTH CONCEPTS

Contemporary philosophy of health is very much focused on the problem of determining the nature of the concepts of health, illness and disease from a scientific point of view. Some theorists claim and argue that these concepts are value-free and descriptive in the same sense as the concepts of atom, metal and rain are value-free and descriptive. Moreover, a disease in a human being can be discovered, according to this line of thought, through ordinary inspection and through the use of scientifically validated procedures without invoking any normative evaluations of the person's body or mind. To say that a person has a certain disease or that he or she is unhealthy is thus to objectively describe this person. On the other hand, it certainly does not preclude an additional evaluation of the state of affairs as something undesirable or bad. The basic scientific description and the evaluation are, however, two independent matters, according to this kind of theory. The most famous protagonist of such a theory is Christopher Boorse (1977; 1997). Others who have thought along these lines are J.G. Scadding (1967) and G. Hesslow (1993).

Other philosophers claim that the concept of health, in particular, but also derivatively the other medical concepts, are essentially value-laden. To establish that a person is healthy does not just entail some objective inspection and measurement. It presupposes also an evaluation of the general state of the person. A statement that he or she is healthy does not merely contain a scientific description of certain features of the person's body or mind but entails also a (positive) evaluation of the person's bodily and mental state. There are now many theorists who think along these lines, although the specific analyses made by them can be quite different: G.J. Agich (1983; 19,97), K.W.M. Fulford (1989), D. Seedhouse (1986), I. Pörn (1993) and myself (1995; 18997; 2000)

It is quite difficult to trace all the intellectual threads in this discussion. (One admirable attempt has been made by Boorse (1997) in his latest defence of his own theory of health.) It is particularly difficult to pin down the various senses of "evaluative" or "normative" that the different theorists have in mind in their characterizations. In the analysis put forward in this paper I will therefore propose a different and simplified strategy for the assessment of the relevant theories of health. My proposal for analysis is helped by the fact that there is a structural similarity between the major "descriptive" theory of health (Boorse's) and some of the "normative" theories of health (Seedhouse's, Pörn's and my own). Both kinds of theories entail a relativization of a state of health to some state of affairs which is called a goal (either a biological goal, a life-goal or some chosen goal). This structural similarity makes it comparatively easy to pinpoint both similarities and differences between the theories. It also sets the stage for an exploration into new variants of concepts of health, since it opens the door to a logical space of possible health concepts. Moreover, as mentioned, this structural similarity between the theories enables us to make a preliminary evaluation of them. Such an evaluation will, however, only be initiated in this paper.

My concrete procedure will entail a comparison between a kind of *biostatistical theory of health (BSTe)* and a *holistic theory of health (HTHe)* (The suffix *e* will be explained below). The former is quite obviously inspired by Christopher Boorse (1977) but the theory is in this paper taken to be applicable to a larger domain than Boorse – at least at present (1997) – claims that it covers. [1] The latter can be viewed as a theoretical model which is some kind of common denominator of Seedhouse's (1986), Pörn's (1993) and my own (1995; 2000) models. It can probably also without too much stretch incorporate essential elements in K.W.M. Fulford's theory of health and illness (1989).

In order to make these comparisons and evaluations in a reasonably clear and expedient way I shall make some simplifying assumptions. These mainly concern the basic concepts used in my essay.

III. BASIC CONCEPTS AND ASSUMPTIONS

A clarification of the concepts of health and disease depend on how they relate to several other concepts, including organ, normal function, person, ability, environment, vital goal, and malady. It also depends on certain issues with regard to the logical priority of health versus illness. In this section, I provide some initial definitions, so that the use of these concepts in later discussion of health concepts will be clear.

The concept of an organ. Boorse's initial version of the *BST* (1977) talks about normal functions of bodily parts. These parts can be constituted by anything from cells or even parts of cells to organs (in the ordinary sense of the word, such as the heart, the lungs and the liver) and in the extreme case to the body as a whole. For the sake of simplification I will here use the term "organ" to refer to any level of bodily organization and any partition of the body. "Organ" will here also cover the concept of mental faculty. Both the

BST and the various versions of *HTH* claim to be theories of health, illness and disease in general, and not just theories of either somatic or mental health.

The concept of a normal function. An organ is assumed to have a function. I will say that an organ O has a function F(g), if and only if, O is directed towards a goal G. For the purpose of clarifying the *BST* I will need the more complicated characterization: Organ O of a particular individual A has a function F(g) if, and only if, there is a state of affairs G and A belongs to a species in the majority of whose members O is directed to G. In the *BST* G is taken to be the survival of the individual or the species. I will say that organ O functions normally in A if, and only if, it makes its species-typical causal contribution to the survival of A. (For very specific functions the goal can also be the survival of the species to which A belongs.) The species-typical contribution is to be assessed statistically with the whole species as the reference class.

The concept of a person is a concept to be distinguished from the concept of a body. A person is an autonomous agent of intentional actions. A person moreover is the bearer of mental attributes such as emotions and moods.

The concept of ability. A person has the ability to perform intentional actions. Ability is a three-place relation. An ability always has a *goal.* When A is able, he or she is able to do *something* or reach *some* state of affairs. This ability can be of a *first-order* kind, i.e. actual ability. A has the ability to do F, if and only if, if A tries to do F, then A does F. The ability can also be of a *second-order* kind, i.e. potential ability. A has the second-order ability to do F, if and only if, if A tries to learn to do F, then A achieves the first-order ability to do F. A third concept to be recognized is that of *capacity,* which is a potential ability of a more indeterminate kind. A has the capacity to do F , if and only if, there is a possibility that A will get the second- or first-order ability to do F. Children have the capacity to do many things which they have neither the second- nor the first-order ability to do.

In other works, I have argued that the ability involved in the concept of health is of a second-order kind (Nordenfelt 1995; 1997). This will not be presupposed here. I will leave the nature of the ability undetermined in the present discussion.

The concepts of environment and circumstance. Organic processes, as well as human actions, always occur in a context. An organic process occurs in an inner milieu as well as in an external environment. The inner milieu and the external environment set the limits of the process and influence the process. An action is always performed in a set of circumstances which constitute the opportunities for the action. This also then holds for abilities. When A is able to perform an action F, then this holds relative to some set of circumstances C. I will use the term "environment" with regard to bodily processes and "circumstance" with regard to abilities. Apart from this convention there is no difference of meaning entailed. A particular feature in the case of ability, though, is that a person's *perception* of circumstances can play a role for the execution of ability. This distinguishes ability of action from functional ability.

A statistically normal environment and standard circumstances. There are different ways of specifying the environment and the circumstances which are presupposed as surrounding an organ or a person. The various versions of the *BST* presuppose that the environment is statistically normal. The *HTH*s, on the contrary, presuppose a set of circumstances that is culturally defined. Standard circumstances are the circumstances that are implicitly presupposed in a particular cultural context when a person's ability is assessed. Such circumstances are not determined with the help of a rigorous statistical method.

The concept of a vital goal. Most versions of the *HTH* try to give a criterion for the healthy person's ability. This is mostly done in terms of a set of actions that the healthy person should be able to perform or a goal that he or she should be able to achieve. I will here use the technical term "vital goal" to signify such a goal. In my own version of the *HTH* I use the concept of a vital goal as a key concept. It has in my theory a precise interpretation. A vital goal of A's is a state of affairs which is a necessary condition for A's minimal happiness in the long run. Other holistic theories specify the goal implied in the concept of health in slightly different ways. Fulford talks about performing "ordinary" doings, i.e. reaching "ordinary" goals, and Seedhouse talks about a person's realistic chosen and biological potential. For reasons of simplicity I will leave the concept of vital goal open here.

The concept of a malady. It has been noted several times in the philosophy of medicine that we need a generic concept of "pathology" covering not only diseases, but also injuries, impairments and defects. Culver and Gert (1982) have suggested the general term "malady" as denoting all these negative medical states. I will adopt this terminology here and use the term "malady" when no more specific reference is needed.

The primacy of the concept of health. It is often presupposed that one of the differences between the *BST*s and the *HTH*s is that in the *BST*s disease is the logically prior concept, whereas in the *HTH*s health is the logically prior concept. This is understandable given the way the theories have normally been formulated. In Boorse's version of the *BST* the notion of disease is first defined and health is said to be *the absence of diseases*. In the various versions of the *HTH,* on the other hand, health is first defined and diseases are viewed as states which reduce health. This difference is crucial for some analyses. It can, however, be disregarded here, since it is possible to define a theory equivalent to the *BST*, where health is the concept first defined, and disease is defined derivatively. This will be the procedure here, where the main purpose is a comparison between the two kinds of theory. Thus health will be the primary object of study through the whole of my analysis.

The concepts of ill health and illness. I choose to use the term "ill health" for the contradictory of health. Sometimes "illness" is used for this purpose. This has a disadvantage, since "illness" is often used as denoting specific types of ill health or even specific diseases.

IV. THE BASIC VERSIONS OF THE BST AND HTH

The characterization of health given in Boorse's *Biostatistical Theory of Health (BST)* (1977; 1997) is the following:

I. A is completely healthy, if and only if, all organs of A function normally, i.e. if they, given a statistically normal environment, make at least their statistically normal contribution to the survival of A or to the survival of the species to which A belongs.

The *BST* admits talk of the health of the organs themselves:

Ia. An organ of A is healthy, if and only if, it functions normally, given a statistically normal environment.

The concept of *malady* in *BST* (1997) is given in the following:

Ib. A has a malady, if and only if, there is at least one organ of A which functions subnormally, given a statistically normal environment. The malady is identical with the subnormal functioning of the organ. (Observe that the term "malady" is not used by Boorse. His term "disease" is, however, intended to cover the range defined above with regard to the concept of malady.)

The characterization of health given in my version of the Holistic Theory of Health (HTH) (1995) is the following:

II. A is completely healthy, if and only if A has the second order ability, given standard circumstances, to reach all his or her vital goals. [2]

My version of *HTH* does not allow talk of an organ's being healthy, except in a derivative sense. Health is a concept which basically pertains to the whole person. The concept of malady in the *HTH* comes out in the following way:

IIa. A has a malady, if and only if , A has at least one organ which is involved in such a state or process as tends to reduce the health of A. The malady is identical with the state or process itself. [3]

The phrase "tends to reduce the health of A" is selected because not all maladies actually compromise health in the holistic sense of being able to realize vital goals. Some maladies are aborted, i.e. disappear before they have influenced the person as a whole; others are latent; yet others are so trivial that they are never recognized by their bearer.

The presented versions of the *BST* and the *HTH* are clearly quite different theories of health. The differences can be summarized thus:

1. In the *BST* health is a function of internal processes in the human body or mind. In the *HTH* health is a function of a person's abilities to perform intentional actions and achieve goals.

2. In the *BST* health is a concept to be defined solely in biological and statistical terms. In the *HTH* the concept of health presupposes extrabiological concepts such as "person," "intentional action," and "cultural standard."

3. In the *BST* health is identical with the absence of maladies. In the *HTH* health is compatible with the presence of maladies. The concept of malady is, however, logically related to the concept of ill health also according to the *HTH*. A malady is defined as a state or process which tends to reduce its bearer's health.

There are, however, also important structural similarities between the *BST* and the *HTH*. These similarities have come out in my presentation. Health in both models has to do with whether a whole person or some of his or her bodily or mental parts are able to achieve certain goals. I will in the following explore this structural similarity in the attempt to visualize alternative conceptions of health.

V. GENERALIZING THE MODELS OF HEALTH

In principle the biostatistical foundation of *BST* could be extended to cover the domain of human abilities in the following way:

III. *The expanded version of the BST: BSTe.* A is completely healthy, if and only if, all organs of A function normally, and A has the ability to perform all actions which are statistically normal, given a statistically normal environment. (The definition of disease is identical with the one given in Ib.) [4] The difference between the expanded version and the original one is that the expanded version also covers the field of human actions.

Conversely, the *HTH* can be extended to cover the biological domain:

IV. *The expanded version of the HTH: HTHe.* A is completely healthy, if and only if, the organic structure of A is such that it enables A to achieve all his or her vital goals, given standard circumstances. [5]

The definition of disease is identical with the one given in IIa. Note however the difference which I suggest between III and IV as to organic functioning. In IV it is not important that all organs function in a satisfactory way. What is of importance is that the sum total of organic functioning is such that A is enabled to achieve all his or her goals. Thus health according to the *HTHe* is, of course, compatible with the existence of maladies, not only in the *BSTe* sense, which we established before, but also in the *HTHe* sense. An organ of A's can be functioning in such a way that it tends to reduce A's health (i.e. there can be a malady in the *HTHe* sense) without A's total organic functioning being reduced. (For instance, the disease can be trivial or the organ in question can be compensated by some other organ.) [6]

By this expansion we can observe the two features which fundamentally distinguish the *BSTe* from the *HTHe.* The first really differentiating criterion concerns the goals of the organism and the person. According to the *BSTe* the only goals that are relevant in the analysis of the health concept are the survival of the individual and the survival of the species. According to the *HTHe* there are further goals. The agent has other possible vital goals than the one of pure survival. (In my own system pure survival *must*, however, be included in the set of vital goals. This is so because survival is a necessary condition for the individual's long-term happiness.) Thus, according to the *HTHe* there are things required in order to be completely healthy other than those required in the *BSTe*. According to the *HTHe* it is not sufficient that you are normal in relation to survival. You must also have resources which are adequate for other vital goals. There are several persons who are healthy according to the *BSTe,* who come out as unhealthy according to the *HTHe*. An example of this is the person who does not have any pathology (as understood in the biostatistical sense) but who has a reduced ability to realize his or her vital goals. The reason for this disability could be a problem of an existential kind.

Conversely, however, there are several persons lacking complete health according to the *BSTe*, who come out as healthy in the *HTHe*. These are the ones who have some clear malady, that, however, does not affect the ability to reach vital goals. The malady may be trivial, latent or very quickly aborted. In the end it is unclear which of the two theories is the more demanding.

The second important differentiating criterion concerns the nature of the circumstances presupposed in the concept of health. The *BSTe* refers to statistical normality. The *HTHe* refers to circumstances which are considered to be standard in a particular cultural context.

VI. ON ASSESSING THE CONCEPTS OF HEALTH

What are the criteria for assessing concepts of health, illness and disease? By what standards can we say that either of the two theories is superior to the other? I shall from now on refer exclusively to the *BSTe* and the *HTHe*. (This means that I shall in effect refer to theories which have, *in toto*, not been formulated by anyone beside me. An advantage of this is that it saves me from making detailed provisos in relation to the various relevant authors.)

There are several possible criteria for assessing the concepts, for instance the following:

a. internal coherence
b. closeness to ordinary language
c. closeness to medical language
d. usefulness for scientific purposes (for instance, health being strictly defined and being measurable)
e. usefulness in medical practice
f. usefulness in public health contexts.

The criteria are not mutually exclusive. Some of them support each other or even presuppose each other. Internal coherence, for instance, ought to be a presupposition in all contexts. Usefulness for scientific purposes is often *eo ipso* usefulness for medical practice, as well as for public health work. Usefulness in medical practice, on the other hand, also presupposes considerable closeness to ordinary language, since medical practice entails communication with laymen.

The scope of this paper does not permit a close analysis along all these lines. I will here only consider the criteria e and f: usefulness in medical practice and in public health contexts. This means that I will consider health as a goal of medicine. Let me then first consider the *medical encounter*, the encounter between a potential patient and a medical practitioner (a doctor, a nurse or a paramedic).

1. A person approaches health care with a problem. John approaches his family doctor with a problem. He says that he has been ill for some time. He has had considerable pain in his stomach and this has prevented him from going to work for a week. He says that he must have some disease. He cannot explain his ill health otherwise. Here we see that John asserts that he is ill. He has not made any inspection of his body in order to establish this fact. He has noticed his pain (a pain which has no immediate external cause) and he observes that he is prevented from going to work. He assumes that there is a disease which is responsible for this problem.

2. The doctor diagnoses the problem and treats the patient. The doctor makes an examination of John. He tries to assess the nature of the problem and when he is convinced about its nature, he seeks the causes of it. Given his medical training he will in the first instance try to find the causes of the problem in the organic functioning of John's body. In short, he seeks some malady. It is however important here to see that he is not seeking a malady for its own sake. He is not seeking any old malady. He wants to find the cause of the patient's problem, primarily in terms of the disease language to be found in medical classifications and textbooks. Having found the malady that he believes to be the cause of the problem he starts treating it *lege artis,* i.e. according to the recommendations of the contemporary art of medicine.

3. The patient is healthy again when he or she no longer has the problem. The medical encounter is considered successful, and John considers himself healthy, when he no longer feels the pain in the stomach and can go to work as usual.

This simple exposition of the typical successful medical encounter indicates to me that the health concept used is a variant of the *HTHe.* The establishment of the fact that John is ill does not presuppose any internal inspection on the organ level. John can himself (at least equally as well as the doctor) determine that he is in a state of ill health. Ill health for John is when he is in pain and unable to do something urgent for him, viz. go to work, given that the circumstances are standard.

Second, it is clear that health as assumed by the patient, as well as by the health care personnel, is a state of affairs *over and above the absence of malady*. Health has not been restored just because a malady has been eliminated, i.e. the disease has been cured. Normally, the patient cannot go to work until after a time of recovery and rehabilitation.This also speaks in favour of a version of the *HTH*. On the other hand the expanded *BSTe,* as I have defined it, also covers the performance of actions, viz. such actions as are statistically normal for a person of John's kind. A full analysis of this idea presupposes a further interpretation of the notion of statistical normality. It may here suffice to note that John can have a very unusual job which partly requires the performance of rather specialized actions, which are not statistically normal in any sense. A full rehabilitation of John, however, presupposes that he will be able to perform also these unusual actions. It seems, then, more appropriate to say that health entails he can realize his vital goals, than that it entails that he can realize what is statistically normal for a person of his type.

The *HTHe* is the theory of health which proves to be in accordance with the ordinary medical encounter. The *HTHe* explains the ordinary use of the medical concepts in this situation better than the *BSTe*.

There are also very different sorts of medical encounters, which do not presuppose an active patient seeking help for an illness that plagues him or her. Let me exemplify this with the case of a woman, Jane, who is brought against her will into a mental hospital after having been diagnosed as psychotic, i.e. we envisage *a case of compulsory treatment of a psychotic patient*. What are the crucial steps in such a medical encounter?

1. Jane's nearest and dearest observe changes in her mood and appearance. They become worried by the fact that Jane has become a changed person. They observe that she has lost her former abilities to take part in social life and fulfill her professional role. She withdraws from company; she often becomes aggressive and sometimes exhibits outrageous behaviour. Jane's family believes that she has become mentally ill. Jane herself rejects this suggestion and refuses to seek help. The family, however, decides to seek psychiatric expertise.

2. The psychiatrist presents a diagnosis of Jane's illness. A psychiatrist consulted finds that Jane suffers from severe schizophrenia claiming emphatically that Jane must be helped and treated. If Jane refuses she must be taken into custody in accordance with the Mental Health Act.

3. Jane is taken into a psychiatric clinic against her will and is treated there. Since Jane refuses to accept that she is ill, she does not seek any medical help on her own. The medical authority therefore makes the decision for her. She is taken against her will to the nearest psychiatric clinic and is treated in accordance with contemporary psychiatric standards.

4. Jane is considered healthy again when she exhibits her previous good mood and can return to her old social contacts and to her work. This more unusual case of medical encounter exhibits some interesting differences from the standard one. It raises, in particular, many ethical issues. From the point of view of the health concept used, the differences are, however, not so significant. The crucial distinction between the cases has to do with who has the *epistemic access* to the problem. In the standard case of medical encounter it is the subject him or herself who knows or believes that he or she is ill and seeks help. In the case where a psychotic person is diagnosed, it is typically someone else who makes the sickness declaration. But do they use different concepts of health? What constitutes ill health in the two kinds of situation? In the example from a standard medical encounter John is in ill health when he is in pain and is unable to do important things in life, for instance go to work. In the case of the psychosis ill health exists when, for instance, the subject is in a depressed mood, is confused, or for other reasons unable to concentrate and therefore unable to go to work. The concept of disability figures equally prominently in both cases. And the disability is in relation to some set of vital goals, not just in relation to survival. [7]

What makes a difference between the two cases of medical care is not the basic concept. The difference is constituted by the fact that in the case of psychosis the subject is impaired also in his or her cognitive functionings and therefore also unable to judge his or her own state of health.

How does the question of disease enter in the case of psychosis? And what kind of disease concept is used? Indeed the psychiatrist must put a disease label on Jane in order to be entitled to call for her compulsory detention. The disease categories will be found in the ordinary textbooks of psychiatry, as well as in, for instance, the *International Classification of Diseases and Related Conditions* (ICD; 1992). The standard mental diseases are characterized in terms of impaired mental functioning. Consider, for instance, some of the defining characteristics of schizophrenia: "The schizophrenic disorders are characterized in general by fundamental and characteristic distortions of thinking and perception, and affects that are inappropriate or blunted... The most important psychopathological phenomena include ... thought insertion or withdrawal ... delusional perceptions and delusions of control ... thought disorders and negative symptoms" (p. 325). In fact, it is significant that the disease concept here already partly connotes a set of disabilities. The disease concept characterizes the respects in which the subject has impaired functionings. (This indicates an interesting difference between the typical somatic and the typical mental disease concepts. This warrants a slight modification in the *HTHe* characterization of diseases.) [8]

Can we say anything about the goals presupposed in the traditional mental disease concepts? Should the mental disabilities be viewed as disabilities in relation to survival or in relation to other goals? The ordinary diagnostic descriptions give us little help on this point. The best answer we can get is probably given in the *Diagnostic and Statistical Manual of Mental Disorders (*DSM *IV*, 1994) which formulates a general characterization of mental disorder: "Each of the mental disorders is conceptualised as a clinically significant behavioral or psychological syndrome or pattern that occurs in an individual and that is associated with present distress (e.g. a painful symptom) or disability (i.e. impairment in one or more important areas of functioning) or with a significantly increased risk of suffering death, pain, disability or an important loss of freedom)" (p. xxi). It is significant that avoidance of death is only one in a large set of goals mentioned in this central paragraph. I conclude that the *BSTe* is wholly inadequate for capturing the concept of mental health and ill health as well as the traditional mental disease concepts.

Consider now a rather different medical context, viz. the one of *health promotion*. Here, there is no patient presenting a concrete problem to be solved. Instead, a health care unit or some other authority has the ambition to improve or protect the health of the population. This can be done by various means. A very salient one is *disease prevention through inoculation.*

The starting point here is not a case of illness for which a cause is to be sought. In this instance medical science must already have identified a particular species of disease or malady irrespective of any present cases of illness. A particular type of process which can occur in a human body or mind must have been identified and given a name. What type of process is this? Has it been identified via the criterion of the *BSTe* or via the criterion of the *HTHe*? In answering this question I will not here make any attempt at giving a historical reconstruction. My aim is rather to see what is logically required.

The *BSTe* provides a clearcut means for the identification of maladies without any reference to human illness in terms of, for instance, suffering and disability. We only have to detect a statistically subnormal human functioning in order to find the malady. This does not seem possible in the *HTHe* model, where the question whether some bodily state or process is a malady is dependent on whether the state or process tends to reduce a person's health. It then seems as if this situation would speak in favour of the *BSTe* model of health and disease. I will, however, contest this rash conclusion here.

Although it is true that a malady can be identified without the presence of an actual state of ill health, it can be disputed whether it can be identified without previous experience of cases of ill health due to this malady. According to the argument of the *HTHe* the logical starting point is always an instance of ill health. The internal cause of this state of ill health is identified. It is then discovered that similar instances of ill health have the same type of cause. This type of cause is given a name, viz. the malady-name "M." As a result the malady M becomes the topic of medical research. M is measured and described from various biological points of view; in short it is given a pathological description. As soon as this happens the malady type M starts living a life of its own. Instances of M can be identified without first establishing a case of illness in a person. One can start looking at the epidemiology of M, and at the pathogenesis of M. When this has been properly done and the knowledge about the epidemiology and pathogenesis of M has been diffused, it becomes possible to try to find the general mechanisms which might prevent the incidence of M. If M is an infection, for instance, the relevant immunological mechanism has to be established. The preventive program can start; the program of inoculation against the discovered disease can start.

Given this reasoning it is possible to identify a type of malady also according to the *HTHe* without first observing a present case of illness which is dependent on the alleged malady. We can perform disease-preventive programs following both the *BSTe* and the *HTHe*. It should be noted, though, that the *BSTe* is of limited use when we wish to prevent a malady that compromises a person's ability to act without entailing any subnormal biological functioning.

Consider now a health-promotive program which does not presuppose the identification of any particular disease. Many health-promotive campaigns such as those concerning healthy eating, physical exercise, moderate consumption of alcohol and abstention from smoking are not directed at the prevention of a specific disease or even a specific range of diseases. How should such a program be characterized according to the two models? And are both equally successful in such a characterization? Let us call this *the case of general health-promotive programs.*

The answer to the question which model fits this situation better is certainly dependent on how the situation is interpreted. A protagonist of the *HTHe* would say that this case clearly speaks in favour of the *HTHe.* General health promotion does not primarily concern prevention of maladies. The primary aim is that the subject should feel hale and hearty and in general be able to achieve the things he or she is aiming for. This goal certainly presupposes the prevention of all serious maladies. It need not, however, presuppose the prevention of all pathology. Being fit and able is clearly compatible with the presence of many trivial diseases or other maladies.

A defender of the *BSTe,* on the other hand, would perhaps argue along the following lines: It may be true that a general health-promotive program need not have identified a particular malady or a range of maladies as its target. From this, however, it does not follow that the goal is not to prevent the incidence of serious malady. In the case where abstention from smoke or alcohol is at issue it is clear that there are some salient maladies that the promoters have in mind. Cardiac and respiratory diseases, as well as a number of cancers, are in focus in the case of smoking. Neurological diseases, liver cirrhosis, and indeed physical accidents are in focus in the case of grave alcohol abuse. And even if physical exercise and healthy eating seem to have a more indefinite target, says the *BSTe* defender, the health- promotive part of these enterprises concerns precisely the prevention of maladies. If they aim for something more they are, strictly speaking, also some other kind of program, for instance, a "fitness program," that is logically separated from the health promotion proper.

My answer to this, in favour of the *HTHe,* is the following. It seems very artificial and implausible to say that broad health-promotive programs, with their very general recommendations concerning people's life style, are aiming at disease prevention and nothing more. To say that the remaining part of the program is logically unrelated to health seems to be a purely theoretical stipulation against the ordinary use of language. The *BSTe* here does not help to explain or reflect either standard medical or lay discourse. To adopt the *BSTe* as the more adequate theory in the case of general health promotion would be to legislate against ordinary language.

VII. CONCLUSIONS

In this paper I have formulated two major competing conceptions of health and related concepts. On the platform of Christopher Boorse's biostatistical theory of health I have formulated an extended theory called *BSTe,* which has been compared with a simplified holistic theory *HTHe* based on a number of contemporary proposals. I have noticed the essential differences and similarities between the two approaches. I have also initiated an assessment of the two conceptions, mainly from the point of view of medical practice and public health. My conclusions from this preliminary assessment are the following:

a. The health concept used in general practice (including both the standard medical encounter and the case of compulsory mental care) is related to vital goals and not just to survival. Moreover, health is something over and above the absence of disease (also when the concept of disease is interpreted in the *HTHe* sense.)

b. The malady concept used in the context of disease prevention could be the one of *BSTe.* I have however argued that it could equally well be the one of *HTHe.* It should be noted that using the two different models will partly yield quite different results.

c. The health concept used in the context of general health promotion is, I argue, much more naturally interpreted along *HTHe* lines than along *BSTe* lines.

These analyses bring me to the conclusion that the *HTHe* is in general a more plausible theory of health then the *BSTe*, when these are assessed from the points of view of medical practice and public health.

Let me conclude this paper by emphasizing the practical and ethical consequences of an analysis like the one I have presented above. If health is the main goal of medicine and health care, then the interpretation of health is of utmost importance for all medical practice and health care. If the objective of medicine stops at the stage of the elimination of diseases, then we end up with a purified and completely medicalized care. If, on the other hand, the objective is instead the person's fitness and ability to realize vital goals, then medicine and health care must take up a number of further roles, including psychological and social rehabilitation. The health care establishment now seems to waver between these standpoints. It is salient, as I have noticed in my analysis, that modern Western medicine entails activities over and above the cure of biological disease. Many representatives of the medical establishment are, however, reluctant to accept a role as broad as the one envisaged in the holistic concept of health. This is understandable, since the typical medical education does not train the doctor for a role beyond the one of the curer and to some extent the carer. A full implementation of the *HTH* model requires a lot of reconstruction of medical care, medical research and, as a result, also of medical education. To discuss this major topic, however, requires a treatise of its own.

Dept of Health and Society
Linköping University
58183 Linköping, Sweden

NOTES

1. Boorse (1997) has clarified that his analysis of the concept of disease solely concerns the concept of pathological disease. He therefore, now, has no claims concerning the application of his notion of disease to, for instance, a clinical encounter. I grant this, and the analysis of the BSTe performed in this paper must therefore not be seen as a criticism of Boorse's position. I think, however, that my analysis is still warranted since there are several supporters of some version of BST who consider that the theory has clinical significance.
2. This definition is simplified in relation to my full-blown characterization (Nordenfelt 2000, p. 93). It is essential in my theory that the ability involved is a second-order ability.
3. The definition of malady in HTHe is simplified in that it solely refers to the agent who is unhealthy. A careful definition of the notion of malady is the following: M is a malady-type in some circumstance C, if and only if, M is a type of physical or mental process, which when instanced in a person P in C, would with high probability cause ill health in P.
4. Perhaps this is the extension made by Norman Daniels in *Just Health Care* (1985), pp.28-35. I think Boorse would accept it, at least if, as I propose, we stick to second-order abilities, i.e. that we do not presuppose that the person has been trained. Health for Boorse must provide the potentials for performing statistically normal actions, as long as "normal" is interpreted in relation to survival.
5. Indeed, this is what is essentially contained in my own version of the HTH. I say that A is completely healthy, if and only if, A is in a bodily and mental state which is such that A has the second-order ability to realize all his or her vital goals.
6. In Nordenfelt, 1995, pp. 110-112, and Nordenfelt, 2000, p. 114. I have proposed variants of health concepts within the HTH framework. One is called the disease concept of health. By this is meant that a person is healthy when his basic health is not compromised by any disease (in the HTH sense of disease).
7. For substantive treatments of the notions of ability, disability, and handicap see my analyses in Nordenfelt (1997; 2000).
8. The observation made here suggests to me that there are perhaps no mental diseases but only mental illnesses. By an illness I then mean a cluster of specific disabilities.

REFERENCES

Agich, G.J.: 1983, ' Disease and Value: A Rejection of the Value-neutralityThesis,' *Theoretical Medicine*, 4, 27-41.

Agich, G.J.: 1997, ' A Pragmatic Theory of Disease,' in Humber, J.M. and Almeder, R.F. (eds.), *What is Disease?,* Biomedical Ethics Reviews, Humana Press, New Jersey, pp.221-246.

Boorse, C.: 1977, ' Health as a Theoretical Concept,' Philosophy of Science 44, 542-573.

Boorse, C.: 1997, 'A Rebuttal on Health,' in Humber, J.M. and Almeder, R.F. (eds.), *What is Disease*?, Biomedical Ethics Reviews, Humana Press, New Jersey, pp. 1-134.

Culver, C.M. and Gert, E.: 1982, *Philosophy in Medicine: Conceptual and Ethical Issues in Medicine and Psychiatry*, Oxford University Press, Oxford.

Daniels, N.: 1985, *Just Health Care*, Cambridge University Press, Cambridge.

Diagnostic and Statistical Manual of Mental Disorders (IV): 1994, The American Psychiatric Association, Washington.

Fulford, K.W.M.: 1989, *Moral Theory and Medical Practice*, Cambridge University Press, Cambridge.

Hesslow, G.: 1993, 'Do we need a concept of disease?,' *Theoretical Medicine* 14, 1-14.
The International Statistical Classification of Diseases and Related Conditions: 1992, Tenth Edition, The World Health Organization, Geneva.
Nordenfelt, L. 1995, *On the Nature of Health*, Second revised edition, Kluwer Academic Publishers, Dordrecht.
Nordenfelt, L. 1997, 'On Ability, Opportunity, and Competence: An Inquiry into People's Possibility for Action,' in Holmström-Hintikka and Tuomela (eds.), *Contemporary Action Theory*, vol.1., Kluwer Academic Publishers, Dordrecht.
Nordenfelt, L. 2000, *Action, Ability and Health: Essays in the Philosophy of Action and Welfare*, Kluwer Academic Publishers, Dordrecht.
Pellegrino, E. and D.Thomasma: 1981, *A Philosophical Basis of Medical Practice: Toward a Philosophy and Ethic of the Healing Professions*, Oxford University Press, Oxford.
Pörn, I.: 1993, 'Health and Adaptedness,' *Theoretical Medicine* 14, 295-303..
Scadding, J.G.: 1967, 'Diagnosis: The Clinician and the Computer,' *The Lancet* 21, 877- 882.
Seedhouse, D.: 1986, *Health: Foundations of Achievement*, Chichester.

RICHARD ZANER

PHYSICIANS AND PATIENTS IN RELATION: CLINICAL INTERPRETATION AND DIALOGUES OF TRUST

I. MEDICINE AS A HUMAN ENTERPRISE

An Agency of Change

Only brief reflection on medicine since its emergence from World War II as a scientific enterprise makes it plain that even modest general practitioners have at their disposal a truly amazing arsenal of regimens, procedures, techniques, technologies, specialties, drugs, anesthesias, and the like – any of which can readily be brought to bear on the merest of complaints. And the cornucopia of medicine's scientific breakthroughs and technological prowess are continuously enhanced.

At the same time, these developments are fraught with serious questions. Clearly, the governing aim of medicine involves performing some action on some unique individual (or, in some situations, on some group or broader collective), an action which is designed to bring about specific sorts of changes in the recipient(s) (usually in the form of benefits, though it may of course also occur in some form of risk, hence harm, or at least unexpected outcome) (Pellegrino, 1979). As profound changes in one or more human lives are sometimes brought about in each clinical encounter (whether from sick to healed, disabled to functional, or otherwise), so medicine itself undergoes at times profound changes. For example, as a direct result of the ability to enable fewer babies to die at birth and more people to stay alive afterwards (and for longer periods), the resultant population explosion helped bring about fundamental changes in the ways people relate with each other and the surrounding world (urban versus rural lifestyles and environments, but also types of disease, accidents, etc.). Similarly, the advent of the 'pill' brought about for the first time in our history a real separation of recreative from reproductive sexual activity, which in turn resulted in wholly new kinds of ethical issues – which, further, brought about changes in family structure, educational policy, child rearing norms and practices, individual family and population planning, and even such a sensitive issue as the right to reproduce.

G. Khushf (ed.), Handbook of Bioethics, 223–250.

All of which is mere preface to recent discoveries in genetics and the consequently awesome questions about privacy, patenting of life forms, and the prospects of plant, animal and human cloning. Enhanced technical prowess brings about fundamental, irreversible alterations in human life and society (Jonas, 1984, pp. 25-50; Bayertz, 1994).

Specifically *modern* technology, Hans Jonas understood, results in decisive changes from what has been hitherto regarded as merely (if at all) possible. Even though in classical times, for instance, Sophocles' *Antigone* gave "awestruck homage to man's powers," that proud celebration of human "violent and violating irruption into the cosmic order" was inevitably accompanied by "a subdued and even anxious quality…[for] man is still small by the measure of the elements…" (Jonas, 1974, p. 5). With modern technology, however, everything changes. For with the drastically altered scale and rush of technological intervention, it is starkly clear that we've also hit upon nature's "critical *vulnerability*…to man's technological intervention—unsuspected before it began to show itself in damage already done" (Jonas, 1974, p. 9). The dimensions of this kind of power and what it brings in its wake is only beginning to be appreciated – the profound changes in the scope and range of human action, for instance, or the expansion and specificity of artificial environments (whether space capsule or a lamb's cellular entity as a veritable laboratory for Wilmott's cloning of 'Dolly' – the mitochondria of the 'parent' cell being left to 'work' with the 'donor' cell's nucleus).

Of equal import are certain social ramifications. It is currently plausible, for example, not only to conceive and practice genetic control of future people ("taking our own evolution in our own hands," in Sir John Eccles' phrase; 1979, p. 120), nor merely to conceive and practice behavior control on individuals and entire populations, using psycho-pharmacological interventions (Clark, 1973, pp. 94-5), but as well it has become plausible to control death (a genetic error in somatic cells) (Burnett, 1978, p. 2). This dramatic enhancement of medicine's reach into individual lives and the social sphere harbors great significance, within and outside of technological cultures. Medicine is a unique enterprise, which has its own special kind of discipline.

A Form of Power

In the first place, medicine is a kind of disciplinary bridge between the natural and social or humanistic sciences (Engelhardt, 1973, p. 451). It is the "most human of the sciences and most scientific of the humanities" (Pellegrino, 1979, pp. 9-15).

In the second place, well beyond Rudolph Virchow's well-known observation that medicine is a "social science," the contemporary forms of the medical art suggest that it is also a kind of social engineering. With contraception, transplantation, genetic manipulation, in vitro fertilization, cloning, even cosmetic surgery or dietary regimens, medicine is effectively an agent of social change and change of individuals. It is, as H. Tristram Engelhardt emphasized more than twenty years ago, engaged in multiple forms of "remaking" human individual and social life:

> Medicine is the most revolutionary of human technologies. It does not sculpt statues or paint paintings: it restructures man and man's life.... In short, medicine is not merely a science, not merely a technology. Medicine is a singular art which has as its object man himself. Medicine is the art of remaking man, not in the image of nature, but in his own image; medicine operates with an implicit idea of what man *should* be.... The more competent medicine becomes, the more powerful it is, the more able it is to remake man, the more necessary it consequently becomes to understand what medicine should do with its competence (1973, p. 445).

It is not only, then, that medicine is involved in value questions; precisely these questions are at its heart and have singular significance.

To express the point as sharply as possible: it lies within the power of modern medicine to alter, perhaps radically and irrevocably, and for better or worse, our very capacity to recognize, reckon with and understand medicine itself – and thereby our ability to understand ourselves and the changes in us due to medicine. Medicine is thus a *critical* discipline, not only in view of the current crises mentioned earlier but rather precisely because *what it is to be human* is *at issue* in every medical encounter and every medical decision – for physicians, of course, but far more for patients. Philosophical issues, as was appreciated for centuries (until our own), are at the heart of medical practice and theory (Buchanan, 1991).

The sheer presence of technological power of itself elicits and encourages the use of that power. Power – especially of the sort modern bio-technology places at the disposal of (or in some form makes available to) every physician – is ineluctably seductive; it withers and is lost if left unused or dormant, and the subjects of its use are, like its users, human beings. "There is no other way of exercising the power," Jonas notes, adding, "Where use is forgone the power must lapse, but there is no limit to the extension of either" (1966, p. 193). Of itself, this calls for serious and judicious rethinking of the place of values in a world of use.

There is another side to the point. To repeat, the power at issue here is defined by its ability to cancel cognizance of its having been used in the first place. The medical enterprise makes available precisely this kind of technological power throughout the settings of its practice, but primarily to those able to understand and use it (other physicians, researchers, etc.), or to those who are able to pay for it (sponsors, funding agencies, etc.). Medicine thereby attracts social prominence and prestige to itself, which further enhances the very power in question. To mention but one facet of the matter, the social enhancement of medicine, its 'presence' as authoritative for how we view ourselves, and not merely our bodies, invariably includes and encourages highly influential ways for those of us outside of it to understand ourselves – children most of all, but others as well – whether as ‘sick’ or ‘well’. Not only does the social prominence of medical understanding, in short, encourage those in the wider culture to view themselves in medical ways, but also to pass these ways on to others – through education, self-promotion, the media, and other indirect ways.

Thus, how and what we are constantly induced to think and to feel about ourselves and others, individually and together, is significantly influenced by the sheer potency and presence of modern medicine. Its influence, moreover, is enhanced and sharpened by the fact that its practitioners frequently have to deal with the ultimate issues of human life: loss, grief, disease, impairment, guilt, failure, life and death – with human beings as regards what *defines humanity*. To be impaired is to suffer a lessening of the ability to choose, of self-image, in short of the freedom to think and act on one’s own.

II. PHYSICIANS AND PATIENTS IN RELATIONSHIP

The Sociality of the Relationship

These considerations prompt reflection on several aspects of the physician-patient relationship.

Consider a typical encounter between physician and patient in a hospital. To make the discussion manageable, we can ignore here many of its features – e.g., the specific ailment, whether the patient is self-admitted or transported by ambulance, whether the physician is on the hospital's clinical staff or only has admitting privileges, and so on. We can concentrate instead on the doctor's efforts to explain the diagnosis and offer therapeutic recommendations.

The situation is still considerably complex. To explain to a patient (and loved ones) what a particular procedure will involve is for specific people to be engaged in a concrete conversational process going on at a specific time and within specific circumstances. The encounter occurs within a particular hospital unit (intensive care, cardiac unit, surgical ward) that includes other providers (nurses, consultants, residents, technicians, ward clerks), which is only one of many such units in the hospital, itself only one of many hospitals in that area, region, state, and the United States. Each of these contexts (units, clinics, hospitals, regions, etc.) operates under certain written and unwritten guidelines, protocols, regulations, and laws, the totality of which lies within the broader society with its own characteristic patterns of prevailing policies with their own specific values (about, among other things, doctors, hospitals, sickness and health).

Each provider has his/her own respective personal biographical situation, including values, beliefs, habits, etc. (Schutz and Luckmann, 1973), and works within a specific profession with its codes and practices. Each professional practices within a specific hospital unit (with its own protocols on resuscitation, accepted therapeutic regimens, written and unwritten rules and codes of conduct, and so on). Beyond this is the hospital as a socially legitimated institution with its complex of rules, committees, policies, etc.; the particular hospital region and state with their body of regulations, licensure policies, laws, etc.; the federal government with its regulations, policies, etc.; the medical profession and specialty and subspecialty organizations with their accepted standards of practice, etc. – all of which are components of the current culture with its complex folkways, mores, laws, institutions, history, etc. There are thus personal, professional, institutional, and prevailing social value-contexts that configure each medical encounter.

On the side of the patient (and loved ones), there is an equal, though less formally organized complexity. On the one hand, each patient has his/her own specific biographical situation with its distinctive values, attitudes, history, linguistic usages, habits, etc. Each patient, moreover, is only rarely without some immediate family or friends (loved ones, circle of intimates), who (implicitly and explicitly) share certain beliefs, values, attitudes, history, and the like, with the patient. Every patient is a member of social, business, political, or religious groups, each of which has its own specific traditions, values, usages, etc. – which in various ways point to and reveal personal characteristics and views that can, on occasion, prove to be quite significant for decision-making. Like every provider, moreover, every patient is part of the same (or at the very least part of some) culture with its prevailing, commonly shared nexus of social values, mores, folkways, etc.

As has long been recognized, prevailing social conditions (fragmentation, specialization, mobility, etc.) mean that, aside from a person's immediate family, circle of intimates, small groups and associations, or (at times) relatively stable neighborhood, people interact for the most part as strangers. They often do not know whether they share values, beliefs, or attitudes – in particular, about health and illness – and thus do not usually know what claims they may legitimately make on each other. When what brings them together is a need for help by one and the claim of being able to help by the other, their relationship can be quite difficult.

When that need is signaled by distress, illness, or injury, the situation is often ripe for trouble – when, for instance, the physician proposes to carry out quite aggressive and intimate actions on patients who are strangers. Yet, for the hospitalized patient, there is unavoidably little choice but to trust in numerous ways: other people (often anonymous, from doctors to manufacturers of drugs), things (equipment, substances), and procedures (protocols, surgical regimens, etc.) (Zaner, 1991). A close and historically informed look into clinical encounters of any type suggests that the "implicit demand for joint decision making" must invariably "confront the painful realization that even in their most intimate relationships, human beings remain strangers to one another" (Katz, 1984, p. xviii). Still, conversations are clearly essential, for they are the sole means by which the physician can at all earn that trust – even if it be only temporary (Lenrow, 1982).

For their part, physicians often have equally little choice but to take care of a patient, even if s/he is regarded as a "gomer" or a "dirtball" (Donnelly, 1986), is seductive or deceptive (Cassell, vol. I, 1985), or even if only mimicking illness. Although there are limits, the doctor also often has little choice but to trust: for instance, that the patient really wants help and is candid, capable and accurate; that tests and radiologic data are correct and equipment functioning properly; that drugs have been properly manufactured and dispensed; etc. In any event, given the plight of the patient who is not only ill but surrounded by strangeness – the illness, people, places, customs, etc. – the physician must bear the responsibility for initiating and sustaining these conversations.

To initiate a conversation with a patient, the physician engages in a conversational (and interpretive) effort to find out "what's going on" in the patient's life. This requires attention to patient talk for its content and intent, which are always framed by a variety of paralinguistic features, physiognomic gestures and situational components that help determine the sense of what is said and done. Within this framework, the physician must also detect the patho-physiology and how it is modified within this particular sick person's embodying organism, personal life, and cultural milieu. That they are strangers, ineluctably influences and shapes their dialogue, especially when recommended regimens, surgeries, or other proposed treatments are aggressive, invasive, risky, and intimate. Diagnosis and therapy – both semiotic activities – require careful attention to a patient's pertinent history, understanding of his/her ability to understand and comply, as well as his/her willingness to accept recommended regimens and act appropriately.

Clinical Hermeneutics: At the Cutting Edge

These considerations suggest that the interpretive aspects of clinical medicine are inherent to it. They also indicate that clinical medicine incorporates a kind of hermeneutics, a clinical-circumstantial probing and understanding whose skills and rationales should manifestly be part of physician training. Several features of that discipline can now be delineated.

To interpret in a clinically appropriate manner requires that the doctor must never *pre*-interpret what a patient displays – neither moans or groans, expressions of pain or comfort, nor what s/he is trying to say ("communicative intent"; Cassell, 1985, vol. I, Ch. 4). The doctor has to be constantly on the *alert* and *oriented* to each patient in ways that are opposed to almost everything in medicine's "millennia-long tradition of solitary decision making" (Katz , 1988, p. 85) – from the deeply-rooted place of authority and presumptions about any patient's ability to understand when in great pain, to the wariness and confusion generated by the serious linguistic gap between medical and everyday terms, and the different forms and conveying of uncertainty.

The physician who initiates serious conversation with patients must invariably become painfully aware of the pervasive principle of everyday life, which Alfred Schutz identified as "taking things for granted" (1967; 1973). This suggests that conversations with patients and their families (and/or circle of intimates) must be governed by what might be termed the first principle of clinical hermeneutics: as far as possible, *take nothing for granted!* While it is easy to state the principle, it can be inordinately difficult to practice while at the same time engaged in dialogue with a patient. This stems not merely from the sheer strength and pervasiveness of this fundamental feature of daily life generally, but also from medicine's long history of solitary decision making. To allow, not to say encourage, patient participation in decisions, however, has precisely this principle as a primary methodological orientation: decisional participation is premised on establishing some shared grounds, which itself requires the effort to take nothing (or little) for granted.

Sick or well, people not only experience and interpret their own bodies (pain, discomforts, hunger, etc.) and themselves (well, sick, poorly, etc.), but also are emotional about these (fear, hope, uncertainty, etc.). It is clear from numerous case studies that sick people want to know, at times fervently, about their illnesses; they want to know "what's going on" and "what can and should be done about it," as they want to know that the people taking care *of* them also care *for* them. Whether their doctors specifically inform them (and whether well, poorly, or not at all) about their ailments, people not only experience but interpret their own illnesses in typical ways: "bronchitis," a "cold," "ulcers," but also "unfortunate," a "damned nuisance," "devastating," and the like.

A second principle of clinical hermeneutics is thus delineated: clinical engagements with patients are invariably *'second-order' interpretations* (Schutz, 1967 I, p. 59), interpretations (physician) of (patient) interpretations, precisely because each physician is always presented with some patient's specific experiences and interpretations (of self, body, world, as well as of physician, nurse, etc.). In somewhat different terms, like language more broadly, illness is intrinsically complex: the disease and the way it is manifested in the patient's body; what being ill means to the patient; what the doctor says and how this is in turn interpreted and understood by the patient (and *vice versa*); what family or friends say; and still other factors always needing to be taken into account. Thus, medical interpretations are inevitably multi-dimensional: the doctor interprets patho-physiology, the way the disease is uniquely manifested in a particular patient, how the disease process is experienced (illness experience) and what its significance is for the sick person, how the patient reacts to the physician's talk, and still other forms.

Physician-Patient Asymmetry: I

A third principle of clinical hermeneutics is suggested: medical interpretations (diagnosis, therapeutic alternatives, prognosis) cannot ignore what things are really like *from the patient's and family's points of view*. While that seems evident enough on the face of it, one significant feature of the physician-patient relationship gives the point a critical moral edge.

The relationship between the helper and the person helped is *asymmetrical*, with power (in the form of knowledge, skills, access to resources, social authorization, and legal legitimation) in favor of the helper (Lenrow, 1982, p. 48). The social organization of professional help within bureaucratically structured institutions further enhances the asymmetry, as the organization itself also carries substantial weight. So far as helper and person helped are most often strangers, the one often engaged in aggressive and quite intimate actions to and on the other, the asymmetry itself is a prime source of immensely difficult issues. The social authority of physicians, however, extends far beyond the usual senses of power. A brief historical remark will be helpful to clarify this.

Even though medical therapies were for many centuries virtually useless and medical understanding often erroneous, the relationship's asymmetry of power has been a fundamental and remarkably unaltered component of medicine's self-understanding almost from its inception. Rarely understood in this way, this is nevertheless an essential component of the Hippocratic tradition. Steeped in an understanding that the relation to patients must be governed by certain fundamental virtues, these ancient physicians had a remarkable insight, I believe, into a key facet of the moral order.

Medical historian Ludwig Edelstein lucidly shows that the principal generic virtues in the Hippocratic Oath are justice (*dike*) and self-restraint (*sophrosyne*). Whatever one may think of the at times barbaric 'treatments' practiced in medicine (almost up to the twentieth century), ancient physicians realized full well that the practice of medicine involved the physician in the most intimate kind of contact with other human beings – and required decisions that could affect the patient and family in profound ways. On the other hand, it was also realized that the patient faced an urgent issue: "How can he be sure that he may have trust in the doctor, not only in his knowledge, but also in the man himself" (Edelstein, 1967, p. 329)? This critical blend of virtues – judicious restraint – was, not unsurprisingly, regarded in the ancient Hippocratic texts as the primary sense of medical wisdom (*On Decorum* and *On the Physician*; in Edelstein, 1967, pp. 6-35).

To be a patient is to be intimately exposed, directly vulnerable. Because of the specific type of knowledge unique to medicine, the healer's possession of drugs and technical skills, and the access to the intimate spheres of patient life (person, body, family, and household), the ancients clearly realized that they were uniquely in position to take advantage of patients while, by contrast, patients were disadvantaged both by illness or injury and by the very asymmetry of the relationship. Precisely this appreciation of the asymmetry of power in favor of the physician led to an understanding of 'the art' as a fundamentally moral enterprise under the guidance of central virtues: justice and restraint.

III. AN INTERLUDE: GYGES AND AESCULAPIUS

A patient once poignantly remarked, "you have to trust these people, the physicians, like you do God. You're all in their hands, and if they don't take care of you, who's going to?" (Hardy, 1978, p. 40). Noting how "overpowering" doctors can be, another emphasized, "They've got an edge on you" (Hardy, 1978, pp. 92-93).

In these plaintive words is the echo of an ancient puzzle – the temptation of having actual power over the existentially vulnerable patient. This puzzle, I am convinced, is at the heart of the Hippocratic tradition in medicine. It is especially plain in light of the Oath's apparent mythic sources with the god Apollo and his progeny, Aesculapius, "the god of doctors and of patients" (Edelstein, 1967, p. 225). Physicians who took the Oath were covenanted to help sick and injured people of all sorts, without bias. They became involved with vulnerable people in the most potent and intimate ways, at times called on to render judgments and make decisions that reached far beyond the application of merely technical knowledge and skills. They thus believed they were entrusted by the gods with a supreme wisdom about afflicted people, committed to be "physicians of the soul no less than of the body" (Edelstein, 1967, pp. 24-25).

The Aesculapian healing-places were open to every sick or injured person, whether the person be slave or free, pauper or prince, man or woman. Following the guidance of Aesculapius – "the god who prided himself most of all on his virtue of philanthropy" (Edelstein, p. 344) – the healer assumed certain fundamental responsibilities. Sarapion laid these out in a poem inscribed on stone in the Athenian temple of Aesculapius: "First to heal his mind and to give assistance to himself before giving it to anyone," and only then to "cure with moral courage and with the proper moral attitude... For we are all brothers" (in Edelstein, 1967, p. 344).

Beneath the covenant is an understanding of social life – including what brought the vulnerable sick person face to face with the healer and powers of the "art." The Oath's covenant invoked a moral vision focused on the healer-patient relationship. It also showed a strong sense of the power inherent in the art, a potential for control and even violence to the patient who was "in the hands" of the physician. Acting on behalf of the sick person and maintaining strict "silence" are as integral to the Oath as certain conducts which were strictly banned (for instance abortions and providing lethal substances for suicide). It thus incorporates that peculiar blend of justice and restraint – and, it seems evident, courage – to govern the relationship. This implies that physicians evidently recognized that they were in a unique position to take advantage of people when they are most vulnerable and accessible. It also strongly suggests a recognition of the central challenge and temptation inherent to the work of physicians, thus demonstrating the emergence of a sophisticated moral cognizance (Zaner, 1988, pp. 202-223). *Vulnerability* – and its correlates, personal *integrity* and *dignity* – were therefore as essential to the art as was acting on behalf of and never harming the patient (beneficence and non-malificence).

Indeed, the Oath lays out for this "sacred art" a "morality of the highest order;" the healer was enjoined to "a life almost saintly and bound by the strictest rules of purity and holiness" (Edelstein, pp. 326-327). To practice medicine was and is deliberately and voluntarily to assume the responsibility for being attentive and responsive to each and every individual who seeks aid – bound by a covenant with each person, but also his or her family, and household.

The moral cognizance at the heart of the Oath is striking but, as noted, forces a searching moral question: What could possibly move any physician *not* to take advantage of the vulnerable patient? Why *not* take advantage, especially when the patient is, precisely, vulnerable? Buried squarely within the Hippocratic tradition is that ancient puzzle. One need only consider another, equally ancient and powerful myth about the temptation of having actual power, to put the puzzle into perspective: the Ring of Gyges story in the Second Book of Plato's *The Republic*.

Having gained the power of the ring (to become invisible when the ring's collet is turned) found in the belly of a bronze horse (uncovered in a crevice by an earthquake), Gyges is then able to do whatever he wishes. And, he does just that: seducing the queen and, with her assistance, slaying the king – becoming the king. The puzzle within the Hippocratic Oath is striking: *having* the advantage, the power, a Gygean physician would surely *take* advantage, just because, given the ring and its power, the patient is vulnerable and readily accessible (as were the queen and king of Lydia). Interpreting medicine from the Gygean myth, the very Oath itself is either nonsense or a façade for the exercise of power.

When people are strangers, there is all the more reason for suspicion and distrust as the basic form of social orientation, since the very grounds for trust in the helping relation are missing, or at the very least are quite problematic. On the one hand, there is no common, enduring, and mutual understanding between strangers: neither the healer nor the one seeking help knows what, if any, values they share nor how values differ. Is the healer trustworthy? Does the patient mean what she says? On the other hand, at the core of the relationship is the asymmetry of power in favor of the healer over against the vulnerability of the one seeking the healer's help. While the healer has the power to influence the patient, often without his/her knowing, the healer does not know how this power is regarded by the patient nor whether s/he is trusted to use power for the patient's benefit.

But if the Gyges myth is alien to the Hippocratic-Aesculapian understanding of medicine, it nevertheless highlights the key moral issue – the power noted, which gives a cutting edge to things in our times. If healers are to be entrusted with such power and intimacies (affecting the patient's body, the person, the family, the household), the crucial question concerns what they must do and be to deserve and ensure that trust (be trustworthy). Why *not* use the asymmetry for the healer's own advantage? The patient must trust precisely while being at the mercy of the physician – the very one who professes and then proceeds to use the power of the art (knowledge, skills, resources, etc.), who proposes and then proceeds to engage in highly intimate, potent, and consequential actions on people when they are most vulnerable, at times bringing about important forms of change in what, even who, they are or hope to be.

These myths invoke contrary visions of the social order, especially the social context of clinical encounters. In both, one with power confronts another at a decided disadvantage. For the Hippocratic, the potencies of the art were clearly appreciated and given expression in its Oath: the injunctions to act always "on behalf of" the sick person, never to take advantage of the patient or his family/household, never to "spread abroad" what is learned in the privacy of the relationship with the sick person. For the Gygean, however, the therapeutic act can make no sense: why engage in helping, since that merely lets the vulnerable become less vulnerable, less open to coercion? But even if *therapeia* is understood as Aesculapian, the grave moral issues do not disappear but become an abiding part of medicine's history: why "first do no harm" then "act in the patient's interest?"

IV. DIMENSIONS OF UNCERTAINTY

Physician-Patient Asymmetry: II

The clinical encounter is a very special sort of relationship because it is inherently haunted by Gyges – the extraordinary temptation to manipulate, control, or otherwise take advantage of the ineluctably vulnerable person, the one whose personal integrity has been compromised by illness. It is in the mythic interplay of these figures that the moral character of encountering the other(-stranger)-as-ill is best understood. While the "medical art" is unable to abide a Gygean reading, Gyges nevertheless nestles snugly within every physician (and every person). It haunts the clinical relationship, making its moral character all the more evident.

As I interpret it, the asymmetry of the physician-patient relationship is tempered by the moral recognition of the governance of judicious restraint in (and the courage not to violate) every relationship with patients. That imbalance of power is a fundamentally moral phenomenon, a sustaining feature of medicine that is poignantly captured in Cassell's remark:

> I remember a patient, lying undressed on the examining table, who said quizzically, "Why am I letting you touch me?" It is a very reasonable question. She was a patient new to me, a stranger, and fifteen minutes after our meeting, I was poking at her breasts! Similarly I have access to the homes and darkest secrets of people who are virtual strangers. In other words, the usual boundaries of a person, both physical and emotional, are crossed with impunity by physicians (1985, vol. I, p. 119).

Because the asymmetry is in favor of the physician – who, unlike the patient, is not (or ought not be) ill – the physician must be the one who encourages and authorizes the patient's experiences and interpretations. The physician thus bears the responsibility of helping patients and loved ones appreciate that their views of the illness are a legitimate part of what's going on – as well as what *can* and *should* be done about it – in the clinical encounter.

This seems to me what Kleinman (1988) means when he remarks that, just as illness "demoralizes," helping the patient understand "remoralizes" – an act that stems from "empathic witnessing," including sensitive talk and alert listening. To correct for the structural asymmetry, in moral terms, there must be what he terms an "existential commitment to be with the sick person and to facilitate his or her building of an illness narrative that will make sense of and give value to the experience" (p. 54). In these terms, the virtues Edelstein takes as constitutive of the therapeutic act – self-restraint and fairness – must not only be expanded to include courage, but these need to be appreciated as ingredient to the relationship itself, not merely a characteristic of the clinician's actions.

The third principle of clinical hermeneutics suggested above therefore has profound moral significance. In view of the essential vulnerability of patients within the asymmetrical relationship, every medical interpretation must be framed so as to be sensitive to, and to capture as exactly and fully as possible, what things are really like from the patient's point of view. The clinical encounter is, when properly executed, a form of affiliation (Zaner, 1988, pp. 315-19), or what in daily life is termed "putting yourself in the other's shoes" –

an act that therefore seems quite fundamental to the constitution of the moral order.

Error and Uncertainty

There is, moreover, a fourth principle of clinical hermeneutics: the physician's work with patients must always include *the constant possibility of diagnostic, therapeutic, prognostic, and other types of interpretive error*, which therefore requires developing concrete plans in the event mistakes occur. This clinical event is, Pellegrino right insists (1983), the "architectonic" of medicine, the governing end of the face-to-face relationship (Schutz and Luckmann, 1973) with a sick or injured person.[1]

To speak of clinical medicine is in the first place to emphasize two crucial aspects of the clinical event. First, as noted, the cultural enterprise of medicine possesses a unique kind of power. Medical actions *change* the particulars with which they are concerned – human beings, individually and collectively. Second, just as its focus is on altering individuals and collectives, so is medicine in this very process *thereby itself changed.* Medicine does not merely study and explain a certain range of phenomena, for internal to its very understanding and knowledge is an inherent thrust to therapeutic practice: it both acts on and alters what is acted-on, and is in turn itself changed by that very action. The fact of medical fallibility and error make this uniqueness apparent.

To emphasize that physicians are fallible is to note that the possibility of mistakes is essential to the clinical event. What distinguishes medicine from most other professions is this "constant possibility of error" – which, moreover, harbors the potential of "doing terrible harm to someone" (Cassell, 1985, vol. II, p. 7). The physician's work with patients must therefore always be alert to the constant possibility of error – whether diagnostic, therapeutic, prognostic, or other – which itself requires having concrete plans at hand in the event of a mistake. The clinical event thus includes the always difficult attempt to talk with patients both about the risks of treatment and the experiential uncertainties associated with every treatment – as patients, after all, must live with the aftermath of every decision. At the same time, physicians must learn to listen for the varieties of experienced uncertainties presented by patients (and their loved ones).

Katz notes, too, how physicians readily and intelligibly converse with one another about uncertainty in theoretical discussions (which at times but not always includes error), yet seem to suppress such talk when they discuss clinical issues, especially with patients and others in the patient's circle of intimates. "The distinguishing characteristic of this mode of thought is that the physician will tell a false or incomplete story not only to his patient but to himself as well" (Katz, 1984, p. 170). This applies to patients as well. To ensure, for instance, that life-threatening events are really not so threatening, and that there are firm supports available to us, all of us engage in some form of denial.

While there is surely talk of "risks" outside the arena of the laboratory and experimentation, it often seems more to ignore, even mask, than confront the uncertainties and ambiguities the patient must face. There is a certain "flight from uncertainty," coupled with a kind of "training for certainty," that begins already in medical school (Katz, p. 184) – and, doubtless, much earlier than that. On the other hand, it might also be noted that those physicians who do try to talk with their patients about the actual and possible errors, the at times terrible almost unbearable uncertainties inherent to many treatments of grievous conditions, inevitably come up against the awesome difficulties of making plain sense to patients about statistical probabilities – about risks, benefits, and ambiguities coupled with those treatments. Clearly, the requirement both *to understand* and *to be understanding* (affiliation) as regards patients and their loved ones takes on a critical dimension at this point, precisely because the way in which patients and families understand and accommodate to the prospect of error and uncertainty most often differs in important ways from those with which physicians are most familiar.

To understand the uncertainties that are inherent to clinical situations, it is essential on the one hand to distinguish medicine from every other science focused on particular affairs, and on the other hand to emphasize that the changes in human life and society brought about by medicine have a unique kind of emotive and valuational force. Like the social and humanistic disciplines, medicine's principal object is the human, individual and social. Unlike being merely an object of study by such disciplines, however, to be brought as a patient to a physician is not only to find oneself observed and explained (well or badly), but *to be changed thereby*, whether trivially or more seriously. Such changes, however, are brought about within the nexus of medicine's *necessary fallibility* (Gorovitz and MacIntyre, 1976, pp. 51-71).

Uncertainty and Dialogue

The effort to optimize decision and action in the face of uncertainty is a basic characteristic of medicine, the "art of conjuncture" (Celsus). It also indicates a logical difficulty intrinsic to clinical judgment: that biomedical principles apply only to a population of patients sharing certain characteristics, which means that one can never know in advance, on the frequency distribution curve for any illness, where any individual patient will be. Uncertainty in this sense, and with it the constant possibility of error, haunts the clinical arena.

At the same time, when people become sick or distressed, they need and seek help precisely because they do not know what is wrong, what can be done about it, or which professed healer can best do whatever can be done, and what should be done. These complex uncertainties texture the clinical situation for both physician and patient (and loved ones).

Two things therefore assume critical significance in clinical situations. On the one hand, uncertainty leads to a decisional strategy that is oriented toward making the least irreversible decisions until greater certainty is at hand. On the other, uncertainty creates a crucial need for listening to and talking with patients and loved ones, for among the most difficult and common moral issues they confront is the need for definitive, even irreversible, decisions whose basis is invariably uncertainty. Clinical decisions for physicians are without fail delicate balancing acts (a) between what is suggested by statistical distribution curves of diseases, and the way in which a particular disease is manifested in the body of each unique individual; and (b) between waiting until greater certainty is at hand, and yet being attentive to the needs and feelings of individual patients. Yet, inevitably accompanying this process is the awareness that decisions often must be made before certainty is or even could be at hand.

The key question concerns how such conversations or dialogues should be conceived. Just here, it seems to me, Pellegrino captures the critical issue: the moral imperative for the physician to "be responsive to the way the patient wishes to spend his life" (1983, p. 165). Given that, how should the conversations between a doctor and a patient be understood? What is the force of this moral imperative to be responsive to the patient?

A person needing help asks, appeals to, a doctor for help; then, the doctor begins to respond, first by asking questions to which the patient in turn responds, followed by further questions and responses, etc., all designed to delineate what is wrong, what can be done about it, and then what should be done. But the person's appeal and responses to the doctor are not trivial; they arise from and constantly refer to the distress, the dis-ease, experienced by the person. In one way or another, they are critical for the person; the more grievous the illness is (or is thought to be by the patient), the more urgent is the appeal. Hence, for the doctor to be responsive, there is nothing for him/her but to seek in every appropriate way for the patient's own ways of experiencing the illness, to probe every clue available to the doctor for what the illness means to the patient. Cassell is right: because "medical care flows through the relationship between physician and patient … the spoken language is the most important tool in medicine" (1985, vol. I, p. 1). Reiser suggests there is more to this: "medical encounters begin with dialogue," in the course of which the patient's experience of illness gets transformed into “subjective portraits” (in Cassell, 1985, vol. I, p. ix) or more accurately, into personal stories or “narratives” (Kleinman, 1988; Frank, 1995).

Merely asking for help does not of itself guarantee a response. In most settings, the doctor to whom the patient appeals may refuse to respond for a variety of reasons (which themselves must be evaluated). Still, if the doctor does respond, a dialogue begins – a form of interpersonal relating with its own internal order, demands and aims. To respond to the patient means both that the doctor believes s/he might be able to help; and, whatever that response may be, it is inherently open to the patient's further queries – since it is the patient who urgently needs to know and asks for help. The doctor who responds is one who claims or “professes” to be able to help. Not only that, for it is essential to the course of dialogue that both physician and patient engage each other as truthfully and amply as possible, and for as long as the relationship continues.

To be sure, as with any conversation, the dialogue may break down for any number of reasons (which themselves require evaluation). In clinical encounters, furthermore, dialogues with patients are intrinsically periodic as well as limited. At various points in the course of the clinical event, for instance, the doctor shifts back-and-forth from the person to the embodying organism, seeking to ascertain the nature of the disease, the way it is experienced and understood by the patient, etc. During the course of the dialogical relationship, however, even when such shifts of attention are necessary, being responsive to the patient means that the doctor should never lose sight of the person being diagnosed and treated, nor the patient's narrative expressing the experience and its meaning. At some point, of course, the doctor's work is over and the patient leaves, temporarily or permanently.

The doctor-patient relationship, thus, is essentially a special form of what Schutz calls *Du-Einstellung*: being-oriented-to-another. He points out that this can be either *unilateral* (as when the other person ignores me) or *reciprocal* (the other is oriented toward me, recognizes that I am a person, too) (1973, pp. 72-88). In the healing relationship, however, the orientation to the other is inevitably more intimate and intense, a form of mutual relationship of trust and care that is at the heart of the dialogue.

V. THE INTERPRETIVE DIMENSION

The Healing Relationship

In every society and historical era, there is a critical common factor in the encounter between professed healers and patients: the need for healing. "Medical thought grows out of, and is governed by, therapeutic experience. Therapeutic theories in all their variety are attempts to makes sense out of the healer's experience with the patient" (Coulter, 1975 I, p. viii). In different terms, to think about the clinical event is to discover "the universal fact that humans become ill and in that state seek and need help, healing and cure" (Pellegrino, 1983, p. 162). No matter how ill, the ill person presents in a

> special state of vulnerability and wounded humanity not shared by other states of human deprivation and vulnerability. ... In no other deprivation is the dissolution of the person so intimate that it impairs the capacity to deal with all other deprivations (Pellegrino, 1982, p. 159).

Arising as a disturbance within that most intimate sphere of relationships between self and its own body, it is this vulnerability and its coordinate appeal for the physician to help, "to restore wholeness or, if this is not possible, to assist in striking some new balance between what the body imposes and the self aspires to" (Pellegrino, 1983, p. 163). If full restoration is not possible, then amelioration, adaptation or coping, palliation, etc., become the ends of the healing relation.

These ends are specific to medicine and distinguish it from other human activities, as well as from other activities in which physicians may also engage. For instance, when the causes and pathogenic mechanisms of disease are sought in the form of a biomedical experiment, the end is primarily knowledge. The end of preventive medicine, on the other hand, is mainly to preserve the well-being of individuals and groups, whereas social medicine seeks the health of an entire population, the public good. Physicians are important for all these, and while the clinical skills usually associated with being a physician surely are important for those other tasks, only within the healing relationship itself is the physician required to brings those skills directly to bear.

The defining moment of medicine is, then, the clinical event, and it is within this direct and always intimate relation with the vulnerable ill person that the physician seeks an action that is both technically right (scientifically sound) and, with the patient (and loved ones), morally good (a healing action).

The Praxis of Medicine

To provide that help, certain questions must be answered with each patient: What is wrong, and what will it do to the patient? What can be done to help? What *should* be done? These converge on the choice of an action that is right and good. Thus, the moment of decision of each healing relation is the centerpiece and "true clinical moment of truth, and in that moment what is most characteristic of medicine comes into existence" (Pellegrino, 1983, p. 163).

Thus, to view medicine merely or mainly as a matter of knowledge is critically inadequate: merely to possess biomedical knowledge does not imply that healing is understood, much less that the patient's interests will be served, error appreciated and discussed, etc. Rather, oriented towards therapy – medical knowledge is inseparable from *praxis*[2] – medical knowledge is essentially directed to and governed by the relationship to the one who is ill, vulnerable, and anxiously appeals for help from the person who has or claims to have knowledge and ability to help. As Henri Bergson insisted before the turn of the century (1953, pp. 12, 57, 66), knowledge is first of all in the service of action, specifically here the interests and needs of the ill person. Hence, clinical medicine is to be understood not by virtue of its abstractively considered epistemic moments, but rather as the clinical event, the moment of clinical truth.

As noted, there is an inherent logical difficulty here: each patient is more than merely an instance of some scientific principle or statistical norm, even while such principles and norms are surely pertinent as regards the patient's condition. In somewhat different terms, understanding the biology of disease requires that disease symptoms and their sundry mechanisms be abstracted from individual patients then generalized into commonly recognizable diagnostic disease patterns (which in ancient medicine was termed "logical classification of diseases"). Diseases are typically expressed in fairly constant ways in cells, organs, or organ systems; similarly, a person's genetic makeup or changes in the immune system can alter his or her biological reaction to diseases. As is suggested by clinical interventions, however, it is equally clear that personal habits, diet, physical conditioning, and the like can also alter that reaction. Each illness

> is unique and differs from every other illness episode because of the person in whom it occurs. Even when a disease recurs in the same individual, the illness is changed by the fact that it is a recurrence... [T]he presentation, course, and outcome of a disease can also be affected by whether the patient likes or fears physicians, "believes" in medication or abuses drugs, is brave or cowardly, "self-destructive" or vain, has unconscious conflicts into which the illness does or does not fit, and so on (Cassell, 1985, vol. I, p. 6).

These concerns have become all the more critical in light of the problems presented by the major diseases of our times: heart disease, cancer, stroke, ulcers, diabetes, even the malignancies associated with AIDS, which stem "primarily from the way we live" (Cassell, 1976, p. 16); thus, treating them requires sensitivity to these modes of actual living, quite as much as do the range of chronic illnesses.

Uncertainty can be reduced to some degree by having the best available information at hand and insuring that meticulous attention is given to the clinical arts: history taking, physical exams, critical use of probabilistic and modal logic, and mastery of the art of clinical listening and dialogue. It is morally imperative for the physician to understand what illness means for the patient. Illness is an experience that challenges, often in a critical and deeply personal way, the meanings of personal life, suffering, relationships with others, and the person's fundamental values. Thus, it is imperative for the physician to elicit, listen for, understand and be understanding of the patient's own experiences and understandings of his/her presented illness.

Illness and Disease

Taking patients' histories and engaging in clinical conversations with them and their families (or whomever they include in their circle of intimates) have thus become increasingly central. Clinical conversations are typically aimed at devising strategies of intervention that must be designed jointly with patients and families. These strategies are necessary, not only because their effective realization depends on the patient's initiative, compliance, and discipline (as well as support and understanding of family and/or significant others) – precisely why courage is a vital virtue in these situations. Beyond this, it has become well-recognized that the patient or legal surrogate is the real authority for decisions – arising mainly from situations involving the initiation or withdrawal of life supports at the end of life.

It is imperative, Norman Cousins argues, for physicians to learn to "strike a sensible balance between psychological and biologic factors in the understanding and management of disease" (1988, p. 1612). Personal and emotional life have too long been regarded merely as "intangibles and imponderables." Instead, there is a "presiding fact" in these inquiries: "namely, the physician has a prime resource at his disposal in the form of the patient's own apothecary, especially when combined with the prescription pad" (p. 1611). Or, as Cassell says, "the illness the patient brings to the physician arises from the interaction between the biological entity that is the disease and the person of the patient, all occurring within a specific context" (1985, vol. II, pp. 4-5).

Patients organize and embody the illness experience most often in narrative formats, deeply personal though often truncated stories (Frank, 1991; 1995); it is thus imperative for physicians not only to recognize each patient's "story," but also to develop and refine their abilities to encourage voicing and, eventually, interpreting them. Frequently, however, neither the patient nor family is able to express their full narrative adequately or accurately – surely a requirement for judging whether s/he is truly informed, uncoerced, and capable of making decisions. Recognizing this, Stanley Joel Reiser emphasizes that, although the "patient's role as narrator in the drama of illness has declined in the twentieth century," the fact is that "medical encounters begin with dialogue." Indeed, he continues, "there are few more important tasks for contemporary medicine" than the cultivation and enhancement of "our communication skills" through which a balance must be sought "between understanding general biologic processes that make us ill and understanding the illness as experienced. ... by the patient" (Reiser, in Cassell, 1985, vol. I, pp. ix-x). These considerations lead to several points bearing directly on disciplining the physician's interpretive intelligence in clinical conversational contexts.

Clinical Semeiotics

Interpretation occurs within specific contexts whose various constituents and multiple interrelations determine the physician's interpretations (Gurwitsch, 1964, pp. 105-154). In the scientific approach to medicine, Kleinman emphasizes, physicians are trained to be "naive realists, like Dashiell Hammett's Sam Spade, who are led to believe that symptoms are clues to disease, evidence of a 'natural' process, a physical entity to be discovered or uncovered" – incorporating a positive tendency to "regard with suspicion patients' illness narratives and causal beliefs" (1988, p. 17). Too few practitioners of scientific medicine truly give credit to the patient's "subjective" account.[3] It is thus not surprising that chronic illnesses are typically regarded as messy and threatening.

Nevertheless, diagnosis, therapeutic recommendations and prognosis are precisely semiotic activities, by which one symbol system (patient *complaints*) is translated into another (*signs* of disease) (Kleinman, 1988, p. 16). Central to clinical encounters, it is therefore imperative for clinicians to become proficient at these symbolic, interpretive translations. For that, it is not enough to attend merely to what are taken to be physical symptoms. One must rather attend to the full context of what each patient presents. As the latter includes bodily experiences (and patient interpretations of them expressed for the most part in common discourse with its socially derived categories), as well as a rich tapestry of personal meanings, the clinician is always faced with symptoms that are contextually determined and configured. Thus, the symptom and its personal and cultural context are, like symbols in a text, mutually determinative and enlightening: the context elaborates the meaning of the symbol, and the symbol crystallizes the context. Kleinman observes that there is

> both sufficient redundancy in the living symbolism of the symptoms and density of meanings in the life text and enough uncertainty and ambiguity in their interpretation to make this aspect of clinical work more like literary criticism or anthropological analysis of a ritual in an alien society than like the interpretation of a laboratory test or a microscopic slide of a tumor (Kleinman, p. 42).

Clinical methods differ importantly from those of physical science; they are, indeed, "closer to the human sciences" (Kleinman, p. 42). But so, too, are the interpretive "methods" of the patient and family – who, like "revisionist historians," "archivists," "diarists," even "cartographers," search their pasts for present meaning, record the most minute difficulties on the map of changing terrain of ongoing illness, and focus on the "artifacts of disease (color of sputum, softness of stool, intensity of knee pain, size and form of skin lesions)" (Kleinman, p. 48).

Compassion as Affiliative Feeling

"To be technically right, a decision must be objective; to be good, it must be compassionate," and thus, Pellegrino advises, the physician must not only be able to stand back, analyze, classify, measure and reason, but also to

> feel something of the experience of illness felt by this patient. He must literally suffer something of the patient's pain along with him, for this is what compassion literally means. Often the physician heals himself while healing the patient; oftentimes he cannot heal until he has healed himself. . .(1983, p. 165).

This needs cautious probing: "compassion" as literally suffering "something of the patient's pain," for there is something subtle and even precarious here.

Sick people frequently, although not always expressly, want the doctor to see things from their point of view: "put yourself in my shoes." The patient is surely asking the doctor to *do* something regarded as quite vital; but what is it? It seems clear that she/he is *not* asking the doctor to think about his/her predicament as if it were in fact his own ("What would you do if you were me?"). It is *not* a matter of a sort of imaginative identification (asking the doctor somehow to *become* the patient and actually feel the very pain felt by the patient). Nor is the patient asking him to consider what he would do if he were faced with the same problem, through a sort of imaginative transposal ("Suppose you faced this dilemma. . .?"). Nor is the patient asking the doctor to be judgmental – to pronounce moral judgments on the person's character or decision. To "put yourself in my shoes," rather, is to see things as the patient experiences them while yet remaining oneself. To be compassionate is *not* to obliterate the always crucial distinction between doctor and patient.

Not to belabor the obvious, such a patient is urgently asking the doctor to see his/her situation *from his/her own point of* view, that is, not only *to understand* but *to be understanding*. For this, the doctor has "to place himself in the lived experience of the patient's illness," to understand the situation "as the patient understands, perceives, and feels it" (Kleinman, 1988, p. 232). Such an act seems not simply to suppose something contrary to fact, but rather seems to be a kind of "fiction deliberately flying in the face of the facts," something "contrary to all normal possibilities" (Spiegelberg, 1986, p. 100), for how could one person – with his own biological wherewithal, experiences, history, frame of mind, etc. – assume the place of another person – with *his* own unique personal, bodily, social, and historical situation?

On the face of it, to "put yourself in my shoes" may seem absurd or at least extravagant. Still, in a sense each of us does this quite often in our daily lives. Thanks to socially derived and typified everyday knowledge, each of us typically knows what it is like, for instance, to be a postman or lawyer even though we are neither; to drive a semi or a tractor even though we do not drive; to use a wrench or operate a crane even though we have done neither; to suffer acute pain or be faced with an urgent dilemma even though we currently experience neither.

As Alfred Schutz has pointed out, our everyday knowledge of the life-world is incredibly rich and detailed even while it is also unevenly distributed into different regions (each of us knows some things better than we know others) (Schutz and Luckmann, 1973). Despite the inadequacies, inconsistencies, and inconstancies of commonsense knowledge and understanding, it is for the most part quite sufficient in the context of daily concerns – we get along for "all practical purposes," as we say. For the most part, when we are asked to "put yourself in my shoes," we typically do not go beyond such taken-for-granted, typified forms of understanding.

At times, though, something more than this typified knowledge and understanding is demanded. The doctor may be urgently asked to understand the patient's dilemma as it is actually faced and experienced: to "feel-with" him/her, as it were, from his/her own perspective and set of moral beliefs, values, etc. In these terms, to "put yourself in my shoes" involves several critical steps: helping the person to articulate and understand what that moral framework actually includes, what values the person has and how they are ordered by him/her; identifying which issues seem most pressing for him/her, given the basic ordering of values and commitments; considering the several alternatives with an eye on respective aftermaths and which seems most consonant with those beliefs, etc. Providing that kind of help, it is clear, requires disciplined self-knowledge: frequently practiced and disciplined reflection intended to delineate one's own feelings, moral beliefs, and social framework, followed by a rigorously disciplined suspension of it, in order to understand what things are like for the other person – a kind of practical, situational distantiation that undergirds the act of compassion or affiliative feeling (Zaner, 1988, pp. 315-319).

VI. CONCLUDING WORD

These considerations make it evident that the fundamental shift in medicine that began in the early 1960's and continued for some years after, is toward the recognition that clinical practice is a complex interpretive or hermeneutic discipline, more akin to the human sciences than to the biological sciences, as Edmund Pellegrino observed long ago (Pellegrino, 1979). Recent events, however – most significantly, what Hans-Jörg Rheinberger terms the "molecularization of medicine," brought into the very heart of medicine and its future by the remarkable developments in genetics (Rheinberger, 1995) – have made the turn toward interpretation (noted by Kleinman and others) quite problematic. While perhaps more deeply appreciated, the new medical paradigm of molecular biology has in effect obfuscated that turn, bringing instead a renewed and quite powerful shift toward genetic science.

This shift has resulted in shifting that insight into interpretation away from the center and more into the margins of medical concerns. For that very reason, however, the interpretive shift constitutes one of the most pressing issues facing the philosophy of medicine and clinical ethics in the forthcoming decades. Other such areas calling for concentrated labors fall generally into two groups: probing clinical encounters as they come further under the sweeping umbrella of genetics; and, on the other hand, exploring and providing needed criticisms of that new paradigm itself.

Briefly, the first area makes imperative several highly significant inquiries: whether or to what degree the new paradigm alters the features of the physician-patient relationship laid out above (e.g. in what ways new concerns for privacy will affect the clinical event); how our understanding and interpretation of "sickness" and the "illness experience" may be altered as there is increased focus on the genetic under-pinnings of disease; and, merely to mention one more area of critical issues, whether the physician's self-understanding might be changed as the understanding of the human body shifts more and more to those genetic sub-layers of human life.

The second area of issue, equally imperative if also more speculative, concerns: in what ways our understanding and experience of our own embodiment will change, as some believe is both necessary and desirable (Churchland,1986); how the molecularization of medicine not only shifts away from traditional paradigms but incorporates a radical new understanding of the basic tasks of medicine (such as genetic engineering) (Rheinberger, 1985), which brings a collapse of the traditional distinction between nature and culture. Merely to mention one other set of profound issues only now being realized (Zaner, 2001a; 2001b), it is essential to probe the fundamental "scandal" implicit in the new paradigm (Rheinberger identifies this as the potential for compromise to the very idea of 'truth'); I have argued that we must also probe more deeply into the ontological as well as ethical implications of the new paradigm, especially in light of what was marked by Sir John Eccles (1979) about "exploiting" human and animal evolution for purposes that remain quite obscure even today.

To conclude the present analysis, I have tried to show that the relationship between patient and physician is unique in a number of ways, and is among the most intimate and certainly most delicate among persons (Zaner, 1990). Because of its inherent inequality (of condition and awareness) and structural asymmetry (of power, knowledge, resources, legitimation), the relationship is especially fragile and exposed to constant dangers and temptations: manipulation and coercion, improper intimacies, and therapeutically compromising forms of remoteness, among others. Whatever else may be said about the interpretive disciplining of medical intelligence, therefore, it must surely include its being understood and practiced as a fundamentally moral discipline.

Center for Clinical and Research Ethics
Vanderbilt University
Nashville, Tennessee

NOTES

[1] It should be noted that quite a few thinkers have provided important insights into the physician-patient relationship – although those cited textually have been most significant for my own work. See, for instance, Tom Beauchamp and James Childress (1994, 4th Ed.); Scott Buchanan (1991/1938), James Childress (1982); H. T. Engelhardt (1986); Stanley Hauerwas (1986); Albert Jonsen and Stephen Toulmin (1988); Pedro Lain-Entralgo (1969); Kerr L. White (1988); Robert M. Veatch (1981).

[2] As Hans Jonas has pointed out, the practical use of medical knowledge (diagnosis, therapeutics, prognosis) is by no means accidental to modern science and theory more generally – and, it seems perfectly clear, to medicine as well to the very extent that it has allied itself with that science. Theory and power are integral to one another, he long ago argued: "the fusion of theory and practice becomes inseparable in ways which the mere terms 'pure' and 'applied' science fail to convey. Effecting changes in nature as a means and as a result of knowing it are inextricably interlocked." Hence "science" is technological by its nature (Jonas, pp. 194-95). At the same time, to the very extent that medicine's theory and practice is ordained to the diagnosis, therapeutic assessment, and prognosis of specific patients, it is a matter of practice as well. Precisely this characteristic was noted by Edelstein in his studies of classical Greek Methodism: this understanding of the close alliance between theory and practice – that practice informs and shapes theory – "is medicine's own creation and…its original contribution" (Edelstein, 1967, p. 201n18).

3. This is one of the anomalies in traditional allopathic medicine (Zaner, 1988, pp. 96-106), which demonstrate the "dominance of technologically centered techniques of medical evaluation in which the views of patients become largely irrelevant, if not obtrusive" (Reiser in Cassell, vol. I, 1985, p. ix).

REFERENCES

Bayertz, K.: 1994, *GenEthics: Technological Intervention in Human Reproduction as a Philosophical Problem*, Cambridge University Press, Cambridge. (Tr from the German by S. L. Kirby, *GenEthik*, Rowohlt Taschenbuch Verlag GmbH, Hamburg, 1987).

Beauchamp, T. L. and Childress, J. F.: 1994, *Principles of Biomedical Ethics*, 4th Ed., Oxford University Press, New York and London.

Bergson, H.: 1953, *Matière et Mèmoire*, 54th Ed., Presses Universitaires de France, Paris.

Buchanan, S.: 1938/1991, *The Doctrine of Signatures: A Defense of Theory in Medicine*, University of Illinois Press, Urbana and Chicago.

Burnett, M.: 1978, *Endurance of Life: The Implications of Genetics for Human Life*, Cambridge University Press, London.

Cassell, E. J.: 1976/1985, *The Healer's Art: A New Approach to the Doctor-Patient Relationship*, J. B. Lippincott Co., Philadelphia, MIT Press, Boston.

Cassell, E. J.: 1985, *Talking With Patients*, two vols., MIT Press, Boston.

Childress, James F.: 1982, *Who Should Decide? Paternalism in Health Care*, Oxford University Press, New York.

Churchland, P. S.: 1986, *Neurophilosophy: Toward a unified Science of the Mind/Brain*, MIT Press, Cambridge, MA.

Clark, K. B.: (1973), 'Psychotechnology and the Pathos of Power,' in F. W. Matson (ed.), *Within/Without: Behaviorism and Humanism,* Brooks/Cole Pub. Co., Monterey, CA

Coulter, H. B.: 1975, *The Divided Legacy*, vol. I, Wehawken Book Co., Washington, D.C.

Cousins, N.: 1988, 'Intangibles in Medicine: An Attempt at a Balancing Perspective,' *Journal of the American Medical Association* 260:11 (September 16), 1610-12.

Donnelly, W. J.: 1986, 'Medical Language as Symptom: Doctor Talk in Teaching Hospitals,' *Perspectives in Biology and Medicine* 30:1 (Autumn), 81-94.

Eccles, J.: (1979), *The Human Mystery: The Gifford Lectures (1977-78),* Springer-Verlag, New York and Heidelberg.

Edelstein, L.: 1967, *Ancient Medicine*, O. and C. L. Tempkin (eds.), The Johns Hopkins Press, Baltimore.

Engel, G.L.: 1988, 'How Much Longer Must Medicine's Science Be Bound by a Seventeenth Century World View?' in K. L. White (ed.), *The Task of Medicine: Dialogue at Wickenburg,* The Henry J. Kaiser Family Foundation, Menlo Park, CA.

Engelhardt, H. Tristram, Jr.: 1986, *The Foundations of Bioethics*, Oxford University Press, New York.

Frank, A. W.: 1991, *At the Will of the Body: Reflections on Illness*, Houghton Mifflin, Boston.

Frank, A. W.: 1995, *The Wounded Storyteller: Body, Illness, and Ethics*, The University of Chicago Press, Chicago, IL.

Gurwitsch, A.: 1964, *The Field of Consciousness*, Duquesne Studies, Psychological Series, Duquesne University Press, Pittsburgh.

Hardy, Robert C.: 1978, *Sick: How People Feel About Being Sick and What They Think of Those Who Care for Them,* Teach'em, Inc., Chicago.

Hauerwas, S.: 1986, *Suffering Presence*, University of Notre Dame Press, Notre Dame, IN.

Jonas, H.: 1966, 'The Practical Uses of Theory,' in Hans Jonas, *The Phenomenon of Life: Toward a Philosophical Biology*, University of Chicago Press, Chicago, pp 194-95.

Jonas, H.: 1974, *Philosophical Essays: From Ancient Creed to Technological Man*, Prentice-Hall, Inc., Englewood Cliffs, NJ.

Jonas, H.: 1977, 'The Concept of Responsibility: An Inquiry into the Foundations of an Ethics for Our Age,' In H. T. Engelhardt, Jr. and D. Callahan (eds.), *Knowledge, Value and Belief*, The Institute of Society, Ethics and Life Sciences, 169-198.

Jonas, H.: 1984, *The Imperative of Responsibility: In Search of an Ethics for the Technological Age*, University of Chicago Press, Chicago, IL.

Jonsen, A.J. and Toulmin, S.: 1988, *The Abuse of Casuistry,* University of California Press, Berkeley and Los Angeles.

Katz, J.: 1988, *The Silent World of Doctor and Patient*, The Free Press, New York.

Kleinman, A.: 1988, *The Illness Narratives: Suffering, Healing and the Human Condition*, Basic Books, New York.

Lain-Entralgo, P.: 1969: *Doctor and Patient*. New York: Pantheon Books.

Lenrow, P. B.: 1982 (1978), 'The Work of Helping Strangers,' in: H. Rubenstein and M. H. Block (eds.), *Things That Matter: Influences on Helping Relationships*, Macmillan Publishing Co., New York, 42-57.

Odegaard, C. E.: 1988, 'Towards an Improved Dialogue,' in K. L. White (ed.), *The Task of Medicine: Dialogue at Wickenburg*, The Henry J. Kaiser Family Foundation, Menlo Park, CA, 99-112.

Pellegrino, E. D.: 1974/79, 'Medicine and Philosophy: Some Notes on the flirtations of Minerva and Aesculapius,' In E. D. Pellegrino, *Humanism and the Physician*, University of Tennessee Press, Knoxville, TN.

Pellegrino, Edmund D. 'The Anatomy of Clinical Judgments: Some Notes on Right Reason and Right Action,' in H.T. Engelhardt, Jr., S.F. Spicker, and B. Towers (eds.), *Clinical Judgment: A Critical Appraisal,* D. Reidel Publishing Company, Dordrecth/Boston/London, pp. 169-194.

Pellegrino, E. D.: 1982, 'Being Ill and Being Healed: Some Reflections on the Grounding of Medical Morality,' in V. Kestenbaum (ed.), *The Humanity of the Ill: Phenomenological Perspectives*, University of Tennessee Press, Knoxville, 157-166.

Pellegrino, E. D.: 1983, 'The Healing Relationship: The Architectonics of Clinical Medicine,' in E. A. Shelp (ed.), *The Clinical Encounter: The Moral Fabric of the Physician-Patient Relationship*, D. Reidel Publishing Company, Boston and Dordrecht, 153-172.

Rheinberger, H-J.: 1995, 'Beyond Nature and Culture: A Note on Medicine in the Age of Molecular Biology,' in *Medicine as a Cultural System*, M. Heyd and H.-JR Rheinberger (eds.), special issue of *Science in Context* 8:1 (Spring), pp. 249-63.

Ruark, J. E., Raffin, T. A., and the Stanford University Medical Center Committee on Ethics: 1988, 'Initiating and Withdrawing Life Support,' *The New England Journal of Medicine* 381:1 (January 7), 25-30.

Schutz, A.: 1967, 'Concept and Theory Formation in the Social Sciences,' in: *Collected Papers*, Vol. I, ed. M. Natanson, Martinus Nijhoff, The Hague, 48-66.

Schutz, A. and Luckmann, T.: 1973, *The Structure of the Life-World*, Vol. I., tr. H. T. Engelhardt, Jr. and R. M. Zaner, Northwestern University Press, Evanston, IL.

Veatch, R. M.: 1981, *A Theory of Medical Ethics*, Basic Books, Inc., New York.

White, Kerr L. (ed.): 1988, *The Task of Medicine*, The Henry J. Kaiser Family Foundation, Menlo Park, CA..

Zaner, R. M.: 1985, 'How the Hell Did I Get Here?' Reflections on Being a Patient,' in: A. H. Bishop and J. R. Scudder, Jr. (eds.), *Caring, Curing, Coping*, The University of Alabama Press, University, AL., 80-105.

Zaner, R. M.: 1988, *Ethics and the Clinical Encounter*, Prentice-Hall, Inc., Englewood Cliffs, NJ.

Zaner, R. M.: 1990, 'Medicine and Dialogue,' in Engelhardt, H. T. (Ed.), Special Issue: 'Edmund Pellegrino's Philosophy of Medicine: An Overview and an Assessment,' *Journal of Medicine and Philosophy* 15: 3, 303-325.

Zaner, R. M.: 1991, 'The Phenomenon of Trust in the Patient-Physician Relationship,' In E. Pellegrino (ed.), *Ethics, Trust, and The Professions: Philosophical and Cultural Aspects*, Georgetown University Press, Washington, D.C., 45-67.

Zaner, R. M.: 1992, 'Parted Bodies, Departed Souls: The Body in Ancient Medicine and Anatomy,' In D. Leder (ed.), *The Body in Medical Thought and Practice,* Kluwer Academic Publishers, Dordrecht & Boston, 101-122.

Zaner, R. M.: 1994a, 'Illness and the Other,' in G. P. McKenny and J. R. Sande (eds.), *Theological Analyses of the Clinical Encounter*, Theology and Medicine Series, Kluwer Academic Publishers, Boston, 185-201

Zaner, R. M.: 1994b, 'Experience and Moral Life: A Phenomenological Approach to Bioethics,' in E. R. DuBose, R. Hamel and L. J. O'Connell (eds.), *A Matter of Principles? Ferment in U.S. Bioethics*, The Park Ridge Center for the Study of Health, Faith, and Ethics, Trinity Press International, Valley Forge, PA, 211-239.

Zaner, R. M.: 1994c, 'Encountering the Other,' in C. S. Campbell & A. Lustig (eds.), *Duties to Others*, Theology and Medicine Series, Kluwer Academic Publishers, Boston, 17-38.

Zaner, R. M.: 1995, 'Interpretation and Dialogue: Medicine as a Moral Discipline,' Essays in Honor of Maurice Natanson, S. Galt Crowell (ed.), *The Prism of Self*, Kluwer Academic Pubs., Dordrecht and Boston, 147-168.

Zaner, R. M.: 2001a, 'Sisyphus Without Knees,' invited lecture on the occasion of the retirement of Professor Kay Toombs, Baylor University, April 2, 2001 (unpublished with publication planned).

Zaner, R. M.: 2001b, 'Thinking About Medicine,' in Kay Toombs (Ed.), *Handbook of Phenomenology and Medicine*, Kluwer Academic Publishers, Dordrecht.

STEPHEN WEAR

INFORMED CONSENT

One would be hard-pressed to identify a realm of practical activity that has received more intense, detailed and comprehensive philosophical attention than contemporary medical practice. The field of bioethics, especially in the last quarter of the twentieth century, has certainly been intensely critical and reflective about this practice in all regards, from focusing on generic issues regarding the proper nature of the clinician-patient interaction, to reflection on a multitude of specific issues, whether in the areas of death and dying, genetics, or access to health care. And perhaps no issue is more fundamental to this extraordinary, ongoing critique than the issue of informed consent. Its fundamental place in bioethics seems plain in a quantitative sense in that a simple tallying of references to it over the last 25 years easily seems to rank it first in the attention it has received (Sugarman, 1999). Equally, it seems quite clear that informed consent is now being offered as an established principle that anchors what has been called the "new ethos of patient autonomy" (McCullough and Wear, 1985), a reasonably pithy phrase for capturing what bioethics has come to advocate as an across-the-board reconstruction of medical care.

For all this comprehensive critique and theorizing about bioethical issues, much of it offered by academically trained philosophers, the core theme of this paper is that bioethics has evolved into a discipline that shares more in common with the "getting the job done" character of medicine than with philosophy. In effect, I submit that we now have what amounts to an orthodoxy within bioethics that is broadly seen as settled, particularly regarding the notion of informed consent and related themes. But, as I will illustrate and argue, much of this orthodoxy merits (and has received) formidable criticism, and should be seen, particularly by philosophers, as providing at best a tentative draft that calls for more searching critique and development. More specifically, this article will argue that this "philosophical" reflection on bioethics, from its infancy onward, has arisen out of a preconceived set of principles and values that was being robustly promoted from the beginning. As philosophers, we will be obliged to dig deeper than bioethics itself has to unearth and provide a proper philosophical critique of informed consent.

I can anticipate the basic thrust of the ensuing argument by signaling two primary features of contemporary bioethics that I see as productive of this

G. Khushf (ed.), Handbook of Bioethics, 251–290.

orthodoxy. First, it is clear that bioethicists came to their study of medicine early on with certain basic and seminal intuitions and concerns regarding informed consent. Generically, the concern about informed consent arose out of what may well have been the guiding, basic intuition of the bioethics movement, viz. the broad perception of the "silence between doctor and patient" (Katz, 1984), and the collateral view that the physician paternalism that produced this silence conflicted with the basic values of a free society (Ramsey, 1970; Baron, 1987). In a crucial sense, it seems accurate to say that most bioethicists were convinced from the start of the unethical character of much of medicine, and were intent on modifying this practice for the better. The second salient feature lies in the fact that certain paradigm situations and concerns operated early on in bioethics. Again regarding informed consent specifically, concerns over the conduct of research on human subjects and the paternalistic tendency to over-treat certain devastated patients against their wills, provided much of the early focus and fuel of bioethics. But, for all the specific merits of such pre-formed convictions and robust paradigms, I will argue that they are now being utilized to support an orthodoxy that begs for extensive philosophical critique and re-construction. In short, I submit, bioethics has presumed to reach closure regarding the issue of informed consent and related notions much too prematurely.

In this essay, I (1) describe the bioethical consensus regarding informed consent, attending both to the sources of this evolution, as well as to its widely embraced result; (2) reappraise this consensus by identifying and critiquing its basic assumptions, with significant aid from the literature of bioethics itself; (3) offer a revised model of informed consent that presumes to be informed by and respond to the previous critique; and then (4) conclude with a further reflection on the philosophical issues that arise within such a revised model and throughout bioethics generally.

I. THE EVOLVED CONSENSUS REGARDING INFORMED CONSENT

If the harsh criticism of medicine found in much of the literature on informed consent is accurate, one might well wonder how physicians managed to attract any patients at all. Most people do not appreciate being lied to or deceived, resent being coerced, and supposedly like to retain control of their own lives. In truth, the usual physician-patient interaction was much more benign. Physicians often saw little point in informing patients, preferring to make recommendations that their patients were supposed to accept and with which they were expected to comply. Such behavior may well not have enhanced freedom and self-determination, but neither was it outrageously paternalistic or disrespectful.

Both parties in the relationship simply felt that the doctor knew best; such decisions were seen as matters of medical expertise and judgment.

Whatever the status of its primal intuitions about the "silence between doctors and patients," a basic and corollary impetus for the new ethos of patient autonomy seems to have developed from other, more specific sources and insights, *not* solely from generic concern over the silence within the usual physician-patient interaction. Two sorts of clinical situations provided much of the fuel that has made the ethics of medicine such a burning issue in recent years: biomedical research and extraordinary cases.

Bioethics' Initial Critique of Twentieth Century Medical Practice

Though much is now said about more positive goals of enhancing patient autonomy, via informed consent and otherwise, the early insights and agendas of the new ethos had a much more negative focus, i.e., that medical practice contains profound threats to both patient freedom *and* well-being. These perceived threats initially arose under two basic rubrics, (1) concerns about research subjects, and (2) extraordinary cases.

A. Concerns about Research Subjects. The initial well-spring of such concerns was biomedical research. Here, though various demands for informed consent had already been made in response to the specter of the Nazi doctors (Lifton, 1986), it was not until the sixties that this area received the attention it merited. Initially highlighted by members of the medical community itself, e.g., in Henry Beecher's article "Ethics and Clinical Research" (Beecher, 1966), the growing perception was that the biomedical research community had become much too zealous in its dealings with research subjects, who were generally provided little or no informed consent to risky and often non-therapeutic procedures. Further, the populations most often studied were among the most vulnerable and least autonomous in the society. Often research subjects were institutionalized patients, such as prisoners, the mentally ill and developmentally disabled, and the elderly.

Particularly for the medical profession, the problem (and the threat) was clear. Biomedical researchers, in their zeal, had sinned against the traditional principle that the physician's primary allegiance must be to the protection and promotion of the best interests of his patients, regardless of the impact of doing so on other considerations, e.g., the advancement of medical knowledge. Such physicians were thus coming to the physician-patient relationship with conflicting interests and agendas. In the research setting the clinician-researcher attempts to benefit not only the patient-subject, but future patients as well. That the emphasis had swung intolerably far away from the interests of the research subjects themselves was apparent in numerous instances, such as the Tuskegee and Willowbrook experiments where, respectively, southern blacks with syphilis were left untreated so that the natural history of the disease could be studied, and developmentally disabled patients were intentionally infected with hepatitis toward the same research goal (Final Report of the Tuskagee Syphilis Study Ad Hoc Advisory Panel, 1977). The medical profession itself seemed to recognize the inappropriateness and dangers in such behaviors and a mandate for informed consent in the research setting gradually evolved, along with the formation of monitoring bodies, i.e., institutional review boards. But such corrections were not accomplished without the growth of a residual distrust within the society that physicians might use patients as "guinea pigs."

B. The Influence of Extraordinary Cases. The new ethos of patient autonomy also has deep roots in a consuming focus on extraordinary types of cases. These cases tended to be drawn from the setting of the large, urban, multi-specialty hospital, and focus on conflicts between physicians and patients in extraordinary life-threatening situations. Students and teachers of bioethics are familiar with such paradigm cases: adult, competent Jehovah's Witnesses being transfused against their wills; severely burned or spinal cord injured patients being sedated or ignored in response to their rejection of aggressive treatment for lives that they do not consider worth living; and adult, competent patients having their lives prolonged in "inhumane" ways, especially in intensive care units, rather than being allowed to "die with dignity," perhaps at home or in a hospice.

We should not miss the images in such cases; they go far beyond mere argument. Whether it be the badly burned Donald Cowart in "Please Let Me Die" (Videotape, 1974), or Richard Dreyfus as a quadriplegic in "Whose Life Is It Anyway?" (Metro-Goldwyn-Mayer, 1981), fuel for moral indignation abounds. In both films, clearly competent and extremely articulate patients are presented as demanding an exit from ghastly and hopeless situations that one might hesitate to wish even on one's worst enemy. And there are the doctors, aloof and arrogant in their white smocks, refusing to honor wishes that any of us might express in such situations, and seeking to circumvent them by drugging their patients, denying their apparent competence, and generally treating them like ignorant, hysterical children.

It is crucial to note at this juncture, however, that the preceding hardly calls for the across-the-board reformation of medical practice that the "new ethos of patient autonomy" promotes and the doctrine of informed consent institutes. Relatively few patients become research subjects, and the previous sorts of extraordinary cases are just that, extraordinary. The latter could be dealt with by insisting on the right to refuse treatment, and the former with an insistence on detailed informed consents, but most of medical practice might well proceed as before. Whence then the new ethos of patient autonomy, if it was not actually called for by the preceding? To answer this question, aside from observing that we may well be faced simply with the philosopher's ancient tendency to generalize too readily from the particular to the general, we must look to the concurrent development of informed consent in the law.

The Evolution of the Legal Doctrine of Informed Consent

The law's focus on informed consent predated the emergence of bioethics, but eventually came to be part and parcel with the evolved consensus that we are seeking to articulate. Prior to the twentieth century, some precedents showed concern for issues such as whether the patient had consented to a procedure (without concern for the provision of any information), whether physicians are obliged to alert patients to basic risks of a proposed treatment, and whether fraud had occurred in the case of a physician's representation of a patient's problem (Faden and Beauchamp, 1986, pp. 114-125). In the past these sorts of issues were usually resolved by referring to the customary practice of members of the medical profession. In effect, the traditional response to our concerns about what informed consent should be was that the medical profession itself was the legally recognized source of guidelines and criteria, not specific legal principles and agendas.

Reliance on the customary practice of the medical profession, i.e. on professional standards, surely makes things simpler for the clinician, at least to the extent that an individual clinician shares a common training, experience and perspective with his peers, and should have a general sense of what is customary. The clinician is thus not also obliged to attempt to fathom the different languages, agendas and formulations of the legal profession. Ironically, as Flexner's reforms of medical education made such a common experience and perspective among clinicians more of a reality, the law tended to shift away from a solely professional standard.

In the early twentieth century, certain principles and agendas that had long been basic to the law but less central to medical practice, began to come to the fore in judicial determinations regarding patient consent and patient-physician interactions. Most basic, at first glance, was the common law's traditional concern for the "bodily integrity of the individual," in effect a concern for patient "self-determination," which at least requires patient consent to treatment, however uninformed.[1] But other legal concerns and agendas also began to assert themselves: (1) the law's prescription against battery, i.e. the "unauthorized touching" of one individual by another (Faden and Beauchamp, 1986, pp.120-122); (2) tort law's concern about the infliction of physical and emotional harm, and legal mechanisms for compensation (Faden and Beauchamp, 1986, pp. 125-132), and (3) the constitutional concern for the individual's privacy, his right to be left alone.[2]

The specific history of the modern emergence of such concerns into matters traditionally seen as relating to medical custom and expertise is quite controversial among legal scholars (Burt, 1979; Faden and Beauchamp, 1986, pp. 53-60; Mesiel, 1988; Szczygiel, 1994). The controversy relates particularly to which principle or branch of law is governing. One option here is to emphasize patient self-determination, which many of the decisions do, at least rhetorically. But strong argument has been advanced that the principle of self-determination has not been the dominant force behind the development of the legal doctrine.[3] Had it been so, courts would have developed more specific disclosure requirements and guidelines. They also might have been much more concerned about the most effective ways to enhance understanding and fight against ritualistic disclosures that barely satisfy the letter, much less the spirit, of such a legal principle. Another way to put this is that if courts were truly concerned about patient self-determination, then an offense against it, e.g., not adequately informing a patient, would in and of itself be treated as an actionable harm, even if no other physical or emotional harm occurred. A lack of *informed* consent could be considered and penalized as a battery, an unauthorized touching, even if a bare consent had been obtained and the intervention was medically successful.

But such developments have not occurred. A lack of informed consent has been actionable, and damages awarded, *only when* its absence could be shown to have caused some *other* emotional or physical harm. By way of example, if a patient was not told of a major risk of a procedure, e.g. the risk of death from general anesthesia, and death did *not* occur, damages would not be awarded simply for the affront to patient self-determination. If, however, such a result did occur, and it could be proved that the patient would not have undergone the procedure had he known of that risk, then damages would be awarded for the injury, even if it was a foreseeable risk of the intervention and no other malpractice was involved. In sum, an inadequate consent is actionable only if some emotional or physical injury also resulted, but there is no legal action for an inadequate provision of information *per se*.

This state of affairs may not seem disconcerting to clinicians, particularly given the current malpractice climate, in that damages are awarded only if there are actual emotional or physical injuries, not simply if an adequate informed consent is lacking in a situation where treatment is otherwise benign or successful. But to those who are concerned that patient self-determination itself be protected *and* fostered in clinical medicine, and who see a lack of an informed consent as a serious affront and harm to a patient's liberty, all this may be regrettable. For our purposes, the significance of the preceding is that much of the development of the legal doctrine of informed consent has instead occurred within the context of tort law, i.e. malpractice actions (Wear, 1998, pp. 9-24).

The broader point to be garnered from this rehearsal of the evolution of legal thinking on informed consent is that the law's rhetorical concern over self-determination does not end up getting the practical emphasis and support that one might well have expected. Legal concerns regarding valid informed consents instead get thrown into the realm of tort law and the sort of self-determination that a philosopher might recognize seems to be allocated to the back burner. That bioethics proper ends up embracing a similar result will be seen as we move to the legal-ethical doctrine that describes the elements of informed consent and the process by which these elements are to be convey

The Elements and Process of Informed Consent

Bioethics and the law eventually part company regarding the doctrine of informed consent in various ways, but there are certain basic principles and intuitions that both share and which are mutually supportive. This is so particularly regarding the *types* of information that informed consents should contain, and the presumption that patients are generally competent to receive and respond to such informed consents. Both clearly end up requiring that informed consents be offered across-the-board in medical practice, whenever there is any possibility of harm in the proposed intervention (to many thinkers, this includes virtually all medical interactions, as even benign, non-invasive testing runs a risk of false positives and then reference to further, less benign, diagnostic and therapeutic responses).

Regarding the basic elements to be disclosed, a legally and ethically valid informed consent should instruct the patient regarding: (1) the problem or diagnosis for which further investigation or intervention is proposed, (2) the recommended intervention coupled with the significant benefits and risks attendant to it, (3) the results or prognosis if no intervention is attempted, and (4) any significant alternative modalities with their attendant risks and benefits. Further, all competent patients must receive such information about any diagnostic or therapeutic intervention except in situations where (a) the patient is threatened with serious harm or death if the intervention is not immediately provided (the *emergency exception*), or (b) the patient voluntarily gives up the right to be so informed and consents, in advance, to what the physician considers the appropriate form of action (the *waiver exception*), or (c) the physician has sufficient reason to believe that disclosure itself would cause serious physical or psychological harm to the patient (the *therapeutic privilege exception*), or (d) the information at issue is too commonly known to merit mention (the *common knowledge exception*) (Meisel, 1979; Wear, 1998, pp. 21-24). Concerning which patients are competent (and thus entitled) to give an informed consent, such competence should be presumed unless sufficient reasons to the contrary are identified, e.g. gross mental deficits or incapacity. And, finally, all this should occur without any coercion or manipulation that undermines the patient's ability to choose.

The preceding may seem straightforward enough, but numerous issues quickly arise as one reflects further. (1) Even though the types of information are clear, we also need more specific criteria for determining which information should actually be disclosed within a given type. This is particularly the case when one gets to the disclosure of the risks of an intervention, as reference to any drug insert advises us that the *possible* risks of any drug are usually quite numerous. So some selection criteria regarding the main or most significant risks are essential, lest risk disclosure degenerate into an undifferentiated list that tends to produce information overload rather than insight in the patient. (2) We need to know what specific criteria are available to determine whether one of the four exceptions to informed consent may be legitimately appealed to in a given case; without clear and specific definition, exceptions tend to destroy the rule (Wear, 1998, pp. 156-170). (3) An operational definition of competence is required so that one can determine who shall or should not be offered an informed consent, i.e. what are "sufficient reasons" to abandon the mandated presumption of competence in a given case? And (4) the issue of how informed consents should be offered comes to the fore, with the law seemingly satisfied with an "event" model of informed consent, where it occurs at a given place and time and is documented in some sort of written, signed informed consent. Many bioethicists have rejected such an event model in favor of a process model where the informing somehow occurs over time, and is keyed to the patient's developing an increased understanding of his or her situation, prospects and options.

The Basic Presuppositions of Informed Consent

From the preceding, we may now summarize the basic presuppositions of the doctrine of informed consent that bioethics has evolved. Bioethics is, first of all, clear as to the basic elements of informed consent.. It is also clear that any medical situation that involves the proposal of a diagnostic or therapeutic intervention that has some sense of risk or potential discomfort to it must be accompanied by an informed consent. Further, bioethics has fully embraced the caveat that a patient's competence should be assumed unless proven otherwise; the bias thus is that patients should, *prima facie*, be considered ready, willing and able to receive and respond to informed consents about their medical care.

What remains unclear, thus far, is what sort of criteria one should use to decide what the actual content of disclosure should be, given that information overload is not a welcome result. Nor are we clear as to how to decide if a given patient lacks adequate competence for informed consent should this issue somehow arise beyond the *prima facie* bias that is commended. Rather than attempt to sort these issues out at this juncture, however, we will be better served if we step back and review various criticisms that have been made of the preceding consensus, criticisms that have arisen within bioethics itself, and which go to the heart of our concern as to what philosophical assumptions, concerns, and commitments are truly operative in bioethics generally and with particular reference to informed consent.

II. A CRITICAL REAPPRAISAL OF THIS CONSENSUS

I believe that if one approached the average bioethicist and asked for his or her sense of the nature and grounds of the doctrine of informed consent, one would receive something like the preceding, certainly regarding the sorts of information an informed consent should include, and equally regarding the presumption of competence regarding patients. Older hands in the field, at least, would also tend to identify the roots of the doctrine in the concerns about the treatment of research subjects and the extraordinary sorts of cases previously noted. What at times worries me about bioethics, and makes me welcome this volume, is that I sense that many bioethicists tend to see all this as settled "doctrine," with little sense that there are any residual issues to deal with; in effect, many bioethicists seem to have concluded that the nature, grounds and goals of informed consent are clear and uncontroversial.

This is, I submit, a major mistake and lies at the heart of my claim that contemporary bioethics, for all its previously noted presumption to philosophical reflection and erudition, may well still be seen as having prematurely concluded that it has reached an adequate closure in this area.

There is much that merits challenge in the preceding, in fact, and much of this challenge can be abstracted out of the literature of bioethics itself. I will now proceed to document this internal critique under a number of headings, all of which will assist us in clarifying the underlying assumptions, concerns, and commitments of bioethics as well as how they are inadequate. These headings are (1) "promoting, restoring and enhancing self-determination" wherein we attempt to clarify the notions of freedom and self-determination operative in bioethics; (2) "empirical studies of informed consent" wherein the large empirical literature regarding patient comprehension of informed consents is reviewed, in part toward the issue of how successful an intervention informed consent actually is at the bedside; and (3) "a clinical critique of the competency assumption" wherein the intuition that illness tends to diminish people's abilities to understand and make decisions about their care is canvassed for its implications regarding the bias in favor of patient competence. Having reviewed these three areas, we will then proceed to restate what may be the more realistic goals of informed consent at the bedside, and then offer a revised model of informed consent that seeks to capture these goals. In passing we will concurrently get a much clearer sense of the ways that bioethics' evolved consensus regarding informed consent is not at all as clear and stable as many take it to be.

Bioethics' Rhetoric Regarding Promotion, Restoration and Enhancement of Self-determination

We have already noted that the law on informed consent, however much it rhetorically emphasizes self-determination, tends to slight it in practice, i.e. by making departures from adequate informed consents punishable within tort law, wherein actual harms, beyond just those to self-determination, must occur for sanctions to be justified. If we move to bioethics proper, we certainly find a similar emphasis on self-determination, but should prudently wonder if this is also just more rhetoric. That is: does bioethics itself keep to a robust notion of patient autonomy, arguably its most basic tenet, or does its commitment to and conception of this notion tend to get watered down in practice, as has happened in the law.

Now bioethicists routinely call for consents to be "full and informed." They regularly emphasize the need for a process model of informed consent over an event model, wherein individual patients can be assisted to gain such "full and informed" consents in their own ways, and at their own individual speeds (Wear, 1998, pp. 91-100). And all this is leavened by further recommendations that (1) informed consents be in laypersons' language; (2) instead of accepting the medical model sort of rendition of a patient's situation and prospects, a narrative model of a patient's own sense of their medical history should be constructed (Hunter, 1991); and (3) to insure that the patient understood what was communicated in the informed consent, some sort of feedback mechanism should be incorporated into the consent process where the patient is encouraged to report what he or she understood from the consent disclosure and inaccuracies or confusions can be responded to by the clinician (Wear, 1998, pp. 122-24).

Now the preceding paragraph surely calls for tactics, e.g. interactive informing, narrative construction, and feedback, that would amount to a quite aggressive stance towards promoting self-determination in patients, via informed consents and otherwise. Such tactics certainly go way beyond purely negative sorts of injunctions against lying to or not adequately informing people. But, as with many practices, we should be leery of accepting the rhetoric of a practice as accurately describing the actual practice and not just of what happens, but particularly regarding what those who offer the rhetoric are willing to accept. And given that this rhetoric is offered regarding *all* informed consents, not just, for example, experimental treatment or extraordinary cases, we should clarify how informed consents regularly occur in the more mundane cases.

It seems quite clear that most consents occur as a single event, with little or no attempt at the construction of the patient's narrative or the solicitation of feedback as to what the patient actually understands, and are completed by a signature on some document that is not very detailed, and certainly not individualized to the particular patient or the specific clinical situation at hand. Moreover, the consent event-process is usually almost entirely one-way, i.e. the physician offers a recommendation that is parsed out in terms of its basic risks and benefits, alternatives, and some sense of what would happen if no therapeutic response is attempted. Finally, in most instances, if the patient remains silent, asks no questions, and simply indicates that he or she accepts the clinician's recommendations, then an adequate process is deemed to have occurred.

Now this may well be all that is usually needed or desired by and for patients. It merits emphasizing, however, that this is surely not the robust sort of "promoting, restoring and enhancing of self-determination" that the bioethicists' rhetoric insists upon. One would, in fact, be naïve to think that such one-way events, and silence by patients, often amounts to anything more than a bare and essentially uninformed *authorization* by patients, not a "full and informed consent." And, as we shall shortly see from reviewing the empirical literature on informed consent, this is the case. Our issue at present, however, regards bioethics' reaction to all this. In sum, does it stick to its rhetoric when it moves to reflect specifically on practice situations?

There are two answers to this question. One is that there are certainly numerous special circumstances where bioethicists emphasize the above tactics for "promoting, restoring and enhancing self-determination," e.g. when non-therapeutic experimentation is being proposed (Lavori, 1999), or when patients are faced with extraordinary treatment situations, and advanced directives or limitations of treatment are at issue. Equally, in situations of extended chronic illness (Kaplan, 1989; Lidz, 1985) or potentially terminal illness where extensive therapeutic response is being contemplated, the importance of nuanced communications, feedback, and attention to the patient's own sense of things is touted, and not just for the sake of so-called "autonomous decision-making," but to respond to the perceived danger of patients becoming diminished in their self-determination by the withering effects of such extended and/or catastrophic illness (Parsons, 1975).

All this is as it should be, and arises out of the early concerns about research subjects and extraordinary cases previously mentioned. It is part of the basic, necessary and crucial contribution that bioethics has actually made to medical practice. But the second answer to our question is that when we move to the rest of medical care, i.e., the vast majority of it, none of this aggressive autonomy promotion is occurring, the silent patient is accepted, and bioethics is itself essentially silent about this.

Why is this so? Grudgingly, I would submit, and only somewhat explicitly, bioethics has come to accept that there are serious practical barriers to its aggressive autonomy promoting tactics, and often insufficient need or desire for them. These practical limiting caveats are roughly the following: (1) in the usual medical situation, clinicians are offering patients interventions that seem to them to be straightforwardly in the patients' best interests, with no sense of controversy about the choice involved; in fact, the clinician would be alarmed if the patient did anything else but agree to proceed; (2) whatever the concerns of the bioethicists, or the much touted "medical consumer" rhetoric, sick folks mostly tend to come to clinicians to be "fixed and reassured," not to be informed and asked to make decisions about issues that they see as mainly a matter of the doctor's expertise, not theirs; and (3) given these first two points, such tactics as feedback, constructing the patient's narrative, or insisting on an extended process of communication and consent, are not only not seen as needed by either party to the clinician-patient relationship, but simply not possible in all or many cases, given the lack of time to pursue them.

And what does the preceding signify? I submit that, as with the law where its rhetorical concern for self-determination evaporated as it made informed consent a creature of tort law, so bioethics, for all its well-taken, nay crucial, emphasis that certain special circumstances in medicine absolutely require such autonomy enhancing tactics, in the usual case bioethics has tacitly accepted that such tactics are often neither needed nor desired. But our point must then be that the "new ethos of patient autonomy" is as much false rhetoric as reality, and its recommendations for medicine as a whole are much more limited. We will need to clarify shortly what this all means for the actual goals and values that bioethics should be proposing for medical practice. But we should first canvass two other grounds for questioning the rhetoric of patient autonomy and medicine's response to it, viz. "empirical studies of informed consent" and the "clinical critique of the competency assumption."

Empirical Studies of Informed Consent

Informed consent may be usefully approached just like any other clinical intervention, i.e. as an intervention that is intended to change outcomes. What outcomes? The primary one would seem to regard the degree of actual understanding that patients have as the result of an informed consent. And this has, not surprisingly, been extensively studied, with certain clear findings.

There are numerous studies of the general effectiveness of informed consent, which measure the percentage of patient recall of the total information provided. Their results vary markedly, but none show a particularly rosy picture. The best results were obtained by Bengler and his colleagues who found that 72 percent of disclosed information was retained immediately after provision (Bengler, 1980). This percentage, not surprisingly, dropped to 61 percent at 3 months post disclosure. At the other extreme, Robinson et al. found only 20 percent recall of information at four to six months post disclosure (Robinson, 1976). The bulk of the studies show recall rates within the 30 to 50 percent range.

There are good reasons to quarrel about the significance of all such studies, depending on one's bias. The four to six month lag time of the Robinson study probably means that it is the result of the natural forgetting process over time that is being documented. This study's 20 percent recall rate hardly proves that patient understanding, *at the point of consent*, was that low. Many of the studies were, in fact, conducted well beyond the point of consent and treatment, and thus erroneously tend to conflate remembering with understanding. Many other studies contain no sense of what was actually said to patients or how it was presented. In short, many such studies offer no sense or reassurance regarding the content and quality of the communication the study subjects received. One study, however, by Morgan and his colleagues, regarding cataract lens replacement surgery (Morgan, 1986), seems to escape such flaws by providing a relatively simple informed consent and a test of patient understanding on the day after surgery. This study found a 37% rate of overall understanding. Risking ageism, one might argue, however, that its one design flaw lies in the fact that its average patient age was 75, with no attempt to identify cognitively diminished patients or to address the issue that perhaps this result is a function of an elderly population which is particularly prone to "let doctor decide."

No study has documented an overall understanding rate of over 72 percent and the average of all of them would probably be around 50 percent. This, at least, indicates that optimism about the effectiveness of informed consent is not warranted, at least in the sense of information acquisition. Further, those studies that tested for specific elements of informed consent reported particularly troublesome findings. In Morgan's study, even if one worries about the number of unidentified incompetents in his elderly population, only 4 percent of the patients recalled more than two of five disclosed risks, the most serious risk of blindness was recalled only by one-third, and 95 percent did not remember three out of the five mentioned complications. Finally, only 50 percent remembered either of the two alternative treatments, and 16 percent could not even recall that they were given an informed consent (Morgan, 1986, p. 42).

The low levels of recall and the absence of any study showing high recall of even the "essentials" surely challenges any optimistic view of informed consent. Morgan, for his part, concluded that the poor results called for an "excessive pursuit of informed consent," but he did not indicate what such an "excessive pursuit" would involve, or why, given his relatively simple and straightforward approach, a more intensive pursuit would be more effective (Morgan, 1986, p. 45).

Aside from the comprehension of information, numerous other empirical findings seem to confirm the anecdotal views of clinicians. Cassileth found that understanding decreased as the degree of illness increased, an expected result that common clinical experience supports (Cassileth, 1980). A number of studies showed that patients do not seem to take the informed consent process seriously (Cassileth, 1980, p. 896; Lidz, 1984, p. 318; Fellner, 1970), a majority not even reading the consent forms carefully, and that "most believed consent forms were meant to protect the physician" (Cassileth, 1980, p. 896; President's Commission, 1983, p. 108). Others found that patient understanding was quite idiosyncratic and related more to the patient's own past experience than to the information the physician provided (Fellner, 1970; Faden, 1980). And a couple of studies documented some degree of "nocebo" effect in that certain side effects seemed to increase in intensity and occurrence as a result of disclosure (Cassileth, 1980, p. 896; Loftus and Fries, 1979; Cairns, 1985).

On the other side of the ledger, however, certain empirical findings contradicted common clinician views. Consistently, studies have found that most patients (usually above 90 percent) definitely wanted to be informed and participate in decision-making (Alfidi, 1971; Harris, 1982), and that physicians routinely underestimated such patient desires (Faden and Beauchamp, 1980). Such opinions were often gained from healthy people or stable patients, who may well, as clinicians often suggest, not be so interested when they are sick and in jeopardy, but the latter tendency has not been empirically substantiated. Further, the concern that informing patients will lead to an increase in the refusal of treatment has been generally disproved. In fact, not only has such an increase not occurred, but it has been found, instead, that treatment refusals tend to increase when patients are *not* informed (Faden, 1978; Leydhecker, 1980; Morgan, 1986). Finally, numerous studies have documented a significant discrepancy between what physicians think their patients want and what those patients actually do desire (Bedell, 1984, p. 1089; Uhlmann, et al., 1988, p. 115).

The problem of increasing anxiety, particularly by mentioning the risks of treatment, has also been extensively studied. While a few studies report that anxiety was increased to some extent (Leeb, 1976), most seem to indicate that anxiety is not increased and often is reduced when patients are better informed (Freeman, 1981; Morgan, 1986). There is also substantial evidence that informed consent is a key factor in patient satisfaction and that informed patients cope with and adapt to situations better (Cassileth, 1980; Wallace, 1986). It has also been shown that anticipating the occurrence of pain tends to decrease its felt intensity, while more anesthesia is needed in the absence of such an anticipation (Wallace, 1986, p. 32). Finally, to the common criticism that informed consent takes too much time, one study found that clinicians overestimated the time spent giving informed consent by a factor of nine (Waitzkin, 1976).

These studies, with their flaws, conflicting findings, and differing contexts, yield something for everyone. Proponents of informed consent can point to enhanced coping and adaptation to situations, the reduction of pain and anxiety, and a higher rate of acceptance of treatment. Further, however short of perfect recall, proponents can call for enhancing the process *and* insist that some degree of understanding is better than none at all.

On the other hand, those who oppose (or are at least luke-warm about) informed consent retain the option of insisting that its effectiveness is shown to be, at best, marginal by the preceding data. They can continue to hold that informed patient participation in decision-making is not an outcome that one should assume is easily gained or that merits striving for. The law's unqualified presumption of patient competence is thus clinically impeached. In effect, clinicians may still be justified in believing it foolish to approach the average patient as a capable co-participant in decision-making. Further, whatever the suggestions from the enthusiasts, there is no evidence that an "excessive pursuit" of informed consent will result in any significant increase in patient understanding. Finally, we must note that such results regarding patient understanding relate only to the bits and pieces of informed consent. Even a strong showing in this regard hardly reassures us that patients will be able to move to the more complex and difficult stage of evaluation and deliberation regarding such "bits and pieces" with a similar level of success.

Our own conclusion must be that neither side wins the day thus far. However much the proponents of informed consent have societal support and the weight of rhetoric on their side, their enterprise surely remains marginal when it comes to actual effectiveness. One earlier comprehensive evaluation of these empirical studies, in fact, concluded that "whether informed consent is or is not feasible is still an open question" (Roth, 1981, p. 2476). Twenty years of studies later, we must still entertain the same conclusion. Though certain benefits seem to clearly derive from informed consent, e.g., enhanced compliance, coping and a diminished experience of anxiety and anticipated pain, the informed patient as decision-maker has just not regularly materialized. And to pursue the former goals does not require belief in the latter.

The converse conclusion that informed consent is a myth (Ravitch, 1978) is no more tenable. There is no clear reason to assume that competent persons attain substantially higher levels of insight in any other area of activity of their lives than the preceding groups of subjects. The empirical studies of informed consent may simply be reporting the level of understanding people bring to all of their endeavors. But we are hardly going to conclude from this that freedom is a bad idea and turn the management of our lives over to experts. Equally, opponents of informed consent are placed in the tenuous position of quarreling with strong patient preferences, basic societal presumptions, and the agendas of its elected representatives (i.e., as seen in informed consent statutes and court findings), to the extent they wish to reject the process of informed consent as not *sufficiently* effective. Such a radical conclusion clearly bears the burden of proof in a free society; at least in this sense, freedom does trump.

The preceding does not, however, constitute the sort of adequate investigation of the clinical experience of patient autonomy that we can and should perform at this juncture. Even if clinician's views are often anecdotal and the results of empirical studies ambiguous, much more needs to be said about certain relevant clinical realities. We should recall Edmund Pellegrino's contention that sickness results in "wounded humanity" (Pellegrino, 1979). This effect of illness creates a need for informed consent as well as allied counseling responses, but it equally raises questions about the actual abilities of the "wounded" to make decisions.

The Clinical Critique of the Competency Assumption

We have noted that the law presumes that the adult patient is competent to participate in decision-making and give informed consent. In one sense, such a presumption serves the important goal of protecting patients' legal status as free citizens who retain control over their bodies and their affairs. This presumption also tends to sustain such status in gray areas, i.e. when the patient's actual cognitive and decision-making capacities are marginal. This is arguably appropriate. Even if the patient lacks a detailed sense of his situation and prospects, he may still be sufficiently in touch with his basic values and life experiences to assess whether the proposed intervention is congruent with them. The low-recall studies are disconcerting; they are not damning. They do, however, surely undercut any facile optimism about patient decision-making, an optimism that should completely evaporate once one attends to the realities of illness and health care, as we shall now do.

As citizens and potential patients, we would tend to subscribe to the presumption of patient autonomy and self-control. For clinicians this presumption can be quite troubling. Many factors that clearly tend to diminish patients' abilities to understand, evaluate, and decide are often present in a given patient. Whether such factors actually place the patient below the threshold of competence is an issue that we will return to later in fashioning an operational notion of competence. For now, our task is simply to identify those factors and reflect upon their potential effect on patients' decision-making abilities. We can certainly quarrel about the degree to which such abilities are to be required, or the significance of any such diminishing factor, especially regarding a summary judgment of whether a given patient is competent or not. It seems undeniable, however, that such diminishing factors are both often present and numerous, thus advising us that the competence the law asks us to presume, or the abilities we hope for, may only be marginally present.

Finally, many of these diminishing factors are commonly present and, alarmingly, often increase in intensity and number as the health of the patient decreases. In other words, competence tends to become especially questionable in exactly those situations when it is particularly important for patients to understand and participate in decision-making.

Before we address such issues, however, we must first identify the diminishing factors involved. Fear, stress, and anxiety are common and surely can diminish a patient's mental abilities, as can pain, drugs, and the confusion often attendant upon illness. Many other sorts of diminishing factors can also be identified.

Metabolic abnormalities, or poor oxygenation of the brain from pulmonary or vascular deficits, have mental sequelae, as is the case with physical discomfort, clinical depression and lethargy. Most clinicians can supply numerous anecdotes of patients with such deficits who, while satisfying minimal competency requirements, refused treatment, but later did not even remember they had said anything of the sort and were glad to have somehow survived. Equally, certain co-morbidities of severe illness, e.g. uremia in renal insufficiency or hypercalcaemia in advanced carcinoma, definitely tend to obtund patients who may still, on the surface, seem to be alert, oriented and able to respond appropriately. Particularly when we think of the "essay mode" of decision-making, i.e. pulling all those "bits and pieces" of the informed consent disclosure together into some sort of comprehensive, summary evaluation and decision, which is surely what we hope patients would attain in decision making, such surface presentations are not reassuring.

The passive nature of the sick role has long been recognized (Peabody, 1927), particularly in the way it often tends to produce an almost child-like state of regression. Such a traditional description should be suspect, however, since paternalistic physicians may have tended to see what their ideology leads them to expect. Equally, paternalistic behavior, i.e., Jay Katz' "silence" at the bedside (Katz, 1984), may well have been as much a cause of as a response to such regression. Any reflection on what actually happens to patients as they enter the health care milieu, e.g. waiting helplessly until seen, asked to remove their clothes, etc., can also be seen as contributory. It at least seems accurate to say that patients have traditionally been given little opportunity, encouragement or assistance to be other than passive, and much routinely occurs that would tend to both shore up and produce such passivity.

But such regression in patients is too common and significant to attribute it solely to clinician behavior or the nature of health care delivery. The previously mentioned needs to be fixed and reassured are clearly not autonomy enhancing factors, however common and 'normal' such responses to the threat of illness are. Such a passive orientation can also clearly result in patients not listening to or being interested in the tasks of informed consent, as well as being overly submissive to or respectful of physician expertise. At times such passive behavior appears similar to the behavior of an animal who "freezes" when faced with or grasped by a predator. Denial, for its part, may well be adaptive and appropriate (or, at least, "normal") at certain junctures, but it remains, by definition, a refusal or inability to recognize the facts of one's situation and prospects (a necessary condition, one would think, of informed consent and competence) (Roth, 1982). Denial might thus be seen as just such a "freezing" in a cognitive sense, as might other passive behaviors commonly witnessed in patients, e.g. not reading the consent forms, or not asking the doctor questions that one has (commonly reported by nurses). Finally, patients can be quite overly or under trusting of physicians, may appear extremely risk averse (or the converse), and may well be keying to past experiences of their own, or of loved ones, that are quite disanalogous to their present situations. One poignant personal example of this was a woman who presented with an eminently resectable, discrete, bowel tumor who, on having received confirmation that she had cancer, concluded that she was in the same situation as her sister who had recently died of an untreatable pancreatic carcinoma. She steadfastly refused counseling and treatment.

The etiology of patient passivity is thus multifactorial, not just iatrogenic, and no modification of clinician behavior or health care delivery is likely to remove it completely. It appears to be a relatively normal response to illness. Moreover, for many patients, it may just be their standard way of dealing with life, not just illness. Many people, in varying degrees, are basically passive and have no urge or habit of decision-making, having been more directed than autonomous in both their work situations as well as their family units. Often such individuals, when asked what they wish to do, will throw the ball back immediately, either at the clinician, or a dominant family member. Equally, some come with unrealistically optimistic or pessimistic views about treatment, or life itself, which, to the onlooker, may be more suggestive of a personality disorder than anything one would tend to see as autonomously held views.

The patient with little education, a short attention span, an impaired memory, a past stroke or history of mental illness, hardly inspires confidence, nor do those who simply do not seem very bright. Patients may come with needs that conflict with autonomy, like the patient who tries to solicit reassurance beyond what the facts of his case afford. Attention-seeking or manipulative behavior can be manifested in refusals of treatment as well as non-compliance. Some patients are clearly being manipulated by family members with their own rather bizarre or conflicting agendas. Equally, a patient may find that he has been abandoned by family and supposed friends precisely at the point when such support systems are most needed. How much easier, then, is it for such people to give way to the counsel of their fears? Or, as an old saying puts it: "a man alone is in bad company."

The preceding excursion should provide substantial credence to the clinical skepticism about decision-making competence in the sick, as should our review of the empirical literature regarding informed consent. These many autonomy-diminishing factors are not that rare, are often conjointly present in patients, and are surely prime contributors to the less than reassuring data from the empirical literature. We may well ultimately stay with the law's presumption of competence, however realism dictates that we should often add the word "diminished" to it. But whatever realism counsels, it also seems fair to say that however minimal our requirements for the status of competence are, it is often going to fall short of what we would hope for and, particularly when the decisions at hand are poignant, may well be the cause of true alarm. Whether and to what extent the law and society should maintain their low threshold presumption about competence will be reviewed shortly. That many patients are not the capable decision makers for which we might hope is surely also the case.

If we are to gain a realistic appreciation of what the "new ethos of patient autonomy" might reasonably aspire to regarding informed consents and otherwise, the guiding assumptions and stated goals must be restated in light of the previous criticisms. Otherwise we are merely dealing with overcooked rhetoric. Without providing the extensive argument and elaboration they otherwise require (see Wear, 1998), I will now offer certain basic restatements of the assumptions about and goals of informed consent that I take to be called for by the preceding.

Taking the bull by the horns, as it were, we might well begin by simply asking whether the preceding does not simply impeach the whole "new ethos of patient autonomy" as not just philosophically incoherent, but practically unrealistic at its core. In sum, the preceding seems to instruct us, aside from certain relatively rare extraordinary cases, that bioethics itself does not insist on any sort of patient autonomy worthy of the name, that the empirical literature suggests that patients may well generally not be up to the task of being autonomous, and that this generic inability may well not stem from silent, paternalistic physicians, but from the autonomy diminishing characteristics of illness. Perhaps, to borrow from T. S. Elliot, what we have here is a "haul which will not bear examination."

There are two basic answers to this line of thinking: first, that however much we retail the preceding "deficiencies" of patient autonomy, it seems completely unlikely that citizens of a free society are going to accept any sort of reversion to physician paternalism. When I have reviewed the preceding considerations for students, although they are willing to consider the implications for others, e.g. elderly parents, they react vehemently against the idea that such considerations could legitimately be used to justify physician paternalism against themselves. In fact, no matter how forcefully arrayed, most of these students seem to explicitly want to keep a quite low threshold notion of competence for themselves; in effect, if they can still respond in some basically coherent fashion, they want to retain the right to throw out the anchor by a simple refusal of treatment.

The second line of response to the preceding criticisms comes from reflecting on what sorts of goods and values might be lost if we were to step back from a commitment to informed consent for apparently competent patients. In effect, the point of this argument is that however marginal patient competence may be, too much is potentially at stake to do anything but assume it in most cases. We may well have to conclude that we are likely to get much less than we hope for in this regard, but most anything is better than nothing. To these many potential goods and values, I now turn.

III. THE HETEROGENEITY OF GOODS AND VALUES

Realism certainly dictates that we see in clinical situations a broad spectrum regarding the nature and extent of the goods and values at stake. At one end of this spectrum, one way clinician informed consents, with the patient being silent beyond a simple affirmation, are ethically adequate. In effect, when the clinician perceives that there is a clear treatment of choice, with no reasonable, truly competing alternatives, and sees inactivity as being simply foolish, and the patient passively agrees, a pro forma informed consent may well satisfy any reasonable ethical requirements. At the other end, the aggressive autonomy promoting and enhancing tactics touted by bioethicists may well be absolutely necessary, when profound patient situations and choices are at hand. Agonizing over how aggressively to respond to the recurrence of a malignancy, or seriously debilitating illness, is thus a luxury only for those who insist on ignoring its significance. In sum, the protection, restoration and enhancement of patient autonomy will in some cases need to be the primary, overriding clinical goal, while in many other cases it merits no serious concern at all. And given that we are speaking of a spectrum here, clearly we are not talking about all-or-nothing, but a range across which clinical concern and response will vary enormously.

A. Goods and Values That Are at Stake in Any Informed Consent Situation. This result will be quite objectionable to some, so let me get somewhat clearer about what I am and am not suggesting. In effect, and at a minimum, I would continue to insist on the provision of informed consent in all clinical situations where risk is present. However humbled we must be about patient autonomy, such across-the-board provision can still, efficiently and effectively, accomplish certain important goals. Most basically, such events give patients the *opportunity* to understand their situation and prospects, to ask questions or drag their feet, and to be told, formally, that there is a choice being made, whether or not they wish to understand the ingredients of that choice or participate in it. Further, at a minimum, we may say that other goals are addressed, such as ruling out hesitancy or ambivalence on the patient's part (which they may signal explicitly or, not seldom, by body language or facial expression). But the point regarding all informed consents is that if the patient does not respond to the opportunity offered, there may often be no reason that he or she should, and the clinician has no obligation to insist that he or she does.

Nor are these the only possible goals in any informed consent. Surely one way to provide patients the reassurance they seek is to have the physician diagnose and explain their condition and any plans for treating it. Simply labeling the disease turns some nameless threat into something that is known and may now be dealt with. The clinician thereby can establish himself as an advocate and guide within the otherwise threatening medical assembly-line. Equally, concern and compassion can be expressed in the specific terms of the patient's actual problems, not just by some abstract hand-holding or similar gestures that have no anchor in the patient's life. What better way to announce that help has arrived? Potentially counter-productive anxiety can thus be diminished, acceptance of and commitment to a treatment regimen gained, compliance, cooperation and self-monitoring stimulated. Even if the clinician sees no sense in which a real choice is at hand, all these are treatment enhancing goals, and one might wonder if there is a better way to pursue them than in terms of a cogent explanation of the problem at hand and the planned response to it.

Such a communication process can thus enhance *any therapeutic encounter* and have value in developing the physician-patient relationship. It will also aid in the pursuit of diagnostic and therapeutic goals. While patient participation in decision-making may be minimal in many interventions, a basic rendition of the risks and benefits could serve a rule-out function by tending to identify the patient who has idiosyncratic fears, misconceptions or hesitations regarding medical care in general. Surely such factors merit identification up front and might well be easily identified and rectified. Finally, it is a commonplace that mention of the significant risks of any intervention will tend to diminish the risk of suit for the clinician, not only by fulfilling the legal requirement but, hopefully, by also instilling in the patient a corresponding sense that he has chosen the intervention and thus feels some personal responsibility for this choice.

B. Goods and Values that are Occasionally But Crucially Present in Certain Cases. At the other extreme of our spectrum, a passive patient "waiver" of the informed consent "opportunity" may well need to be as vigorously opposed as the most foolish rejection of beneficial therapy. The patient who, for example, does not want to hear the gory details and is simply waiting to hear how the clinician intends to fix a recurrent cancer or remove the debilitating effects of a major stoke, does not need a "daddy" to tell what he would do if he were in the patient's shoes (which he is not), This patient needs a counselor who steadfastly refuses to give any quick fix and resolutely insists that the patient has certain profound and personal choices to make. There are, in sum, times when the waiver of informed consent that the law allows must not be accepted by the clinician.

Special needs, at times far more important than any treatment regimen, also arise in certain types of cases, especially where chronic or terminal illness threatens the patient. A number of these variable needs have already been noted: (1) the need for the patient to understand, ponder and pass judgment on major, eminently personal implications of their treatment and further care; (2) the patient's need to grasp and adapt to profound effects on his life style and expectations, given the presence of chronic or terminal illness; and (3) assisting the patient, who because of pain, abandonment or situational depression, has tended to give up hope and control. Institutionalized patients come particularly to mind with their oft-noted child-like deterioration. Such deterioration can be particularly counter-productive in situations where the difference between success or failure may turn on the patient's coping and adaptive responses to a reduced quality or expected duration of life. *In sum, the protection or restoration of patient autonomy may, at times, need to occupy center stage and be preparatory and foundational to all other therapeutic goals.* It does little good to enhance function or extend life if the patient is indisposed to value or take advantage of it. And at times the primary goal will relate to objectives that only the patient can address: active participation in rehabilitation, major modifications and acceptance of a life style imbedded in chronic illness, or the performance of "last things" to the extent terminal illness threatens. Few of us do such things well; many stumble and falter. Informed consent, far beyond the legal ritual, allied with referrals to self-help groups, community assistance, and counseling services, can make all the difference.

Other needs often arise in such situations: (1) A knowledgeable patient will be better able to reflect on the degree of aggressiveness of treatment they desire, and how and where they wish to spend their remaining time. They will also tend to more effectively address the issues involved in formulating a living will or designating a proxy for the possibility that such treatment might be either limited or discontinued at a certain point. The patient who has been kept informed from the beginning, however modest or circumscribed the initial information, is better prepared to deal with such profound matters. One must question the so-called "compassion" that keeps physicians from sharing their deeper fears with patients early on. Such "compassion" is false in two senses: (a) it appears to be done more toward the goal of the physician avoiding a disagreeable discussion, and (b) it often results in much more suffering down the line, either because the patient was not given the chance to rule out aggressive treatment that he would not have wanted, or is unprepared to do so from lack of anticipation. (2) Some patients with self-destructive life styles, e.g. nicotine or alcohol abuse, seem much more approachable at the point of their first major hospitalization. The clinician can sometimes generate an about-face in behavior that television messages or advice in the office setting cannot. (3) The role of families can also be addressed, enlisting them in patient counseling and the provision of alternative care in the home situation, and assisting them to see the often significant effects on their own lives that chronic or terminal illness in a loved one may bring.

C. More Commonly Present Goods and Values. In between these two extremes, various other goods and values will be variably present and merit varying degrees of response. To the extent that patient compliance with an ongoing regimen, or self-monitoring for complications as an outpatient, are crucial to the success and safety of a proposed therapy, the one-way, essentially silent patient consent scenario must be leavened with more conversation, as well as confirmation that the patient has gained sufficient insight. The perception of confusion and ambivalence on the patient's part may well also require further intervention, however much the recommended treatment is clearly indicated. Other variably present goods and values that an informed consent might address include the following: (1) It could assist in developing the physician-patient relationship beyond the state of "moral strangers" (Engelhardt, 1986; Rothman, 1991). Even if such an enhanced relationship is unlikely to be significant in the current situation, it might well be at another time, for example, when the need to address advanced directives presents itself. (2) To the extent the benefits of and prognosis with treatment may fall short of the ideal of full and timely restoration of function, the patient can be led to a more realistic appreciation of his situation and prospects. Such an appreciation would also tend to protect the viability of the physician-patient relationship if the results are, in fact, less than what had been hoped. (3) Even the abstract presentation of the choice between treatment and no intervention at all might serve to instruct the patient that such an activity on his part can be important within the clinical encounter. Perhaps some diminishment of the urge to be passive and assume that doctor always knows best could occur. Such a result could be of value at another time when patient understanding and participation *is* important. And (4) particularly if interactive, the patient being queried as to what characteristics and effects his problem has, the communication process could also: (a) assist the physician in seeing what personal meanings the patient's problem has in his life, (b) assist in identifying the subtly but functionally incompetent patient, (c) identify and respond to any significant misconceptions, false hopes or fears that the patient may have, and (d) stimulate the patient to be more forthcoming with information, thus enhancing the quality of the history and workup in terms more specific to that particular patient.

Other positive, variably present needs may also be noted: (1) that the anticipation of expected pain and discomfort tends to diminish its effect; (2) to the extent the problem stems from self-destructive patient behaviors, preventive goals can be pursued; (3) the patient can be appraised of possible chronic effects to his life style and assisted to plan for these; and (4) the clinician can often enhance the possibility of successful treatment by generating a knowledgeable, committed, even optimistic patient, which in turn can yield dividends in patient compliance, cooperation and self-monitoring.

Negatively, an interactive process can identify fear, confusion, anxiety and misconceptions on the patient's part. These factors will not necessarily be uncovered by a ritualistic, one-sided offering of informed consent. But they surely merit identification and response since they may well cause the patient more suffering than the treatment or disease will produce. Further, patients' initial perceptions of their problem and its potential resolutions are *likely* to be wide of the mark. They may be overly apprehensive in a situation where treatment will probably be successful. Conversely, they may fail to appreciate the significance of the threat, or of the effort required to remove it. The clinician might often dispel such sources of counter-productive and harmful behavior in patients, particularly before they have time to fester.

In sum, one size does not fit all in medicine. In some cases, the "promotion, restoration and enhancement of autonomy" may well be the most basic and primary goal of the therapeutic encounter. Conversely, in many cases, patient freedom and self-determination may not be meaningfully at issue or threatened at all, and a pro forma event that addresses neither is not objectionable. We would all do well to remember, in this regard, Eric Cassell's point that often the best way to protect or restore the patient's autonomy is to cure the patient (Cassell, 1982). At some point in much of this, the supposedly competent patient simply gets to choose if such a pro forma event is sufficient. If he or she says otherwise, simple politeness, among other things, calls for the clinician's response. Lacking this, the enhancing tactics of the new ethos should be called into play by the clinician to the extent the case calls for them.

IV. RETHINKING THE COMPETENCY ASSUMPTION

My argument thus far is that there are potentially so many important goods and values at stake in clinical encounters that the presumption of competence must be generally maintained so that at least some attempt will be made to capture them. Realism here must grapple not just with how “competent” patients actually are, but also with what would be lost if the threshold was raised.

Fair enough. But now we must grab the other horn of the bull and get clearer as to when and how the presumption of competency should be questioned and an individual patient’s competency specifically assessed. And given that entire books have been written on the subject of competency (Cutter, 1991; White 1994), we can surely address only so much of the topic in a sub-section of this article. I will proceed by stating, rather than arguing for, a conceptual solution to this issue, along with directing the reader elsewhere for more detailed reflections (Abernathy, 1984; Appelbaum, 1981; Baumgarten, 1981; Drane, 1984 and 1985; Gert, 1981; Kopelman 1990; Miller, 1981; Morreim, 1983 and 1991; Moreno, 1989; Roth, 1977; Sherlock, 1984; Tancredi, 1982; Wear, 1998, pp. 126-155). I begin by commending a basic distinction regarding the notion of competency that efficiently takes us to the heart of the matter.

A. Competence as a Dual Status and Capacity Notion. Much of the writing on the notion of competency suffers, I submit, from a failure to distinguish two intertwined but distinct features of it, viz. *status* and *capacity*. On the one hand, competence is seen as a *status* concept. Both the law and the new ethos tend to instruct us that it should generally be presumed, and such generally unassessed status confers numerous privileges and rights on patients who are presumed to have it, e.g. the right to informed consent and the right to refuse treatment. To presume competence in patients is thus protective of patient freedom and autonomy. And if the new ethos of patient autonomy has anything right, it is that such protection is needed. This presumption also advances our concern for efficiency because it rejects the need to perform a detailed assessment of the average "alert and oriented" patient as to whether he is capable of performing, or has actually performed, the cognitive and participatory tasks already detailed. It is surely not even remotely feasible to do competency assessments on even a small minority of patients, and many patients would rebel if any such broader review was attempted. Finally, competence in the "status" sense is clearly an either-or sort of notion. Patients either have such status or they do not.

On the other hand, competence is also seen as an *ability* or *capacity* notion in that a competent patient is deemed to have sufficient ability or capacity to "participate in medical decision making.” This, in effect, is what the law and the bioethicists are presuming in the usual case. This sense of competence, particularly given our previous reflections on the clinical factors that tend to produce "diminished competence,” is clearly a spectrum concept. Competent patients fall within a wide range, from the marginally competent, to those who are particularly informable, knowledgeable, and reflective about their situation and prospects. A *capacity* sense of competence thus tends to remain up for grabs however much we are instructed to presume competence in the usual patient, absent significant counter-indications. This is so because the patient might still fail to have or exercise sufficient ability in any given instance of information acquisition or decision-making.

Now these two senses have led to an extraordinary amount of dispute and confusion in the literature, as well as at the bedside. The presumption of competence runs the risk of false-positives, i.e. patients who are treated as competent to make certain decisions who actually are not. Conversely, an over-emphasis on the capacity sense of competence can lead to an invasive monitoring and assessment of patient's abilities that has no analogy in the free world and can hardly be attempted even in a small minority of cases. The account offered here avoids opting for one sense of the notion of competence over the other, as previous accounts tend to do. Instead, I propose a notion of competence that incorporates both aspects, viz. *status* and *capacity*.

To briefly summarize, this account proposes to retain the efficient, freedom protecting presumption of competence that its *status* sense entails. We will concurrently, however, need to pursue a more detailed account, keyed to the *capacity* sense of competence, of when and how such a presumption may be defeated in favor of a more detailed investigation and assessment of an individual patient's actual capacity for and performance of cognitive and decision-making functions. A basic feature of this account will regard what sort of clinical factors should "trigger" a capacity oriented investigation beyond the usual presumption of competence. Arguably, such an investigation should be conducted when the seemingly "alert and oriented" patient is refusing an intervention that the clinician sees as "clearly indicated" for that patient, or when the patient has a history of mental illness. But, given that such factors hardly prove incompetence, we must be as concerned with how to assess such data as with which data legitimately initiates such an inquiry. Past mental illness does not prove present incapacity, nor does disagreement with the clinician as to where one's best interests lie. Our solution will give such factors status only as *triggers*, eliminating them as factors within the competency assessment itself.

B. Triggering and then Assessing Competence. The issue of what should initiate a competency assessment is surely problematic given our previous reflections. In sum, we must now realize that the presumption of competency often masks a reality of marginal patient capacity and performance. But if we are going to accept this from the many, how are we going to fairly and legitimately not accept it from the few? How this all tends to work out at the bedside has been well described by Loren Roth and his colleagues:

> It has been our experience that competency is presumed as long as the patient modulates his or her behavior, talks in a comprehensible way, remembers what he or she is told, dresses and acts so as to appear to be in meaningful communication with the environment, and has not been declared to be legally incompetent. In other words, if patients have their wits about them in a layman's sense it is assumed that they will understand what they are told about treatment, including its risks, benefits, and alternatives. This is the equivalent of saying that the legal presumption is one of competency until found otherwise. The Pandora's box of the question of whether and to what extent the patient is able to understand or has understood what has been disclosed is therefore never opened (Roth, 1977, p. 282).

This seems to be an accurate description of current practice and does, as noted, honor the prescription that competence should be presumed unless proven otherwise. In sum, the basic answer to our initial question about the nature of competence, arrives not in the form of a specific conceptual formulation, but rather in the sense that competence, whatever it might be, is a thing that is assumed when we are faced with "patients who have their wits about them in a layman's sense." The Roth et al. article goes on to focus on the patient whose competence becomes an "issue" during the informed consent process. Roth et al. spend the bulk of their article providing a sliding scale of tests for competence, ranging from the simple test of whether the patient can "evince a choice" (i.e. say yes or no to the recommendation offered), to the most elaborate test of whether the patient has a detailed "actual understanding" of the elements of the decision at hand. In between, tests regarding whether the patient's choice was based on "rational reasons," or tests assessing the patient's "ability to understand" are sometimes used, depending on the situation.

Depending on *what* about the situation? Their answer to this constituted their basic contribution, a contribution that 20 years later enjoys fairly wide-spread scholarly support. The crucial move that they make is to contend that the selection of which test to use, from the simple and easily met "evincing a choice" test, to more complex and challenging tests, turns on the degree to which the patient's decision is "favorable" or not. In sum, the more "favorable" the decision, the lower the hurdle, and conversely.

How is this "favorability" to be determined? Roth et al., and others, have offered numerous comments and distinctions in this regard, but the most often expressed ground of this determination relates to the "risk-benefit ratio" of the treatment at issue, what Roth et al. refer to as the "valence" of the decision (Roth, 1977, p. 282). Thus to the extent the risk-benefit ratio is favorable, and the patient consents to it, a low threshold test may be used. To the extent such a favorable treatment is rejected by the patient, a relatively more stringent test will be used. Conversely, if the patient refuses an unfavorable or questionable treatment, the test will be low threshold, high if he consents to or demands an unfavorable or questionable treatment (Roth, 1977).

And who determines whether a given treatment is favorable or not, questionable or not? Roth et al. reply that it is "the person determining competency" (Roth, 1977, p. 283), whose identity and training they do not specify. But one assumes they are thinking either of the patient's physician or a psychiatrist consulting to this individual, the psychiatrist basically dependent on the former for the judgment of what treatments are "favorable" or not. Favorability, then, is seen as a matter of objective medical judgment.

There is much to recommend this approach. It is surely the efficient way to proceed, assuming competence in most cases where the patient appears basically "alert and oriented" *and*, as is again usually the case, where he accepts his physician's recommendation. This also honors the law's insistence that competence be presumed unless proven otherwise. Further, it is surely an improvement over the older tradition of assuming that patients are generically incompetent to make medical decisions. It also appropriately rejects the traditional notion that the patient is likely to be incompetent if he refuses the physician's recommendations. At least in the latter situation, Roth et al. give the patient the opportunity to prove that his "unfavorable" decision is based on a competent choice, however idiosyncratic, tragic, ill-advised, or self-destructive it might appear to the clinician.

Other benefits are claimed for such an approach: the empirical studies we previously reviewed suggest both the presence of significantly diminished competence in many patients, as well as a lack of comprehensive understanding. Roth et al. point out that this Pandora's box need not be opened in the usual case of the generically "alert and oriented" patient who accepts recommended treatment. Further, the authors also emphasize the advantage of being able to employ "a low test of competence .... to find a marginal patient competent so that his or her decision can be honored." Conversely, when a favorable treatment is refused, "even a somewhat knowledgeable patient may be found incompetent so that consent may be sought from a substitute decision maker and treatment administered despite the patient's refusal" (Roth, 1977, p. 283).

At some point, however, we must object to all this. The last point particularly, where a "somewhat knowledgeable" patient's unfavorable decision is not honored seems especially alarming if we value and respect patient autonomy. The whole enterprise starts to look much too expedient. Roth et al. themselves note that "in theory competency is an independent variable that determines whether or not the patient's decision to accept or refuse treatment is to be honored" (Roth, 1977, p. 282). But they are surely being accurate as a descriptive matter when they point out that "favorable decisions are more likely to be accepted at face value", whereas "unfavorable" ones are likely to trigger "further investigation" (Roth, 1977, p. 282).

Now it must be emphasized that this focus on "consequences" in competency assessments enjoys a substantial consensus among contemporary writers in bioethics. Perhaps surprisingly, the majority of the authors in a published edition regarding competency determinations support this sort of move (Cutter and Shelp, 1991). I say "perhaps surprisingly" since these authors, as well as numerous others, are all devotees of the new ethos of patient autonomy; the list includes such names as Edmund Pellegrino, James Knight, Tom Beauchamp, and John Robertson. And it seems clear that the "appeal to consequences" move fundamentally contrasts with the whole thrust of the new ethos (Wear, 1991). Aside from ignoring the pluralism of values that would undercut the idea that "favorability is an objective medical determination," this whole enterprise could just as easily be seen as only a slight improvement over traditional paternalistic medicine. In effect, the Roth et al. view seems to say: "we will respect your autonomy without question as long as you agree with standard medical judgment, even if you otherwise do not know if you are on foot or horseback. Depart from or disagree with this canon and you will be forced to jump a substantially higher hurdle to retain such respect." It would appear that what we actually have here is the old paternalistic wolf dressed up in sheep's clothing, fangs and claws intact.

The way out here, I submit, is to sharply distinguish between what legitimately "triggers" a competency assessment, and what is legitimately a part of that assessment once it is triggered. "Favorability" is the best example of such a trigger. My suggestion is that unfavorable decisions by patients legitimately trigger competency assessments, but should not be allowed to effect the nature of the resulting assessment in any way, e.g. making the hurdle higher the more unfavorable the patient's decision is, as Roth and his colleagues, and many other bioethicists suggest. Directing the reader on to more detailed arguments by myself and others regarding this point (Morreim, 1983 and 1991; Moreno, 1989; Wear, 1998, pp. 137-142), I will merely suggest that this resolution provides the proper balance. We should be loathe to routinely disenfranchise patients simply because they do not agree with what their clinician has in mind, but must equally be concerned that the canon of what clinicians tend to recommend often has strong empirical support, and we are already well enough aware that the root cause of the rejection by the patient may well be the patient's own lack of understanding, confusion, or some other illness related factor diminishing capacity.

The further balancing proviso here is that once the competency assessment is triggered, it is to be accomplished by ascertaining the degree of actual understanding the patient has about the specific clinical situation and issue at hand. This, in effect, rules out doing a status sort of competency assessment, e.g. a mini-mental status exam, an exam that many patients who choose favorably might not pass. Rather, the patient's right to informed consent, and the refusal of treatment, come to turn on the patient's actual performance regarding the specific issue at hand. In sum, does the patient understand the basic features that a reasonable informed consent would provide?

There are other triggers and the same distinction can be used to address their role in determinations of competency; namely, they are allowed to trigger the competency assessment, but should not be allowed to be a part of it. The presence or a history of mental illness is another common trigger. But one can anticipate a quick rejoinder from mental health advocates that we must not go back to the dark days when such a factor quickly and effectively resulted in mentally ill patients being routinely disenfranchised. So again the proposed solution is that mental illness, past or present, can trigger a competency assessment. But with the proviso that such patients get assessed in terms of their actual understanding, not their mini-mental status or erroneous beliefs about which planet they are on, or what the CIA is doing. An extreme but pertinent example here would be a patient with paranoid schizophrenia who gets involuntarily committed for a suicide attempt. Once committed, the suggestion here is that if the patient then refuses psychiatric medications, e.g. haldol, in the process explaining that he or she does not want to get tardive dyskenesia (however colloquially the patient describes this), then he or she might well be deemed to have competently exercised decision making regarding the issue at hand (Munetz, 1982).

Which brings us full circle to a final issue in this discussion, viz. what, in sum, is the basic informed consent that, among other things, will be the focus when such "triggering" patients actually get their competence assessed? We know the types of information, e.g. risks and benefits, but how much of each, how presented, and how selected?

V. A THREE STAGE MODEL OF INFORMED CONSENT

My own generic, revisionist proposal (Wear, 1998, pp. 100-125) is that informed consent should be conceived of as having three distinct stages, each aimed at quite different, though complementary goals: (1) the *comprehensive disclosure* stage, which will roughly approximate the detailed presentation of risks, benefits and alternatives of a given intervention required by American law, but will be aimed at more modest goals, e.g. providing the patient the opportunity to rule in or out regarding hesitancy, ambivalence or misconceptions; (2) the *core disclosure* stage, which will attempt to counter the information overload tendency of the first stage and give patients something relatively simple and structured which they might minimally react to and evaluate, viz., the essential choice at hand; and (3) the *assessment, clarification, and patient choice* stage which will be the only necessarily interactive part of the informed consent event (unless the patient spontaneously chooses this mode at any other point), and which will key to the specific patient's level of understanding and other concerns. This stage will proceed by probing into the patient's understanding of the information provided in the preceding two stages and will respond to this with appropriate clarification of the patient's developing sense of the issues at hand.

For the sake of our discussion in this article, the following basic caveat should be emphasized: given our prior recognition and acceptance that many informed consents occur in a one way fashion with little or no questioning of and by patients, the third "*assessment, clarification, and patient choice*" stage of the above model will occur only so often and then, usually, more out of the clinician's perception that important goods and values merit it, than any tendency of patients to call for it. More generally, how much of this stage needs to occur at all will be a function of the degree the various goods and values we have previously identified are actually at stake in the specific clinical encounter.

Having made this point, and leaving off a detailed explication and argument for this three stage model, I submit that the final task for us here is to clarify what the second stage "core disclosure" should involve. Aside from the fact that this is clearly the core of what this account takes to be involved in legitimate, necessary and effective informed consents, it takes us to the pivotal issue that remained at the end of our previous section, i.e. when we are in a position where we actually must assess an individual patient's competency, how much does a given patient need to know to pass this test? Clearly an assessment based on the comprehensive disclosure stage would be detailed enough to disenfranchise many patients who, if only they had made a "favorable" choice, would not have been assessed in the first place. So, to be fair and balanced, what is the minimal, core amount of information upon which they can be assessed?

A. The Notion of Transparency. The basic goal of the second "core disclosure" stage is to present the essentials of the choice at hand to the patient in an approachable and palatable fashion. This choice will often involve no more than asking the patient to choose to agree with the treatment recommended by the physician when, as anticipated, no reasonable alternative is seen to exist, significant risk is either not present or relatively insignificant, and the option of non-treatment is accurately seen as unattractive. At the other extreme, the patient may need to be presented with a layperson's rendition of various competing treatment options, as well as the non-treatment option, with no clear recommendation provided. As already noted, there is a wide spectrum of the sense of choice here, and it is this actual, specific choice that the physician should emphasize, absent much of the potentially stultifying detail of the comprehensive disclosure stage. The aim of this stage should thus be seen as keying to the activity of choice, not the presentation of information, the latter already having been provided in detail. The nature of such a presentation, particularly in the sense of the perspective it will offer, has been previously addressed by the physician-philosopher Howard Brody.

Sharing many of the same concerns that we have previously struggled with, Brody has proposed a "transparency" model of informed consent (Brody, 1989). Though Brody offered his model with specific reference to primary care situations, it might be proposed for all informed consent situations. And, as will now be seen, it might be particularly attractive to practicing clinicians, as well as respond elegantly to our own concerns.

The essential point of Brody's model is that, rather than have physicians provide the usual complex and content-filled informed consents, they avoid such potentially stultifying detail in favor of "making transparent" to the patient why they prefer and are recommending a given therapy. Such a presentation might well mention a number of risks, or even discuss an alternative modality, not for the wider informative purposes that informed consent usually has, but rather because they are seen by the physician as significant factors in the physician's own decision-making process.

Brody argues that the advantages of this approach are substantial: (1) rather than informed consent being some sort of alien body patched on to medical practice, it asks no more than that "the typical patient-management thought process" be arrayed for the patient, "only do it out loud in language understandable to the patient" (Brody, 1989, p. 8). (2) It provides a sense of informed consent that has clear criteria as to what is involved and when the process is adequately completed. And (3) it avoids hyper-informing the patient in favor of a structured communication of the basic factors and issues at hand.

B. Problems with the Notion of Transparency. As a generic model for the informed consent event, I submit that we must reject Brody's strategy, and for a variety of reasons: (1) It tends to falsely assume that a physician can legitimately formulate a recommendation for a given patient, in *all* cases, without any prior input from the patient; such an assumption can proceed only by ignoring the fact of the diversity of values. (2) It ignores the possibility that profound and eminently personal choices may be at hand to which only the patient can speak, and where a recommendation, at any point, would go far beyond the physician's expertise. (3) It would run the real risk of having the physician ignore certain risks or complications that he sees as routine or inconsequential, but which might well make the patient pause, or even decide differently. And (4), it would tend to diminish the possibility that the patient express hesitancy, ambivalence or misconceptions. The "weight" of detail in the first comprehensive disclosure stage is thus basic to triggering additional assessments important in the broader informed consent process.

However, there seems to be no reason why Brody's model could not incorporate these concerns. In fact, one of the results that we should hope for is that clinicians, as part of their "patient-management thought process," should also be considering the possibility that a recommendation must be preceded by prior discussions with the patient regarding the personal preferences and values that he may or may not have. As a general, pre-consent tactic, an enterprise such as getting a "values history" from the patient could be utilized, as some have suggested, although these have been focused on the more specific issues surrounding aggressiveness of treatment. This thought process should also become more patient centered in the sense of what patients, not physicians, tend to find significant, as well as key to when "similarly situated" patients tend to choose differently, or when "profound and eminently personal" choices are at hand. Thus, it is not clear that we need disagree with Brody on anything more than emphasis.

In the end, I believe that Brody's suggestion makes the correct and crucial emphasis in trying to correlate the informed consent event with the actual process of medical decision-making. To the extent that we are hoping for effective informed consents, surely the closer the correlation, the more likely they will be provided. Further, as long as we insist that "patient-management thought processes" be more concerned with and keyed to a patient-centered perspective, with an emphasis on the real choice for the patient, then at least with regard to our second "core disclosure" stage, the orientation is just right. After all, in a majority of situations, the choice actually presented to the patient is whether to agree to his physician's assessment of his situation and prospects and the response being recommended. In this sense, the "core disclosure" is appropriately offered in the sense of "this is how I see your situation and the reasons I have for wanting to respond to it in a certain way." The exact form of a given "core disclosure" will vary from case to case, of course, and the previous discussion of Brody's recommendation at best gives a generic perspective to guide it. And it gives sufficient conceptual clarification of our own present issue, i.e. what amount of information can a patient who has triggered a competency assessment be held to.

VI. FOOD FOR THOUGHT FOR PHILOSOPHERS

There is surely much additional work that needs to be done by philosophers, whether they focus specifically on the model of informed consent sketched out above, or more broadly on the "new ethos of patient autonomy" that underlies it. By way of concluding, I will identify what I see as a number of potentially quite valuable avenues for philosophical reflection.

One avenue is roughly the following: clearly the preceding strongly commends a notion of clinician discretion and beneficence that conflicts with the early bioethics intuition that beneficence is merely a creature of physician paternalism and must be discredited in favor of a thoroughgoing commitment to patient autonomy. But, at least in the situation of illness, the robustness and effectiveness of patient autonomy is surely suspect, and a beneficence based attention by clinicians regarding when patients should be stimulated to address more than they would otherwise must be part of the picture. The alternative that all such potential goods and values be routinely pursued for patients in all clinical encounters is not just not remotely feasible, it is not desired by most parties to such clinical events. In this sense, autonomy may well continue to trump whatever beneficence concerns are at hand, but beneficence will be the touchstone and trigger of much that actually needs doing in these encounters. So philosophical reflection on the nature of beneficence, within the realm of freedom and autonomy, is requisite.

Other philosophical issues certainly commend themselves here, perhaps most basically the philosopher could focus on the notion of autonomy, a very rich philosophical notion that the above discussion has facilely treated as roughly equivalent to the notion of competence, an equivalence that surely evaporates once one tries to more precisely define either notion. Issues keyed to the notion of autonomy might include: first, how minimally robust does self-determination have to be to be worthy of the name, and to what extent does actual understanding need to be present to credibly assert its presence? Second, and analogous to the traditional issue of negative and positive rights, how do we conceptualize situations where autonomy is not so robust and seems to crucially need assistance from others for its adequate exercise and protection? And third, the new ethos seems to have entailed a sort of procedural ethics where all we have is a meeting of "moral strangers" who do not necessarily share any values in common. Given the preceding, however, it appears that such mere proceduralism must give way to a system where beneficence is a pivotal consideration in any encounter, and the assumption that it is merely a meeting of moral strangers, often factually wrong and potentially disastrous morally.

How do we address this issue of shared values?

By way of concluding, I offer what I think is a valuable heuristic, though it is at best a caricature of both the history of medicine and bioethics, as well as the philosophy of Hegel. Taking a watered down sense of Hegel's notion of thesis-antithesis, we might see the history of medicine as one where, until roughly the 1970s, the notion of beneficence held sway with little or no concern for patient autonomy, or the recognition of the diversity of values. Then, thanks to the bioethics movement in the last quarter century, the notion of autonomy triumphed, coupled with a corresponding rejection of the idea of beneficence as in any way a guide to clinical behavior and decision making.

The preceding discussion suggests that a purely autonomy based view of informed consent and related notions is completely inadequate to the realm of practice that we are investigating. Far from being a meeting of moral strangers, various common values between clinicians and patients must be assumed both to account for the common situation where clinicians and patients routinely agree to proceed, as well as to ground our perception that the clinician must be alert to various goods and values that may well be important to patients but which, without the clinicians' counsel and encouragement, patients might well not tend to address, particularly given the diminishing factors present in the situation of illness. Equally, if "favorability" is going to be a pivotal consideration in determining when the presumption of competence is to be jettisoned in any particular case, as all seem to agree that it must be, then the notion of beneficence is back, seated front and center.

So a synthesis of the notions of autonomy and beneficence are required for an adequate account of how the clinician-patient interaction occurs and should occur. And the settled orthodoxy of the "new ethos of patient autonomy" must evaporate in favor of a more nuanced account of the interaction of these two principles. Food for thought for philosophers, indeed!

NOTES

1. The use of the term "self-determination" may be questioned here, as it does not appear in the early cases. I concur with Faden and Beauchamp's argument that this contemporary term best captures the sense of much of the early language of the courts in this regard (Faden and Beauchamp, 1986, p.121ff). Mohr v. Williams, for example, speaks of the "free citizen's first and greatest right...the right to himself" and expresses its concern about "violating without permission the bodily integrity of the patient...without his knowledge or consent" (95 Minn. 261, 104 N.W. 12 [1905]). The landmark Schloendorff case states: "every human being of adult years and sound mind has a right to determine what shall be done with his own body" (211 N.Y. 128, 105 N.E. 93 [1914]).
2. An early use of the appeal to the right to privacy is found in the Quinlan decision, a principle later applied in Roe vs. Wade regarding abortion; see Faden and Beauchamp, 1986, p. 39ff.

3. For a particularly detailed and forceful presentation of this view, see Katz, 1984, pp.48-84.
4. For a fairly exhaustive list of most of the earliest existing studies, see the bibliography of Meisel, 1981. For whatever reason, the bulk of the studies that actually document specific levels of patient understanding to informed consent occurred in the 1970's and early 1980's; beyond that point there are few such studies (none of which give much detail as to what was asked, how it was tested, etc.), with more attention to either patient perceptions of informed consent or patient satisfaction, neither of which help us regarding actual understanding. An excellent updated bibliography of the empirical literature is offered by Jeremy Sugarman and his colleagues (Sugarman, 1999).

REFERENCES

Abernethy, V.: 1984, 'Compassion, control, and decisions about competency,' *American Journal of Psychiatry* 141 (1), 53-60.
Alfidi, R.: 1971, 'Informed consent: A study of patient reaction,' *Journal of the American Medical Association* 216 (8), 1325-29.
Applbaum, P. *et al.*: 1981, 'Empirical assessment of competency to consent to psychiatric hospitalization,' *American Journal of Psychiatry* 138, 1170-76.
Appelbaum, P. and L.H. Roth.: 1981, 'Clinical issues in the assessment of competency,' *American Journal of Psychiatry* 138 (11), 1462-66.
Baron, C.: 1987, 'On knowing one's chains and decking them with flowers: Limits on patient autonomy in "The Silent World of Doctor and Patient",' *Western New England Law Review* 9, 31-41.
Baumgarten, E.: 1980, 'The concept of " competence" in medical ethics,' *Journal of Medical Ethics* 6, 180-84.
Bedell, SE and Delbanco, TL: 1984, 'Choices about cardiopulmonary resuscitation in the hospital,' *New England Journal of Medicine* 310, 1089-92.
Beecher, H.: 1966, 'Ethics and clinical research,' *New England Journal of Medicine* 274 (24), 1354-60.
Bergler J., C. Pennington, M. Metcalfe, E. Freis.: 1980, 'Informed Consent: How Much Does The Patient understand?' *Clinical Pharmacology and Therapeutics* 27 (4), 435-440.
Brody, H.: 1989, 'Transparency: Informed consent in primary care,' *Hastings Center Report* 19, 5-9.
Burt, R.A.: 1979, *Taking care of strangers: The rule of law in doctor-patient relations*, New York: Free Press.
Cairns, JA: 1985, 'Aspirin, sulfinpyrazone, or both in unstable angina,' *New England Journal of Medicine* 313, 1369-75.
Casseleth B.R., et al.: 1980, 'Informed Consent – Why Are Its Goals Imperfectly Realized?,' *The New England Journal of Medicine*, 896-900 .
Cassell, E.J.: 1982, 'The nature of suffering and the goals of medicine,' *New England Journal of Medicine* 306, 639-45.
Cutter, M., E.E Shelp (eds.).: 1991, *Competency: A study of informal competency determinations in primary care*, Kluwer Academic Publishers, Dordrecht.
Doukas, D.J. and L.B. McCullough: 1991, 'The values history,' *Journal of Family Practice* 32, 145-50.
Drane, J.F.: 1984, 'Competency to give informed consent,' *Journal of the American Medical Association* 252, 925-27.
Drane, J.F.: 1985, 'The many faces of competency,' *The Hastings Center Report* 15, 17-21.
Engelhardt, H.T.: 1986, *The foundations of Bioethics*, Oxford University Press, New York.
Faden, R. and A. Faden: 1978, 'Informed consent in medical practice: With particular reference to

neurology,' *Archives of Neurology* 35, 761-4.
Faden, R. and T. Beauchamp: 1980, 'Decision-making and informed consent,' *Social Indicators Research* 7: 313-36,
Faden, R. and T. Beauchamp: 1986, *A History and Theory of Informed Consent*, Oxford, New York.
Fellner, CH and Marshall, JR: 1970, 'Kidney donors – the myth of informed consent,' *American Journal of Psychiatry* 126, 1245-51.
Final Report of the Tuskagee Syphilis Study Ad Hoc Advisory Panel: 1977, In *Ethics in Medicine*, Reiser, J., et al. (eds.), MIT Press, Cambridge, MA., pp. 316-321.
Fries, J.F. and E.F. Loftus: 1979, 'Informed consent: Right or Rite?' *CA-A Cancer Journal for Clinicians* 29, 316-8.
Gert, B. and C. Culver: 1981, 'Competence to consent: A philosophical overview,' In *Competency and Informed Consent,* ed. N. Reating, National Institutes of Mental Health, Rockville, MD, pp. 12-31.
Harris, L. and Associates: 1982, 'Views of informed consent and decision making: Parallel surveys of physicians and the public,' In *Making Health Care Decisions.* Volume 2, Presidents Commission for the study of Ethical Problems in Medicine and Biomedical and Behavioral Research, Washington, DC, pp. 17-316.
Hunter, K.M.: 1991, *Doctors stories: The narrative structure of medical knowledge*, Princeton University Press, Princeton, NJ.
Kaplan, S.H., S. Greenfield and J.E. Ware: 1989, 'Assessing the effects of physician- patient interactions on the outcomes of chronic disease,' *Medical Care* 27, 110-27.
Katz, J.: 1984, *The silent world of the doctor and patient*, Free Press, New York.
Kopelman, L.M.: 1990, 'On the evaluative nature of competency and capacity judgments,' *International Journal of Law and Psychiatry* 13 (4), 309-29.
Lavori P., J. Sugarman, M. Hays, J. Feussner: 1999, 'Improving Informed Consent in Clinical Trials: A Duty to Experiment, *Controlled Clinical Trials* 20, 187-193.
Leeb, D., D.G. Bowers and J.B. Lynch: 1976, 'Observations on the myth of informed consent,' *Plastic Reconstructive Surgery* 58, 280-2.
Leydhecker, W., E. Gramer and G.K. Krieglstein: 1980, 'Patient information before cataract surgery,' *Ophthalmologic Base* 180, 241.
Lidz, C.W., *et al.*: 1984, *Informed consent: A study of decisionmaking in psychiatry*, The Guilford Press, New York.
Lidz, C.W., A. Meisel, and M. Munetz: 1985, 'Chronic disease: The sick role and informed consent,' *Culture, Medicine and Psychiatry* 9, 241-55.
Lifton, R.J.: 1986, *The Nazi Doctors*, Basic Books, New York.
McCullough, L. and S. Wear :1985, 'Respect for autonomy and medical paternalism reconsidered,' *Theoretical Medicine* 6, 295-308.
Meisel, A.: 1979, 'The "exceptions" to the informed consent doctrine: Striking a balance between competing values in medical decision-making,' *Wisconsin LawReview* 1979 (2), 413-88.
Meisel, A.: 1981, 'The "exceptions" to informed consent,' *Connecticut Medicine* 45, 27-32.
Meisel, A.: 1988, 'A dignitary tort as a bridge between the idea of informed consent and the law of informed consent,' *Law, Medicine and Health Care* 16(3-4), 210-18.
Meisel, A.and L.H. Roth: 1981, 'What we do and do not know about informed consent,' *Journal of the American Medical Association* 246 (21), 2473-77.
Metro-Goldwyn-Mayer: 1981, Whose Life is it Anyway? Los Angeles, California.
Miller, B.L.: 1981, 'Autonomy and the refusal of lifesaving treatment,' *The Hastings Center Report* 11, 22-28.
Moreno, J.D.: 1989, 'Treating the adolescent patient: An ethical analysis,' *Journal of Adolescent Health Care* 10, 454-59.
Morgan, L.W. and I.R. Schwab: 1986, 'Informed consent for senile cataract extraction,' *Archives of*

Ophthalmology 104, 42-45.
Morreim, E.H.: 1983, 'Three concepts of patient competence,' *Theoretical Medicine* 4, 231-51.
Morreim E.H.: 1991, 'Competence: At the intersection of law, medicine and philosophy,' In *Competency: A study of informal competency determinations in primary care,* Eds. M.A. Cutter and E. Shelp, Kluwer Academic Publishers, Dordrecht, pp. 93-125.
Munetz Mark R., Roth Loren H., Cornes Cleon L.: 1982, 'Tardive Dyskinesia and Informed Consent: Myths and Realities,' Bulletin of the AAPL, Vol. 10, No. 2, 77-88.
Parsons, T.: 1975, 'The sick role and the role of the physician considered,' *Milbank Memorial Fund Quarterly* 53, 257-77.
Peabody, F.W.: 1927, 'The care of the patient,' *Journal of the American Medical Association* 88, 877-82.
Pellegrino, E.D.: 1979, 'Toward a reconstruction of medical morality: The primacy of the act of profession and the fact of illness,' *Journal of Medicine and Philosophy* 4, 35-46.
President's Commission for the Study of Ethical Problems in Medicine and Biomedical and Behavioral Research: 1982, *Making Health Care Decisions: The Ethical and Legal Implications of Informed Consent in the Patient- Practitioner Relationship*. U.S. Government Printing Office, Washington, DC.
Ramsey, P.: 1970, *The patient as person*, Yale University Press, New Haven.
Ravitch, M.: 1978, 'The myth of informed consent,' *Surgical Rounds* 1, 7-8.
Robinson, G. and A. Merav: 1976, 'Informed consent: Recall by patients Tested Postoperatively,' *Annals of Thoracic Surgery* 22, 209-12.
Roth, L.H. *et al.*: 1977, 'Tests of competency to consent to treatment., *American Journal of Psychiatry* 134, 279-84.
Roth, L. and A. Meisel: 1981, 'What we do and do not know about informed consent,' *Journal of American Medical Association* 246, 2473-7.
Roth, L., et al.: 1982, 'The dilemma of denial in the assessment of competency to refuse treatment,' *American Journal of Psychiatry* 139, 910-13.
Roth, L. H., et al.: 1982, 'The Dilemma of Denial in The Assessment of Competency To Refuse Treatment,' *American Journal of Psychiatry*, 139 (7), 910-913.
Rothman, D.J.: 1991, *Strangers at the bedside.*, Basic Books Inc., New York.
Sherlock, R.: 1984, 'Competency to consent to medical care: Toward a general view,' *General Hospital Psychiatry* 6, 71-76.
Sugarman, J., D. McCrory, D. Powell, A. Krasney, B. Adams, E. Ball, C. Cassell: 1999, 'Empirical Research on Informed Consent: An Annotated Bibliography,' *Hastings Center Report*, January- February.
Szczygiel, A.: 1994, 'Beyond informed consent,' *Ohio Northern University Law Review* 21, 171-262.
Tancredi, L.: 1982, 'Competency for informed consent,' International Journal of Law and Psychiatry 5, 51-63.
Uhlmann, RF, Pearlman, RA, and Cain, KC: 1988, 'Physicians' and spouses' predictions of elderloy patients' resuscitation preferences,' *Journal of Gerontology* 43, 115-21.
Videotape in the Library of Psychiatric Disorders series: 1974, Department of Psychiatry, University of Texas Medical Branch at Galveston.
Waitzkin, H.: 1984, 'Doctor-patient communication: Clinical implications of social scientific research,' *Journal of the American Medical Association* 252 (17), 2441-46..
Wallace, L.M.: 1986, 'Informed consent to elective surgery: The therapeutic value, ' *Social Science and Medicine* 22, 29-33.
Wear, S.: 1991,' Patient freedom and competence in health care,' In *Competency: A study of informal competency determinations in primary care,* Eds. M.A.G. Cutter and E.E. Shelp, Kluwer Academic Publishers, Dordrecht, pp. 227-236.
Wear S.: 1991, 'The Irreducibly Clinical Character of Bioethics,' *The Journal of Medicine and*

Philosophy 16, 53-70.
Wear, S.: 1998, *Informed Consent: Patient Autonomy and Clinician Beneficence within Health Care (2nd Edition)*. Georgetown University Press, Washington, DC.
White, B.: 1994, *Competence to Consent*, Georgetown University Press, Washington, DC.

CASES AND STATUTES

Canterbury v. Spence: 464 F.2d 772, 781 (U.S. Ct. App. D.C. Cir. 1972).
Cobbs v. Grant 502 P.2d 1 (Cal. 1972).
Mohr v. Williams 95 Minn. 261, 104 N.W. 12 (1905).
Natanson v. Kline 350 P.2d 1093 (Kan. 1960) reaff'd 354 P.2d 670 (Kan. 1960).
Salgo v. Leland Stanford Jr. University Board of Regents 317 P.2d 170 (Cal. App. 1 Dist. 1957).
Scholoendorff v. Society of New York Hospitals 105 N.E. 92 (N.Y. 1914)
Wilkinson v. Vesey 295 A.2d 676 (R.I. 1972).

CYNTHIA B. COHEN

PHILOSOPHICAL CHALLENGES TO THE USE OF ADVANCE DIRECTIVES

I. INTRODUCTION

The power of medicine to sustain life increased dramatically in the 1960s as new technologies promised to keep people alive beyond all previous expectations. Early enthusiasm for these novel medical measures, however, was soon tempered by the realization that they could not only keep people alive, but could also prolong their dying. To combat a rising concern about the medical overextension of the lives of those who could no longer express their choices, “living wills” were developed in the 1970s. These written documents enable competent persons to give instructions about how life-sustaining treatments should be used on their behalf should they become terminally ill and unable to speak for themselves. Durable powers of attorney for health care soon followed. In these directives, persons appoint a proxy to make medical decisions for them should they become unable to make their own decisions. Proxy directives apply not only to treatment decisions near the end of life, but to all medical care decisions. These two sorts of advance directives have also been combined so that people can appoint a proxy and indicate what sorts of treatments they do or do not want in certain kinds of situations in one document. The term "advance directives" has been coined to cover the entire repertoire of measures available to those who wish to guide the use of life-sustaining medical treatment for them should they become decisionally incapacitated.

Initially, advance directives were heralded as a means of enhancing individual autonomy for formerly competent persons who were now incompetent and for whom difficult decisions about the use of life-sustaining treatment had to be made. Many states enacted legislation giving legal recognition to written “living wills” that conformed to a certain format and also gave legal force to durable powers of attorney for health care. Indeed, in 1990, the Supreme Court stated that advance directives are important documents that might help resolve difficult questions about the withdrawal of treatment from patients (*Cruzan*, 1990). In 1991, Congress passed the Patient Self-Determination Act (U.S. Congress, 1991) to encourage greater knowledge about and use of these instruments. The law requires that patients be asked upon admission to a health

G. Khushf (ed.), Handbook of Bioethics, 291–314.

care institution that receives Medicare or Medicaid funds whether they have developed written advance directives and, if they have not, that they be given information about how to formulate them.

This rosy acceptance of advance directives began to fade, however, as difficulties were encountered in their use and questions were raised about their ethical foundations. At least three interrelated challenges to advance directives have been raised since their inception: to their meaning, their applicability, and their basic philosophical and ethical grounding.

The earliest challenges focused on ambiguities of meaning that could be found in individual "living wills." Critics contended that these written documents often contained only vague declarations that provided little specific guidance to families and health care professionals about how to treat the patient in the current circumstances. For instance, the refusal of a life-sustaining technology in one context did not indicate whether all other forms of life-sustaining treatment were also to be withheld in that context. Moreover, challengers claimed, durable powers of attorney did not provide proxies with sufficient information to allow them to make decisions that expressed the wishes of those they represented. Too often, critics declared, professionals and proxies had to read between the lines to discern what the authors of these declarations had intended.

Challenges to the applicability of advance directives soon followed. Physicians and families often could find no accepted standard for determining that a patient had become decisionally incapacitated and that his or her advance directive had therefore been triggered. Some critics maintained that proxy directives did not delimit the scope of authority of proxies and therefore provided no way to assess when they were carrying out their charge properly. Such challenges to the applicability of advance directives increased in number and intensity as empirical studies indicated that some authors of these documents change their minds and their preferences over time (Everhart and Pearlman, 1990) and that others are loathe to have their directives followed to the letter (Sehgal *et. al.*, 1992). Such studies raised serious questions about whether directives should unfailingly be followed.

More philosophical challenges to advance directives have been enunciated in recent years as questions about the meaning and ethical weight of individual autonomy, the role of interests, and the nature of personal identity have been raised in relation to advance directives. Thus, some question whether these directives can meet the same standards of informed, autonomous choice that undergird contemporaneous treatment decisions (Brock, 1991, p. S5; Buchanan, 1988), an important requirement if they are to be given the same ethical weight. Others assert that people's interests do not remain static over the course of a life-time, but change as they move from competence to incompetence. When advance directives no longer express the interests of persons who have lapsed into incompetence, according to some critics, they can be overridden (Dresser and Robertson, 1989). An even more radical philosophical challenge to advance directives has been leveled: that those who are incompetent not only have different interests from those they had while competent, but that they have become new and different persons. The very process that robs individuals of competence and brings their advance directive into play can often destroy the conditions necessary for them to retain the same personal identity. In such instances, the advance directives of their former selves do not apply to their new selves and should be disregarded (Dresser, 1995; Dresser, 1994a; Dresser, 1990; Dresser, 1989, Dresser, 1986).

Such philosophical challenges to the foundations of advance directives threaten to undermine them in an even more lethal fashion than do the challenges to their meaning and applicability. Should they go unmet, advance directives would be shorn of their ethical grounding and ultimately have to be abandoned. Should they be successfully rebuffed, the philosophical framework and arguments developed in the process will assist in addressing the challenges of meaning and applicability that have also been thrust at advance directives.

In this article, I will set out and analyze the philosophical challenges that have been raised to advance directives, and will consider responses and counterarguments that have been and could be given to them. I will also highlight future philosophical challenges to advance directives that are beginning to take form and indicate how these might shape future directions for utilizing these documents.

II. ADVANCE DIRECTIVES AND UNCERTAINTY ABOUT THE FUTURE

According to the accepted view, instruction directives, or "living wills," have the same ethical standing as contemporaneous treatment decisions and are to be followed out of respect for the autonomy of their authors (Beauchamp and Childress, 1994, p. 179). Competent persons should be allowed to shape their dying process during possible subsequent incompetence in ways that accord with their deepest values and preferences, much as they shape their lives when they make contemporaneous treatment choices. Yet treatment decisions expressed in instruction directives prior to illness are often made when their authors are in a state of uncertainty about the future and therefore, it is arguable, cannot meet the requirements of competent informed consent.

In order to give competent informed consent or refusal to a proposed intervention, individuals must have substantial understanding of facts that are usually considered material to reaching a decision about it (Beauchamp and

Childress, 1994, p. 147). They ordinarily have this sort of understanding when they make contemporaneous treatment decisions, for they know the medical circumstances and condition at issue, the parameters of the available treatment options, and the risks and outcomes of each. Here is the rub with advance directives, critics maintain. Those who create these instruments often cannot know just what circumstances they will face in the future when their directives take effect (King, 1996, pp. 6-7; Cantor, 1993, pp. 26-27; Buchanan and Brock, 1989, p. 104; Buchanan, 1988). Therapeutic options may have changed and, along with them, the outlook for recovery. Consequently, it is arguable that the authors of these declarations often lack sufficient understanding and information about their future circumstances at the time that they formulate them to meet the standards for informed consent.

Furthermore, some patients indicate that they have not discussed their treatment preferences with health care professionals, family members or proxies (Uhlmann *et al.*, 1988; Cohen-Mansfield *et al.* 1991; Gamble *et al.*, 1991). This suggests that informal safeguards provided by others who might restrain uninformed or imprudent contemporaneous choices about the future may not be present to those developing advance directives (Cantor, 1993, p. 26; Buchanan and Brock, 1989, pp. 106-107). At the time that individuals compose these documents:

> [t]he decision to forego life-sustaining treatment is a remote and abstract possibility, the mere contemplation of which is unlikely to elicit the same protective responses that are provoked in family members and health care professionals when they are actually confronted with a patient . . . (Buchanan and Brock, 1989, p. 107).

The inability of the authors of advance directives to know about their future circumstances and the absence of scrutiny by others of the decisions they take in advance directives limit the ethical authority of these documents to direct the course of treatment of incompetent persons, according to these critics.

Proxies are not the answer to these challenges, it would seem. They may know a patient's general values, but be mistaken about the way the patient would want them to guide the current situation (Seckler *et al.*, 1991). Often they must make significant decisions when they are uncertain about how to interpret the values and goals set out by the patient in an advance directive or expressed by the patient earlier in conversation. If the patient did not draw his or her values together into a coherent whole and express them clearly in an advance directive – or if the patient found this project too daunting – proxies would have little basis for understanding what that patient would want done at this later time. Thus, it appears that individuals cannot give informed consent to treatment or its refusal in future situations and that their proxies may not know how to apply their values when those future situations are realized.

May presents an even more formidable philosophical argument against the applicability of advance directives to patients in the future (May, 1997; May, 1999). Autonomous decisions, he states, "reflect an agent's assessment of the balance of reasons for action, 'reasons' here construed in the broadest possible sense" (May, 1997). He claims that advance directives cannot extend individual autonomy to later situations of diminished competence because they provide only a formal strategy for self-binding in the future. They do not allow their authors to reconsider this strategy and make a specific judgment that determines action at the time that a treatment decision must be made. That is, the authors of these declarations have bound themselves to a certain plan of action in the future, but once they have reached that future point, their own evaluations of what should be done in that context do not control the decision taken. They have put themselves on "automatic pilot" in a way that mimics the structure of voluntary slavery. Consequently, he maintains, advance directives are inconsistent with their authors' autonomy. His argument, if successful, would undermine the ethical authority of these directives.

In response to these varied criticisms of advance directives, it can be argued that advance directives are often developed by persons familiar with a specific illness who know the clinical course of their condition and can tailor their instructions to it. Thus, Lynn and Teno observe that

> [t]he increase of chronic disease in the population also grants patients the opportunity to become knowledgeable about the expected course of illness and the merits of alternative plans of care for anticipated future situations, thus enabling their giving informed directions about future choices (1995, p. 572).

Advance directives in many such instances can closely approximate the exercise of informed contemporaneous choice. Moreover, if advance directives are reviewed and modified periodically, they will be more likely to take into account newly developing medical conditions and novel therapeutic options that might affect their authors. Thus, they can reflect their knowledge of their possible later condition and prognosis and serve as close approximations of contemporaneous judgments (King, 1996, p. 74; Buchanan and Brock, 1989, p. 104).

It also can be argued that third parties often *do* have the opportunity to influence a mistaken or imprudent decision set out in an advance directive. Research shows that advance care planning catalyzes important and therapeutic discussions between patients, providers, and family members about difficult issues (Miles *et al.*, 1996, p. 1066; Singer *et al.*, 1998). The process of composing advance directives can spark discussions about their authors' underlying values and reveal whether they want various sorts of life-sustaining treatment should they linger near death in the future. Such conversations can give proxies named in advance directives a sense of those values that are important to those whom they represent and of how to interpret their choices should they later lapse into incompetence. Moreover, proxies can step into the breach when individuals have not anticipated their later condition or treatment options and make decisions for them that reflect the ends, beliefs, and preferences that such individuals have expressed earlier. Thus, sound treatment decisions can be, and are made by others on the basis of a sense of a patient's overarching interests, values, and beliefs (Dworkin, 1993, pp. 191-192).

Commentators who defend advance directives against the criticism that they are vitiated by their authors' lack of information about the future, however, do not award them absolute authority over treatment at the time that they take effect. In general, they maintain that there is a presumption in favor of accepting such declarations as binding unless some overriding reason can be produced to indicate that their authors have given flawed consent (Cantor, 1993, p. 28). Such reasons include that an author has not stated his or her preferences clearly and in a way relevant to the situation in the directive (Beauchamp and Childress, 1994, p. 39), that there is evidence that the directive does not represent what its author wanted done in the present circumstances (Brock, 1991, p. S5), that the author did not understand his or her likely future medical context and circumstances (Buchanan and Brock, 1989, p. 105; Dworkin, 1986, p. 14), and that available treatments have changed markedly from when the directive was created (Buchanan and Brock, 1989, p. 104). These grounds for overturning an advance directive suppose that the choice expressed in them by their author was mistaken or not clearly expressed and try to remedy this in order to honor patient autonomy.

These responses provide grounds for addressing the criticism that advance directives cannot overcome the significant ethical gap between giving consent when competent for treatment or non-treatment in the near future and giving such consent for treatment or non-treatment in the far future. Whether this gap is as great as it seems to critics of advance directives and whether it is inconsistent with individual autonomy, as May claims, is open to question.

Consent is given to an action or project under a certain description. When individuals provide contemporaneous consent to a certain treatment or non-treatment, they may understand their current situation, but be unaware of further implications of what they have consented to, of its corollaries and presuppositions. Because they do not and cannot have complete knowledge and understanding of all that will ensue, they cannot consent to all aspects of all proposed treatments. There would always be some additional aspect of the therapeutic situation that called for a specific judgment on the part of patients. Should surgery be carried out with one technique or another to realize their chosen ends? How much sedation should they receive for a moderately uncomfortable procedure in which they might need to make a particular decision? Were individuals required by the value of patient autonomy to make such specific judgments about steps taken over the course of treatment, no treatment could ever be provided to any patient. Not only advance directives, but *all* informed consent, whether contemporaneous or long-term, would be placed in jeopardy.

When seeking contemporaneous informed consent, "[t]he most we can ask for is consent to the more fundamental proposed policies, practices and actions," (O'Neill, 1984, p. 176), not to very detailed information about treatment options, risks, and outcomes. Individuals can articulate certain ends and goals grounded in advance directives that are relevant across a wide spectrum of medical situations without giving particular evaluations of what should be done in specific instances, as May requires. If we accept the opacity of contemporaneous consent, then we must also accept the opacity of consent to future treatment given in advance directives, viewing it as differing only in degree, not in kind.

Furthermore, it seems peculiar to claim that individuals who indicate in advance directives that certain ends that are to be pursued on their behalf in the future should they become incompetent thereby place themselves in the position of voluntary slaves. They are not passive victims of a broad strategy they chose earlier that prohibits them from taking into account specific reasons at the time of decision. The "master" that will determine their decisions at that later time will be the very ends and reasons they have enunciated in their advance directives as these apply to their specific situation. Respect for individual autonomy would seem to involve just that – making decisions that are guided by the chosen ends of the individuals involved, even when those individuals can no longer speak for themselves. The proxy who honors these ends acts on behalf of the now-incompetent author of the advance directive, making specific judgments according to the author's values and ends as these were expressed earlier. This differs from the situation of the voluntary slave for whom decisions are made on the basis of the values and ends of an external master. The person appointed in a proxy directive who acts according to the wishes and values of the now-incompetent person is acting for that person. If there were strong reasons for a proxy to question whether the individual's advance directive led to a decision that would honor that person's chosen ends, that proxy would be responsible for reforming or even abandoning that decision. The proxy would be obligated to decide according to the latter's values, filling in as best he or she can on the basis of knowledge gained through their personal relationship. Thus, contrary to May, advance directives can be used as a way of preserving individual autonomy even after their authors can no longer make specific treatment decisions for themselves.

Future efforts need to be directed toward developing ways of overcoming the opacity of informed consent in both current and future contexts. One such attempt involves a new "living will" format termed the "Medical Directive" (Emanuel and Emanuel, 1989). This comprehensive instruction directive enables individuals to decide in advance of possible future incompetence about what specific medical interventions they would want to have used in 48 hypothetical clinical situations. This document covers many of the clinical circumstances that could arise in the future related to life-sustaining care and allows individuals to express their preferences and values about their treatment in each of these scenarios. It also allows its authors to name a proxy to make decisions for them in those circumstances that are not considered in the directive. Professional caregivers are expected to assist patients in developing this document, and this, in turn, gives physicians and others an opportunity to respond protectively to decisions made by patients that seem mistaken or imprudent.

Brett (1991, p. 826), however, has expressed reservations about using "a checklist of interventions" to direct the care of patients who can no longer choose for themselves. He maintains that this "runs the risk of promoting the selection or rejection of interventions because of their inherent characteristics, rather than as appropriate means to ends that the patient would have wanted." Patients' goals should be the primary consideration in deciding about the use of life-sustaining treatment, Brett argues. This is an important concern, but it does not indicate that patients' ends and goals are necessarily lost by the use of more detailed and comprehensive advance directives. As Cantor notes, it is not possible to acknowledge all possible future situations, given the limits of human imagination (Cantor, 1993, p. 27). This means that the values and ends that individuals express in such directives can be taken by proxies as guides for their care in clinical contexts that they have not specifically addressed. Continued explorations of ways to address the opacity of individual consent given in advanced directives to future treatment will perhaps never completely overcome this difficulty, but can serve to make these declarations express more reliably their authors' intent.

III. ADVANCE DIRECTIVES AND CHANGES OF INTERESTS

While there may be ways to ameliorate the factual and decisional uncertainty surrounding decisions made in advance directives about an unknown future, some critics maintain there is no way to conquer uncertainty about what their authors' future interests as incompetent persons will be. Some argue that those who lapse into incompetence may undergo major changes in the way they think, feel, and act that they could not have foreseen when they were competent (Dresser, 1994b, p. 611; Brock, 1991, p. S5; Dresser and Robertson, 1989, p. 236). At the time when they composed their advance directives, they were necessarily in a state of uncertainty about how they would respond should they become incapacitated and were therefore poorly equipped to forecast what their needs and interests would later be (King, 1996, pp. 76-81; Dresser, 1994a, p. 35). For instance, it might have seemed demeaning to them when competent to live in a debilitated state, and yet once they became incompetent and entered into that very state, they might have found it brought certain satisfactions that they could not have anticipated when competent (Dresser, 1986, p. 379; Dresser and Robertson, 1989, p. 236). If people cannot accurately envision their future needs and interests, but can only guess about them, there seems little reason to honor the decisions that they make in their advance directives, these challengers maintain.

As the lives of those who become decisionally incapacitated fade into incomprehension and they can no longer be aware of their life projects and concerns, their significant interests fade as well, these critics hold. This is because to have an interest, they contend, requires that its possessor be capable of consciously appreciating that interest. Those who cannot comprehend what it means to have dignity or privacy can have no remaining interest in their own dignity or privacy (Dresser, 1990, p. 431; Dresser and Robertson, 1989, p. 238). Incompetent persons who do not grasp their interests therefore lose them (Dresser, 1994b, p. 695; Dresser, 1990, p. 431; Dresser and Robertson, 1989, p. 238). They "can receive no present benefit from treatment decisions in accord with their former preferences" (Dresser, 1986, p. 381). In their radically changed circumstances, Dresser maintains, their advance directives may not protect their current well-being. When this is the case, these directives should be abandoned.

What standard, then, should govern the care of those who become incompetent after developing an advance directive? In the course of her extensive discussion of this issue, Dresser appears to adopt three different views. At times she maintains that the current interests expressed by incompetent persons should prevail. We can learn about these interests by attempting to "ascertain their point of view, their perspective on what is to be decided" (Dresser, 1994b, p. 613). At other times, however, Dresser recommends that the choices of the formerly competent person and the now-demented person be given equal weight. Thus, she maintains that "[e]ven if the patient is incompetent, the present request ought to carry at least as much weight as the preferences expressed in the past" (Dresser, 1989, p. 163). And at still other times, she holds that neither the choices of patients while competent nor incompetent should be honored. Instead, a "modified best-interest principle" should be used "that both protects patients' experiential welfare and permits proxy decisionmakers to choose from an array of reasonable treatment options for dementia and other impaired patients" (Dresser, 1994b, p. 617). Since Dresser awards the fullest discussion to this third view in her writings, it will be presumed to represent her core position.

On this third view, that a best interest principle should be used, primacy is given to the experiential welfare of incompetent patients, rather than to their past autonomous choices or currently stated preferences. Decisions about whether to treat incompetent patients should be made on the basis of "shared judgments and values, . . . choosing as most people would for themselves" (Dresser, 1994b, p. 658). This involves utilizing "a conception of what experiences are burdensome and beneficial to human beings, as well as . . . a broader vision of when life constitutes a 'good' for people and when it does not" (Dresser, 1994b, p. 639). This modified best interest principle should be used, Dresser holds, because the current interests of cognitively-impaired persons cannot be separated from a more general conception of what would be best for most people in their particular situation (Dresser, 1994b, p. 662).

In assessing the value of the diminished lives of those who are cognitively impaired, according to Dresser, decisionmakers should filter through the socially accepted standard of what constitutes a life worth living the subjective world of individual cognitively-impaired persons. Since the interests of these individuals are grounded in their experience, it is important to delve into their consciousness, gathering behavioral and neurological data about them. Their current awareness of such states as pain, distress, and pleasure and their physical movements and interactions with people and surrounding objects all serve to indicate the nature of their current interests (Dresser, 1994a, p. 37; Dresser, 1994b, pp. 638-639; Dresser, 1990, p. 430; Dresser, 1986, pp. 390-393).

This method, Dresser finds, yields the conclusion that life is a good for individuals when they have "the capacity to interact with their environment" and are free from excruciating and uncontrollable pain or other negative experiences, such as those that are created by permanent restraints or sedation (Dresser, 1994b, pp. 661, 665, 708-713; Dresser, 1990, p. 433; Dresser and Robertson, 1989, p. 240). Dresser would mandate treatment for incompetent patients who meet this criterion, regardless of provisos in their living wills to the contrary (Dresser, 1990, pp. 428-9). Incompetent persons who are permanently unconscious and cannot relate to their environment do not have an interest in continued life because they "can gain none of the goods that make human life valuable to an individual" (Dresser, 1990, p. 428). Hence, treatment would be withheld or withdrawn from them, no matter what they had indicated earlier in their advance directives (Dresser, 1994b, pp. 665-666, 695-697). On Dresser's view, the irremediable uncertainty of the competent authors of advance directives about what their interests will be during future incompetency requires that socially shared values and judgments about individual well-being should govern the treatment of incompetent persons.

To disregard the past autonomous choices of now-incompetent persons does not show disrespect for them or an intent to injure them, according to Dresser. Since they cannot understand what it means to annul their earlier choices, they can experience no injury if this is done (Dresser, 1986, p. 381). She maintains that "[w]e would show more genuine respect for incompetent patients by treating them in accord with what rational maximizers of self-interest would want: the outcome that is most beneficial and least burdensome to them in their current situations" (Dresser, 1990, p. 389) than by honoring their advance directives.

This far-reaching challenge to advance directives is grounded in the supposition that the authors of these instruments are, by and large, ignorant about how they would think, sense, and feel should they later become demented. Dresser asks about those with dementia, "How could we ever know what it would be like to be someone like that?" (Dresser, 1994b, p. 667) This may not be as difficult as she maintains, however. As those in Western society live longer and consequently enter states of dementia with greater frequency, many are becoming familiar with the nature of dementia through their experience of family and friends affected by it. On the basis of the intimate knowledge of the progression of dementia that they derive, those close to the person with dementia can develop an evidence-based understanding of what it is like. As a consequence, when they develop their own directives, the choices that competent individuals make about their care during future dementia may be based on knowledge that is as reliable as our knowledge of any other condition that can be observed but has not been experienced.

Advance directives, taken as a whole, will not reveal one monolithic view of personal well-being. Some individuals may have decided to receive life-sustaining treatment for as long as possible should they become demented out of the conviction that dementia is not as awful as the general view makes it seem. Others may have elected to refuse such treatment should they become demented because they do not wish to have their lives dragged out by "extraordinary" means. Still others, concerned about the burden to their family and friends, may elect to have life-sustaining treatment withdrawn at a certain point as they approach death. And yet others may have decided to receive such treatment because they know that it is important to their family to feel that they have provided care for them as they move toward death. Dresser's approach does not permit giving effect to these differing conceptions of well-being and to these altruistic motivations, but instead imposes a putative social consensus about well-being upon all those with dementia.

Dresser maintains that the interests expressed by competent persons in advance directives vanish when they become demented and that their interests are now to be defined in terms of that of which they are consciously aware. Thus, they can have no interests in dignity and privacy. Indeed, Kuhse, arguing along similar lines, holds that those who are severely demented are no longer persons because they are not conscious of themselves as existing over time. Therefore, they no longer have an interest in their own continued existence (Kuhse 1999). Other commentators maintain, however, that a person's interests can survive incompetency, even though the person can no longer appreciate them (Dworkin, 1993, pp. 226-229; Buchanan and Brock, 1989, pp. 100, 128-129, 162-164; Buchanan, pp. 286-287).

Ronald Dworkin, a prominent advocate of this view, distinguishes between two sorts of interests: "experiential" and "critical" interests. "Experiential" interests are those that people have in carrying out certain activities just for the experience of doing so (Dworkin, 1993, p. 201). These must be experienced by those who have them to qualify as bona fide "experiential" interests. They include any projects that individuals find pleasurable or exciting as experiences, such as playing softball, cooking, or walking in the woods in the fall.

"Critical" interests embrace the self-chosen values and goals that give overarching meaning to people's lives, regardless of what sorts of experiences result from fulfilling them. Such aims as having close friendships, a warm relationship with one's children, and success in one's work qualify as "critical" interests because they are based on critical judgments about what makes a life good. They differ from experiential preferences in that they identify the significance that people find in their lives as a whole (Dworkin, 1993, p. 202-204). Dworkin explains that:

> [p]eople think it important not just that their life contain a variety of the right experiences, achievements, and connections, but that it have a structure that expresses a coherent choice among these – for some, that it display a steady, self-defining commitment to a vision of character or achievement that the life as a whole, seen as an integral creative narrative, illustrates and expresses (1993, p. 205).

Individuals identify themselves with their critical, rather than their experiential interests, for the former incorporate life-shaping commitments that give coherence to their lives and define their view of their own well-being. The latter, in contrast, are more fleeting and relative to time and place. When people make a commitment to a certain sort of life as the good life, to certain critical interests, it is essential to their integrity and dignity that they be allowed to live – and die – according to that commitment (Dworkin, 1993, p. 224). Thus, the point of what Dworkin terms "precedent autonomy" is to allow individuals to shape their lives "according to some coherent and distinctive sense of character, conviction, and interest," to define what well-being would consist of for them (Dworkin, 1986, p. 8). This is the sort of autonomy that can be expressed in advance directives.

Persons who develop dementia often cannot appreciate the critical interests that they were able to grasp while competent. Dworkin agrees with Dresser that many "have no sense of a whole life, a past joined to a future, that could be the object of any evaluation or concern as a whole" (Dworkin, 1993, p. 230). Indeed, their critical interests may conflict with their experiential interests at this time. Nevertheless, Dworkin maintains, contrary to Dresser, that the values and choices they have made in the past do not evaporate when incompetence clouds their memory and they are no longer aware of them. They retain their critical interests, such as those in dignity and privacy, and these take precedence over their experiential interests (Dworkin, 1993, pp. pp. 219-220, 229-230, 237). What happens to these persons in their current state of incompetence, Dworkin contends, has ethical import for the value of their lives as a whole. To override their critical interests by refusing to adhere to choices they made in their advance directives while competent constitutes a form of paternalism that denies their continued ethical standing as persons with dignity (Dworkin, 1993, pp. 231, 236). Dworkin declares that "[m]aking someone die in a way that others approve, but he believes a horrifying contradiction of his life, is a devastating, odious form of tyranny" (Dworkin, 1993, p. 217).

As Post observes, at some point in the progression of dementia, the individual reaches a threshold where continuity with the past is so diminished that the self appears to live almost entirely within the present (Post, 1995). Jaworska grapples with this reality in her presentation of an alternative to Dworkin's view, arguing that even though they have lost considerable memory of the past, many with dementia retain the "capacity to value" and are therefore still capable of making autonomous decisions (Jaworska, 1999). She provides several vignettes of individuals with Alzheimer's to illustrate that they have not completely lost this capacity, even though they cannot recall their life story, and goes on to maintain that they therefore retain authority over their well-being. For Jaworska, autonomy does not require a grasp of the narrative of one's whole life plan, but only the capacity to value; well-being consists of acting according to one's values. Thus, she finds that those who can no longer recall their earlier lives can retain their autonomy during dementia and that respect for their immediate interests is not contrary to their autonomy or well-being. Yet it is not clear what this rechristening of autonomy as the capacity to value gains and how it differs from Dresser's ultimate approach. It surely promotes respect for the interests and concerns of those with dementia, which is a laudable pursuit. However, it leaves unanswered the question whether to honor the values expressed by patients with dementia when these conflict irresolvably with those that they have expressed in advance directives.

A significant philosophical question is at issue in this debate: How should we think of the self in assessing the interests of those who are decisionally impaired? Should we take into account only a person's present situation and capacities or should we consider previously expressed values that take into account the entire sweep of his or her life?

Dresser's way of responding to this question takes what she calls a "subjective" turn, focusing on the demented person before her and bracketing the person he or she once was. This leads her to mount an heroic attempt as an onlooker, akin to that of Jaworska, to pierce the altered consciousness of the individual incompetent person in order to understand his or her experience and the value life has for that person. Dworkin, in contrast, adopts a more historical approach, urging us to consider the entire course of the demented person's life in deciding how we should think of him or her. He mounts an attempt that Dresser would regard as equally heroic: to apply the values that competent persons stated earlier were important to their life plans when providing them with care in their current situation of dementia.

Dresser's is a more immediate empirical endeavor: she asks what do these persons with dementia, in fact, currently experience? If we are to make decisions for them as they are in the present, we must attempt to ascertain what they are currently going through and how society would tend to decide what should be done, given their experience. Dworkin's is a more long-range endeavor: he asks what over-all values guide the lives of these formerly competent persons? If we are to make decisions about them as persons, we must consider their past interests, projects, and relationships, as well as their current situation.

Dresser's approach can be criticized because it divorces persons with dementia from their life-long views about the good life, their beliefs about what they owe to others, their relationship with their family, and their significant religious beliefs and communities. Thus, Rhoden questioned whether standards for treating those who are vulnerable and at risk of death can be established by isolating persons who have become demented from their past history and interests, from "all that makes humans special"(Rhoden 1988, p. 378). She declared that "a perniciously impoverished view of humans" lurks behind the drive to restrict the interests of incompetent persons exclusively to what they can currently experience (Rhoden, 1988, p. 408-11).

Yet Dworkin's view can also be criticized on grounds that it ignores the current welfare of vulnerable incompetent persons. Dresser contends that we must care for the incompetent person as best we can in the present, recognizing the central role that compassion should play in shaping our decisions. We are ethically required to respect the lives of disabled people and to protect them from the harm that could result from following their advance directives (Dresser, 1994b, p. 628). Consequently, when treatment choices dictated by competent persons in advance directives would be harmful to them in a future state of dementia, we owe it to them to overrule their earlier autonomous choices for the sake of their current well-being. Harm is to be identified by others who evaluate what outcome would be most solicitous of the present interests of incompetent persons. It is not defined by what currently incompetent persons viewed as harm when they were competent, which was grounded in radically different interests (Dresser, 1995, p. 35; Dresser, 1994a, pp. 159-160, 163; Dresser, 1990, p. 426; Dresser, 1989, pp. 159-160; Dresser, 1986, p. 379).

Some commentators offer mediating views of the ethical force of advance directives that accommodate Dresser's concern about directives that appear to violate the well-being of those who are incompetent and yet respect the choices made by persons when competent. They would maintain a presumption in favor of honoring advance directives, but would discount such directives when human compassion demands this. Thus, both Dworkin and Cantor maintain that an advance directive should be set aside when it appears to require care for an incompetent person that would generally be considered "inhumane" (Cantor, 1993, pp. 114-115; Dworkin, 1986, p. 13). Cantor illustrates this with the example of Person B who believes:

> both that life should be preserved to the maximum extent possible and that suffering is preordained and carries redemptive value in an after-life. B prepares an advance directive in which all possible life-extending medical intervention is requested and all pain relief is rejected. . . . Subsequently, B suffers from cancer, which both affects his brain, rendering him incompetent, and causes him to suffer excruciating pain. Further medical treatment such as radiation will extend B's life but will not itself relieve the pain or cause any remission in which competence would return. . . . [B is] not only groaning in agony but also thrashing to the point at which continuous physical restraint is necessary to keep his life-preserving medical intervention in place (1993, pp. 114-115).

Here we reach the ethical boundary beyond which advance directives should not be honored, Cantor holds, for "basic humanity" demands that we provide B with pain relief, despite his carefully considered advance directive. He maintains:

> Just as a self-determination prerogative cannot morally encompass self-mutilation or consent to slavery, . . . the prospective imposition of an utterly degrading status on an incompetent person is simply beyond tolerable bounds. . . . Only when actual suffering reaches the level most people would label intrinsically inhumane should the patient's directive be overridden (1993, p. 115).

Thus, Dworkin and Cantor would nullify advance directives of decisionally incapacitated persons only in such extreme cases of great suffering on grounds of human compassion. Dresser would do so more broadly when such individuals have some level of conscious awareness of their surroundings and are free from pain and other negative experiences and yet earlier in an advance directive elected to refuse treatment in such circumstances and also when individuals are permanently comatose and yet earlier in an advance directive elected to receive treatment.

Future efforts will need to focus on the question whether the goal of advance directive is to protect their authors' well-being during incompetence or to express their autonomous choices about future treatment or non-treatment. If, with Dresser, we conclude that advance directives should be set aside in order to protect patient well-being during incompetence, we will have to conceptualize more fully the nature of that well-being and address the apparent conflict between the view of competent persons about what their own well-being requires during incompetence and that of society. If, with Dworkin, we maintain that we should respect patient autonomy and well-being as these were defined by the competent person in an advance directive, we will have to explain how the competent person's notion of well-being is to be weighed in the balance against the implied view of well-being of that person when incompetent.

IV. ADVANCE DIRECTIVES AND CHANGES IN PERSONAL IDENTITY

An even more profound and radical philosophical challenge to advance directives has been posed. When incompetence sets in, it is arguable that a new person with only tentative connections to the earlier person he or she once was appears on the scene. Since that person's advance directive was authored by a self that no longer exists, to honor it would be paternalistically to impose on a vulnerable, incompetent person the interests of an entirely different person. The long-standing philosophical question of personal identity, or what it is to be the same person persisting over time, reemerges with this argument.

Dresser, its foremost advocate, maintains that many of those who suffer from dementia experience such significant cognitive changes that they become altogether different individuals (Dresser, 1995, p. 35; Dresser, 1994a, pp. 157-158; Dresser, 1990, pp. 432; Dresser, 1989, pp. 158-162; 1986, pp. 379-381). In developing her position, she refers to a theory of personal identity propounded by Derek Parfit (Parfit, 1984, pp. 204-205, 223-226). According to Parfit, personal identity is based on "psychological continuity" or the strength of the connectedness of an individual's mental events, such as memories, desires, beliefs, and intentions. Personal identity varies over time, depending on the degree of psychological continuity that ties a person's experiences together over segments of time. To have psychological continuity, persons need not recall every experience they have had, but must recall a significant number of their overlapping psychological features from previous days and years to be called the same person (Parfit, 1982, pp. 204-206). A person who cannot do this becomes a different person. In Parfit's view, questions of personal identity are not about a single unitary entity that persists over the course of a life-time, but about a series of successive selves, each of which is expressed by that part of a life that exhibits psychological continuity with earlier and future selves.

Dresser, taking this view of personal identity as her starting point, claims that the person who suffers from dementia experiences a chain of psychological events that is broken at many points. Consequently, no psychological continuity can be found between an earlier competent self and the current incompetent self, and the latter must be considered a new, different person. Moreover, the earlier self is not ethically warranted in determining the fate of the new post-competent self. Advance directives, on this argument, do not have the ethical authority to govern the treatment of those who author them when competent and subsequently succumb to dementia (Dresser, 1995, p. 37; Dresser, 1989, pp. 161, 164-165).

To illuminate her view, Dresser borrows a case from Sanford Kadish (Dresser, 1994b, pp. 624-625; Kadish, 1992, pp. 871-872). We are to suppose that world-famous Composer Then instructs her son in a combined instruction-proxy directive that if she becomes incompetent and "permanently unable to experience music in any way," no life-saving medical treatment should be administered to her should she need it. Years later, as Composer Now, she becomes demented and incurs a life-threatening, but curable disease. At this later time, Composer Now still has considerable awareness of the world and gains many pleasures from life. She likes to sit in the garden, enjoys being attended, smiles at her grandchildren, and shows certain preferences for foods and television programs. When asked if she prefers to die, she becomes agitated and says "No".

Dresser maintains that the treatment refusal expressed by Composer Then in her advance directive clearly conflicts with Composer Now's current interest in continued living. Therefore, she concludes, the document should not be honored. To allow the advance directive of a competent person to govern the care of that person once incompetent can constitute an ethically unacceptable form of "self-paternalism" in which the best interests of the later self are defined by a different, earlier individual (Dresser, 1989, p. 160). Such self-paternalism is ethically appropriate only when it furthers the later self's best interests or coincides with the later self's preferences.

Dresser's view of personal identity does not define the degree of memory loss or character change that would mark the transformation of a person into a new person. Consequently, it leaves us with no way to know the point of discontinuity that distinguishes between these different selves and to determine that an advance directive made by a former self is inapplicable to the current self (Beauchamp and Childress, 1994, p. 132; Cantor, 1993, p. 110; Buchanan and Brock, 1989, p. 158). Indeed, in actual clinical scenarios, DeGrazia points out, even if the person with dementia could not remember life as a competent person, he or she would probably have many overlapping chains of memory that extend bit by bit from the present back to that earlier time (De Grazia, 1999). Moreover, that person would likely retain many beliefs and skills and some desires and character traits from his or her earlier competent days. Consequently, it would be difficult to assess with any theoretical confidence when a currently demented person has so little psychological continuity with an earlier competent person that it can be said that he or she is now a different person.

Dresser's argument can be extended to those who experience decisional incapacity of any kind, such as those with moderate brain injuries who experience some forgetfulness or those who are comatose. The earlier self in such instances does not survive, on this argument, because there is only "weak connectedness" between its former and current mental events. Should the person recover from the brain injury or emerge from the coma, however, would he or she now be yet a third person due to serious psychological discontinuities with the brain-injured or comatose earlier self and also the self that preceded that self even earlier? Ultimately, Dresser's argument can be used to deny that advance directives should ever apply to *anyone*, for, as she states, "observers can never be sure that the person who executed the directive is the same person whose medical treatment is presently at issue" (Dresser, 1989, p. 162).

Those, like Dresser, who focus on the person in his or her subjective immediacy tend to hold that personal identity is constituted by psychological factors. This view owes a debt to David Hume, who stated, "I may venture to affirm . . . that [we] are nothing but a bundle or collection of different perceptions which succeed each other with an inconceivable rapidity, and are in a perpetual flux and movement" (Hume, 1978, p. 399). The individual who has no important connections of memory, character, and personality traits with the person who existed earlier, even though they share the same body, becomes a different person on this view. There is no substratum, no self, that persists over the course of a person's life, but only a series of mental events. Hume admitted that this left him unable to explain personal identity and to declare that he had fallen into a labyrinth from which he could find no escape.

Certain other philosophers (Butler, 1897; Reid, 1785; Chisholm, 1976, Swinburne, 1976) maintain that psychological continuity and other criteria of personal identity that have been proposed, such as persistence of body and brain, are criteria of personal identity only in the sense that they provide evidence for it. They do not in themselves constitute personal identity. Indeed, there is nothing else that personal identity consists in, these thinkers hold, for it is an "ultimate unanalysable fact, distinct from everything observable or experienceable that might be evidence for it" (Noonan, 1989, p. 19). On this view, persons are enduring subjects of experience who are distinguishable from their experiences, not mere labels for thoughts, sensations, and memories that occur over a slice of time. A person need not hold the same beliefs, desires, or memories over a long span of time in order to be the same person, according to these philosophers, but may change these over the course of a life-time and yet remain the same.

Some of these thinkers, such as Joseph Butler, find that the view of personal identity as psychological continuity is viciously circular in that it *presupposes* personal identity (Butler, 1898, p. 329). One must first be in possession of the concept of the personal identity of a specific individual and determine that it applies to that same individual to be in a position to say that the memories, beliefs and desires of that individual have changed – and on that basis to deny that the same individual is the same individual! Thus, the very proposition that Composer Now does not remember certain things, such as the values, beliefs, and desires "she" expressed in Composer Then's advance directive, presupposes that Composer Now and Composer Then are the same person – Composer. We make this distinction by an appeal to the underlying personal identity of the individual under discussion.

Finally, even those who are sympathetic to the Dresser view of a succession of selves over the course of a life-time, find that ethical and policy considerations provide a barrier to accepting the position that the person with advanced dementia has become a different person. This is because personal identity is a necessary condition of being accorded rights or having responsibility for what one does. Only if a person is the same person over time can he or she be held heir to the merit or blame of the earlier person. On the Dresser view, the person who makes a commitment at one point in time may undergo major changes in intentions, desires, and beliefs at another time and thus become a different person with no responsibility for what that first person has promised. This view of personal identity would destroy the ethical authority of promises, contracts, or any other practices in which we make commitments to bring about certain events in the future (Rich, 1997). As Buchanan and Brock observe,

> Some of our most important social practices and institutions – those dealing with contracts, promises, civil and criminal liability, and the assignment of moral praise and blame – apparently presuppose a view of personal identity according to which a person can survive quite radical psychological *discontinuity*. If this is so, then since these practices and institutions are so valuable, we would have to have extraordinarily weighty reasons for giving up the view of personal identity upon which they are founded (1989, p. 174).

Such policy considerations led Rhoden to maintain that we should adopt the "notion that a person is one person, and one person only, from birth through old age, despite whatever changes and vicissitudes she might undergo" (Rhoden, 1988, p. 414). Consequently, Dressser's challenge to the validity of advance directives, while intriguing, seems to fall by the wayside for reasons both of critical philosophy and of public policy.

V. FUTURE DIRECTIONS

Puzzles and problems created by the use of advance directives are bringing to light significant philosophical questions with import beyond their use. These difficulties spur us to reconsider such basic philosophical issues as the meaning of the self, the nature of personal identity, and the role of the community in new ways. The most important of these has to do with an issue that has appeared and reappeared consistently throughout the discussion of advance directives: What conception of the self and of self-determination is necessary to sustain the value of respect for the choices that competent individuals have made in advance directives? What is the relationship between the self and its choices and acts? How are we to understand the nature of the underlying self when incompetence sets in? Does that self become a different self or even a non-self?

If it is difficult to maintain that the self is merely a certain degree of connectedness between segments of the experience of an embodied entity, so, too, is it arduous to hold that the self is an ultimate unanalyzable fact, distinct from everything observable or experienceable, divorced from its history and narrative. For this reason, Korsgaard has made the post-Kantian suggestion that if we regard persons primarily as agents, rather than as the locus of experience, we will find that as a matter of practical necessity, personal identity is a prerequisite for coordinating action and carrying out plans (Korsgaard, 1996). We need not ponder the significance of the narrative of our whole life, but need only have a practical identity, a normative concept of ourselves.

Others have also criticized the Kantian conception of a self that is independent of choices it might make, unaffected by them or by its history and tradition, but have given greater prominence to a rational plan of life through which persons define themselves (Buchanan and Brock 1989, May 1999, Tollefsen, 1998). We develop desires and interests as we move into the future through the pursuit of ends in an order specified in our comprehensive plan. Thus, the self develops through time and is constituted mainly through its choices (Tollefsen, 1998). We shape ourselves according to our commitments and character and act to make that person a reality. Our earlier decisions as competent persons affect who we are during incompetence, determining who we will be at that time and what will be done for us. Thus, on this view, advance directives function in a way similar to any other choices of the self: they determine our future character now.

Blustein has approached the question of whether to honor the advance directives of decisionally impaired patients by proposing that it is the narrative self-conception of persons that makes them the individuals who they are (Blustein, 1999). When those overtaken by dementia can no longer give narrative sense to their lives, it is the role of proxies to step in to assist them to continue their life stories. Thus, it is not a list of values that defines what should be done for incompetent patients, but the narrative that they have constructed over the course of their life and expressed in their advance directives. This view, however, retains some of the problems of the psychological continuity view, such as how high a degree of coherence a narrative must retain to constitute a self. How can those who continue a person's narrative fill out an incomplete story when an advance directive does not tell them how the individual would have brought it to a close? By what moral authority do advance directives have priority in such situations?

Some maintain, in response to these sorts of difficulties, that our selves are not individual, but social creations, affected by the community in which they develop (Kuczewski, 1994, 1999; Nelson, 1995). Our lives can be conceived of as stages in a history that plays over the course of time during which we discover our values, ends, interests, and preferences as we interact with the community. Thus, Kuczewski states that "I conceive of myself objectively as part of a larger group that contains a 'part' of me that transcends my individual consciousness and psychological continuity" (Kuczewski, 1994, pp. 42-44). Psychological continuity is not a necessary condition of personal identity on this communitarian view of the self. Yet the lives of those who are no longer competent to make their own treatment decisions still need to be considered as a unity. This requires the community to complete their personal identity by continuing to tell their story from a vantage point outside their stream of consciousness in ways that are in general accord with their values and ends. Thus, the interests and values of the formerly competent person, according to this social view of human selfhood, survive in the deliberative process of the proxy who takes their character into account and thereby fulfills their trust.

What we mean by the self and self-determination also underlies consideration of whether the goal of advance directives is to protect their authors' current interests during incompetence or to ensure respect for competent choices they made that were grounded in earlier, sometimes different, interests. Pursuing this inquiry will require us to explore more fully what is meant by the "interests" of the competent self in order to understand whether those interests extend to the now incompetent self or have changed radically. This, in turn, will require us to clarify how the notion of "interests" relates both to the well-being and the autonomy of the self. If interests are reduced to current preferences, then the interests of at least some incompetent selves will differ sharply from those they had while competent. Yet there is a sense in which interests are not equivalent to what we are "interested in," but have to do with what would promote our good, with what is "in our interests." Thus, it is in our interest to be healthy, although we currently have a preference for a remarkably unhealthy diet and no exercise. Our interests in this second sense are directed toward the values by which we define our well-being as whole persons over time. These interests can outlive our competence and cannot be ignored without risking mutilation of our selves.

Bioethics was not founded on overarching ethical theories, but on values and concepts of significance in our culture, such as self-determination, respect for the dignity of persons, beneficence, and personal identity over time. The importance of certain of these values and concepts is sharply challenged in some of the philosophical attacks launched against the use of advance directives. To meet these challenges, we must continue to explore philosophical issues that have been at stake in bioethics with even greater intensity and vigor.

In the process of doing so, it will be important to recognize that the policies that we come to adopt about the treatment of the demented have not only philosophical, but also symbolic import. If we define those who suffer from dementia as non-selves who are outside the compass of the human community, we risk demeaning and abandoning them (Post 1995; Jaworska, 1999). If we make it easier to terminate their treatment (Kuhse, 1999), contrary to their earlier directions, we risk devaluing lives that could be lived tolerably well because of our social loathing of a life burdened with that condition (Callahan, 1995). And yet if we insist on a policy of retaining those with severe dementia in existence by means of life-support, in conflict with their advance directives, we risk treating the self as an entity that must be preserved alive at any cost. Thus, should we depart from the values set out by individuals in advance directive because we wish to avoid slavish adherence to the wishes of a formerly competent person, we do so at social peril. This social peril must also be taken into account when we grapple with the difficult philosophical questions that surround the use of advance directives.

Kennedy Institute of Ethics
Georgetown University
Washington, D.C., U.S.A.

REFERENCES

Beauchamp, T. and Childress J.: 1994, *Principles of Biomedical Ethics*, Fourth Edition, Oxford University Press, New York and Oxford.

Blustein J.: 1999, 'Choosing for others as continuing a life story: the problem of personal identity revisited', *Journal of Law, Medicine & Ethics* 27, 20-31.

Brett, A.S.: 1991, 'Limitations of listing specific medical interventions in advance directives', *Journal of the American Medical Association* 266, 825-828.

Brock, D.W.: 1991 'Trumping advance directives,' *Hastings Center Report* 21, s5-6.

Buchanan, A.E. and Brock, D.W.: 1989, *Deciding for Others: The Ethics of Surrogate Decision Making*, Cambridge University Press, Cambridge and New York.

Butler, J.: 1898, 'Of personal identity,' Appendix to *The Analogy of Religion*, London, George Bell and Sons, pp. 328-334.

Callahan D.: 1995, 'Terminating life-sustaining treatment of the demented,' *HastingsCenter Report* 25, 25-31.

Cantor, N.L.: 1993, *Advance Directives and the Pursuit of Death with Dignity*, Indiana University Press, Bloomington and Indianapolis.

Chisholm, R.: 1976, *Person and Object,* Allen and Unwin, London.

Cohen-Mansfield J. et al.: 1991, 'The decision to execute a durable power of attorney for health care and preferences regarding the utilization of life-sustaining treatments in nursing home residents', *Archives of Internal Medicine* 151, 289-294.

Cruzan v. Director, Missouri Department of Health, 110 S. Ct. 2841 (1990).

DeGrazia, D.: 1999, 'Advance directives, dementia, and "the someone else problem" ,' *Bioethics* 13, 373-391.

Dresser, R.: 1995, 'Dworkin on dementia: elegant theory, questionable policy,' *Hastings Center Report* 25, 32-38.

Dresser, R.S.: 1994a, 'Advance directives: implications for policy,' *Hastings CenterReport* 24, S2-S5.

Dresser, R.S.: 1994b, 'Missing persons: legal perceptions of incompetent patients,' *Rutgers Law Review* 46, 609-719.

Dresser, R.: 1990, 'Relitigating life and death,' *Ohio State Law Journal* 51, 425-437.

Dresser, R.S.: 1989, 'Advance directives, self-determination and personal identity,' in C. Hackler, R. Moseley, and D. E. Vawter (eds.) *Advance Directives in Medicine*, Praeger, New York, pp. 155-170.

Dresser, R.: 1986, 'Life, death, and incompetent patients: conceptual infirmities and hidden values in the law,' *Arizona Law Review* 28, 373-405.

Dresser, R.S. and Robertson, J.A.: 1989, 'Quality-of-life and non-treatment decisions for incompetent patients: a critique of the orthodox approach,' *Law, Medicine & Health Care* 17(3) 234-244.

Dworkin, R.: 1986, 'Autonomy and the demented self,' *Milbank Quarterly* 64, Supplement **2**, 4-16.
Dworkin, R.: 1993, *Life's Dominion: An Argument about Abortion, Euthanasia, and Individual Freedom*, Knopf, New York.
Emanuel, L.L. and Emanuel, E.J.: 1989, 'The medical directive: a new comprehensive advance care document,' *Journal of the American Medical Association* 261, 3288-3293.
Everhart, M.A. and Pearlman,R.A.: 1990, 'Stability of patient preferences regarding life-sustaining treatments,' *Chest* 97, 159-164.
Gamble E.R. et al.: 1991, 'Knowledge, attitudes, and behavior of elderly persons regarding living wills,' *Archives of Internal Medicine* 151, 277-280.
Hume, D.: 1978, *A Treatise of Human Nature*, ed. L.A. Selby-Bigge and P.H. Nidditch, Oxford, Clarendon Press.
Jaworska, A.: 1999, 'Respecting the margins of agency: Alzheimer's patients and the capacity to value,' *Philosophy and Public Affairs*, 28, 105-138.
Kadish, S.H.: 1992, 'Letting patients die: legal and moral reflections,' California Law Review 80, 857-888.
King, N.M.P.: 1996, *Making Sense of Advance Directives*, Georgetown University Press, Washington, D.C.
Korsgaard, C.:1996, *The Sources of Normativity*, Cambridge University Press, Cambridge, 90-130.
Kuczewski, M.G.:1994, 'Whose will is it anyway? A discussion of advance directives, personal identity, and consensus in medical ethics,' *Bioethics* 8, 27-48.
Kuczewski, M.G. 1999: 'Commentary: Narrative Views of Personal Identity and Substituted Judgment in Surrogate Decision Making,' *Law Medicine & Ethics*, 27, 32-36.
Kuhse, H.: 1999, 'Some reflections on the problem of advance directives, personhood, and personal identity,' *Kennedy Institute of Ethics Journal* 9, 347- 364.
Lynn, J. and Teno, J.M.: 1995, `Death and dying: euthanasia and sustaining life: III. Advance directives', *Encyclopedia of Bioethics*, Vol. I, 572-577.
May, T.:1997, 'Reassessing the reliability of advance directives,' *Cambridge Quarterly of Healthcare Ethics* 6, 325-338.
May, T.: 1999, 'Slavery, commitment, and choice: do advance directives reflect autonomy?, *Cambridge Quarterly of Healthcare Ethics* 8, 358-363.
Miles, S. *et al.*: 1996, 'Advance end-of-life treatment planning: A research review,' *Archives of Internal Medicine* 156, 1062-1068.
Nelson J.L.: 1995, 'Critical interests and sources of familial decision-making authority for incapacitated patients,' *Journal of Law, Medicine & Ethics*, 23, 143-148.
Noonan, H.W.: 1989, *Personal Identity*, London and New York, Routledge.
O'Neill, O.: 1984, 'Paternalism and partial autonomy,' *Journal of Medical Ethics* 10,173-178.
Parfit, D.: 1984, *Reasons and Persons*, Oxford University press, New York.
Post, S.: 1995 'Alzheimer disease and the "then" self,' *Kennedy Institute of Ethics Journal* 5, 307-321.
Reid, T.: 1941, *Essays on the Intellectual Powers of Man*, ed. A.D. Woozley, Macm London.
Rhoden, N.K.: 1988, 'Litigating life and death,' *Harvard Law Review*, 102, 375-446.
Rich, B.A.: 1997, 'Prospective autonomy and critical interests: a narrative defense of the moral authority of advance directives,' *Cambridge Quarterly of Healthcare Ethics*, 6, 138-147.
Seckler, A.B. *et al.*.: 1991, 'Substituted judgment: how accurate are proxy predictions?', *Annals of Internal Medicine* 115, 92-98.
Sehgal, A., *et al.*: 1992, 'How strictly do dialysis patients want their advance directives followed?,' *Journal of the American Medical Association* 267, 59-63.
Singer P.A. *et al.*.: 1998, 'Reconceptualizing advance care planning from the patient's perspective,' *Archives of Internal Medicine*, 158, 879-884.
Swinburne, R.G.: 1976, 'Persons and personal identity,' in H.D. Lewis, (ed.), *Contemporary British Philosophy*, Allen and Unwin, London, 219-238.

Tollefsen, C.: 1998, 'Advance directives and voluntary slavery,' *Cambridge Quarterly of Healthcare Ethics* 7, 405-413.

Uhlmann, R.F. *et al.*.: 1988, 'Physicians' and spouses predictions of elderly patients' resuscitation preferences,' *Journal of Gerontology* 43 (suppl.), M115-M121.

U.S. Congress, Omnibus Budget Reconciliation Act of 1990, Pub. L. No. 101-508, secs. 4206, 4751.

MARK G. KUCZEWSKI

ETHICS COMMITTEES AND CASE CONSULTATION: THEORY AND PRACTICE

I. INTRODUCTION

Healthcare ethics committees (HECs) are possibly the most enigmatic aspect of biomedical ethics.[1] Interest in biomedical ethics is strong and ethics committees have begun to capture popular interest through their portrayal in non-fiction books (Belkin, 1994) and television shows. Ethics committees have also proliferated rapidly during the 1990s with many hospitals and other healthcare institutions chartering such mechanisms for the first time. With this rapid development and expansion, some even speak of the "ethics committee movement" (Wilson, 1998, p.366).

In this vein, some have begun to worry that ethics committees are gaining too much momentum and wielding too much power. Some of these concerns are in response to the enthusiasm courts and governmental commissions have shown for ethics committees as a more appropriate forum for decision making than the legal arena, and to proposed, implied, or statutory immunity protections conferred on these committees by state legislatures (Fleetwood and Unger, 1994; Meisel, 1995, pp. 285, 331-334; Wilson, 1998). These concerns generally focus on whether committees are equipped to carry out a judicial review function in lieu of the courts.

Philosophically speaking, there is also a concern with the legitimacy of the power possessed by committee members who conduct bedside consultation in difficult cases. These concerns vary from basic issues regarding the uniformity of training and skills that such consultants should have (Fletcher and Hoffman, 1994; Aulisio, et al., 1998) to the legitimacy of such an institutional representative potentially imposing his or her values on patients (Scofield, 1993). These issues overlap with questions regarding the philosophical importance of consensus. Consensus seems to be the great hope for moral inquiry in a pluralistic society. If consensus can be produced on a moral issue, we need not be overly concerned with the legitimacy of imposing values on another.[2] But how can committees be sure that the answer produced by consensus is necessarily a morally appropriate one?

On the practical level, the problems that committees encounter seem to be the opposite of those that concern legal and philosophical scholars. For instance,

G. Khushf (ed.), Handbook of Bioethics, 315–334.

it is not the abuse of their power that is often at issue. Rather, defining a mission for the committee, designing ways to address that mission, finding some way to achieve credibility with the medical staff and to become a positive influence on clinical practices all tend to consume the energy of the committee (Ross, et al., 1993).

One of the problems with sketching a philosophy of ethics committees is that it is not clear whether one should do a philosophy of ethics committees as they might be some day if they manage to obtain their greatest ambitions or a philosophy of ethics committees in their mundane existence. The former would largely be a philosophy of power and efficacy, the latter a philosophy of moral or professional education. I will try to address both sets of concerns. However, a philosophy of ethics committees as they currently exist will usually take precedence.

I will examine the philosophical issues that underlie the structure and functions of ethics committees. It is a commonplace that ethics committees provide education, case consultation, and policy review for their institutions. Of course, any particular committee will set its own priorities among these functions. I will argue that there has been far too much emphasis, both philosophically and practically, on concurrent bedside consultation and that this focus has generally hindered the functioning of ethics committees and has minimized their potential positive impact.

II. ETHICS COMMITTEES: PURPOSE AND RATIONALE

Ethics committees have been heralded as the way to resolve clinical ethical issues without involving the courts. The courts have been disparaged as the mechanism for case resolution for a number of reasons, including that court proceedings can create unnecessary expense for institutions and patients, that the members of the judiciary system lack medical knowledge, and that the adversarial nature of the court system is ill-suited to resolving patient care problems. As a result, it has been thought that members of ethics committees could help to resolve problems by bringing to bear knowledge in a way that is much more sensitive to the needs of patients and providers. This way of conceiving the role of ethics committees emphasizes concurrent case consultation as its central mission.

An emphasis on concurrent case consultation is natural because much of the original motivation for the creation of ethics committees comes from a desire to find a replacement for judicial proceedings. This case consultation function received additional impetus from other sources such as the Joint Commission on the Accreditation of Healthcare Organizations (JCAHO). Many hospitals seek JCAHO accreditation and spend a good deal of time, money, and energy in ensuring that their hospital meets the standards set by the commission. In 1992, the JCAHO issued standards related to ethics and patient rights. The importance of this event cannot be overestimated as it has resulted in an unprecedented proliferation of ethics committees.[3] At the same time that it led to their proliferation, the standards also seemed to imply that their mission was case consultation. The standards require that the healthcare organization have "structures to support patient rights" (JCAHO, 1997, Intent of Standard RI.1) to resolve ethical conflicts in patient care. Because the JCAHO is concerned with patient care, it is natural that they should be interested in resolving problems as they arise in the delivery of patient care. Although these standards allow for administrative mechanisms to resolve ethical problems, many hospitals have interpreted them as requiring an ethics committee that performs concurrent case consultation.

Perhaps the main reason that concurrent case consultation has received so much attention is simply that it "fits" with the culture of medicine (Ross, et al., 1993, p. 91). Much of medical care is delivered in terms of expert knowledge brought by specialized medical advice that is solicited through consultation of the appropriate medical service. If a primary care physician encounters an unusual problem with her patient's heart, she calls the consultation service that is staffed by the cardiology department. Similarly, if she had an ethical issue arise in his case, she is likely to want the ethics service to provide a bedside consultation. In each instance, one would expect the service to bring expert knowledge to the bedside that can result in effective treatment options. Of course, one could easily ask whether ethics consultation can meet the expectations of this model.

We can also conceive of the mission of an ethics committee quite differently. Rather than seeing it as the repository of specialized knowledge, its main mission might be to disseminate knowledge about a dimension of clinical practice that accompanies virtually all medical encounters. We can see this possibility by analogy to a medical specialty such as infectious disease control. A physician wants to be able to call an infectious disease specialist when he or she encounters an exotic and puzzling disease. But every physician needs to have a basic knowledge of infectious diseases and to understand and practice good hygiene and other prophylactic measures that prevent their spread. There is good reason to suspect that ethics is similar in nature.

The fundamental value in contemporary secular medical ethics is respect for the autonomy of the patient. This value motivates the informed consent procedure that should precede every invasive medical intervention. It seems likely that the better the informed consent procedure, the less likely that there will be communication problems, misunderstandings, and unresolvable value conflicts later in the treatment process. Of course, clinicians do not call specialists in ethics to conduct the informed consent process. Each clinician needs to become adept at the subtleties and nuances of informed consent. If ethics is analogous to infectious disease control in this way, then the fundamental task of ethics committee members should be educating their colleagues and developing their skills in these preventive ethics measures (Forrow, et al., 1993). However, this aspect of ethics committees has received little attention by comparison to the case consultation function.

III. THE EDUCATIONAL MISSION OF ETHICS COMMITTEES

The notion that ethics committees should primarily direct their efforts toward education has practical as well as theoretical reasons for its advocacy. For one thing, education sounds far less threatening to medical staff than case consultation or case review. Medical staff members are often worried that ethics committees will become the "ethics police" who constantly second guess treatment decisions (Pinkus, et al., 1995). But, saying that the ethics committee exists for the purpose of education can defuse these fears. Second, there is simply a lack of good evidence that case consultation is especially effective. This may be due to the nascent state of the art, but could also be due to the nature of ethics. If our analogy between ethics and infectious disease is plausible, it may be that the most effective strategies are preventive rather than curative. Such reasoning provides a theoretical reason in favor of education as the focus of ethics committees.

When education is the main focus of a committee's activities, two questions must be answered. First, what is the content that should be learned and taught? Second, how is ethics best taught? Let us look at each question in turn.

Biomedical ethics is clearly a multidisciplinary field, or perhaps, an interdisciplinary field that borrows from several fields. The thinking of bioethicists is certainly shaped by philosophical frameworks, legal precedents, and clinical experience. As a result, every ethics committee member[4] will need to be familiar with a certain amount of philosophical, legal, and clinical concepts.[5] For instance, committee members need not be philosophers but should be able to identify typical philosophical considerations such as consequentialist and nonconsequentialist concerns. Similarly, they need to be able to speak the language of the four principles of biomedical ethics and have some familiarity regarding how these concepts might be balanced when they conflict. This is not because ethical problems are typically settled by "applying" principles and concepts to cases (Beauchamp, 1984; Kopelman, 1990). Rather, these concepts constitute the language of medical ethics and a set of considerations that suggest potential strategies for resolution.

Similarly, committee members should know something about classic legal cases such as those of Karen Ann Quinlan and Nancy Cruzan. Perhaps a certain familiarity with the U.S. Supreme Court cases involving physician-assisted suicide would also be apropos (Annas, 1997). Of course, the kind of patients for whom the hospital cares will be important in determining what legal knowledge is likely to be helpful. For instance, members of an ethics committee at a rehabilitation hospital should know a bit about disability law; those at a children's hospital have an additional set of classic cases that are worth perusing. The committee members should be familiar with any appellate court decisions that have particular impact in their state. Once again, this knowledge is not being acquired in order that these committee members are able to understand legal nuances in the way an attorney does. They need to understand the law philosophically. That is, what members of ethics committees need to understand is how our legal system has attempted to balance the same concerns for patient autonomy (self-determination), the responsibility of the care giver for the patient's well being (beneficence), and the integrity of the medical profession. By obtaining such a knowledge, committee members at once free themselves from common but gross misunderstandings of the law and obtain additional access to aspects of our societal and cultural narrative that form the backdrop of clinical ethics issues.

Perhaps of greatest importance is a need for “clinical” knowledge. This includes familiarity with clinical realities, experience with cases in which ethical issues have arisen in the past and knowledge of how they were handled and resolved. This should involve not only classic dilemmas that have become part of the local lore but also observation of cases that are considered routine and in which the ethical dimension is barely noticeable. Control of ethical problems, like that of infectious diseases, involves knowledge of good practice, not just the handling of disasters.

If we accept that these are the major content areas with which ethics committee members need to become familiar, we must ask how the committee members can acquire such knowledge and how they may become effective peer educators.[6] The most obvious answer to the knowledge acquisition question involves formal study. Graduate programs that confer Master of Arts degrees in biomedical ethics are increasingly common. There are also scaled back versions of these offerings, often called "certificate" programs. Of far less intensity are other kinds of ethics conferences such as weeklong "intensive courses" that offer introductions to the basics of these content areas. Of course, as one moves away from full-fledged degree programs to less weighty endeavors, the possibility of genuinely acquiring the knowledge base one needs diminishes, especially if the educational program provides no opportunity for a clinical practicum. As a result, conferences and other kinds of such training should probably be seen as ways of acquiring materials and resources for committees to use for their self-education. Here we come to the crux of the matter.

Clinicians on an ethics committee may find "book learning" relatively adequate if they have been around other practitioners whose standard practices are ethically sensitive. The committee members would then be used to seeing good practice. Ethics education can provide them with reasons why such practices are good and help them to extend this reasoning to increasingly problematic scenarios. It is evident that such committee members are learning through a combination of discursive and observational strategies. It is probably by such strategies that clinical education routinely proceeds.

"Watch one, do one, teach one" has long been the motto of medical education. This is the strategy by which ethics committee members seem to learn biomedical ethics and is likely the manner in which they will teach it. Ideally, they will point out good clinical practice and find ways to distill the values, processes or ethical frameworks that underpin such practices. This can happen in the clinical setting or through more structured teaching sessions. From what we have said, it seems that good clinical ethics teaching will usually involve a case, whether it be one at hand or a hypothetical one. Furthermore, this kind of teaching ultimately involves role modeling.

Once again, following our analogy of infectious disease control, we know that physicians can learn to wash their hands by reading studies and data regarding the efficacy of hand washing. However, it is possible that they learn best from seeing peers model this behavior around them. They can see the behavior of these peer educators, learn the rationale from them, and then illustrate and narrate this behavior for others. This epitomizes medical education's motto. If this is the case, then the most effective means for ethics committee members to achieve their goal is by becoming good role models. This is accomplished through effective self-education. In sum, self-education could be the main task for ethics committee members (Ross, et al., 1993, pp. 186-187).

Philosophically speaking, a phenomenology of ethics committee functioning suggests that its educational function is accomplished in a manner described by the Aristotelian tradition, particularly its contemporary progeny, virtue ethics, communitarianism, and casuistry. Ethics is learned by attention to the particular right act (*to eschaton*) accompanied by an account (*logos*). Virtue ethics and casuistry make clear that such particulars are variable and are likely to be most ably identified by those who have developed the proper habits of character (Baylis, 1994, pp. 36-41; Pellegrino and Thomasma, 1993; Jonsen, 1991). Similarly, communitarianism emphasizes that the clinical practices that develop these states of character are ultimately a function of the virtue of the community (MacIntyre, 1984, 193). The relationship of virtue and the community shows the dialectical character of moral education. On the one hand, the ethics committee members must educate themselves to become role models in order to change the community of practitioners in their institution. On the other hand, the acquisition of virtue by these members seems to be dependent on the good practices of the community of practitioners of the institution. But, as we have already seen, this need not be a vicious circle.

Clinicians have not usually had to wait for ethicists to arrive in order to practice morally sound medicine. The physician-patient relationship has historically embodied certain virtues as a quintessential fiduciary relationship. Making thematic the ethical dimension simply extrapolates these moral concepts and values and makes it easier for clinicians to make their practices even more virtuous by extending these values to areas where they may have traditionally been excluded. In this way, medical ethics makes it more likely that good practitioners will "hit the right mark" (Aristotle, *Nicomachean Ethics,* 1.2, 1094a24).

We can draw two additional lessons for ethics committees from this neo-Aristotelian phenomenology. First, if education takes place through role modeling and discursive illustration, committee members should utilize a variety of formats for their peer education efforts. It has become quite common in hospitals for most ethics education efforts to take the form of one-hour medical staff presentations or lunchtime sessions. Although there is nothing wrong with such an approach as long as the sessions are case-oriented, we might wonder why educational efforts need be structured in this way. I have suggested that ethics can be learned on the shop floor and so, we might wonder whether ethics committee members have underestimated the power of "ethics rounds" in which committee members routinely discuss cases with the clinicians even though no consultation has been requested (Dowdy, et al., 1998). Second, if virtue is ultimately a function of the community, perhaps educational efforts should more commonly be focused on potential patients and members of the broader community. It has been shown that merely providing practitioners with information does not necessarily change the level of physician-patient dialogue (The SUPPORT Principal Investigators, 1995). But, empowering patients with information may elevate the dialogue between them and their physicians. This is a natural corollary of the notion that the community, not the individual practitioner, must foster virtue and good practice.

IV. CASE CONSULTATION

Most of the literature on ethics committees focuses on their consultative function. Several debates have raged in the literature. The legitimacy of ethics consultants as healthcare decision makers has fueled most of controversies. In particular, some have asked from where ethics consultants derive their mandate and of what their expertise consists. Of course, these questions presuppose that we know what an ethics consultant does. This is not nearly as obvious as it might seem. There are at least two different visions of the role of ethics consultation.

A. The Legalistic Functions of Ethics Consultation

The role ethics consultants play is likely to vary depending on whether the patient has the capacity to consent to treatment or not. The role of an ethics consultant is more readily defined when the patient lacks decision-making capacity than when he or she does not. In fact, the President's Commission for the Study of Ethical Problems in Biomedical and Behavioral Research (1983, pp.160-170) originally envisaged the role of ethics consultants in terms of the decisionally incapacitated.

In such cases, ethics committees can help to identify the appropriate surrogate decision maker, clarify the ethical standards of surrogate decision making, facilitate the interpretation of advance directives, or even act as a surrogate decision maker. This kind of consultation represents an application of the kind of ethico-legal frameworks for which contemporary bioethics is famed. That is, the case is reviewed in terms of the principles of surrogate decision making as they've evolved from the doctrine of informed consent. The principles, e.g., the surrogate who is chosen should be someone who is likely to know the wishes of the patient and to safeguard his or her best interests, guide the review of the specifics of the case. This modus operandi is clearly quasi-legal and can be seen either as a substitute for judicial review or propadeutic to judicial review.

The case for ethics consultation or committee review prior to judicial review is fairly clear and does not engender controversy. Ethics committee members can bring a multidisciplinary perspective, ethical, legal, and philosophical knowledge, skill in facilitating the decision-making process, and experience with similar cases to bear on a problematic case. This is likely to improve the quality of the decision making and should help the decision to bear up under judicial scrutiny.[7] Some related current controversies center on whether the proceedings of the ethics committee or the ethics consultants should be documented in the patient's medical record or should be discoverable in a court of law. The arguments against documentation and discovery are based on the claim that ethics consultation is a kind of quality assurance function that will improve patient care and therefore, should be encouraged among healthcare professionals who might not utilize such a service if they feared a potential negative impact on their liability. However, such fears are probably unfounded and liability exposure is likely to be reduced by demonstrating that a careful decision-making process took place rather than shrouding such a procedure in secrecy (Meisel, 1995, pp. 299-301).

Far more controversial is the suggestion that review by an ethics committee should replace judicial proceedings and any recourse to the courts (Wilson, 1995, pp. 365-366). Such an argument can be based on the greater familiarity of ethics committee members with clinical medical realities and/or the less cumbersome nature of ethics committee procedures. This can be clarified through an example.

Suppose an indigent patient is in the care of a hospital. His condition is such that he is likely to be permanently unconscious and dependent on technologies such as artificial nutrition and hydration. Even with such technology, the patient is unlikely to live very long as he has a degenerative condition. He has no advance directive, no next of kin, and no friends willing to serve as a surrogate decision maker. The caregivers do not want to perform CPR on this patient should he go into cardiac arrest and they would like to discontinue all aggressive treatment in favor of palliative care. This sentiment is based on their belief that most persons would make this same choice for themselves.

The caregivers believe that most of their patients in similar situations want aggressive treatment when there is a chance that their condition will improve but would decline it in favor of being made more comfortable if such improvement were not possible. But, the usual procedure is for the patient or family member to make such decisions in concert with the caregivers. Furthermore, the caregivers know that if they seek a court-appointed guardian to make the decision, the court will appoint someone from that state's department of social services. No such guardian has ever refused any life-prolonging treatment, including CPR, for a patient in this particular jurisdiction. The department has an explicit bias in favor of attempting prolongation of life, no matter how burdensome and unlikely to be efficacious the treatments when the patient has not left explicit instructions. If such scenarios occur repeatedly, it might become the consensus (i.e., among the courts, the hospital, the state department of social services, and even the state legislature) that patients would be better served by an informal procedure in which an ethics committee or consultant acted as the guardian for the patient and made the treatment decisions. Decisions might be made for patients in a way that was more likely to correspond to their wishes if they could speak for themselves. This case is bolstered if such decisions are based on treatment guidelines that are developed with local or regional community support (Emanuel and Emanuel, 1993).

This is a very strong argument. The major controversy would seem to surround whether such a proceeding should be immunized against potential liability. On a philosophical level, the case for such immunity is fairly evenly balanced. That is, a grant of immunity from liability would foster the implementation of these ethics committee processes and can provide a powerful stimulus to the development of guidelines to inform such processes. The potential for improved quality of decision making in such instances is clear. Of course, redress to the courts is one of the primary safeguards of patient rights and should not be discarded lightly. For instance, we do not know whether committee members can implement such guidelines evenhandedly or whether certain classes of vulnerable persons will be harmed due to implicit biases in the determination of their wishes or interests.[8] And, because there is little quality assurance regarding the training and performance of ethics committee members, conferring immunity should, at best, be postponed until standards in this area are instituted (Fletcher and Hoffmann, 1994). It is exactly this argument that motivated the recent laudable efforts by the professional societies in bioethics to issue a report on the core competencies needed by bedside consultants (Task Force on Standards for Bioethics Consultation, 1998). These standards are strictly voluntary but are likely to influence the requirements for accreditation of the JCAHO.[9]

B. Ethics Committees and the Competent Patient: "Qualified Facilitation"

The role of ethics consultants in cases that involve patients who possess decision-making capacity has expanded in recent years. It has evolved far beyond the role that the President's Commission envisioned in which committees primarily safeguard the rights and interests of the incapacitated patient. The logic of this evolution has been guided by two factors that, ironically, were acknowledged by the commission. First, the line between competence and incompetence is fluid. Ethics consultants cannot simply be called when patients lack decision-making capacity because the capacity of the patient is often what is at issue. As a result, the ethics consultants are often involved in assessing the capacity of the patient and in helping to tailor information and facilitate dialogue so that patients with marginal capacity may participate in their treatment decisions. Second, informed consent has come to be seen more as a process than an event (Lidz, et al., 1988; Kuczewski, 1996). It is increasingly acknowledged that patients come to understand medical information over time through both cognitive and experiential means. As a result, we cannot simply dismiss ethics consultants from a case in which the patient possesses capacity by saying that the patient should be informed of his or her options and then make a choice. Rather, a skillful facilitator can aid the informed consent process. As a result, ethics consultants can have a potentially important role in virtually any case

The role of ethics consultants as facilitators has been championed by the Task Force on Standards for Bioethics Consultation of the American Society for Bioethics Consultation (ASBH). In their report, they argue against any authoritarian approach to ethics consultation in favor of consensus building. But they also recognize that mere consensus is not the goal of consultation. Consensus among the parties to a case can sometimes be found outside established ethical boundaries and is, therefore, wrong. They cite the example of a consensus achieved among the members of a healthcare team and an incapacitated patient's family members to ignore a valid advance directive (Task Force on Standards for Bioethics Consultation, 1998, p.6). As a result, they endorse a process model or ethics facilitation model of consultation that is *qualified* by the moral frameworks of patient rights that we have noted to be the currency of contemporary bioethics. It is a facilitation model of consultation because it aims to clarify the ethical issues and values at stake and to achieve consensus among parties to the case. But, it is qualified by certain moral boundaries that limit the range of acceptable outcomes.

The role of consensus is clearly important. On the one hand, the building of consensus is a way to assuage concerns about the consultant imposing moral solutions by fiat rather than moral expertise (Moreno, 1995, pp. 9-16). Because moral absolutes may only be clear and widely accepted in uncomplicated, paradigmatic cases or at the boundaries of our moral frameworks, there may be a variety of morally acceptable resolutions in particular cases. Which one is best is probably the one that is chosen by the consensus of the involved parties. Of course, as the ASBH Task Force indicated, one cannot simply take consensus as an end-in-itself.

For a particular consensus to be of moral import, it must be a consensus that is achieved in a certain way. Namely, it must be achieved through a process in which the values of each party with standing are clarified and options chosen collaboratively. Often, there is a gap between values and particular choices such as treatment preferences. For instance, a patient has certain values such as respect for life and a desire to be free of unnecessary suffering. How these two values will be balanced given a patient's prognosis and the options available are something that can only be determined through an appropriate process of shared decision making. Clearly, there are some pragmatic assumptions behind this view of moral consensus.

In particular, the relevant parties, i.e., those with moral standing, must be identified and made comfortable with the decision-making process. They must not simply be interested in asserting their rights but must come to believe that the process will foster their interests in a way that would not have been possible for them apart from the group process (Caws, 1991). As a result, some of the knowledge and skills that ethics consultants need are of the group process variety (Task Force on Standards for Bioethics Consultation, 1998, pp. 14-16). This approach can be interpreted as presupposing that persons typically possess a kind of moral sense that can determine an appropriate outcome through a process of dialogue (Moreno, 1995, pp. 10, 106-125). Nevertheless, even if this assumption should prove problematic, the boundaries of the ethico-legal frameworks provide a safeguard that should help if the group processes or the moral sense fail to produce acceptable results.

C. Consultation as Policy Formation: The Example of Futility

It is typically the function of ethics committees to assist in the creation of policies related to patient rights and choices that may involve readily defined ethical quandaries. These policies are often related to informed consent, end-of-life decision making, reproductive choices, and most recently, organizational ethics. Ethics committees seldom create policies unilaterally. Instead they often draft a policy that is then modified and adopted by the appropriate institutional authority, e.g., administrative officer, medical staff, or legal counsel. The drafting of policies is not often a signal event in the life of an ethics committee but may provide an opportunity for the self-education of members. Because they wish to make policy recommendations that accord with the ethico-legal framework of contemporary biomedical ethics, the members usually wish to educate themselves in the basic frameworks of contemporary biomedical ethics.

Most patient's rights policies such as those related to informed consent and end-of-life decision making have become rather standardized and no longer generate much interest or research. However, two kinds of policies, in particular, have generated interest from a philosophical point of view: organizational ethics policies and "futility" policies. Organizational ethics policies are being proliferated by recent JCAHO requirements and concern such things as truthful disclosure in billing, ethical advertising practices, and disclosure of potential conflicts of interest. These items seem to loosely fit the ideal of informed consent, i.e., they are concerned with the disclosure of information to patients. Because these matters are far more administrative and business-oriented than the patient care issues that ethics committees usually confront, they raise important questions regarding ethical expertise and its multidisciplinary nature. That is, is the ethical skill developed in clinical case analysis applicable to business issues? And, if the clinical ethics skill of committee members needs to be supplemented by business expertise, does that mean that a number of business administrators should be added to ethics committees? If so, should they also participate in the deliberations of clinical ethics cases?

More pertinent to our present investigation is the wave of interest in policies regarding the cessation of "futile" medical treatment. The first wave of development of clinical ethics policies guaranteed that when a physician proposed a treatment, e.g., CPR or mechanical ventilation, the patient had a right to refuse it. But, the question naturally followed concerning what should happen if the physician did not want to propose a treatment such as CPR. For various historical reasons, the administering of CPR at the end of life has become the norm in acute care, emergency settings, and many long-term care institutions despite startlingly low success rates for resuscitation measures with many classes of persons (Murphy, 1988). As more healthcare professionals began to attend to the outcomes of life-sustaining treatments and to appreciate that their duty to sustain life may be outweighed by the harms done and violations of dignity they are perpetrating, they have asked whether they must always offer these treatments. Many ethics committees have answered "No, not even if the patient or his surrogate asks for them."

This debate has raged in the literature for over ten years and much of it has surrounded what exactly is meant by saying that that a treatment is "futile" (Youngner, 1988). Among the possible things that futility can mean are (1) that a treatment won't work, e.g., CPR will not restore cardiac function, (2) that there is a very low probability that a treatment will work, e.g., it is very unlikely that CPR will restore cardiac function, or (3) that a treatment will achieve its goal but that outcome is undesirable, e.g., CPR will restore cardiac function but will leave the patient with a very low quality of life. Philosophical analysis of these definitions reveals that the first definition has some potential for policy making.

Our society has never really required that physicians provide all treatments a patient might want when there is no reason to suspect efficacy, e.g., physicians are not required to provide antibiotics for a patient with a viral infection even if the patient asks for them. Some ethics committees have crafted policies that allow physicians to unilaterally write Do-Not-Resuscitate orders for a patient if they believe that CPR simply will not work (Waisel and Truog, 1995). However, since it is very difficult to make such determinations absolutely, one suspects that, in practice, such unilateral decision making is usually based on a low probability of the efficacy of the treatment and may be improperly influenced by implicit quality-of-life judgments.

I have made this excursus into this particular policy development because it raises several interesting philosophical issues. First, it is interesting that many ethics committees have been quick to place their blessing on a decision-making model that runs counter to its ethos of patient autonomy. This raises the suspicion that bioethicists, despite their traditional reluctance to pass quality-of-life judgments on patients, may actually sometimes be concerned with such judgments and are occasionally willing to sacrifice their principle of respect for patient autonomy. But, they have generally tried to base such policies on the physician's judgment that a treatment just won't work. By so doing, they try to convert futility questions into medical judgments. Of course, the medical judgment that is required is that the treatment just won't work while, in reality, the best the physician could usually say is that it is unlikely to work (Rubin 1998; Halevy, et al., 1996)

It is surprising that ethicists and ethics committees have been willing to adopt futility policies based on this line of argument and it is important to try to understand what motivates this willingness to do so (Truog, et al., 1992).[10] This brings us to our second philosophical consideration. It is unlikely that ethics committees can be completely impartial mechanisms for safeguarding the rights and interests of patients. Members of the committee are implicitly likely to try and please medical staff members. Despite the fact that ethics committees bring a multidisciplinary perspective to their work, the disciplines that are represented are mainly healthcare disciplines. In the early days of medical ethics, it was thought that philosophers and those from the humanities and social sciences would counterbalance the medical perspective. But as these humanists become more at home in the clinical setting, i.e., become bioethicists, they lose some of the perspective of the "outsider" (Chambers, 2001). Thus, when a unilateral consensus is sought among ethics committee members on an issue such as futility, that consensus may not be as morally significant as a bedside consensus that includes the patient and/or family in the consensus-building process. Similarly, a futility policy will likely have more legitimacy if there is some type of community consultation involved in its development.

Our answer as to why ethics committees countenance futility policies seems to be two answers. On the one hand, the committee is made up of medical insiders. Therefore, committees may reason in the same way that medical staff members reason. But, I have also suggested that committees may wish to please the medical staff members. This desire to please comes largely from the fact that ethics committees are very unlikely to be effective without significant cooperation from physicians. It is perhaps in response to this lack of power that proposals to legislate or grant vast powers to ethics committees arise. But, implementing such proposals could be problematic since it might empower ethics committees to pursue their biases rather than to correct them. As a result, an array of empirical questions regarding appropriate policies and safeguards in the work of ethics committees stands before us.

V. THE FUTURE OF ETHICS COMMITTEES?

It is one of the great ironies of contemporary biomedical ethics that hospital ethics committees are routinely reported to be struggling to find their mission and purpose. This is ironic because it comes at a time in history when biomedical ethics is considered important and worthy of public attention. At the same time, bedside ethics consultation is proliferating. These two observations are probably related and represent a tension for the future of ethics committees.

When ethics committees focus on concurrent case consultation, it often becomes the main job of the committee to review the work of the bedside consultants on a regular basis. For instance, the consultants may present the cases on which they were consulted at a monthly meeting of the entire committee. The committee then assumes a quality control function over the consultants. This presentation of the cases and group discussion may dominate the agenda of the committee. When this is the case, many of the members who do not conduct bedside consultation may begin to lose interest in the committee and they become increasingly marginalized and the committee will become an appendage to the ethics consultants. We saw earlier that this need not be the case. The entire committee can have a vibrant existence if the members understand their mission in terms of their self- and peer-education. Of course, there are important philosophical differences between these two approaches.

The main difference concerns what kind of knowledge ethics ultimately is. If it is highly specialized knowledge that can only be the province of a formally trained elite, then case consultation would seem to be where emphasis should be placed. Consultation only requires that a small number of people be trained in the esoteric knowledge and they bring this knowledge to bear in difficult cases. Attempts to educate other healthcare practitioners would be a secondary priority because few could attain a high level of expertise. Or, this model at least assumes that the most teachable moment for a healthcare professional comes during a moment of crisis when a specialist transmits the information. The model of consultation we saw earlier, that of qualified facilitation, seems to deny these very premises since it does not conceive the consultant's role as transmission of esoteric knowledge but in terms of facilitating a consensus-building process that is bounded by certain easily cognizable frameworks. Nevertheless, even the ASBH Task Force views the bedside consultant as attaining an advanced stage of knowledge while they believe most committee members attain only basic knowledge (Task Force on Standards for Bioethics Consultation, 1998, pp. 15, 20).

In tension with this consultative model of clinical ethics is that of the educational model of ethics. An educational model assumes that ethics is most effectively taught by the routine role-modeling behavior of peers and that self-education by a large number of committee members is do-able since the kind of knowledge at issue is not so esoteric. Such an approach sees the life of the ethics committee as primary and that of bedside consultation as, at best, secondary.

VI. RESEARCH DIRECTIONS

Clearly, the effectiveness of ethics committees and consultation services are in great need of evaluation (Fox et al., 1996). However, this task is intrinsically difficult as it is not so clear what counts as success in consultation. Patient and physician satisfaction provide some insight but it is also possible that facilitating an ethical resolution can leave some parties dissatisfied. Another possibility is to view case consultation as hastening the decision-making process. By facilitating decision making, consultation may reduce the length of stay of the patient in the hospital and thereby prove cost effective. Little data has been produced regarding ethics consultation to date and much of what has been produced rests on optimistic assumptions that remain to be substantiated and reproduced (Bacchetta and Fins, 1997; Heilicser, et al., 2000). It is likely that the efficacy of case consultation will only be determined over time through a series of quantitative and qualitative research employing a variety of endpoints.

Similarly, there is a need to determine what approaches to education and policy development are effective within health-care institutions. For instance, institutions need to insure that patient wishes are respected through informed consent, advance directives, and surrogate decision making as appropriate. In most cases, these ethical standards should be routinely respected without recourse to the ethics committee or consultation service. But, there are not widely accepted indicators of quality decision making and consensus regarding how to improve clinical performance. As a result, there is a growing movement to think of ethical behavior as something that can be measured and bettered through the language and methods of continuous quality improvement (CQI) management strategies (Feldstein, et al., 1997). While this approach has a prima facie appeal, much research is needed to see if clinical ethics will benefit from this paradigm.

Of course, for the foreseeable future, bioethicists are likely to continue to debate the same issues that they have been debating for a number of years, e.g., the legitimacy of the power of ethics committees and consultants, the benefits or horrors of credentialing bedside consultants, the extent of liability for giving or following the advice of ethics committees, etc. We have seen that the culture of the clinic fosters an interest in these questions. It is also likely that the culture of academia fosters this interest as well. That is, one major determinant of academic debate is simply what academics have been debating. However, it is important that bioethicists begin to conceptualize the ethical dimensions of the changing health-care landscape.

Ethics services within hospitals were developed when the hospital was much more of the focal point of health-care delivery. By being developed within this context, virtually all questions identified as ethical issues involved medical treatment decisions for in-patients. However, as hospitals have come under pressure to reduce the length of stay of patients, decision making conflicts increasingly involve placement or disposition decisions rather than treatment decisions (Kuczewski, et al. 1999, pp. 85-87). Furthermore, as medical care is now delivered along a continuum of providers, the site of ethical issues is increasingly in home care, rehabilitation care, ambulatory care, or skilled nursing facilities. (Arras, 1995; McCullough, et al., 1995; Barnard et al., 2000; Kuczewski, 2001). Although bioethicists have made some effort to address these new horizons of ethics, much more research is needed to determine how to effectively support health-care providers in settings where the traditional committee and case consultation model may not be easily applied.

Neiswanger Institute for Bioethics and Health Policy
Stritch School of Medicine
Maywood, Illinois, U.S.A.

NOTES

1. A terminological clarification is in order. The vast majority of ethics committees are found in acute care hospitals. However, in recent years, they have also proliferated in other settings such as nursing homes, hospices, group practices, and rehabilitation facilities. As a result, they are generically referred to as Institutional Ethics Committees (IECs) or Healthcare Ethics Committees (HECs). The latter abbreviation has become increasingly popular, probably due to its use by the President's Commission for the Study of Ethical Problems in Biomedical and Behavioral Research (1983) and the increasing popularity of the journal known as the *HEC Forum* that is targeted at members of these committees.
2. Consensus has many meanings. In this context, consensus means something like "near unanimity that a course of action is at least minimally morally acceptable." When a consultant approaches a clinical case, there may be a range of morally acceptable outcomes. The goal will be to come to agreement among the parties on one that though it may not be anyone's first choice, does not violate the fundamental values of anyone who is party to the case. Similarly, on a societal level, consensus on a course of social policy usually takes the form of a course of action that the vast majority of persons find at least minimally acceptable. Unfortunately, a society as large as ours will always find it difficult to produce the kind of unanimous consensus that is possible in particular clinical cases. Consensus, nevertheless, makes potentially divisive social problems manageable by demonstrating a concern for widely shared values rather than simply adjudicating matters in terms of rights.
3. The growth of ethics committees and their vitality is difficult to map because there have been very few attempts at a comprehensive national study of ethics committees (This lacuna is probably due more to a lack of interested funders than interested researchers). But, we do know that the growth of ethics committees has been very rapid over the course of the last 20 years. A study done for the President's Commission (Youngner, et al., 1983, p.446) found that only about 1% of hospitals had such committees at that time. Estimates by the American Hospital Association (AHA Statistics 212, 1994) place the number of acute care hospitals with more than 200 beds that have an ethics committee at over 80%. Virtually every study finds that smaller institutions (acute care hospitals with fewer than 200 – 250 beds) are less likely to have an ethics committee (Hoffmann, 1991). However, it is probably a safe inference to conclude that far more than half of all healthcare institutions now have an ethics committee and that the number may well be approaching 80% (Wilson, 1998, p. 357).

4. In a sense, I am following the procedure of the ASBH Task Force on Standards for Bioethics Consultation that prescribed a certain basic level of education that every committee member should have. However, I am not necessarily following their teleology that sees bedside consultation as the pinnacle for an ethics committee member and therefore requires advanced levels of knowledge and skills for such members.
5. I am omitting explicit discussion of sociological and anthropological knowledge required by ethics committee members. Rather, I am assuming that in acquiring "clinical" and "philosophical" knowledge, the committee members will learn a good deal about the socio-economic circumstances of the patient population that the institution serves and the religious and other belief systems commonly encountered among this clientele (see DeVries and Subedi, 1998).
6. Again, my content list does not contradict the areas identified by the ASBH Task Force. However, they parse the content more finely and add a set of process skills necessary to bring such information to bear in case consultation.
7. Note that this kind of case consultation can take place before any real decision is made in a case, e.g., to withdraw treatment from an incapacitated patient with no identifiable surrogate, or to review this decision after the healthcare team has made such a decision but prior to its implementation. The latter approach is probably what the courts initially viewed as the procedure ethics committees would follow. The former approach has come to dominate recent thinking such as that of the ASBH Task Force on Standards for Bioethics Consultation. In each case, the decision is not implemented until an ethics consultant has reviewed it, and in the scenario we are considering, by a court of law. However, significant differences between these approaches may emerge regarding how one sees the role of the ethics consultant. For instance, if the review takes place before the healthcare team makes any decision, the role of the ethics consultant might be more likely to be seen as informational or facilitational. If the review is after a decision has been reached, the review may have a more judgmental function. However, this is not necessarily the case since even in the latter approach, the role of the consultant or ethics committee might be to provide information so that the healthcare team has a chance to reconsider its position prior to a court proceeding.
8. This argument is similar to some of the slippery slope arguments against physician-assisted suicide. In each case, the important point seems to be whether adequate safeguards can be created to prevent a slide down the slope. Because recourse to the courts is an important mechanism by which to prevent such a slide, this right should probably not be removed. But this is an important area for future debate since slippery slope arguments are not easily evaluated.
9. Although the recommendations of the Task Force are strictly voluntary, there is the possibility that they may form the basis for future licensing or credentialing of consultants. The efforts of the Task Force were directed to outlining the content and process skills that consultants need. Once methods of reliably producing these skills are documented, the final step of requiring consultants to partake of these methods might be justified (Aulisio, et al., 1998, pp. 492-493).
10. It is not entirely clear how widespread the creation and implementation of futility policies are and opinions vary as to whether such policies are proliferating or waning. It is clear that clinicians are fairly prone to the use of the term “futility” and favor policies that expand their decision-making capability in this regard (Dickenson, 2000). However, some believe that such policies have been losing favor (Helft, et al., 2000) while others believe that these policies are becoming the standard of care (Schneiderman, et al., 2000; Fine, et al. 2000).

REFERENCES

Annas, G.J.: 1997, ‘The bell tolls for a constitutional right to physician-assisted suicide,’ *New England Journal of Medicine*, 337, 1098-1103.

Arras, J., 1995, *Bringing the hospital home: Ethical and social implications of high-tech home care*, Johns Hopkins University Press, Baltimore, MD.

Aulisio, M.P., Arnold, R.M., Youngner, S.: 1998, 'Can there be educational and training standards for those conducting healthcare ethics consultation?' in D. Thomasma and J. Monagle (eds.), *Healthcare Ethics: Critical Issues for the 21st Century*, Aspen Publishers, Gaithersburg, MD.

Bacchetta, M.D. and Fins, J.J.: 1997, 'The economics of clinical ethics programs: A quantitative justification,' *Cambridge Quarterly of Healthcare Ethics* 6, 451-460.

Barnard, D., Towers, A., Boston, P., Lambrinidou, V.: 2000. *Crossing over: Narratives of palliative care*, Oxford University Press, New York.

Baylis, F.E. (ed.): 1994, *The Healthcare Ethics Consultant*, Humana Press, Totowa, NJ.

Beauchamp, T.L.: 1984, 'On eliminating the distinction between applied ethics and ethical theory,' *Monist*, 67, 514-532.

Belkin, L.: 1994, *First, Do No Harm*, Fawcett Crest, New York.

Caws, P.: 1991, 'Committees and consensus: How many heads are better than one?' *Journal of Medicine and Philosophy*, 16, 375-391.

Chambers, T.: 2001, 'Theory and the Organic Bioethicist,' *Theoretical Medicine and Bioethics*, 22(2), 123-134.

DeVries, R. and Subedi, J. (eds.): 1998, *Bioethics and Society: Constructing the Ethical Enterprise*, Prentice Hall, Saddle River, NJ.

Dickenson, D.L.: 1998, 'Are medical ethicists out of touch? Practitioner attitudes in the US and UK towards decisions at the end of life,' *Journal of Medical Ethics*, 26(4), 254-260.

Dowdy, M.D., Robertson, C., Bander, J.A.: 1998, 'A Study of Proactive Ethics Consultation for Critically and Terminally Ill Patients With Extended Lengths of Stay,' *Critical Care Medicine*, 26, 252-259.

Emanuel, L.L. and Emanuel, E.J.:1993, 'Decisions at the End of Life: Guided by Communities of Patients,' *Hastings Center Report* 23, 6-14.

Feldstein, B.D., Ogle, R.D.: 1997, 'Satisfaction, managed ethics, and the duty to design,' *Hospital Ethics Committee Forum*, 9(4), 333-354.

Fine, R.L., Mayo, T.W. :2000, ' The rise and fall of the futility movement,' [Letter] *New England Journal of Medicine*, 343(21), 1575-1576.

Fleetwood, J. and Unger, S.S.: 1994, 'Institutional ethics committees and the shield of immunity,' *Annals of Internal Medicine* 120, 320-325.

Fletcher, J.C. and Hoffmann, D.E.: 1994, 'Ethics committees: Time to experiment with standards,' *Annals of Internal Medicine* 120, 335-338.

Forrow L., Arnold, R.M., Parker, L.S.:1993, 'Preventive ethics: Expanding the horizon of clinical ethics,' *Journal of Clinical Ethics* 4, 287-294.

Fox, E., Tulsky, J.A.: 1996, 'Evaluation research and the future of ethics consultation,' *Journal of Clinical Ethics* 7(2), 146-149.

Halevy, A., Neal, R.C., Brody, B.A.: 1996, 'The low frequency of futility in an adult intensive care unit setting,' *Archives of Internal Medicine* 156(1), 100-104.

Helft, P.R., Siegler, M., Lantos, J.: 2000, 'The rise and fall of the futility movement,' *New England Journal of Medicine*, 343(4), 293-295.

Heilicser, B.J., Meltzer, D., Siegler, M.,:2000, 'The effect of clinical medical ethics consultation on healthcare costs,' *Journal of Clinical Ethics*, 11(1), 31-38.

Hoffmann, D.E.,:1991, 'Does legislating hospital ethics committees make a difference? A study of hospital ethics committees in Maryland, the District of Columbia, and Virginia,' *Law, Medicine, and Healthcare*, 19, 105-119, 1991.

Jonsen, A.R., :1991b, 'Casuistry as a methodology in clinical ethics,' *Theoretical Medicine* 12(4), 295-307.

Joint Commission on the Accreditation of Healthcare Organizations: 1997, *Accreditation Manual for Hospitals*, Vol. I, *Standards*, The Commission, Oakbrook Terrace, Ill.

Kopelman, L.A.: 1990, 'What is applied about 'applied' philosophy?' *Journal of Medicine and Philosophy* 15, 199-218.

Kuczewski, M.G.: 1996, 'Reconceiving the family: The process of consent in medical decisionmaking,' *Hastings Center Report* 26, 30-37.

Kuczewski, M.G., Pinkus, R.L.: 1999, *An Ethics Casebook for Hospitals: Practical Approaches to Everyday Cases*, Georgetown University Press, Washington, DC.

Kuczewski, M.G.: 2001, 'Disability: An agenda for bioethics,' *American Journal of Bioethics* 1(3).

Lidz, C.W., Appelbaum, P.S., Meisel, A..: 1988, 'Two Models of Implementing Informed Consent,' *Archives of Internal Medicine*, 148, 1385-1389.

MacIntyre, A.: 1984, *After Virtue*, 2nd ed., University of Notre Dame Press, Notre Dame, IN.

McCullough, L.B., Wilson, N.L.: 1995, *Long-term care decisions: Ethical and conceptual dimensions*, Johns Hopkins University Press, Baltimore, MD.

Meisel, A.: 1995, *The Right to Die*, John Wiley & Sons, Inc., New York.

Moreno, J.D.: 1995, *Deciding Together: Bioethics and Moral Consensus*, Oxford University Press, New York.

Murphy, D.J.: 1988, 'Do-not-resuscitate orders: Time for reappraisal in long-term care institutions,' *Journal of the American Medical Association* 260, 2098-2101.

Pellegrino, E.D. and Thomasma, D.C.: 1993, *The Virtues in Medical Practice*, Oxford University Press, New York.

Pinkus, R.L., et al..: 1995, The consortium ethics program: An approach to establishing a permanent regional ethics network, *HEC Forum*, 7, 13-32.

President's Commission for the Study of Ethical Problems in Biomedical and Behavioral Research: 1983, *Deciding to Forego Life-Sustaining Treatment*, U.S. Government Printing Office, Washington, DC.

Ross, J.W. et al.: 1993, *Healthcare Ethics Committees: The Next Generation*, American Hospital Publishing, Chicago, IL.

Rubin, S.B.: 1998, *When Doctors Say No: The Battle Ground of Medical Futility*, Indiana University Press, Indianapolis, IN.

Schneiderman, L.J., Capron A.M.: 2000, 'How can hospital futility policies contribute to establishing standards of practice?' *Cambridge Quarterly of Healthcare Ethics*, 9(4),524-531.

Scofield, G.R.: 1993, 'Ethics consultation: The least dangerous profession,' *Cambridge Quarterly of Healthcare Ethics* 2, 417-426.

The SUPPORT Principal Investigators: 1995, 'A controlled trial to improve care for seriously ill hospitalized patients: The study to understand prognoses and preferences for outcomes and risks of treatments (SUPPORT),' *Journal of the American Medical Association* 274, 1591-1598.

Task Force on Standards for Bioethics Consultation: 1998, *Core Competencies for Healthcare Ethics Consultation*, American Society for Bioethics and Humanities, Glenview, IL.

Truog, R.D. Brett, A.S., Frader, J.:1992, 'The problem with futility,' *New England Journal of Medicine* 326, 1560-1564.

Waisel, D.B. and Truog, R.D.: 1995, 'The cardiopulmonary resuscitation-not-indicated order: Futility revisited,' *Annals of Internal Medicine* 122, 304-308.

Wilson, R.F.: 1998, 'Hospital ethics committees as the forum of last resort: An idea whose time has not come,' *North Carolina Law Review* 76, 353-406.

Youngner, S.J., et al.: 1983, 'A national survey of hospital ethics committees,' in President's Commission for the Study of Ethical Problems in Biomedical and Behavioral Research, *Deciding to Forego Life-Sustaining Treatment*, U.S. Government Printing Office, Washington, DC.

Youngner, S.J.: 1988, 'Who Defines Futility?' *Journal of the American Medical Association* 260, 2094-2095.

SECTION IV

THE PUBLIC POLICY CONTEXT

BARUCH BRODY

THE ETHICS OF CONTROLLED CLINICAL TRIALS

Clinical research in the last fifty years has been dominated by controlled clinical trials. Ever since the 1948 British Medical Research Council clinical trial of streptomycin to treat pulmonary tuberculosis (Medical Research Council, 1948), controlled clinical trials have emerged as the best way of determining the safety and efficacy of new clinical interventions. They have also given rise to a wide variety of ethical issues. This chapter will analyze these issues and possible resolutions of them from a philosophical perspective.

There is, however, one preliminary task. Controlled clinical trials involve research on human subjects. As such, they are subject to the general ethical requirements on research involving human subjects. We must, therefore, analyze from a philosophical perspective those more general requirements before we turn to the analysis of the specific issues raised by clinical trials.

I. THE BACKGROUND REQUIREMENTS OF RESEARCH ETHICS

A remarkable consensus has emerged throughout the world on the requirements for morally acceptable research involving human subjects (Brody, 1998). There is general agreement that independent review of research protocols should insure that the following standards are met: risks are limited, research subjects are selected equitably, prospective informed consent is obtained from the subjects or their surrogates, and privacy of subjects and confidentiality of data are adequately protected. These seem to be reasonable requirements, although their implementation can often be problematic from both a conceptual and a practical perspective.

The requirement of independent review grows out of the reflection that researchers face a conflict of interest between their desire to conduct their research in the most expeditious and promising fashion and their obligation to protect the interests and rights of the subjects of the research. This conflict of interest may lead to inadequate attention to those rights and interests. The task of the independent review is to offer that protection by those who do not face that conflict of interest and who can focus their attention on the protection of the research subjects. The difficulty with this requirement is finding mechanisms for

G. Khushf (ed.), Handbook of Bioethics, 337–352.

independent review that are truly independent and that do focus on providing protection to research subjects. If the independent review is primarily provided, as it usually is, by researchers from the same institution, the potential for its focusing on the needs of the researcher rather than on the interests and rights of the subjects is always present. One of the ongoing debates in this area is how to provide a truly independent review (McNeill, 1993).

The first of the standards used in the independent review is that the research has limited the potential risks to the subject. This standard has two components. One is the requirement that risks be minimized by conducting the research in the scientifically sound manner that minimizes the exposure to risks. The other is the requirement that the resulting minimized risks be reasonable in relation to the anticipated benefits to the subjects and/or to society (there must be a favorable risk-benefit ratio). The difficulty with this second component is related to the uncertainties that are inherent to research: the potential risks and benefits are difficult to define, much less to quantify, so the determination of the risk-benefit ratio is often problematic.

The second of the standards used in the independent review is that there should be an equitable selection of research subjects so that the benefits and burdens of the research are shared equitably. As originally developed, this standard was intended to protect vulnerable research subjects from being exploited by researchers conducting risky research. Prisoners and poor people (whose participation might be less than fully voluntary) and children and cognitively impaired adults (whose participation might be based on a lack of understanding) were among the potential subjects needing protection from exploitation, protection that was often provided by excluding them from participation in research. More recently, as the potential benefits to subjects from participation in research have been better appreciated, this standard has also been interpreted to mean that subjects should not be denied these benefits by being arbitrarily excluded from promising research. An ongoing debate in research ethics is about how to properly balance the protection of vulnerable potential subjects and the inclusion of those who might benefit from participation in research (Kahn, 1998).

The third of the standards used in the independent review is that prospective informed consent must be obtained from the subjects, if they are competent to provide that consent, or from the surrogates of the subjects, if the subjects are incompetent to provide that consent. This requirement offers further protection to the interests of the subjects, as they decide whether or not it is in their interest to participate. It also protects their rights to autonomously decide whether or not they should be part of a research protocol. Difficult problems arise, however, in the determination of the information to be provided to research subjects: should the informational requirement be the same as the requirement in the normal therapeutic setting, or does the research setting require greater disclosures to potential subjects? Further problems arise as one confronts the case of the possibly incompetent subject: Who should determine whether or not the potential subject is competent to consent and what standard of competency should be employed in that determination? Who should serve as the surrogate decision maker if the potential subject is incompetent and what standards should be used by that surrogate in deciding whether or not to consent to the participation? What role, if any, remains for the incompetent individual in the decisional process?

The last of the standards used in the independent review is the protection of subject privacy and confidentiality of research data. Normally, this poses few problems, as research reports need not identify the subjects and research data can be stored in ways that prevent linkage to identifiers of the subjects.

There is a standard account of the philosophical basis of these standards. It was first offered in the Belmont Report (National Commission, 1979). The report claimed that there were three relevant ethical principles: the principle of respect for persons, the principle of beneficence, and the principle of justice. The principle of respect for persons leads to the requirement of prospective informed consent, the principle of beneficence leads to the requirement of a favorable risk-benefit ratio, and the principle of justice leads to the requirement of an equitable selection of research subjects. The additional requirement of protecting privacy and confidentiality can also be justified by an appeal to the principle of beneficence.

Principalism (Beauchamp and Childress, 2001) is an approach to the philosophical foundations of bioethics which insists that bioethical requirements need not be grounded in some fundamental moral theory such as utilitarianism or Kantianism which employs a single basic moral standard. It is sufficient, says principalism, to ground the bioethical requirements in one or more of a wide variety of moral principles, each of which incorporates a different moral standard. Principalism represents a form of moral pluralism (Brody, 1988), in opposition to the moral monism so common in the major ethical theories developed in the history of philosophy. Defenders of principalism often point to the consensus about the requirements for human subjects research, grounded in the appeal to the three principles of the Belmont Report, as a prime example of the successful employment of principalism as the foundation of a set of bioethical requirements. Critics of principalism argue that the principalist is not able to properly specify the precise meaning of the bioethical requirements or deal with conflicts between them. The critics feel that these complex tasks can only be accomplished if the bioethical requirements are grounded in some more fundamental moral theory. Moreover, says these critics, the principalist cannot justify these requirements to critics who insist that the requirements and the principles are nothing more than the cultural beliefs of particular countries. One of the most important philosophical questions about research ethics is whether the appeal to the principles of the Belmont Report offers an adequate foundation for the general requirements on research using human subjects, or whether it needs to be replaced by some more fundamental philosophical appeal.

II. THE SPECIAL ETHICAL ISSUES OF CONTROLLED CLINICAL TRIALS

The easiest way to understand the special ethical issues raised by controlled clinical trials is to review the major features of these trials. As we shall see, each of eight features of such trials generates its own set of ethical issues. Controlled clinical trials are prospective trials of (a) well defined plausible interventions in (b) a well-defined population. The subjects, (c) all of whom have consented to participation, are (d) randomly assigned to receive the intervention or (e) be part of a control group. Both the subjects and the researchers (f) are blinded to that assignment. The trial is designed to have adequate power to determine whether there is a significant difference in the occurrence of (g) well defined end points between the intervention group and the control group. In the course of the trial, results are monitored to determine (h) whether the trial should be stopped because of interim data about the safety or the efficacy of the intervention being tested. We shall review the issues raised by each of these features separately.

A. The intervention being tested must be plausible

Unless there is preliminary evidence suggesting that the intervention being tested is likely to produce sufficiently favorable results, the clinical trial in question should never be run. This is partially a question of resources. Why invest in a clinical trial, which is an expensive activity, unless it is plausible to suppose that the intervention to be tested will have a favorable risk benefit ratio? It is also a question of ethical standards. Why impose the risks of participation in a trial on subjects unless there is the requisite preliminary evidence? However, if there is truly adequate evidence that the intervention has a favorable risk benefit ratio, there would be no reason to run the clinical trial and it would be inappropriate to deny the intervention to a control group. So there is a limited opportunity for running a controlled clinical trial: there must be enough evidence to support running the trial but not enough evidence to support the general use of the intervention in question. Defining that limited period of opportunity more carefully, both from a conceptual and an ethical perspective, has turned out to be very difficult.

Charles Fried (Fried, 1974) first introduced the term "equipoise" to refer to the state of uncertainty that must exist in order for a clinical trial to be justified. As he used that term, it referred to a lack of a reason, taking into account risks and benefits, for preferring the new intervention over the standard treatment. This account of equipoise does not define the relevant situation. If there truly is no reason to support the use of the new intervention, why run the trial at all? Benjamin Freedman (Freedman, 1987) suggested instead that the relevant equipoise is a clinical equipoise, where there is disagreement about the use of the new intervention in the relevant clinical community, even though there is evidence supporting its use. Freedman argued that the opportunity for running a clinical trial was when there was evidence supporting the use of the new intervention but there was still clinical equipoise about its use.

Freedman's approach is a sociological approach; the justification for running the trial is the remaining disagreement about the merits of the new intervention in the relevant clinical community. This approach suffers from conceptual ambiguity. How much disagreement in what community is required in order for the trial to be justified? More crucially, its normative basis is problematic. If the remaining disagreement in the relevant clinical community is merely due to conservatism based on inertia and/or ignorance of the results of earlier research, rather than a justified feeling that the existing evidence is inadequate, why should the intervention be withheld from the needed control group?

I (Brody, 1995) have suggested an alternative approach, based upon some earlier work done by Paul Meier (Meier, 1979). It rests upon the assessment, in a thought experiment, of the already existing evidence by a rational person with a normal amount of both self interest and altruism. If such a person would be willing to be randomized into a trial of the intervention, because the evidence of benefit is sufficiently modest so that the sacrifice of self interest is also sufficiently modest, then the existing evidence is insufficient and the proposed clinical trial is justified. If such a person would not be willing, there is no justification for running the trial. Put more bluntly, the criterion is that researchers should not run clinical trials, and independent review boards should not approve them, unless the researchers and the independent reviewers would be willing to be randomized into the trial.

There are obvious conceptual issues faced by this proposal. Defining the appropriate balance of self interest and altruism is very difficult. Defining honest responses to these types of thought experiments is equally problematic. But the normative basis of this proposal seems to be acceptable. The justification of the clinical trial is the quality of the evidence supporting the new intervention, rather than the willingness of the relevant clinical community to support new interventions.

B. The intervention must be tested in a well defined population

The definition of the population, both by inclusion criteria and exclusion criteria, determines the applicability of the results of the clinical trial. Strictly speaking, the trial results apply only to those future users of the intervention who would have been eligible to participate in the trial. Applying the trial results to other potential users of the intervention represents an extrapolation which may or may not be justified.

This observation has a direct impact on the issue of the equitable selection of research subjects. If members of various groups are excluded from participation in clinical trials, physicians will not know on the basis of the trials whether the interventions tested in those trials should be used in the excluded populations. Those populations are thereby denied some of the benefits of the research. In addition, individuals in those populations are denied the immediate benefits, if any, of participating in the research.

These observations have led to a rethinking of the exclusion of children and cognitively impaired adults from clinical trials. As noted above, these groups were in the past perceived as vulnerable to exploitation in research and were often protected from that exploitation by being excluded from clinical trials. The emphasis today is on a more balanced approach to their inclusion in clinical trials, keeping in mind the benefits as well as the risks of participation both for the individual subjects and for the vulnerable groups.

Women of child bearing potential and pregnant women were in the past often excluded from participation in clinical trials out of fear that the intervention being tested might have a negative impact upon fetuses. This denied the potential women subjects the benefits, if any, of participating in the trials. It also meant that much of the data from clinical trials involving men was not directly applicable to women, denying women and their physicians a firmer basis for deciding about their use of new interventions validated only for men. The emphasis today (DeBruin, 1994) is on a more balanced approach, one which includes women in clinical trials subject to their being adequately informed about fetal risks and subject to adequate use of pregnancy testing and birth control (in the case of women of child bearing potential).

There are an important set of philosophical issues raised by this new approach. Consider the case of research on women using interventions that may be harmful to fetuses but beneficial to the women in question. Should women be included in such trials, so long as they are adequately informed about the risks? Or should women be excluded from such trials, even if they wish to participate, when the risks to the fetuses are too great? These questions cannot be answered without considering a complex set of philosophical questions about the moral status of fetuses, about the responsibility of pregnant women to their future children, about the decisional autonomy of pregnant women and of women of child bearing potential, and about the responsibility of researchers and the research enterprise.

C. All of the subjects must consent to participation

This requirement is, of course, one of the requirements for all research on human subjects. It can be met by obtaining the prospective informed consent of the research subject, or, if the subject is incompetent, of the subject's surrogate.

There are certain very important clinical trials which would have great difficulty meeting this requirement. These are trials of promising new interventions to deal with life threatening emergency medical problems. Many of these interventions must be used very rapidly after the patient presents if they are to have any chance of working. There may not be time to get anyone's informed consent for participation in the research. Moreover, the subjects (after, for example, a myocardial infarction or a stroke) may not be competent and their surrogate may not be present in the relevant time period. Some (Abramson, 1986) have suggested getting retrospective deferred consent when the patient becomes competent or when a surrogate is available. This suggestion makes little sense; how, for example, can one retrospectively refuse to participate if one's participation is complete. Others (OPRR, 1993) have suggested allowing such research to proceed so long as the risks are minimal, but many of these promising interventions carry significant risks.

A very subtle set of philosophical issues are raised by these trials. If the requirement of obtaining prospective informed consent is viewed as an absolute requirement, allowing for no exceptions, then many of these promising interventions could never be tested in clinical trials. This may be unfortunate both from the perspective of future patients and from the perspective of the subjects in the active treatment group (both of whom would benefit from the use of these interventions if the trial shows that they work). Allowing for some of this research to be conducted to obtain these benefits requires adopting a non-absolutist approach to this requirement, one which insists that other moral values can take precedence over the requirement of obtaining informed consent. But then one must specify the circumstances under which these other values take precedence.

One plausible approach (FDA, 1996) insists that the requirement for informed consent can be waived only if the following conditions are met: (1) the medical condition must be life threatening and the standard treatments must be unsatisfactory; (2) the preliminary evidence must strongly suggest, without conclusively demonstrating, that the intervention being tested will have a favorable risk benefit ratio; (3) prospective consent cannot be obtained because potential subjects cannot be identified in advance and there is no possibility at the time of the emergency of getting consent from the subjects or their surrogates. The first condition introduces the social need of conducting the research. The second condition introduces the likelihood that the subjects who receive the intervention will personally benefit. The third condition clarifies why consent cannot be obtained. While this approach is quite plausible, a full philosophical justification of it would require a better theoretical understanding of the conditions under which important moral requirements can be overridden by other pressing moral values.

There is a further complicating concern having to do with the third condition. In many of these trials, a certain percentage of potential subjects will present quickly enough so that there is time to obtain consent, and either they will be competent to make a decision or they will be accompanied by a surrogate who can make the decision about participation. In such cases, the third condition is not literally satisfied. The trial can be run using only those subjects for whom prospective consent can be obtained. However, as these are only a limited subset of the potential subjects, the trial will take much longer to complete (denying society for a longer period of time the vitally needed data about the merits of the new intervention) and many potential subjects who might personally benefit from participation will not be able to participate. May the research proceed without prospective consent in such cases? This is an even more complex balancing of values, and our lack of a philosophical theory of balancing values is particularly troublesome in such complex cases.

D. The subjects must be randomized to receive the intervention or to be part of the control group

This randomization is necessary to insure that any differences in the outcomes are due to the intervention rather than to baseline differences between the intervention group and the control group. As part of the informed consent process, prospective subjects must understand and accept the idea that the treatment they receive will depend on the randomization process.

It is this last point that has given rise to problems. Many prospective subjects are very troubled by the idea that their treatment is determined by randomization. This is particularly true when the medical condition is life threatening and the different treatments to which people are being randomized are radically different. For example, in a trial of total versus segmental mastectomy, with or without radiation therapy, for breast cancer, significant opposition to randomization was expressed by many potential subjects and enrollment was initially very poor

(Fisher, 1985).

Two alternatives have been proposed by Zelen (Angell, 1984). The first is the prerandomization strategy, in which potential subjects, without being notified, are randomized to the intervention group or the control group, consent is obtained only from those receiving the new intervention, and all the subjects are followed to determine comparative outcomes. The second is the double-consent randomized strategy, in which potential subjects, without being notified, are randomized to the intervention group or the control group and are then asked for their consent to receive the treatment to which they were randomized. The second strategy has the advantage of insuring that all subjects in the trial are aware that they are in the trial and have consented to the treatment which they actually receive. Neither strategy has anyone consenting to being randomized, and this is their main attraction from the perspective of increasing enrollment. But there remains the concern that investigators, knowing the group to which the subject has been assigned, will bias their presentation of information in favor of the assigned treatment to increase enrollment.

There is, moreover, a broader philosophical issue about informed consent raised by these strategies. Is it sufficient that people consent to the treatment which they then receive, or must they also consent to the randomization process? Defenders of these new strategies insist that the former is all that is required, while opponents insist that the latter is required as well. What does autonomy actually require?

E. There must be an appropriate control group

The existence of some control group is essential for the scientific validity of the clinical trial. It is the comparison of the outcomes in the group receiving the intervention versus the control group that enables us to determine the risks and benefits of the new intervention, as opposed to the natural history of the disease itself. The trial can be a historically controlled trial, in which case the treatment group is compared to some earlier group that did not receive the intervention in question. The use of a historical control group precludes, of course, randomization; it also raises concerns about whether other differences over time are responsible for differences between the treatment group and the control group. These are the scientific reasons for preferring the use of a concurrent control group, so that the treatment group is compared to a group currently not receiving the intervention in question. The members of the concurrent control group may be receiving some other intervention (active controlled trial) or they may just be receiving a placebo (placebo controlled trial). For a variety of statistical and trial design considerations, placebo controlled trials are easier to run and easier to interpret. (FDA, 1985)

These scientific considerations are not, however, the only considerations. There are also a variety of ethical issues raised by the control group. When is it ethical to deny promising new interventions to members of the control group? Is it sufficient that they knowingly consent to being in a control group? Or are there cases in which it is wrong to deny them those interventions despite their consent? Moreover, if there are already proven therapies of some value, can a placebo control group be used or must we use an active control group, because it is wrong to deny the proven therapy to the control group? Once more, is the issue settled by the consent of the members of the control group, or are there cases in which it is wrong to deny them the proven intervention despite their consent?

From a practical perspective, guidelines (AMA, 1996) have been offered: (a) the more serious the disease process and the more beneficial existing therapies, the more problematic is a placebo controlled trial. Investigators may consider measures such as early rescue or limitation of study duration that would make the placebo controlled trial ethically acceptable. If they are not adequate, then an active controlled trial is probably required; (b) similarly, when there is a serious disease process with no beneficial existing therapies, investigators may be required to run a historically controlled trial unless the above mentioned measures make a concurrent placebo controlled trial ethically acceptable.

These guidelines presuppose that the consent of the control group is not sufficient to justify the trial; their whole point is to indicate the conditions under which the use of a placebo control group is wrong, regardless of any consent by potential subjects. The opponent of such guidelines see them as inappropriately paternalistic. The proponents of these guidelines see it differently. The consent of the members of the placebo control group would certainly mean that the use of such a group violates none of the rights of those who have consented. But, from this perspective, other moral issues are relevant. There are many wrongs we can do even if we violate no rights. One of those wrongs is imposing what we see as excessive harms on others, even if they consent to having the harms imposed, because they see them differently. It is a matter of the moral integrity of the investigators.

F. The subjects and the researchers should be blinded regarding the group into which the subject has been randomized

Blinding the subjects helps prevent excessive dropout from one group as opposed to the other and helps prevent differential seeking of alternative concomitant therapies. Blinding the researchers helps prevent differential assignment of concurrent therapies and helps prevent biased assessment of outcomes. In all of these ways, blinding contributes to the scientific validity of the clinical trial. At the same time, it may impose considerable burdens on the subject. The ethical issue becomes that of balancing the scientific gains from blinding against the burdens imposed on the subjects and deciding when the burdens are too great.

It is of some interest to note that in the 1948 British Medical Research Council trial of streptomycin, where the drug was being injected for an extensive period of time, blinding would have required injecting the control group subjects with sham injections for the equivalent period of time. That trial was not run as a blinded trial, precisely because the investigators decided that the burden on the subjects required to maintain the blind was unethical. They would not have objected if it was just a question of ingesting placebo pills externally indistinguishable from the active medications being tested (which is a common practice); it was the need to inject both the streptomycin and the sham which led the investigators to conclude that the burden was too great.

This feeling has made it very difficult to run blinded trials of surgical techniques. Although a number of such trial were run in the 1950s, including one in which the chests of the placebo subjects were opened, the general consensus has been that this imposes excessive burdens on the subjects and is unethical. This is not surprising; if sham injections are unethical, sham incisions are even more unethical. Nevertheless, in recent years, the issue of sham surgery to maintain the blinding of subjects and investigators has reemerged. In a famous study of transplantation of fetal tissue for Parkinson's Disease (Freeman, 1999), all subjects, including those in the control group, underwent the initial burrhole penetration of the skull in order to maintain blinding. In light of the difficulties in objectively assessing whether the treatment resulted in symptomatic improvement and in light of a real possibility of placebo effect benefit, it was felt by the investigators that the need to maintain the blinding was so great that it justified the sham surgery.

As in the case of the control group issue, there are those who would insist that the whole issue is just a question of adequate informed consent. If all the subjects know that they may be assigned to the control group and know that this will result in sham injections or sham incisions, and if they agree to be randomized, then there is no problem in using sham injections or doing sham surgery. Others (Macklin, 1999) do not see it that way. From their point of view, investigators, as part of maintaining their own moral integrity, should not impose excessive burdens on the subjects in the placebo group, even if the subjects give their informed consent. Adopting this second approach requires the investigators to do their own balancing of the scientific benefits against the burdens of maintaining the blind.

These last two issues raise therefore a crucial philosophical issue about the moral responsibility of investigators. Is it sufficient that they not violate the rights of the research subjects? If that is the limit of their responsibility, then they can use whatever control group and whatever blinding techniques are most scientifically beneficial, providing that they can get subjects to give their voluntary informed consent to participate in the trial as designed. Alternatively, are they responsible to not impose burdens on subjects which the investigators judge to be excessive? If their responsibility extends that far, then they must balance the scientific benefits of certain trial designs against the burdens those designs impose upon subjects, and they must not conduct certain trials for which they could find subjects who would give their voluntary informed consent to participate.

G. The trial must be powered to be able to determine whether there is a significant difference between the intervention group and the control group in terms of the chosen endpoints

The goal of the trial is to determine whether there is a sufficiently favorable risk-benefit ratio to justify using the new intervention in general clinical practice. The risks are the negative endpoints chosen to be studied in the trial and the benefits are the positive endpoints chosen to be studied in the trial. Traditionally, the positive endpoints are reductions in mortality and morbidity which hopefully will be produced by the intervention. In recent years, however, a new approach has emerged, in which clinical trials are based on the use of surrogate endpoints. This approach raises many questions, both from a scientific perspective and an ethical perspective (Fleming and DeMets, 1996).

The point of departure of this new approach is that clinical trials are used to do more than settle scientific questions; they are also used to justify the regulatory approval of new drugs or devices, making them generally available for use. When a new intervention promises to provide help to those suffering from conditions for which there are few available treatments, there is great pressure to make that new intervention generally available as soon as possible. Classical clinical trials, using traditional clinical endpoints, may take a long time to run (in part because the clinical endpoints may take a long time to occur), thereby delaying the approval and availability of the new intervention. This has led to the suggestion that clinical trials should study surrogate endpoints, laboratory endpoints which could be used to predict the later occurrence of the clinical endpoints. If the intervention produces a favorable result in terms of the surrogate endpoints, we might reasonably expect that it will produce the same favorable result in terms of the clinical endpoints, and we could approve the intervention on the basis of the quicker trial using the surrogate endpoints. Thus, many have advocated the approval of new AIDS drugs on the basis of trials showing that they reduce viral load and/or increase CD4 counts, without waiting to see how much they extend the period of time in which HIV positive subjects do not have an AIDS defining clinical event.

There are, of course, many scientific questions raised by this approach. The first has to do with whether improvements in the surrogate endpoint truly predicts ultimate clinical benefit. The second has to do with the possibility that the clinical benefit predicted by the improvement in the surrogate endpoint may be outweighed by unfavorable long term side effects of the new intervention. There are many examples showing that these are not merely theoretical possibilities. For both of these reasons, there are significant concerns about the scientific validity of drawing conclusions about the risk-benefit ratios of new interventions on the basis of clinical trials employing surrogate endpoints.

This observation leads us directly to the ethical and philosophical issues posed by the use of surrogate endpoints. Suppose that the sufferers from the disease in question concede that there are scientific problems with trials using surrogate endpoints and that we do not know for sure that the intervention is worth using just because of a favorable result in a surrogate endpoint trial. Suppose that they say that they want the intervention approved for general use anyway, because of the lack of good alternatives for treating their disease. Why shouldn't they be allowed to take their chances with these new interventions that have some support, they argue, so long as they recognize the risks they are taking? Isn't the insistence on stronger evidence before the intervention is approved for general use an unacceptable form of medical paternalism?

From a philosophical point of view, the issue being raised is even more complex (Brody, 1995). The whole system of not allowing patients to use new medical interventions until society approves of their general use through a drug/device regulatory mechanism is difficult to justify. Using new medical interventions is the very sort of behavior that classical liberal theory, expounded by such authors as Mill, believes should be unregulated by society and left to the free choices of individuals. Even if some justification can be found for such a regulatory mechanism, it is unclear that it could extend to refusing to approve new interventions justified by clinical trials using surrogate endpoints. It is difficult to develop clear conclusions on this question until the philosophical/ethical basis of the regulatory scheme is clarified. But it certainly seems clear that we need to distinguish the scientific benefits of getting firmer evidence about new interventions by running clinical trials with clinical endpoints from the regulatory benefits of allowing access to new interventions on the basis of clinical trials employing surrogate endpoints.

H. The trial should be monitored to determine whether the interim data about safety or efficacy is sufficient to justify stopping the trial

In clinical trials, data about safety and efficacy accumulate over a period of time. The investigators, being blinded to the assignment of the subjects, do not know whether the accumulating data prove that the new intervention is unsafe (so that the trial should be stopped to protect future subjects) or that it is clearly efficacious (so that the trial should be stopped and the new intervention made available to all who could benefit from its use). Some unblinded group independent of the investigators must be empowered to monitor the interim data and to make recommendations about continuing or stopping the trial. Similar recommendations may be required as data becomes available from other sources; in that case, the investigators could make the decision, but there is a concern about their objectivity, and an independent group might be more objective. For these reasons, many clinical trials now have an independent Data Safety and Monitoring Board (DSMB) which monitors data both from the ongoing trial and from other sources and make recommendations as to whether or not the trial should continue. The practice began in the United States at the National Heart Institute in the 1960s, and it has since become very common (DeMets, 1987).

It is often thought that the main issues are statistical, and that statisticians should therefore dominate on such boards. The idea behind this suggestion is based on the fact that the more often one looks at interim data, the more likely one is to find by chance data of efficacy that reaches the usual level of significance (p value less than .05). For example, if you check for efficacy five times while the trial is being conducted, you have a 14% (rather than the desirable 5%) chance of finding a result of efficacy at the usual level of significance. It is necessary therefore to adopt a plan for interim monitoring for efficacy of the intervention that takes this statistical problem into account. These plans adopt one of a variety of available interim stopping rules. So, it might be suggested, the heart of the task of a DSMB is dealing with these statistical problems by adopting an appropriate plan for interim monitoring and then carrying it out. These, it might be suggested, are statistical issues to be dealt with by statisticians.

This suggestion fails to understand the ethical complexities in making decisions in monitoring interim data. It misses three crucial points: (1) one of the other roles of a DSMB is monitoring for safety. If safety concerns about the new intervention emerge, a decision has to be made as to whether the safety risks are too great to allow the trial to continue even if there are benefits from the use of the new intervention. These judgements about the balancing of risks and benefits are complex value questions, and statistical stopping rules are not even relevant to their resolution; (2) the statistical stopping rules relate only to the data emerging from the trial in question. They do not address the issue of data emerging from other sources while the trial is being conducted. The DSMB, considering the data from the other trials, must make a value judgment as to whether or not sufficient equipoise remains to justify continuing the trial. These judgments about remaining equipoise are also complex value questions, and statistical stopping rules are not even relevant to their resolution; (3) statistical stopping rules are relevant to analyzing data of efficacy, but they are not sufficient to make the decision about stopping the trial. Even if benefits have been established according to the rule in question, it may still be appropriate to continue the trial to settle questions of long term safety or efficacy. And even if the benefits have not been established with sufficient certainty according to the stopping rule, it may still be appropriate to stop the trial when the likely benefits are very great and the risks minimal. These additional considerations call for delicate balancings of values, and not just an appeal to a prespecified statistical stopping rule. So even when those rules are relevant, they do not settle the issue of continuing the trial. In short, the monitoring of interim data calls for ethical as well as statistical analysis. For that reason, ethicists now sit on many DSMBs.

What ethical guideline should be used by a DSMB? It seems to me that the following principle should be used. It is ethical to continue the trial, based on the data that has become available from interim monitoring and from other sources, only if it would have been ethical to begin the trial if that data had been known. Referring to our earlier discussion of that issue, we can say that it is ethical to continue the trial only if sufficient equipoise remains (in the normative sense we developed earlier, rather than in the Fried-Freedman sociological sense).

We have in this section reviewed eight major ethical issues about the conduct of clinical trials. We have seen that they raise broader philosophical issues about the role of sociological versus normative judgments in determining the ethics of proposed trials, the decisional autonomy of pregnant women, the criteria for balancing of conflicting values, the extent of the need for subject consent, the role of subject consent versus researcher integrity, and the role of the state in limiting access to desired therapies. Each of these issues needs to be further explored in future research.

Center for Medical Ethics and Health Policy
Baylor College of Medicine
Houston, Texas, U.S.A.

REFERENCES

Abramson, N, Meisel, A, Safar P: 1986, 'deferred consent,' *JAMA* 255, 2466-71

AMA: 1996, 'Ethical use of placebo controls in clinical trials,' *www.ama-assn.org/ama/pub/category/5494.html#a96* (Visited Aug 9, 2001)

Angell, M: 1984, 'Patients' preferences in randomized clinical trials' *New England Journal of Medicine* 310, 1385-7

Beauchamp, T, Childress, J:2001, *Principles of Biomedical Ethics*, Oxford University Press, New York

Brody, B: 1988, *Life and Death Decision Making*, Oxford University Press, New York

Brody, B: 1995, *Ethical Issues in Drug Testing Approval and Pricing*, Oxford University Press, New York

Brody, B: 1998, *The Ethics of Biomedical Research, Oxford University Press*, New York

DeBruin, D: 1994, 'Justice and the inclusion of women in clinical studies' *Kennedy Institute of Ethics Journal* 4, 117-46

DeMets, D: 1987: 'Practical aspects in data monitoring' *Statistics in Medicine* 6, 753-60

FDA: 1985, 'Adequate and well-controlled studies' 21 CFR 314.126

FDA: 1996, 'Exception from informed consent requirements for emergency research' 21 CFR 50.24

Fisher, B, Bauer, M, Margolese, R, et al :1985, 'Five year results of a randomized controlled trial comparing total mastectomy and segmental mastectomy with or without radiation in the treatment of breast cancer,' *New England Journal of Medicine* 312, 665-73

Fleming, T , DeMets, D: 1996, 'Surrogate end points in clinical trials,' *Annals of Internal Medicine* 125, 605-13

Freedman, B: 1987, 'Equipoise and the ethics of clinical research,' *New England Journal of Medicine* 317, 141-5

Freeman TB, Vawter DE, Leaverton PE et al: 1999, 'Use of placebo surgery in controlled-trials of a cellular-based therapy for Parkinson's Disease,' *New England Journal of Medicine* 341: 988-92

Fried,C: 1974, *Medical Experimentation*, North Holland, Amsterdam

Kahn, J, Mastroianni,A, Sugarman,J (eds): 1998, *Beyond Consent: Seeking Justice in Research*, Oxford University Press, New York

Macklin, R: 1999, 'The ethical problem with sham surgery in clinical research,' *New England Journal* of Medicine, 341, 992-6

McNeill, P: 1993, *The Ethics and Politics of Human Experimentation*, Cambridge University Press, Cambridge

Medical Research Council: 1948, 'Streptomycin treatment of pulmonary tuberculosis,' *British Medical Journal*, 769-82

Meier, P: 1979, 'Terminating a trial-the ethical problem,' *Clinical Pharmacology and Therapeutics* 25, 633-40

National Commission: 1979, "The Belmont Report,' Federal Register 44 (April 18)

OPRR: 1993, OPRR Reports #3 (August 12, 1993)

DAN BROCK

ETHICAL ISSUES IN THE USE OF COST EFFECTIVENESS ANALYSIS FOR THE PRIORITIZATION OF HEALTH RESOURCES

Resources to improve health are and always have been scarce, in the sense that health must compete with other desirable social goals like education and personal security for resources.[1] It is not possible to provide all the resources to health, including health care and health care research, that might provide some positive health benefits without great and unacceptable sacrifices in other important social goods. This should go without saying, and in other areas of social expenditures resource scarcity is not denied, but in health care many people mistakenly persist in denying this fact. It follows from resource scarcity that some form of health care rationing is unavoidable, where by rationing I mean some means of allocating health care resources that denies to some persons some potentially beneficial health care. That rationing may take many forms. In most countries with a national health system it is done through some form of global budgeting for health care. In the United States much rationing is by ability to pay, but in both public programs like the Oregon Medicaid program and in many private managed care plans more systematic efforts to prioritize health care resources have been carried out.

To many health policy analysts it is an unquestioned, and so generally undefended, assumption that in the face of limited health care resources, those resources should be allocated so as to maximize the health benefits they produce, measured by either the aggregate health status or disease burden of a population. Cost effectiveness analysis (CEA) that compares the aggregate health benefits secured from a given resource expenditure devoted to alternative health interventions is the standard analytic tool for determining how to maximize the health benefits from limited resources. Natural, even self-evident, as this maximization standard may appear to many health policy analysts and economists, it assumes a utilitarian or consequentialist moral standard. More specifically, it assumes a utilitarian standard of distributive justice, which is widely and I believe correctly taken to be utilitarianism's most problematic feature.

I hope to show in this paper that bringing together critical philosophical work on utilitarianism with the issues that arise for resource prioritization in health care that employs cost effectiveness analysis has benefits both for moral and political philosophy as well as for health care resource prioritization. The critical philosophical work on utilitarianism's account of distributive justice can

G. Khushf (ed.), Handbook of Bioethics, 353–380.

deepen health policy analysts' understanding of the ethical issues and disputes in the use of cost-effectiveness analysis for health care resource prioritization. The moral and political philosopher can benefit from understanding better the full array of distinct and precise equity issues that arise in health care resource prioritization. Much philosophical work in moral and political philosophy on justice has been at too high a level of generality to provide determinate implications for many of the equity issues that arise in health care (or other) resource prioritization. Coming to understand and address these issues will enrich and deepen philosophical theories of justice as well as help make them more useful in practical policy contexts.

Cost effectiveness analysis comparing alternative health interventions in the quality-adjusted life years (QALYs) produced from a given level of resources constitutes a quantitative method for prioritizing different interventions to improve health. There are many unresolved technical and methodological issues in QALYs and CEA, none of which will be my concern here. My concern will be instead with the ethical issues in the construction and use of CEAs for the prioritization of health care resources. The specific issues that I shall briefly discuss below all constitute potential ethical criticisms of CEA as a normative standard, specifically criticisms concerning justice or equity, and so one might hope concerns for justice or equity could be integrated into these quantitative methodologies. There are at least two reasons, however, for caution, at least in the near term, about the possibility of integrating some of these ethical concerns into cost effectiveness models and analyses. First, although a great deal of work in economics and health policy has gone into the development and validation of measures of health status and the burdens of disease, as well as of cost effectiveness methodologies, much less work has been done on how to integrate concerns of ethics and equity into cost effectiveness measures, although I shall mention one means of doing so later. The theoretical and methodological work necessary to do so remains largely undone. Second, each of the issues of ethics and equity that I take up below remain controversial. Since no clear consensus exists about how each should be treated, there is in turn no consensus about what qualifications or constraints they might justify placing on the cost effectiveness goal of maximizing health.

This second difficulty is not likely to be solely a near term limitation, awaiting further work on the ethical issues that I will identify. Instead, most of these issues represent deep divisions in normative ethical theory and in the ethical beliefs of ordinary people; I believe they are likely a permanent fact of ethical life. As I understand and shall present these ethical issues, in most cases there is not a single plausible answer to them. Even from within the standpoint of a particular ethical theory or ethical view, these issues' complexity means that different answers may be appropriate for a particular issue in the different contexts in which CEAs are used. Thus, what is necessary at this point is work developing more clearly and precisely the nature of the issues at stake, the alternative plausible positions on them together with the arguments for and against those positions. Until much more of this work is done, we will not know how deep the conflicts go and the degree to which any can be resolved.

Norman Daniels and James Sabin have recently argued that because ethical theories and theories of justice are indeterminate and/or in conflict on some of these issues, we must turn to fair procedures to arrive at practical solutions to them for health policy (Daniels and Sabin, 1997, pp. 303-350). As practical policy matters that need resolution now they are no doubt correct, and a single quantitative measure or model of equity and justice for health care resource prioritization is certainly not possible now, if it will ever be. But that is not to deny that much important work remains to be done on the substantive issues of equity in health care, and that work should inform the deliberations of those taking part in the fair procedures that we will need to reach practical resolutions and compromises on these issues in real time. What then are some of the main issues of equity raised by cost effectiveness approaches to resource allocation of health care?

I. FIRST ISSUE: HOW SHOULD STATES OF HEALTH AND DISABILITY BE EVALUATED?

Any CEA in health care requires some summary measure of the health benefits of interventions designed to improve the health status and reduce the burden of disease of a given population. Early summary measures of the health status of populations and of the benefits of health interventions often assessed only a single variable, such as life expectancy or infant mortality. The usefulness of life expectancy or infant mortality rates is clearly very limited, however, since they give us information about only one of the aims of health interventions, extending life or preventing premature loss of life, and they provide only limited information about that aim. They give us no information about another, at least as important aim of health interventions, to improve or protect the quality of life by treating or preventing suffering and disability.

Multi-attribute measures like the Sickness Impact Profile (Bergner, Bobbitt, Carter and Gibson, 1981, 787-805) and the SF 36 (Ware and Sherborune, 1992, pp.473-483) distinguish different aspects of overall health related quality of life (HRQL). A particular population can be assessed on these different dimensions, and an intervention assessed for its impact on these different dimensions of health, or HRQL. This type of multi-attribute measure, however, merely distinguishes different aspects of HRQL, but does not assign a quantitative measure of the relative value or importance to the different aspects or attributes of HRQL; consequently, it does not provide a single overall summary measure of HRQL. Thus, if one of two populations or health interventions scores higher in some respect(s) but lower in others, no conclusion can be drawn about whether the overall HRQL of one population, or from one intervention, is better than the other. Much quantitative based resource prioritization requires a methodology that combines in a single measure the two broad kinds of benefits produced by health interventions – extension of length of life and improvements in various aspects of HRQL (Brock, 1992).

Typical summary measures of the benefits over time of health interventions that combine and assign relative value to these two kinds of benefits include QALYs and Disability-Adjusted Life Years (DALYs); for example, a health intervention that extends a patients life for 10 years, but with a less than full quality of life of .75 measured on a zero to one scale, produces a benefit of 7.5 QALYs. QALYs and DALYs require a measure of the health status of individuals and in turn populations at different points in time. Typical measures are the Health Utilities Index (HUI) (Torrance, et.al., 1996) and the Quality of Well-Being Scale (QWB) (Kaplan and Anderson, 1988, PP. 203-235), so as to be able to measure the health benefits in terms of changes in HRQL and length of life produced by different health interventions; the HUI is reproduced in Tables 1and 2 for readers unfamiliar with these types of measures. The construction of any measure like the HUI requires a two step process: first, different states of disability or conditions limiting HRQL are described (Table 1; Torrance, et al., 1996, p. 706); second, different relative values or utilities are assigned to those different conditions (Table 2; Torrance, et al., 1996, p. 711).

TABLE 1. Health Utilities Index Mark 2 Multiattribute Health Status System

Attribute	Level	Description[a]
Sensation	1	Able to see, hear, and speak normally for age.
	2	Requires equipment to see or hear or speak.
	3	Sees, hears, or speaks with limitations even with equipment.
	4	Blind, deaf, or mute
Mobility	1	Able to walk, bend, lift, jump, and run normally for age.

	2	Walks, bends, lifts, jumps, or runs with some limitations but does not require help.
	3	Requires mechanical equipment (such as canes, crutches, braces, or wheelchair) to walk or get around independently.
	4	Requires the help of another person to walk or get around and requires mechanical equipment as well.
	5	Unable to control or use arms or legs.
Emotion	1	Generally happy and free from worry.
	2	Occasionally fretful, angry, irritable, anxious, depressed, or suffering "night terrors."
	3	Often fretful, angry, irritable, anxious, depressed, or suffering "night terrors."
	4	Almost always fretful, angry, irritable, anxious, depressed.
	5	Extremely fretful, angry, irritable, anxious, or depressed usually requiring hospitalization or psychiatric institutional care.
Cognition	1	Learns and remembers school work normally for age.
	2	Learns and remembers school work more slowly than classmates as judged by parents and/or teachers.
	3	Learns and remembers very slowly and usually requires special educational assistance.
	4	Unable to learn and remember.
Self-care	1	Eats, bathes, dresses, and uses the toilet normally for age.
	2	Eats, bathes, dresses, or uses the toilet with difficulty.
	3	Requires mechanical equipment to eat, bathe, dress, or use the toilet independently.
	4	Requires the help of another person to eat, bathe, dress, or use the toilet.
Pain	1	Free of pain and discomfort.
	2	Occasional pain. Discomfort relieved by nonprescription drugs or self control activity without disruption of normal activities.
	3	Frequent pain. Discomfort relieved by oral medicines with occasional disruption of normal activities.

	4	Frequent pain, frequent disruption of normal activities. Discomfort requires prescription narcotics for relief.
	5	Severe pain. Pain not relieved by drugs and constantly disrupts normal activities.
Fertility	1	Able to have children with a fertile spouse.
	2	Difficulty in having children with a fertile spouse.
	3	Unable to have children with a fertile spouse.

[a] Level descriptions are worded here exactly as presented to respondents in the HU1:2 preference survey.

TABLE 2. Measured Values for Levels Within Attributes

Attribute	Level	VAS (n=203) Mean ±SD[b]	95% Confidence Limits
Sensation	1	1.00	
	2	0.59±0.25	0.56,0.62
	3	0.36±0.21	0.33,0.39
	4	0.00	
Mobility[a]	1	1.00	
	2	0.68±0.22	0.65,0.71
	3	0.34±0.22	0.31,0.37
	4	0.17±0.19	0.14,0.20
	5	0.00	
Emotion	1	1.00	
	2	0.58±0.24	0.55,0.61
	3	0.33±0.19	0.30,0.36
	4	0.18±0.15	0.16,0.20
	5	0.00	
Cognition	1	1.00	
	2	0.58±0.22	0.55,0.61
	3	0.38±0.22	0.35,0.41
	4	0.00	
Self-care	1	1.00	
	2	0.56±0.24	0.53,0.59
	3	0.29±0.21	0.26,0.32
	4	0.00	

Pain	1	1.00	
	2	0.72±0.21	0.69,0.75
	3	0.45±0.21	0.42,0.48
	4	0.21±0.17	0.19,0.23
	5	0.00	
Fertility	1	1.00	
	2	0.45±0.24	0.42,0.48
	3	0.00	

VAS, visual analogue scale; SD, standard deviation
[a]The mobility results have been corrected for confounding with self-care
[b]The extreme levels of each attribute were assigned values of 0 and 1.

The determination of a person's or group's different health related conditions in terms of the various areas of function on the HUI both before and after a particular health intervention is an empirical question, which should be answered by appeal to relevant data regarding the burden of a particular disease and the reduction in that burden that a particular health intervention can be expected to produce. Needless to say, often the relevant data are highly imperfect, but that is a problem to be addressed largely by generating better data, not by ethical analysis.

The second step of assigning different relative values or utilities to the different areas and levels of function described by a measure like the HUI is typically done by soliciting people's preferences for life with the various functional limitations. This raises the fundamental question of whose preferences should be used to determine the relative value of life with different limitations in function and how they should be obtained. The developers of the DALY used the preferences of expert health professionals, in part for the practical reason that they are more knowledgeable about the nature of different health states, but the degree to which various conditions reduce overall HRQL is not a matter to be settled by professional expertise. Moreover, health professionals may have systematic biases that skew their value judgments about quality of life from those of ordinary persons. Other measures like the HUI use the value judgments of a random group of ordinary citizens to evaluate different states of disability or limitations in function. The utilities so determined for different functional attributes and their levels in the HUI are shown in Table 2.

A central issue concerning whose evaluations of different states of disability or functional limitation should be used arises from the typical responses of individuals to becoming disabled: adaptation, that is improving one's functional performance through learning and skills development; coping, that is altering one's expectations for performance so as to reduce the self-perceived gap between them and one's actual performance; and adjustment, that is altering one's life plans to give greater importance to activities in which performance is not diminished by disability (Murray, 1996). The result is that the disabled who have gone through these processes often report less distress and limitation of opportunity and a higher quality of life with their disability than the non disabled in evaluating the same condition. If the evaluations of disability states by the non disabled are used for ranking different states of health and disability, then disabilities will be ranked as more serious health needs, but these rankings are open to the charge that they are distorted by the ignorance of the evaluators of what it is like to live with the conditions in question. Moreover, those valuations will assign less value to extending the lives of persons with disabilities. If the evaluations of the disabled themselves are used, however, the rankings are open to the charge that they reflect a different distortion by unjustifiably underestimating the burden of the disability because of the process of adaptation, coping, and adjustment that the disabled person has undergone. Moreover, they will assign less value to prevention or rehabilitation for disability because of the results of this process. The problem here is to determine an appropriate evaluative standpoint for ranking the importance of different disabilities which avoids these potential distortions (Brock, 1995, pp. 159-184).

Since the preferences for different states of disability or HRQL used to determine their relative values should be informed preferences, it is natural to think that the preferences of those who actually experience the disabilities should be used. Because they should have a more informed understanding of what it is actually like to live with the particular disability in question, we can hope to avoid uninformed evaluations. But this is to miss the deeper nature of the problem caused by adaptation, coping, and adjustment to disabilities.

Fundamental to understanding the difficulty posed by adaptation, coping, and adjustment to disabilities for preference evaluation of HRQL with various disabilities is that neither the nondisabled nor the disabled need have made any mistake in their different evaluations of quality of life with that disability. They arrive at different evaluations of the quality of life with that disability because they use different evaluative standpoints as a result of the disabled person's adaptation, coping, and adjustment. Disabled persons who have undergone this process can look back and see that before they became disabled they too would have evaluated the quality of life with that disability as nondisabled people now do. But this provides no basis for concluding that their pre-disability evaluation of the quality of life with that disability was mistaken, and so in turn no basis for discounting or discarding it because mistaken. The problem that I call the perspectives problem is that the nondisabled and the disabled evaluate the quality of life with the disability from two different evaluative perspectives, neither of which is mistaken. It might seem tempting to use the non-disabled's preferences for assessing the importance of prevention or rehabilitation programs, and the disabled's preferences for assessing the importance of life-sustaining treatments for the disabled, but this ignores the necessity of a single unified perspective in order to compare the relative benefits from, and prioritize, the full range of different health interventions.

Moreover, what weight to give to the results of coping with one's condition may depend on the causes of that condition, for example disease or injury that are no one's fault as opposed to unjust social conditions. Most measures of HRQL include some measure of subjective satisfaction or distress, a factor that is importantly influenced by people's expectations. In a society which has long practiced systematic discrimination against women, for example, women may not be dissatisfied with their unjustly disadvantaged state, including the health differences that result from that discrimination. The fact that victims are sufficiently oppressed that they accept an injustice as natural and cope with it by reducing their expectations and adjusting their life plans should not make its effects less serious, as measures of HRQL with a subjective satisfaction or distress component would imply.

When measures like the HUI or QWB are applied across different economic, ethnic, cultural, and social groups, the meaningful states of health and disability and their importance in different groups may vary greatly; for example, in a setting in which most work is manual labor, limitations in physical functioning will have greater importance than it does in a setting in which most individuals are engaged in non-physical, knowledge-based occupations, where certain cognitive disabilities are of greater importance. Different evaluations of health conditions and disabilities seem to be necessary for groups with significantly different relative needs for different functional abilities, but then cross-group comparisons of health and disability, and of the relative value of health interventions, in those different groups will not be possible. The health program benefits will have been measured on two different and apparently incommensurable valuational scales. These differences will be magnified when summary measures of population health are employed for international comparisons across very disparate countries.

Some of this variability of perspective may be avoided by a focus on the evaluation of disability instead of handicap, as these are traditionally distinguished, such as in the 1980 International Classification of Impairments, Disabilities and Handicaps (ICIDH). The ICIDH understands disabilities as "any restriction or lack (resulting from an impairment) of ability to perform an activity in the manner or within the range considered normal for a human being," whereas handicap is "a disadvantage for a given individual, resulting from an impairment or disability, that limits or prevents the fulfillment of a role that is normal (depending on age, sex, and social and cultural factors) for that individual." There will be greater variability between individuals, groups, and cultures in the relative importance of handicaps than of disabilities since handicaps take account of differences in individuals' roles and social conditions that disabilities do not. But it is problematic whether these differences should be ignored in prioritizing health resources for individuals, groups, and societies, that is, whether disabilities or handicaps are the correct focus for evaluation.

II. SECOND ISSUE: DO ALL QALYs COUNT EQUALLY?

QALYs standardly assume that an additional year of life has the same value regardless of the age of the person who receives it, assuming that the different life years are of comparable quality. A year of life extension for an infant, a forty-year-old, and an eighty-year-old all have the same value in QALYs produced, and in turn in a cost effectiveness analysis using QALYs, assuming no difference in the quality of the year of life extension. This is compatible, of course, with using age-based quality adjustments for interventions affecting groups of different age patients to reflect differences in the average quality of life of those different groups; for example, if average quality of life in a group of patients of average age 85 is less than that of patients of average age 25, a year of life extension for the 25 year old would have greater value in QALYs than would a year of life extension for the 85 year old.

In the World Bank Study, *World Development Report 1993; Investing in Health* (World Bank, 1993), the alternative DALY measure was developed to measure the burden of disease in reducing life expectancy and quality of life. Probably the most important ethical difference between QALYs and DALYs is that DALYs assign different value to a year of life extension of the same quality, depending on the age at which an individual receives it; specifically, life extension for individuals during their adult productive work years is assigned greater value than a similar period of life extension for infants and young children or the elderly. The principal justification offered for this feature of DALYs was the different social roles that individuals typically occupy at different ages and the typical emotional, physical, and financial dependence of the very young and the elderly on individuals in their productive work years (Murray, 1994).

This justification of age-based differences in the value of life extension implicitly adopts an ethically problematic social perspective on the value of health care interventions that extend life, or maintain or restore function, that is, an evaluation of the benefits *to others* of extending an individual's life, or maintaining or restoring his or her function, in addition to the benefit to that individual of doing so. This social perspective is in conflict with the usual focus in clinical decision making and treatment only on the benefits to the individuals who receive the health care interventions in question. Typical practice in health policy and public health contexts is more ambiguous on this point, since there benefits to others besides the direct recipient of the intervention are sometimes given substantial weight in the evaluation and justification of health programs; for example, treatment programs for substance abuse are argued to merit high priority because of their benefits in reductions in lost work days and in harmful effects on the substance abusers' family members. This social perspective is ethically problematic because it gives weight to differences between individuals in their social and economic value to others; in so doing, it discriminates against persons with fewer dependencies and social ties, which arguably is not ethically relevant in health care resource allocation. The social perspective justifying the DALY measure is therefore ethically problematic, in a way the alternative QALY measure is not, if the value of health benefits for individuals should focus on the value to the individuals treated of the health benefits, not on the social value for others of treating those persons. The ethical difficulty here is briefly explored further in the section below on what costs and benefits should count in a CEA.

Giving different value to life extension at different ages, however, might be justified ethically if done for different reasons. For example, Norman Daniels has argued that because everyone can expect to pass through the different stages of the life span, giving different value to a year of life extension at different stages in the life span need not unjustly discriminate against individuals in the way giving different weight to life extension for members of different racial, ethnic, or gender groups would unjustly discriminate (Daniels, 1988). Each individual can expect to pass through all the life stages in which life extension is given different value, but is a member of only one race, ethnic group, and gender. Thus, all persons are treated the same at comparable stages of their lives regarding the value of extending their lives, and so the use of DALYs would not constitute unjust age discrimination comparable to gender, ethnic or racial discrimination.

Moreover, individuals, and in turn their society, might choose to give lesser weight to a year of life extension beyond the normal life span than to a year of life extension before one has reached the normal life span based on a conception of what equality of opportunity requires, or on what Alan Williams calls the "fair innings argument" (Williams, 1997, pp. 117-132). People's plans of life and central long term projects will typically be constructed to fit within the normal life span, and so the completion of these central projects will typically require reaching, but not living beyond, the normal life span (Daniels, 1988; Brock 1989).

III. THIRD ISSUE: WHAT COSTS AND BENEFITS SHOULD COUNT IN COST EFFECTIVENESS ANALYSIS OF HEALTH PROGRAMS?

It is widely agreed that cost effectiveness analyses in health should reflect the direct health benefits for individuals of their medical treatment, such as improving renal function or reducing joint swelling, and of public health programs, such as reducing the incidence of infectious diseases through vaccination programs. The direct costs of medical treatment and public health programs, such as the costs of health care professionals' time and of medical equipment and supplies, should also be reflected. But medical and public health interventions typically also have indirect non-health benefits and costs. For example, some disease and illness principally affects adults during their working years, thereby incurring significant economic costs in lost work days associated with the disease or illness, whereas other disease and illness principally affects either young children, such as some infectious diseases, or the elderly, such as Alzheimer's dementia, who in each case are not typically employed and so do not incur lost wages or lost work time from illness. Should an indirect economic burden of disease of this sort be given weight in a cost effectiveness analysis used to prioritize between different health interventions?

From an economic perspective, as well as from a broad utilitarian moral perspective, indirect non-health benefits and costs are real benefits and costs of disease and of efforts to treat or prevent it, even if not direct health benefits and direct treatment costs; they should be reflected in the overall cost effectiveness accounting of how to use scarce health resources so as to produce the maximum aggregate benefit. A possible moral argument for ignoring these indirect non-health costs and benefits in health resource prioritization is grounded in a conception of the moral equality of persons. Giving priority to the treatment of one group of patients over another because treating the first group would produce indirect non-health benefits for others (for example, other family members who were dependent on these patients) or would reduce indirect economic costs to others (for example, the employers of these patients who incur less lost work time) could be argued to fail to treat each group of patients with the equal moral concern and respect that all people deserve; in particular, doing so would fail to give equal moral concern and weight to each person's health care needs. Instead, giving lower priority to the second group of patients simply because they are not a means to the indirect non-health benefits or cost savings produced by treating the first group of patients gives the second group of patients and their health care needs lower priority simply because they are not a means to these indirect non-health benefits or cost savings to others. It would violate the Kantian moral injunction against treating people solely as means for the benefit of others.

In public policy we often use a notion of "separate spheres," which in this case could be used to argue that the purpose of health care and of public health is health and the reduction of disease, and so only these goals and effects should guide health care and public health programs (Kamm, 1993; Walzer, 1983). There are obvious practical grounds for the separate spheres view associated with the difficulty of fully determining and calculating indirect benefits and costs. But the Kantian moral argument could serve as a principled moral basis for ignoring indirect benefits and costs in a cost effectiveness analysis to be used to prioritize health resources and interventions that serve different individuals or groups.

IV. FOURTH ISSUE: SHOULD DISCOUNT RATES BE APPLIED TO HEALTH CARE BENEFITS?

It is both standard and recommended practice in cost effectiveness analyses, within health care and elsewhere, to assume a time preference by applying a discount rate to both the benefits and costs of different programs under evaluation, although the reasons for doing so and the proper rate of discount are controversial (Gold, 1996, Ch.7). It is important to separate clearly the ethical issue about whether health benefits should be discounted from other economic considerations for discounting, as well as to be clear why the issue is important for health policy. It is not ethically controversial that a discount rate should be applied to economic costs and economic benefits; a dollar received today is worth more than a dollar received 10 years from now because we have its use for those ten years, and there is a similar economic advantage in delaying the incurring of economic costs. The ethical issue is whether a discount rate should be applied directly to changes in life extension and well-being or health. Is an improvement in well-being, such as a specific period of life extension, a reduction in suffering, or an improvement in function, extending, say, for one year of substantially less value if it occurs twenty years from now than if it occurs next year?

Future benefits are appropriately discounted when they are more uncertain than proximate benefits. Proximate benefits, such as restoration of an individual's function, also are of more value than distant benefits if they make possible a longer period of, and thus larger, benefit by occurring sooner. But neither of these considerations require the use of a discount rate – they will be taken account of in the measurement of expected benefits of alternative interventions. The ethical question is whether an improvement in an individual's well-being is of lesser value if it occurs in the distant future than if it occurs in the immediate future, simply and only because it occurs later in time. This is a controversial issue in the literature on social discounting and my own view is that no adequate ethical justification has been offered for applying a discount rate directly to changes in health and well-being, though I cannot pursue the justifications offered by proponents of discounting here. The avoidance of paradoxes that arise when no discount rate is applied or when different discount rates are applied to costs and benefits, has influenced many economists to support use of the same discount rate for costs and benefits (Keeler and Cretin, 1983, pp. 300-306), but I believe these are properly dealt with not through discounting, but rather through directly addressing the ethical issues they raise, usually about equity between different generations.

The policy importance of this issue is relatively straightforward in the prioritization of health care interventions. Many health care and public health programs take significantly different lengths of time to produce their benefits. Applying a discount rate to those benefits leads to an unwarranted priority to programs producing benefits more rapidly. It results in a program that produces benefits in health and well-being say twenty years into the future being given lower priority than an alternative health care program that produces substantially less overall improvement in health and well-being, but produces that improvement much sooner. Many public health and preventive interventions, for example, vaccination programs and changes in unhealthy behavior, reap their health benefits years into the future. If those benefits are unjustifiably discounted, they will be given lower priority than alternative programs that produce fewer aggregate benefits. The result is a health policy that produces fewer overall health benefits over time than could have been produced with the same resources.

V. FIFTH ISSUE: WHAT LIFE EXPECTANCIES SHOULD BE USED FOR CALCULATING THE BENEFITS OF LIFE SAVING INTERVENTIONS?

In calculating QALYs it is standard practice to take account of differences in the average ages and in turn life expectancies of patients served by different health care programs; for example, a treatment for a life-threatening childhood disease would produce more QALYs than a comparable treatment for a life-threatening disease affecting primarily the elderly. Similarly, accurate estimates of the expected QALYs from different interventions would adjust for differences in the average life expectancies of patients caused by diseases other than those treated by the interventions; for example, an intervention that improved the quality of life of patients with cystic fibrosis, who have a much lower than average life expectancy as a result of their disease, would produce fewer QALYs than an intervention with a comparable improvement in lifetime quality of life for patients with average life expectancies undiminished by disease. This latter case raises difficult issues about discrimination against people with disabilities that

I take up later. But there are other differences in the life expectancies of different groups that an accurate estimate of QALYs produced by health interventions serving those groups would seemingly have to reflect; for example, there are significant differences in the life expectancies between different genders, racial and ethnic groups, and socio-economic groups within most countries. Internationally, the differences in life expectancies between different countries are often much larger. Should these differences affect calculations of the QALYs gained by health care and public health interventions that extend life or improve quality of life? An accurate estimate of the additional life years actually produced by those interventions should not ignore differences in life expectancies that the health care interventions will not affect, but the result will be that it is less valuable to save the life of a poor person in an underdeveloped country than a rich person in a developed country.

The differences in life expectancies between different racial, ethnic, and socio-economic groups within a single country, as well as the very large differences between life expectancies in economically developed and poor countries, are often principally the result of unjust conditions and deprivations suffered by those with lower life expectancies. It would seem only to compound those injustices to give less value to interventions that save lives or improve quality of life for groups with lower life expectancies caused by the unjust conditions and deprivations from which they suffer. Differences in life expectancies between the genders, on the other hand, are believed to rest in significant part on biological differences, not on unjust social conditions. Whether the biologically based component of gender differences in life expectancies should be reflected in measures like QALYs or DALYs is more controversial. For example, on the one hand, the lower life expectancy of men does not result from any independent injustice, but, on the other hand, it is explicit public policy and required by law in the United States to ignore this gender-based difference in most calculations of pension benefits and annuity costs so as to avoid gender discrimination. The developers of the DALY explicitly chose to use a single uniform measure of life expectancy (except for the biological component of the gender difference), specifically that observed in Japan which has the highest national life expectancy, to measure gains from life saving interventions. They justified their choice in explicitly ethical terms as conforming to a principle of "treating like events as like," although the reasoning was not pursued in any detail (Murray, 1994, p. 7). How this issue is treated can have a substantial impact on the priorities that result from the cost effectiveness analysis, especially at the international level where country differences tend often to be greater than group differences within specific countries.

Each of the preceding five ethical issues can be considered issues in the *construction* of a cost-effectiveness analysis in health care. The other issues I want to briefly note can be considered issues in the *use* of cost effectiveness analysis in health resource prioritization. They are each issues of distributive justice or equity raised by the fact that a cost effectiveness analysis is insensitive to the distribution of health benefits and the costs of producing them. Yet people's beliefs about equity and justice directly affect the relative priority they assign to different health interventions. One standard response to this point is that a CEA can only be an aid to policy making in general, and health resource prioritization in particular, and that policy makers must take account of considerations of equity in final policy decisions and choices. But as with the ethical issues in the construction of CEAs, much work remains to be done to clarify and assess alternative positions on these issues of equity so the policy choices on them can at least be better informed, even if they remain controversial. Here, there is only space to state four of the main equity issues in the use of CEAs and some of the principal ethical considerations supporting different positions on them (Daniels, 1993, pp. 224-233). After doing that, I shall mention an alternative quantitative methodology that, unlike CEA, incorporates considerations of equity within the quantitative analysis.

VI. SIXTH ISSUE: WHAT PRIORITY SHOULD BE GIVEN TO THE SICKEST OR WORST OFF?

It is a commonplace that most theories of distributive justice require some special concern for those who are worst off or most disadvantaged; for example, it is often said that the justice of a society can be measured by how it treats its least well off members. In the context of health care allocation and the prioritization of health interventions, the worst off with regard to need for the good being distributed might reasonably be thought to be the sickest patients. In many cases, the sickest will be given priority by a CEA comparing treating them as opposed to less sick patients; the sickest have greater possible improvements in HRQL because they begin from a lower HRQL, and so, for example, in comparing fully effective treatments those for the sickest will produce the greater benefits. But in other cases giving priority to the sickest will require a sacrifice in aggregate health benefits. An abstract example makes the point most concisely. Suppose Group A patients have a very serious disease that leaves them with a health utility level of .25 as measured by the HUI, and this would be raised only to .45 with the best available treatment because no treatment is very effective for their disease; for example, patients with severe chronic obstructive pulmonary disease or with severe chronic schizophrenia that is largely resistant to standard pharmacological treatments. A similar number of Group B patients have a health utility level of .60 because they have a considerably less serious disease, but since treatment for their disease is more effective, although no more costly, it would raise their health utility level to .90; for example, patients with asthma, or with milder forms of pulmonary disease or schizophrenia that both leave them less disabled without treatment and are more responsive to treatment. Should we give priority to treating Group B because doing so would produce a 50% greater aggregate health benefit at the same cost, as the CEA standard implies, or to treating Group A who are the sickest? In some empirical studies, both ordinary people and health professionals prefer to sacrifice some aggregate health benefits in order to treat the sickest patients, although the degree of sacrifice they are prepared to make is variable and not statistically reliable (Nord, 1993, pp. 227-238).

One difficulty raised by this issue is determining what weight to give to this particular aspect of equity – concern for the worst off. Virtually no one would prefer to treat the sickest, no matter how costly their treatment and how small the benefit to them of doing so, and no matter how beneficial and inexpensive treatment for the less sick might be. However, there seems no objective, principled basis for determining how much priority to give the sickest, that is, how much aggregate health benefits should be sacrificed in order to treat or give priority to the sickest. Instead, the most one can say is that most people and many theories of distributive justice have a concern both for maximizing overall benefits with scarce health resources and for helping the worst off or sickest, but there is a large range of indeterminacy regarding the proper trade off between these two concerns when they are in conflict.

One issue in understanding this concern for the worst off important for health care priorities is whether it should focus on who is worst off at a point in time or instead over an extended period of time, such as a lifetime. When choosing between patients to receive a scarce resource, such as in organ transplantation, it is often plausible to focus on lifetime well being, since otherwise we may give priority to the patient who is worst off at the time the distributive choice is made, but whose lifetime level of well being is far higher than the other patient. Frances Kamm has defended a notion of need in this context according to which the neediest patient is the patient whose life will have gone worst if he or she does not get the scarce resource, such as an organ transplant (Kamm, 1993, ch. 8). However, some justifications for giving priority to the worst off may support focusing on the sickest here and now.

What are the ethical justifications for giving priority to the worst off? I can mention only two possibilities here. One is that we must give priority to the worst off in order to avoid increasing the already unjustified disadvantage or inequality they suffer relative to those better off. But it is worth noting that a concern for the worst off is not always the same as a concern to produce equality in outcomes. In the example above of Groups A and B, equality could be achieved by what Derek Parfit has called "leveling down," that is by bringing B's health utility level down to that of A's instead raising A's level up to that of B (Parfit, 1991). If equity here is equivalent to equality in outcomes, then if it were not possible to raise A's level above .40 with treatment, equity would seem to support not treating Group B and letting their condition deteriorate until it reached the lower level of Group A. The fact that no one would defend doing this suggests that this aspect of our notion of equity or justice is best captured by the idea of giving priority to improving the condition of the worst off, rather than by a simple concern for equality in outcomes. A different justification for giving priority to treating the sickest, offered by some participants in Nord's research, is that it would be subjectively more important to the sickest to obtain treatment, even if the health benefits they receive from treatment are less than those that would go to the less sick; this justification might support focusing on who is worst off at the point in time at which the decision about who to treat is made, not whose lifetime well-being will be lowest (Nord, 1993, 227-238).

One further issue concerning the priority to the worst off should be mentioned. In the context of health resource prioritization in health policy it seems natural to understand the worst off as the sickest. But this may not always be correct. At the most fundamental ethical level in our general theories of equity and distributive justice, our concern should be for those who are overall or all things considered worst off, and they will not always be the sickest. It could be argued that giving priority to the worst off in health resource prioritization sometimes requires giving priority to those with the lowest levels of overall well-being, even at some cost to aggregate health benefits produced *and* at the cost of not treating sicker persons whose overall well-being is much higher. A preference for health interventions that raise the level of well-being of those who are worst off in overall well-being, instead of giving priority to the sickest, might be justified in order not to increase the unjustified disadvantage suffered by those with the lowest overall level of well-being. If, instead, the priority to the worst off in health resource prioritization should focus only on health states and so on the sickest, a justification of this narrowed focus is needed.

VII. SEVENTH ISSUE: WHEN SHOULD SMALL BENEFITS TO A LARGE NUMBER OF PERSONS RECEIVE PRIORITY OVER LARGE BENEFITS TO A SMALL NUMBER OF PERSONS?

Cost effectiveness and utilitarian standards require minimizing the aggregate burden of disease and maximizing the aggregate health of a population without regard to the resulting distribution of disease and health, or *who* gets what benefits. The issue about priority to the worst off focuses on who gets the benefits. A different issue concerns *what* benefits different individuals get. Some would argue that health benefits are often qualitatively different and so cannot all be compared on a single scale like the HUI, or in turn by a single measure like QALYs, but that is not the issue of concern now. In its most general form the issue about aggregation concerns what ethical limits there are, if any, on aggregating together different size benefits for different persons in comparing and prioritizing different health interventions; CEA accepts no such limits. There are many forms in which this issue can arise which cannot be pursued here (Kamm, 1993, Part II), but the version that has received the most attention, and which Daniels has called the aggregation problem, is when, if ever, large benefits to a few individuals should take priority over greater aggregate benefits to a different and much larger group of individuals, each one of whom receives only a small benefit. This issue arises when a very serious disease or condition for those affected that is also very costly to prevent or treat is compared with a much more prevalent disease or condition that both has a very small impact on each individual affected and is very inexpensive to treat or prevent in any one individual. Applying cost effectiveness or utilitarian standards, preventing or treating the very prevalent but low impact disease or condition at a given cost will receive higher priority when doing so produces greater aggregate benefits than using the same funds to treat or prevent the disease or condition that has a very great impact on each individual affected. The example that received considerable attention in the United States arose in the Oregon Medicaid priority setting process where capping teeth for exposed pulp was ranked just above an appendectomy for acute appendicitis, a potentially life-threatening condition. Because an appendectomy is approximately 150 times as expensive as capping a tooth for exposed pulp, the aggregate benefit of capping a tooth for 150 patients was judged to be greater than the benefit of an appendectomy for one patient. Since Medicaid coverage decisions were to be made according to the list of treatment/condition pairs ranked in terms of their relative cost effectiveness, it could have turned out, depending on the overall level of resources available to the Medicaid program, that tooth capping would have been covered but appendectomies not covered.

This result, and other less extreme cases like it, was highly counter-intuitive and unacceptable to most people, whose intuitive rankings of the relative importance or priority of health interventions are based on one-to-one comparisons, for example of one tooth capped as opposed to one appendectomy performed. In the face of these results Oregon made a fundamental change in its prioritization methodology, abandoning the cost effectiveness standard in favor of a standard that did not take account of differences in costs. This was not a minor problem requiring tinkering at the margins of the CEA standard, but a fundamental challenge to it and so required a fundamental revision in it.

Yet it is by no means clear that no such aggregation can be ethically justified. The very case that precipitated Oregon's Medicaid revision was a 12 year old boy in need of a bone marrow transplant as the only effective chance to save his life. Oregon denied coverage under its Medicaid program on the grounds that it could do greater good by using its limited resources to improve prenatal care for pregnant women, in this case giving higher priority to small benefits to many over a potentially much larger benefit to a few. Moreover, many public policy choices appear to give higher priority to small benefits to many over even life saving benefits to a few; for example, governments in the United States support public parks used by tens or hundreds of thousands of persons, while reducing funding for public hospitals resulting in quite predictable loss of life.

The cost effectiveness or utilitarian standard that permits unlimited aggregation of benefits might be defended by distinguishing between the clinical context in which physicians treat individual patients and the public health and health policy context in which health resource allocation decisions are made that will affect different groups in the population. In the clinical context, physicians forced to prioritize between individual patients typically will first treat the patient who will suffer the more serious consequences without treatment, or who will benefit the most from treatment, even if doing so will prevent her treating a larger number of less seriously ill patients. But from a public health or health policy perspective, it could be argued that the potential overall or aggregate effects of alternative interventions on population health is the appropriate perspective. However, the Oregon experience makes clear that even when allocating public resources for interventions to improve the health of a population, it is ethically controversial whether always giving priority to producing the maximum aggregate benefits, even when that is done by giving small benefits to many at the cost of forgoing large benefits to a few, is justified.

Just as with the problem of what priority to give to the worst off, part of the complexity of the aggregation problem is that for most people some, but not all, cases of aggregation are ethically acceptable and equitable. The theoretical problem then is to develop a principled account of when, and for what reasons, different forms of aggregation satisfy requirements of equity and when they do not (Kamm, 1993). There is no consensus on this issue either among ordinary persons or within the literature of health policy or ethics and political philosophy. As with the problem about priority to the worst off, the complexities of this issue have received relatively little attention in bioethics and moral and political philosophy, and there is much difficult but important work to be done.

VIII. EIGHTH ISSUE: THE CONFLICT BETWEEN FAIR CHANCES AND BEST OUTCOMES.

The third ethical issue in the use of CEA for health resource utilization that I will mention here has been characterized as the conflict between fair chances and best outcomes (Daniels, 1993, pp. 224-233). The conflict is most pressing when the health intervention is life saving and not all those whose lives are threatened can be saved, but it arises as well when threats are only to individuals' health and well-being. In the context of health care, this issue first received attention in organ transplantation where there is a scarcity of life saving organs such as hearts and lungs resulting in thousands of deaths each year of patients on waiting lists for an organ for transplant; an abstract example from transplantation can illustrate the issue most clearly and succinctly (Brock, 1988).

Suppose two patients are each in need of a heart transplant to prevent imminent death, but there is only one heart available for transplant. Patient A has a life expectancy with a transplant of ten years and patient B has a life expectancy with a transplant of nine years (of course, precise estimates of this sort are not possible, but the point is that there is a small difference in the expected benefits to be gained depending on which patient gets the scarce organ), with no difference in their expected quality of life. Maximizing health benefits or QALYs, as a CEA standard requires, favors giving the organ to patient A, but patient B might argue that it is unfair to give her no chance to receive the scarce heart. Just as much as A, she needs the heart transplant for life itself and will lose everything, that is her life, if she does not receive it. It is unfair, B might argue, to give the organ to A because the quite small increment in expected benefits from doing so is too small to justly determine who lives and who dies. Instead, she argues, each of them should receive a fair chance of getting the organ and having their health needs met; in this case, that might be done by giving each an equal chance of receiving the transplant through some form of random selection between them, or by a weighted lottery that gives the patient who would benefit more some greater likelihood of being selected to receive the organ, but still gives the patient who would benefit less some significant chance of getting it instead (Broome, 1984, pp. 38-55; Kamm, 1993, Part III; Brock, 1988).

Most prioritization and rationing choices arise not from physical scarcity of the needed health resource, as in organ transplantation, but from economic scarcity, limits in the money society devotes to health care. Will this issue of equity arise in health resource prioritization and allocation choices forced by economic scarcity? Two considerations will often mitigate the force of the ethical conflict between fair chances and best outcomes there. First, allocation of resources in health care is typically not an all or nothing choice, as in the case of selecting recipients for scarce organs, but is usually a matter of the relative priority for funding to be given to different health programs or interventions. That one health program A promises a small gain in aggregate health benefits over a competing program B need not entail that A is fully funded and B receives no funding, but only that A should receive higher priority for, or a higher level of, funding than B. Persons with the disease or condition that A treats will have a somewhat higher probability of being successfully treated than will those who have the disease or condition that B treats; in the case of prevention, those at risk of A will have a somewhat higher probability of successful prevention than will those at risk of B. When there is significant resource scarcity this will involve some sacrifice in aggregate health benefits that might have been produced by always preferring the more cost effective alternative. But doing so means that individuals who are served by B have no complaint that the small difference in expected benefits between programs A and B unfairly prevents them from having their health needs met at all. Instead, the small difference in expected benefits between programs A and B need only result in a comparably small difference in the resources devoted to A and B; it is not obvious that this is unfair to those patients served by B, whose needs are somewhat less well served than patients in program A because of B's lower priority and level of funding.

The second consideration that may mitigate some of the conflict between fair chances and best outcomes in health resource prioritization forced by economic scarcity is that often, probably usually, the diseases and health problems to be treated or prevented are not directly life threatening, but instead only impact on individuals' quality of life, and often for only a limited period of time. In these cases, the difference in health benefits between individuals who receive a needed health intervention that is given a higher priority and individuals who do not receive a needed health intervention because their condition is given lower priority, is much less, making the unfairness arguably less compelling.

These two considerations may mitigate, but they do not fully avoid, the conflict between fair chances and best outcomes in prioritization decisions about health interventions forced by economic scarcity. When a more cost effective health program is developed for one population instead of a different less cost effective health program for a different population, individuals who would have been served by the second program will have a complaint that they did not have a fair chance to have their needs served only because of a small gain in the benefits that are produced by the first program. The fair chances versus best outcome conflict will arise in prioritizing health interventions in health policy; how this conflict can be equitably resolved is complex, controversial, and unclear.

IX. NINTH ISSUE: DOES USE OF CEA TO SET HEALTH CARE PRIORITIES UNJUSTLY DISCRIMINATE AGAINST THE DISABLED?

In several contexts using CEA to set health care priorities will result in assigning lower priority to both life extending and quality of life improving treatment for disabled than nondisabled persons with the same health care needs (Brock, 1995, pp. 159-184; 2000, pp 223-235). Here are five such contexts. First, since already disabled persons have a lower HRQL from their disability than nondisabled persons, treatment that extends their life for a given number of years produces fewer QALYs than the same treatment that extends the life of a nondisabled person for the same number of years. Second, if two groups of patients with the same HRQL have the same need for a life sustaining or quality of life improving treatment, but one will be restored to normal function and the other will be left with a resultant disability, more QALYs will be produced by treating the first group. Third, persons with disabilities often have a lower life expectancy because of their disability than otherwise similar nondisabled persons. As a result, treatments that prevent loss of life or produce lifetime improvements in quality of life will produce fewer QALYs when given to disabled than to nondisabled persons with the same health care needs. Fourth, disabilities often act as comorbid conditions making a treatment less beneficial in QALYs produced for disabled than for nondisabled persons with the same health care needs. Fifth, the presence of a disability can make treatment of disabled persons more difficult and so more costly than for nondisabled persons with the same health care needs; the result is a lower cost effectiveness ratio for treating the disabled persons.

In each of the five cases above, disabled persons have the same medical and health care need as nondisabled persons, and so the same claim to treatment on the basis of their needs. But treating the disabled person will produce less benefit, that is fewer QALYs, *because of their disability* than treating the nondisabled. Thus, their disability is the reason for their receiving lower priority for treatment. This at least arguably fails to give equal moral concern to disabled persons' health care needs and is unjust discrimination against them on grounds of their disability. Indeed, United States Health and Human Services Secretary Louis Sullivan denied Oregon's initial request for a waiver of federal regulations for its proposed revisions to its Medicaid plan on the grounds that Oregon's method of prioritization of services was in violation of the Americans with Disabilities Act (ADA).[2] Sullivan cited some of the five kinds of cases I noted above in support of that position, and Oregon in turn made essentially ad hoc revisions in its ranking to avoid the putative violation of the ADA.

Disabled persons charge that in cases like the first I cited above concerning life saving treatment, the implication of use of CEA to prioritize health care is that saving their lives, and so their lives themselves, have less value than nondisabled persons' lives. They quite plausibly find that implication of CEA threatening and unjust. There are means of avoiding these problems about discrimination against persons with disabilities, but they involve abandoning fundamental features of CEAs. For example, one response to the first case cited above would be to give equal value to a year of life extension, whatever the quality of that life, so long as it is acceptable to the person whose life it is (Kamm, 1993, Part I). But that has problematic implications too since, for example, a small percentage of persons in surveys say they would want their lives sustained even if they were in a persistent vegetative state. I cannot pursue the issues further here, but I believe the problem of whether CEA unjustly discriminates against the disabled is a deep and unresolved difficulty for use of CEA and QALYs to prioritize health care.

The sixth, seventh, and eighth issues above all raise possible criticisms of the maximization standard embodied in CEA; in each case, the claim is that equity requires attention to the distribution of health benefits and costs to distinct individuals. Steadfast utilitarians or consequentialists will reject the criticisms and hold fast to the maximization standard. But most people will accept some departure from the maximization standard of CEA; there are two broad strategies for how to do so. The first and probably most common is to propose CEA as an aid to policy makers who must make prioritization and allocation choices in health care, but then to remind those policy makers that they must take account of these considerations of equity as well in their decision making; this may be, but usually is not, accompanied by some guidance about alternative substantive positions, and reasons in support of them, on the equity issues. Moreover, some use of CEA in health policy and health program evaluation does not raise these last three issues of equity; for example, CEA of alternative treatments that each have uniform but different benefits for a group of patients with a particular medical condition. And outside of a CEA, either QALYs or DALYs can be used for evaluating alternative interventions, or for monitoring changes over time in health status or the burdens of disease, in a given group or population.

The second strategy for responding to concerns about equity seeks to develop a quantitative tool that measures the specific weight people give to different equity concerns in comparing interventions that raise issues of distributive justice because they serve different individuals or benefit individuals differently. The most prominent and promising example is the "person trade-off" approach which explicitly asks people how many outcomes of one kind they consider equivalent in social value to X outcomes of another kind, where the outcomes are for different groups of individuals (Nord, 1999). For example, people can be asked, as in our earlier example, to compare treatment A for very severely ill patients who are at .25 on the HUI without treatment and who can be raised only to .45 with treatment, with treatment B of less severely ill patients who are at .60 and can be raised to .90 with treatment; filled out detailed examples, of course, will make the comparisons more understandable. Respondents are then asked how many patients treated with A would be equivalent in social value to treating 100 patients with B. Answers to questions of this form will tell us in quantitative terms how much importance people give to treating the sickest when doing so conflicts with maximizing aggregate health benefits.

The person trade-off approach is designed to permit people to incorporate concerns for equity or distributive justice into their judgments about the social value of alternative health programs. There has been relatively little exploration and use of this methodology in health care evaluation in comparison with the mass of methodological work on and studies of aggregate QALYs and CEAs, in part because many health policy analysts and health economists assume, often with little or no argument, that the social value of health programs is the sum of the individual utilities produced by the program. As I noted in the introduction to the paper, the early stage we are now at in the development and use of the person trade-off approach is a reason for caution at the present time about using it to settle issues of equity in health resource prioritization. While the utilitarian assumption in CEA is rejected in most philosophical work on distributive justice, as well as in the preferences most ordinary people express for different health outcomes and programs, I also noted in the introduction a second more important reason for caution about bringing considerations of equity into health policy decision making through a quantitative methodology like the person trade-off methodology – the issues of distributive justice that must be addressed by equitable health resource prioritization represent deep and long-standing divisions in moral and political philosophy about which there is not now, and may never be, anything approaching consensus. There is a strong case to be made, though I cannot pursue it here, that important value conflicts about justice of this sort should be addressed in public, democratic political processes, or in fair, participatory and accountable procedures within private institutions like managed care organizations (Daniels and Sabin, 1997, pp. 303-350). The person trade-off method can be a useful aid to those deliberative decision making processes in providing more structure and precision to different people's views about equity in health care resource prioritization and trade-offs, but it is not a substitute for that deliberation. Despite these briefly noted reservations, I do emphasize that for purposes of resource prioritization and allocation, the person trade-off approach is the proper perspective, in comparison with CEA, because it correctly reflects that the choices are typically about how health benefits and costs are distributed to different individuals.

X. CONCLUSION

I have distinguished above nine distinct issues about equity and justice that arise in the construction and use of cost effectiveness analysis to minimize the burdens of disease and to maximize health outcomes. In each case the concern for equity is in my view valid and warrants some constraints on a goal of unqualified maximization of health outcomes. There has not been space here to pursue at all fully any of these nine issues regarding equity and justice – each is complex, controversial, and important. In each case, my point has been that there are important ethical and value choices to be made in constructing and using the measures; the choices are not merely technical, empirical, or economic, but moral and value choices as well. Each requires explicit attention by health policy makers using CEA. In a few cases I have indicated my own view about how the potential conflict between equity and utilitarian maximization might be resolved, but in other cases I have simply summarized briefly some arguments for giving the particular concern about equity some weight when it conflicts with maximization of utility. For some of these issues, the literature and research is at a relatively early stage and one cannot be confident about how the issues should be resolved or even about the range of plausible positions and supporting reasons on them. However, this is not grounds for ignoring the issues, but instead for getting to work on them and for ensuring that they receive explicit attention and deliberation in decisions about health resource prioritization and allocation.

Department of Philosophy
Brown University
Providence, RI 02912, U.S.A.

NOTES

1. This paper draws heavily on my "Considerations of Equity in Relation to Prioritization and Allocation of Health Care Resources," in *Ethics, Equity and Health for All*, eds Z. Bankowski, J.H. Bryant and J. Gallagher (Geneva: CIOMS, 1997) and "Ethical Issues in the Development of Summary Measures of Population Health States" in *Summarizing Population Health: Directions for the Development and Application of Population Metrics* (Washington DC: National Academy Press, 1998).
2. Interventions that would improve health should be understood broadly, and in particular extend substantially beyond health care. It is widely agreed that other factors such as improved sanitation and economic conditions have contributed more to the health gains of the past century than has health care. However, in this paper I shall largely confine myself to health care interventions
3.Unpublished letter from Secretary of Health and Human Services, Louis Sullivan, to Oregon Governor Barbara Roberts, August 3, 1992.

REFERENCES

Bergner, M., Bobbitt, R.A., Carter, W.B., and Gibson, B.S.: 1981, *'The Sickness Impact Profile Development and Final Revision of a Health,' Medical Care* 19, 787-805.

Brock, D.W.: 1988, 'Ethical issues in recipient selection for organ transplantation,' in *Organ Substitution Technology: Ethical, Legal, and Public Policy Issues* (ed.), D. Mathieu, Westview Press, Boulder and London,

Brock, D.W.: 1989, 'Justice, Health Care, and the Elderly,' *Philosophy & Public Affairs* 18(3), 297-312.

Brock, D.W.: 1992, 'Quality of Life Measures in Health Care and Medical Ethics,' in A. Sen and M. Nussbaum (eds.), The Quality of Life, Oxford University Press, Oxford.

Brock, D.W.: 1995, 'Justice and ADA: Does Prioritizing and Rationing Health Care Discriminate against the Disabled?,' *Social Theory and Policy* 12, 159-84.

Brock, D.W.: 2000, 'Health Care Resource Prioritization and Discrimination Against Persons With Disabilities,' in L. Francis and A. Silvers (eds.) *Americans with Disabilities,* Routledge, New York.

Brock, D.W.: 2002, 'Priority to the Worst Off in Health Care Resource Prioritization,' in M. Battin, R. Rhodes, and A. Silvers (eds.), *Medicine and Social Justice,* Oxford University Press, New York.

Broome, J.: 1984, 'Selecting People Randomly,' *Ethics* 95, 38-55.

Daniels, N.: 1988, *Am I My Parents' Keeper? An Essay on Justice Between the Young and the Old,* Oxford University Press, New York.

Daniels, N.: 1993, 'Rationing Fairly: Programmatic Considerations,' *Bioethics,* 7/2-3, 224-233.

Daniels, N. and Sabin, J.: 1997, 'Limits to Health Care: Fair Procedures, Democratic Deliberation, and the Legitimacy Problem for Insurers, *Philosophy and Public Affairs* 26(4), 303-50

Gold, M.R. et. al.: 1996, *Cost-Effectiveness in Health and Medicine*, Oxford University Press, New York.

International Classification of Impairments, Disabilities and Handicaps: 1980, World Health Organization, Geneva .

Kamm, F.M.: 1993, *Morality/Mortality. Volume One. Death and Whom to Save From It,* Oxford University Press, Oxford.

Kaplan, R.M. and Anderson, J.P: 1988, 'A General Health Policy Model: Update and Applications,' *Health Services Research,* June 23, 203-35.

Keeler, E.B. and Cretin, S.: 1983, 'Discounting of Life-Saving and Other Nonmonetary Effects,' *Management Science* 29, 300-306.

Murray, C.J.L.: 1994, 'Quantifying the Burden of Disease: the Technical Basis for Disability-Adjusted life years,' in *Global Comparative. Assessments in the Health Sector: Disease Burden, Expenditures and Intervention Packages,* eds. C.J.L. Murray and A.D. Lopez ,World Health Organization, Geneva.

Murray, C.J.M.:1996, 'Rethinking DALYs, in *The Global Burden of Disease: A Comprehension Assessment of Mortality and Disability From Disease, Injuries, and Risk Factors in 1990 and Projected to 2020*, World Health Organization, Geneva.

Nord, E.: 1999, *Cost-Value Analysis in Health Care: Making Sense of QALYs*, Cambridge University Press, New York.

Nord, E.: 1993 'The trade-off between severity of illness and treatment effect in cost-value analysis of health care,' *Health Policy* 24, 227-38.

Parfit, D.: 1991, 'Equality or priority,' The Lindley Lecture. Copyright: Department of Philosophy, University of Kansas.

Torrance, G.W. *et al.*: 1996 'Multi attribute preference functions for a comprehensive health status classification system.' *Medical Care* 34:7, 702-722.
Reprinted with permission of Lippincott Williams & Wilkins.

Walzer, M.: 1983, *Spheres of Justice,* Basic Books, New York .

Ware, J.E. and Sherbourne, D.C.: 1992, "The MOS 36-item short form health survey," *Medical Care* 30, 473-83.

Williams, A.: 1997, 'Intergenerational Equity: An Exploration of the "Fair Innings' Argument,' *Health Economics* 6(2), 117-32.

World Bank: 1993, *World Development Report 1993: Investing in Health,* Oxford University Press, Oxford.

CHRISTOPHER TOLLEFSEN

SIC ET NON: SOME DISPUTED QUESTIONS IN REPRODUCTIVE ETHICS

I. INTRODUCTION

Reproductive ethics is concerned with human reproduction insofar as its various aspects are under the control of human agents. Broadly, there are three such aspects: conception, gestation, and birth. In each of these three areas, agents have considerable control, as, for example, whether or not to conceive, whether or not to gestate, whether or not to give birth. And agents have further control over the manner of conceiving or not conceiving, gestating or not gestating, and giving birth.

Most of the issues of reproductive ethics may thus be seen as involving one or more of these stages of reproduction, and as attempting to settle normative questions about what can and cannot be done at these stages. Thus, questions of contraception surround the decision not to conceive; many questions concerning 'assisted reproduction' surround the decision to conceive. Finally, while the decision not to give birth is in most cases a decision to abort, still there are questions about the mode of birth, as, for example, when a cesarean section is recommended but turned down by a mother for religious reasons; and there are questions as to the appropriate birth environment, including the proper role of doctors and nurses, and family members.

This does not exhaust the many issues of reproductive ethics, however. For decisions regarding one aspect of reproduction often give rise to further possibilities which must be chosen or refused. If, for example, it is permissible to conceive children outside of sexual intercourse, whether by in vitro fertilization or by cloning, for the purpose of subsequent implantation and gestation, is it also permissible to conceive them for the purposes of medical research, or as a source of donor tissue? If a fetus is aborted, may the tissue remains be used for these purposes? If the fetus can be partially gestated outside of the biological mother's womb, is this permissible? What if the fetus can be completely brought to term outside of a womb? Such questions arise in response to new technological possibilities conjoined with prior decisions concerning the agent's role in reproduction.

G. Khushf (ed.), Handbook of Bioethics, 381–413.

Finally, although again without exhausting all possible issues, these various possibilities give rise to questions and options which, while not directly concerned with conception, gestation, and birth, are nonetheless intimately tied with such activities. On the side of reproduction, questions about the circumstances under which reproduction is and is not permissible give rise to questions about the relation between reproduction and sexual intercourse, and about the nature and role of marriage and the family. On the other end, questions concerning the relationship between expectant mothers and their families, nurses, and doctors, give rise to questions concerning the proper role of doctor and patient, 'medicalization' of normal conditions, and so on. And throughout, there are issues concerning the relationship between science and technology, on the one hand, and nature, morality, and politics, on the other.

Here, then, are a number of issues implicated by the terms 'reproductive ethics.' Their very multiplicity provokes methodological difficulties: what common concerns and issues are there in this multiplicity which can provide a key for a systematic treatment of some, or all, of these difficulties? In the absence of such a key, reproductive ethics would seem not even to be its own distinct field of research within bioethics. On the other hand, two problems are associated with the attempt to introduce large scale systematization into this area.

First, not all those working in the field agree that such systematization would be desirable. A case by case 'situational' approach (Fletcher, 1966), or a broader 'principlism' (Beauchamp and Childress, 1994), is sometimes advocated. On the former approach, no general principle or principles, no issue or distinct set of issues, is capable of doing justice to the variety of issues and concerns which characterize human involvement in an area of such importance. On the latter view, general principles play a role in the arbitration of dispute, but no one principle, no one problem, can be set down as that which is to guide all, or even most, considerations. Proponents of both positions view the attempt to introduce full systematicity into consideration of reproductive ethics with suspicion.

There are difficulties with such views. The extreme situational approach, and the form of principlism described above, run a risk of being no more than masks for subjective judgments and expressions of feelings, giving the appearance of objectivity where there is none (Geach, 1956; Clouser and Gert, 1990). It appears reasonable both to look for general principles of universal applicability, and to engage in the dialectical work necessary for 'specificatory premises', i.e., premises which indicate that a certain action type does, or does not fall under the scope of the principles (Donagan, 1977a). At the same time, contemporary objections to 'theoretical' ethics may be credited with bringing to light the need for attention to those aspects of ethics, including reproductive ethics, which may not be rigorously formulable, as, for example, concern with the virtues, and the role of non-rule guided judgment in particular situations (Clarke and Simpson, 1989; MacIntyre, 1980).

Even among those willing to introduce some unity into ethical consideration of reproduction, however, there is considerable dispute as to where the source of such unity should lie. Difficulties such as the tension between Kantianism and Utilitarianism are well known in this regard. But even apart from controversy over moral *principles*, there is dispute over what the relevant *object* of consideration really is.

Consider, for example, the following claims. Reproduction is centrally concerned with conceiving, bringing to term, and giving birth to babies. Hence the object of reproductive ethics is, above all, that entity which is conceived, carried, and given birth to, an entity variously described as the conceptus, zygote, embryo, fetus, and newborn. It is the nature of this entity, and the demands that that nature makes upon us as agents, which should properly guide all, or most, of our considerations in the field of reproductive ethics. Hence questions concerning that nature, and the demands it makes upon us, should be addressed first.

We should note that this view is, at least initially, compatible with more than one approach to the principles which will guide our reflections. But it does isolate a subject matter for investigation which, if systematically addressed, would serve to unify a number of considerations, for example, concerning abortion, fetal testing and research, fetal therapy, assisted reproduction, surrogacy, and others.

Such a claim, however, is far from universally accepted. Rosemarie Tong writes, for instance, that feminist bioethics "...should center not on the question of whether fetuses are the moral equivalent of adult persons but, rather, on the fact that fertilized eggs develop into infants inside the wombs of women"(Tong, 1997, p. 129). Generalized, this approach sees the key to reproductive ethics not in that which is reproduced, but in those who, traditionally and biologically, have been most involved, and perhaps most burdened by reproduction.

We may further note that within feminist bioethics, the claim that concerns of women form the primary focus of our consideration is, like the claim that the fetus should focus our concerns, susceptible of multiple approaches in terms of the principles brought to bear. Depending upon how the concern's of women are conceptualized, different considerations, and indeed, multiple considerations, may have a considerable bearing on our approach. Thus we have, for example, the feminist-utilitarian approach of Laura Purdy (Purdy, 1996), or the care-based approach of feminists such as Carol Gilligan and Nel Noddings (Gilligan, 1982; Noddings, 1984).

There are of course, those who opt for an intermediate approach, one that focuses both upon women and upon the child. Carson Strong, for example, sets out the following criteria for an "acceptable ethical framework for reproductive" ethics:

> First, it should explore and assess the significance of reproductive freedom. Although reproductive freedom is one of the central values in these issues, it has not been examined adequately. Second, because the interests of offspring are among the main values, an ethical framework should address the question of the importance to be attached to those interests. Thus, it should put forward a view concerning the moral status of offspring during the preembryonic, embryonic, fetal, and postnatal stages of development and discuss whatever obligations procreators might have during these stages. Third, it should advocate an approach to the problem of assigning priorities to conflicting values (Strong, 1997a, pp. 4-5).

Strong's insistence that reproductive ethics address issues of reproductive freedom parallel's Tong's claims about the focus of feminist bioethics; his claim that the status of the offspring must be addressed parallels the claim put forth above in contrast to the feminist position. And his urging that a framework is needed to address the conflicts between the two is an appeal for moral principles that will tell us how to approach, normatively, the object of our consideration.

We see, then, implicitly, the variety of approaches to reproductive ethics that could be taken. When we factor in multiple and competing moral principles over multiple and competing objects of moral interest, there is little hope that a brief survey of reproductive ethics could do justice to such a broad field.

A more efficient manner of proceeding, though not entirely satisfactory, will be to establish at the outset a commitment to one of the several approaches in regards to the subject matter of reproductive ethics, and pursue that, highlighting, at various points, how one's approach, both in terms of principles, and in terms of specificatory premises, to the treatment of that subject matter could differ. The first issue to address on this approach, then, concerns the proper object of reproductive ethics: the offspring, the mother, both in equal measure, or some other entity or interest, such as the advancement of scientific progress, or the future needs of medical patients generally (Maynard-Moody, 1995).

It is generally agreed, however, that neither scientific progress nor medical advances are immune to moral considerations, and the most obvious of these considerations concerns the means by which such progress is attained. Freedom of inquiry is especially considered limited where it runs up against the value and dignity of persons. Thus, no longer is it acceptable practice to conduct experiments on persons without obtaining their informed consent, for to do so is to violate their autonomy as rational agents (Donagan, 1977b). To the extent that this claim is accepted, as it is nearly universally in Western society now, it follows that the question of the moral status of conceptus *must* be addressed before an informed judgment can be made as to whether, for example, it is legitimate to experiment upon the embryo for non-therapeutic purposes, or to use deliberately cultivated embryonic tissue for the medical treatment of third parties.

A similar claim may be made with respect to feminist reproductive ethics: without denying that the procreator is a legitimate and important subject of moral concern, it seems clear that what sorts of concern the procreator is due, and hence what rights and responsibilities the procreator *qua* procreator has, depend upon what the nature of the entity procreated is, for our rights and responsibilities are widely acknowledged to be limited by whether we are acting upon or with other human persons, or upon or with subhuman materials or animals.

In this essay, then, I will take it that the first cluster of questions to be addressed concerns the moral status of the conceptus. Again, while this is not to deny the importance of scientific research, medical advancement, or reproductive autonomy, the nature of the conceptus, as an entity immediately affected by scientists, medical researchers, and procreators, will limit or fail to limit what may permissibly be undertaken by these agents.

We may hope for more than this, however. We should note, in Strong's characterization of the necessary moral framework, that the interests of the offspring, and the significance of reproductive freedom, are to be addressed independently of one another, thus setting up the possibility of radical conflict in need of resolution. Since reproduction is inherently related to that which is reproduced, however, we may hope to find some insight into the nature and value of reproductive freedom by investigating the object of the reproductive capacity. Strong himself, in defending the claim that reproductive freedom is humanly significant, refers to the values of creation of persons, the relation such creation has to love and intimacy, and the experiences of pregnancy, childbirth, and child-rearing. Considerations about the offspring thus play an important, although not exclusive role in his account of the value of reproductive freedom.

Three immediate questions, I believe, should guide our consideration of the conceptus. First, is that entity a human person? Second, whether it is a human person early, or late, what constitutes respectful treatment of the person that is or is to be? Third, what is the value of procreative activities? In what follows, I address each question in turn, showing how answers to these questions would contribute to progress in the field of reproductive ethics, and showing as well where conflicting approaches will lead to conflicting answers. There are thus three sections of the paper to follow. The first, which raises the issue of the moral status of the conceptus, deals with issues of killing, as these are most seriously implicated by a discussion of the status of the conceptus. An example of an issue discussed in this context is abortion. The second section addresses questions of proper treatment of the conceptus which do not, intrinsically, involve killing. Examples here would be cloning or in vitro fertilization with a view to implantation, gestation, and birth. Finally, the third section will address the nature and value of reproduction and reproductive freedom. Among other things, the ramifications of views of reproductive freedom to issues of sex, marriage, and family will be discussed, as well as issues concerning the proper relation between the pregnant woman and the medical community.

II. KILLING AND THE MORAL STATUS OF THE CONCEPTUS

What, then, is the moral status of the conceptus? The answers typically given are these: first, the conceptus is, by its nature, both a human being and a person. Second, the conceptus is by its nature a human being, but is only a person by achievement or development. Third, the conceptus is neither a human being nor a person, but is eventually both by achievement or development.

Until recently, few have argued without qualification for the third approach.[1] If a human being is an individual with membership in a certain species, then any entity may be identified as a human if it is both an individual and may be genetically identified as human. Human gametes – sperm and egg cells – and human somatic cells – skin cells, for example – are genetically human, but not individual members of a species type. A fertilized human egg, on the other hand, is genetically continuous with a recognizable future individual, and genetically distinct from its parent individuals, and thus appears to be a human being from the moment of conception. While the claim that the fetus is merely maternal tissue may be politically effective, the fetus is genetically distinct from all other maternal tissue, and, unlike such tissue, is continuous with a later, and separate, individual. Thus, were there rights that derived to human beings as such, fetuses would be subject to such rights from conception on.

Recently, however, such claims have been attacked: not only, so the objection goes, is the pre-embryo[2] not a *person*, it is not *a human being* although it is clearly human. The pre-embryo is not a human being, and hence not a person, because it is not an individual until its cells cease to be totipotent. For up until such a stage, the pre-embryo has the potential to divide into twins, or to be fused with another pre-embryo to form a chimera.

As George Khushf has pointed out (Khushf, 1997), this strategy seeks to circumvent philosophical or theological questions concerning the various necessary and sufficient conditions for personhood to focus on one necessary condition for personhood – individuality – which science apparently reveals to be absent. The primary focus of this discussion concerns issues of non-therapeutic research on the pre-embryo, including the creation of such pre-embryos for research purposes. A secondary debate focuses on the status of the morning after pill, or emergency contraception medicines. So arguments against individuation at this stage serve to divorce the issue of pre-embryonic creation and research, as well as certain forms of birth control, from the more divisive issue of abortion by isolating a possible area of overlapping consensus.

Two possible responses may be suggested to this approach. First, as Khushf suggests, it is not clear that the conception of individuation at work in these arguments is strictly scientific, or universally held, or independent of prior beliefs concerning the nature of personhood (Khushf, 1997). Second, the arguments against individuation can be directly addressed. Patrick Lee, in a discussion of this delayed individuation thesis (Lee, 1996), addresses several such arguments.[3] He then criticizes the anti-individuation thesis itself, on the grounds that it creates an explanatory gap where there previously was none.

As Lee points out, it seems difficult to deny that the *single* cell conceptus is anything but an individual organism. Nor do proponents of the delayed individuation thesis deny that at a later point there is a single multicellular organism. Thus, argues Lee,

> In effect the proposal is that, first, the unitary, single-celled zygote is formed by the fusion of the sperm and ovum, then it splits into several independent organisms. No explanation is provided for what guides this process. Nothing seems to happen at the point of the appearance of the primitive streak...that might account for the sudden appearance of unity among the previously manifold cells. Only at fertilization, with the fusion of sperm and ovum, is there any event which could be construed as imposing unity on what was previously manifold. In effect, the hypothesis amounts to saying that fertilization is not completed until the primitive streak stage. But...[the] evidence indicates that prior to the primitive streak stage there is already a regularly occurring, predictable, orderly sequence of events of division, differentiation, and growth, beginning with the one-celled organism and leading to an organism with a clear precursor of a brain. I conclude that the more reasonable position is the one taken by the majority of embryologists, that the beginning of the life of the new individual human occurs at the fusion of the sperm and egg (Lee, 1996, p. 102).

Lee's and Khushf's claims are unlikely, at this stage of the debate, to convince the proponents of delayed individuation (Strong, 1997c; Shannon, 1997b). For example, it may be open to proponents of the delayed individuation thesis to deny, *pace* Lee, that they are committed to holding that the pre-embryo "splits into several independent organisms." It might be, rather, that the pre-embryo at this stage is simply not *fully* individuated, rather than being multiply individuated. Such a response, however, would surely make Khushf's claim that there is no strictly scientific, and universally held, criterion of individuation at work here more plausible. Some account must be forthcoming of what it means for something to be neither a single entity, whether composed of parts or not, nor a collection of single entities. It is thus reasonable to predict, as well as to recommend, that more work be done on the issues of individuation, persons, and the status of the pre-embryo.[4]

Many argue that rights derive only to persons, and not to humans, even human individuals, as such. Persons, in such arguments, are typically defined as possessing some set of characteristics that are typical of rational agents. On Mary Ann Warren's view, for instance, persons have at least some of the following five characteristics: consciousness, developed capacity for reasoning, self-motivated activity, capacity to communicate, and self-awareness (Warren, 1973).

Proponents of the claim that the conceptus is a person from conception can accept that persons are rational agents, that rational agents typically manifest Warren's five characteristics, and that it is because they are rational agents that persons are to be respected. Where there is disagreement, typically, is over whether personhood is best viewed as an achievement, as in Warren's view, or as conferred, as in Carson Strong's view (English, 1975; Benn, 1984; Strong, 1997b), or as a status, which individuals have in virtue of their *capacity* to achieve rational agency, a capacity identified on the basis of species membership. On this latter view, when something may be identified as an individual member of a species, the individuals of which have a capacity to achieve rational agency, then that individual is a person in virtue of that capacity.[5]

Disagreement over these positions deeply affects numerous issues in reproductive ethics, beginning, most obviously, with the issue of abortion. For one of the most obvious and widely accepted norms regarding our treatment of persons is that persons should not be deliberately killed. Thus, if the conceptus is a person, then abortion, under this norm, will likely be ruled out as a form of killing; whereas if the conceptus is not a person, but only something which could, if allowed to develop, eventually achieve personhood, then abortion will not be impermissible under the norm concerning killing. (Even here, there is dispute, however. First, utilitarians, and some rights theorists, can hold that direct killing of persons is not always wrong. Second, an argument for the permissibility of abortion made popular by Judith Jarvis Thomson seems to depend upon the claim that abortion is not direct killing, even if the fetus is a person. These positions will be addressed in due course.)

Whether the conceptus is a person is likewise directly related to consideration of the morality of: freezing and disposing of embryos, deliberate creation of embryos for research purposes followed by disposal of embryos, and the deliberate creation of embryos for medical purposes, as when a developing embryo is stimulated at the totipotent stage in such a way as to cease development as a human individual and to continue development only as one or another form of human tissue. The issue of killing is indirectly related to questions of the use of tissue from already aborted fetuses for research or medical purposes. Finally, the status of the fetus as a person or not is important to decisions concerning conflicts of interest between fetus and mother, and subsequent medical decisions to treat the fetus as a patient or not. As H. Tristram Engelhardt argues, the view that personhood is an achievement which comes with rationality and self-awareness is consistently accompanied by the view that many things may be done to the fetus, and indeed, to the neonate and infant which could not morally be done to a fully functional adult (Engelhardt, 1996).

It is perhaps worthwhile here to point out that the answer to this question of personhood, and, as well, the question of individuation discussed above, does not merely determine further answers to questions already acknowledged to be a part of the field of bioethics, but can significantly affect what we conceive to be within that field, specifically in regards to issues of justice.

Suppose, with Warren, one determines that the fetus is not a person. This will have a significant bearing on the questions raised above. Take, for example, the question of the generation of fetal tissue for research and therapeutic purposes. At present, it is not especially resource efficient to create embryos, although there do exist many thousands of frozen embryos in the US. Increasingly, however, as cloning technology has developed, it has become attractive to many scientists to manipulate somatic cells in such a way that they become, effectively, embryonic cells from which stem cells might be extracted. Such stem cells could then be stimulated to develop only into one or another tissue type, to be used in tissue transplants.

If the pre-embryo is determined to be a person, such 'stem cell technology' will, under most moral principles, be ruled out as impermissible: extraction of the stem cell from the embryonic cell cluster would seem to be killing (Mirkes, 2001). If the embryonic cluster of cells is not a person, however, not only will such activities be permissible; rather, there will be significant questions concerning what demands justice makes upon the medical and scientific community, and upon society, to promote the necessary research into these developments, and to ensure that the benefits of such research are adequately distributed. These questions of distributive justice will seem to be increasingly important questions of this area of bioethics (Warnock, 1985; Harris, 1998; Buchanan, et. al., 2000). For those who believe that the pre-embryo, or embryo, is a person, however, these questions of justice will not even arise.

This pattern is typical of the field of reproductive ethics, and contributes to its sometimes fragmented character. Few of the activities which come under scrutiny in the field are entirely neutral from a moral point of view. For any given activity, then, if it is held to be permissible, this is likely to be, at least in part, a result of the special value attributed to that activity. For those who support abortion rights, for example, self-determination and the integrity of one's own body are crucial to the well-being of women. For those who support various forms of assisted reproduction, the values of self-determination, family, propagation of one's genetic line, parenthood, etc., will all play a justificatory role. While these goods might be thought to be merely personal or subjective, they are often taken to be of wider value. In consequence, arguments for the permissiblity of these various activities often depend upon the claim that these goods are either of objective worth, or that they have a broad instrumental value in the achievement of human flourishing. Both views generate the possibility of a strong claim, in justice, that the common good be socially promoted, especially by means of increased, and even subsidized, availablity of the procedures in question (Buchanan, 1995; 1996).

Issues of distributive justice are, of course, important for those who find much of the recent reproductive technology morally problematic. For example, even if some form of assisted reproduction is held to be impermissible, still, infertility as such is likely to be viewed as an obstacle to the achievement of a significant good, a good which the state might reasonably play a role in promoting (Congregation for the Doctrine of the Faith, 1987). In general, however, opponents of much of the reproductive techology of the past forty years have been primarily concerned with issues of commutative justice – that part of justice "in which neither the requirements or incidents of communal enterprise nor the distribution...of a common stock are directly at stake, but in which there can be a question of what is fitting, fair, or just as between the parties to the relationship" (Finnis, 1980, p. 178). This will make sense inasmuch as such thinkers accept the existence of a 'party to the relationship' not accepted by their opponents, viz., the fetus.

While this division is imperfect, it does reflect a trend in much recent work: there are those who, giving some token attention to issues of commutative justice move straight on to distributive issues, and those whose attention is held almost exclusively by commutative issues. At the same time, however, it is clear that, from a societal perspective, if there were duties in commutative justice of parents to unborn offspring, then failure, in a society, of individuals to meet such duties would at the same time be damaging to the common good of the society. As John Finnis (1980, p. 184) has written of a society in which individuals fail in their duties of commutative justice to other individuals, "How can a society be said to be well-off in which individuals do not respect each other's rights?" For a society to allow extensive and serious violations of such duties in commutative justice therefore amounts to a failure of distributive justice.

Returning to the issue of stem cell technology, there is perhaps a limited area of possible rapprochement between adherents to the two positions. Recent work has suggested the possibility that stem cells taken from non-embryonic tissue, or even from non-human animals, might provide many of the significant medical benefits hoped for from embryonic stem cell technology. Given the moral issues surrounding the human conceptus, it would seem reasonable to suggest that these non-controversial sources of stem cells be rigorously investigated before arguments for the "necessity" of embryonic stem cell technology be proferred.

A final area in which the issue of personhood would be of significant importance is one which has been relatively underdeveloped. The new reproductive technologies have made possible, or, in some cases, promise to make possible, transactions with the zygote or embryo which do not initially appear to involve killing, as there will remain after the transaction a live entity. An example of this would be germ-line enhancement of an in vitro embryo. This is generally thought to be of a piece with genetic work done on the parental gametes, e.g., genetic enhancement of sperm. This genetic work on the gametes is also referred to as "germ-line engineering." However, if the conceptus is held to be a person, it seems possible that the two types of germ-line work are different in kind.

Consider again the reasons for considering the conceptus a human being: it is an individual, and, as determined genetically, both distinct from its mother and continuous with a later identifiably human individual. At its early stages, the cells of the conceptus are totipotent, that is, they have not yet been differentiated into the various types of cells which go to make up different organs and perform different tasks in the human body. At this stage, unlike later stages after cell differentiation, it is, at least in theory, possible to change the entire genome of the individual, its entire genetic code; this has been done with mice embryos (Anderson, 1994). Somatic cell therapy, by contrast, is local to the area in which genetic information is introduced, removed, or repaired. So this later, medically altered individual, is still recognizably genetically continuous with the earlier physical organism which resulted from conception.

In germ line engineering of the early embryo, however, such continuity exists only between the modified individual and the later individual. As identified genetically, the individual which came to be in conception no longer exists. It thus appears possible that an individual has been killed, without there being any fewer living entities than before the individual's death. [6]

Nor is this possibility unique to germ-line interventions of this sort. The type of fusion of pre-implantation embryos which occurs naturally in the formation of chimeras can be done in the laboratory. On the earlier argument, such fusion would be a form of killing, whether the cells fused were both human, or whether the chimera was an 'interspecies' mix. Similar considerations would have relevance to genetic testing of a cell removed from a totipotent embryo.

This would be relevantly similar to removal of one of two preimplantation twins for the purpose of experimentation and subsequent disposal. [7]

I have so far outlined, of course, only a variety of implications of views regarding the personhood of the conceptus. Whether this entity is best viewed as a person or not has not been addressed; rather, I have argued that an answer to that question remains central both for answering a number of questions common to all bioethicists, and to determining the structure of further inquiry. As an earlier parenthetical remark noted, however, the picture is not quite so simple.

The presence of utilitarian bioethicists is the cause of one wrinkle in the guided map I have provided. For utilitarians do not accept the normative claim that persons are never to be directly killed. Rather, depending on what they conceive of as the various forms of goodness, utilitarians urge that the good be maximized, or the bad minimized. On such a view, even if the fetus were a person, numerous activities would be permissible if they promised to promote the greater good. So, to return to an example from above, if many lives could be saved or enhanced with the knowledge provided by research on laboratory generated fetuses, the utilitarian may urge that such research be performed.

The second difficulty is suggested by the possibility that in many cases, including, but not limited to abortion, what one does to the fetus is not done directly, or intentionally, but is accepted as the by-product of an otherwise permissible, and sometimes laudable, activity. Thus, as Judith Jarvis Thomson argues, if it would be permissible to unplug a violinist from one's kidneys, not meaning to kill her, but merely to disentangle oneself from an incumbrance which the violinist had no claim upon one to accept, then similarly, to discharge a fetus from one's womb, not meaning to kill, but merely to refuse it housing which it had no claim on one to give would likewise be acceptable (Thomson, 1971).

If we dispense with the violinist analogy, and simply work with the distinction between intended killing and accepted side-effects, Thomson's analysis could presumably be extended to other areas of reproductive ethics where the concern is with the well-being of a fetus. For example, in certain forms, at least, of fetal experimentation, the intention might not be to kill, but merely to gain knowledge, although, given the ex-utero environment, and the invasive procedures in use, death of the fetus was to be expected.

The existence of utilitarian and violinist arguments indicates the overlap, in two key areas, between bioethics and other areas of moral philosophy. First, the plausibility of the utilitarian approach in bioethics depends upon the success of utilitarianism in the larger scrum of normative ethics. And it should be pointed out that, while it will not go away, utilitarianism has consistently received a beating from which it has not fully recovered. The problems of ignorance of consequences (Donagan, 1974), of incommensurability of goods (Grisez, 1978;

Finnis, 1980; Raz, 1986), of failure to treat persons as individuals (Rawls, 1972), of the incompatibility of utilitarianism with freedom (Finnis, 1991), and others remain difficulties which utilitarianism has not, in my view, successfully answered.

Second, violinist (and trolley[9]) problems indicate the need both for a sound theory of action, which allows us to distinguish between intended actions and side-effects, and for normative considerations which guide our judgment of intentions *and* side-effects. This last is often ignored, the assumption being that any traditional ethics which forbids intentional killing has nothing to say about bad side-effects.[10] So, for example, even when a disvalued consequence is a mere side-effect, and not an intended end, norms are necessary to determine when it is legitimate to accept such a consequence. Clearly, not all side-effects, however much outside the intention, are acceptable.

One such norm, which could be helpful in resolving some of the difficulties of reproductive ethics, is that of fairness (Boyle, 1980; 1991). Even in those cases in which the norm against intentional killing is not violated, it remains the case that some side-effects constitute a different form of injustice against those who suffer them. With a norm advocating fairness in the acceptance and distribution of side effects, the following difficulty, raised by Thomas Nagel against the theory of double effect, would seem to be addressed:

> In Indo-China, for example, there is a great deal of aerial bombardment, spraying of napalm, and employment of pellet- or needle-spraying antipersonnel weapons against rural villages in which guerillas are suspected to be hiding...The majority of those killed in these aerial attacks are reported to be women and children, even when some combatants are caught as well. However, the government regards these civilian casualties as a regrettable side-effect of what is a legitimate attack against an armed enemy (Nagel, 1979, pp. 60-1).

Although such casualties might indeed be mere side effects, if they are unfairly inflicted, then they should be ruled out. And fairness can, in this instance, be determined by asking what one would oneself by willing to accept as casualties were the tables turned.

Introducing the norm of fairness, I think, is damaging to the violinist examples, in a way which later commentary on Thomson's article has reflected. Detachment from the violinist seems fair: we would not ourselves seek to demand use of someone else's body if we were ill, we would not ourselves kidnap someone against their will, the hostage has no relation to the violinist, and so on. But, assuming, as Thomson does for the sake of the argument, that the fetus is a person, the situation is completely different with the fetus: it does not aggress upon us, as does the violinist. It is there by voluntary actions on our own part, and its presence does not make the same sorts of extreme demands as the violinist: few are bedridden, for the entire nine months of pregnancy, nor, when a woman is bedridden, is the fetus' presence a gross invasion of privacy, as the violinist's presence is. To take the fetus's life, if it is indeed a person, thus seems, in light of these arguments, unfair.

This is reflected in Warren's view that Thomson's article can only justify abortion, if the fetus is a person, in cases of rape (Warren, 1973); and thus brings us back to the need to determine whether or not Warren's claims, and the claims of those who follow her, about personhood are true. The issues of personhood and killing are central to reproductive ethics. [9]

A final point about the need for a theory of action which is adequate to the problem of intention. Such a theory might make more intelligible what, to date, has been mostly an in-house skirmish of Roman Catholic ethicists over the morality of contraception. Traditional Catholic pronouncements on contraception have likened that act to abortion, without, of course, claiming that anyone is actually killed in contraception. The most rigorous of recent work in the Catholic tradition attempting to make sense of this claim has focused on the nature of the intention of a contracepting couple, and asked whether, in the 'contra-life' intention of such a couple, there is something akin to the intention at work in abortion (Grisez, Finnis, Boyle and May, 1988; Smith, 1993).

III. RIGHT TREATMENT OF A CONCEPTUS

Although issues of killing are central to reproductive ethics, technological developments have made available numerous relations to the zygote, embryo, and fetus which are not forms of killing, but which have raised moral questions. In particular, a host of issues concerning the artificiality of various reproductive techniques have been raised. Thus, there are disputes over the commodification of women and children in surrogacy arrangements, of the artifactuality of the child in cases of cloning, or genetic enhancement. In general, the severance of reproduction from any 'natural' act, and its importation into the laboratory has led many ethicists to ask whether there is a significant moral difference between reproduction, and (mere) production, between being begotten and being made.

In this section, I will, for the most part, concentrate on the nature of what seems to many to be the most radical form of production of persons, human cloning. The structure of many arguments in favor of cloning parallels closely utilitarian arguments for abortion, fetal research, and other technologies which appear to involve killing, but with a twist: not only are the various reproductive technologies discussed in this section of great benefit, social and individual, and not only do they not even involve killing, but the entity which is putatively violated by 'productive' techniques is, in fact, benefited, by being made to live, and in a way which would only have been possible as a result of these techniques. If such arguments are legitimate, they have far reaching consequences for most techniques by which children are brought to life. On the other hand, as we will see, rejection of cloning threatens to have equally far reaching consequences for other, more generally accepted, forms of assisted reproduction.

One qualification to the discussion must be noted. A major strand of argument in favor of various assisted reproduction technologies concerns the value of reproductive freedom and autonomy. To some extent, such arguments tend towards a consequentialist structure: reproductive autonomy is a value which weighs more heavily than other values in certain circumstances. On the other hand, reproductive autonomy could be taken to have 'trump' value – the claims of such autonomy would then be viewed as defeating of other considerations, but not through a consequentialist weighing. To the extent that autonomy considerations concerning assisted reproduction are formally consequentialist, they fall under the considerations already raised about consequentialism. The value of reproductive autonomy in itself, and the ways in which it may or may not provide defeating reasons, will be considered in the next section.

The distinction I wish to make use of in the following discussion of cloning is between that which *happens*, and that which one *does* (Nagel, 1979). Around this distinction may be grouped, generally, those types of ethical concerns labeled consequentialist and deontological, respectively. The former emphasis served to justify, for example, terror bombings of villages intended to force enemy surrender, or, going back, the fire bombing of Dresden, or the atomic bombing of Hiroshima. On this approach, what matters is *results*.

The latter approach, by contrast, in concentrating on what one does, requires that, in acting, one stand in certain sorts of relations to other persons, and avoid other sorts of relations. Historically, the nature of the necessary relation has been explicated in terms of notions such as respect for the dignity of the other as, variously, an end in itself, a rational creature, or a person of absolute, intrinsic, or unconditioned worth (Immanuel Kant, 1985). Such formulations are not without difficulty, for they stand themselves in need of interpretation, a fact that has been used against such approaches in the context of the ethics of cloning. John Harris, for example, in an essay in support of cloning published shortly after reports of the first cloned sheep, writes that "Appeals to human dignity, while universally attractive, are comprehensively vague" (Harris, 1997, p. 353). But, in theory, if we could identify some relation in which we could stand to other persons, as one violating respect, or dignity, etc., then concern for how one acts in relation to others would serve to rule out that possibility. In what follows,

I indicate how concern for what happens and concern for what one does can chart different paths in the ethics of cloning.

The most common arguments, in any moral field, which center around concern for what happens, are utilitarian arguments, but they are not the only sorts. In addition to utilitarian arguments, two others seem important in debate about cloning, what I shall call arguments from dissolution, and the hostage argument.

The utilitarian argument is straightforward: cloning will result in great reproductive benefits, benefits which outweigh any possible negative effects. A few examples will suffice: the infertile will be enabled to reproduce; medical possibilities will be created for those in need of, e.g., bone marrow; the narrow constraints of the traditional family structure will be overcome.

These arguments are, of course, substantially the same as those used to justify other forms of assisted reproduction, such as donor insemination, surrogacy, or in vitro fertilization (IVF), as in the statement by the Ethics Committee of the American Fertility Society, regarding IVF, that "the benefits provided outweigh the risks both to the couple and to the offspring produced" (The Ethics Committee of the American Fertility Society, 1992, pp. 301). Further, as mentioned, they play a role in other areas of reproductive ethics. I have already questioned their success.

Other arguments in the cloning debate which concern themselves with what happens are not straightforwardly utilitarian, however. Still, to the extent that one's moral consciousness has been informed by the underlying concern for what happens in the utilitarian argument, these further arguments might be persuasive.

What I call the argument from dissolution is a response to numerous concerns about the 'unnaturalness' of cloning. Such concerns often run as follows: cloning robs us of our genetic identity or our uniqueness or our individuality; or, cloning means that there is only one parent, eliminating the natural two parent relationship. The response to such objections, and others, which response I am calling the dissolution argument, may be summed up in one word: twins. Richard Lewontin, for example, argues that concerns with individuality are the result of a faulty conception of genetic determination (which he blames on Richard Dawkins) which is readily corrected by looking at twins, who are, in fact, more closely related genetically than clones would be.[11] The one parent argument is also subjected to the twin argument: a clone would actually be the genetic child of the donor's two parents, and the twin of the donor, giving the clone as many parents, and the same biological relations as the donor.

The argument from unnaturalness can generate concerns about twins separated in time, but another form of the dissolution argument could be called upon: if one twin were put on a faster than light spaceship and allowed to travel a bit, when she emerged, she would be considerably younger than her twin, so much so, in fact, that the older twin might gain custody and care for the younger as her child. Who could think this an intrinsic evil for either twin?

The strategy of such arguments is to induce agents to look at the results of cloning purely as happenings – not as doings. For when we look at them in this way, we see that the results of cloning are similar or even identical to natural occurrences which are not regarded as spooky, creepy, or unnatural. Divorced from the perspective of agents, viewing the world , *sub specie aeternitatis*, we are robbed of any deontic point of view which would allow us to worry about how we, as opposed to nature, had gotten to this or that point. Our first person, indeed, human perspective is 'dissolved' in favor of a third person, and therefore purportedly more 'natural' standpoint. And if what nature does is not good enough for us, Lewontin gives an extension of the dissolution argument that I believe rhetorically works the same way although with the addition of irony:

> the Creation business...is both seductive and frightening. Even Jehovah botched the job despite the considerable knowledge of biology that he must have possessed, and we have suffered catastrophic consequences ever since. According to Haggadic legend, the Celestial Cloner put a great deal of thought into technique. In deciding on which of Adam's organs to use for Eve, He had the problem of finding tissue that was what the biologist calls 'totipotent,' that is, not already committed in development to a particular function. So he cloned Eve not from the head, lest she carry her head high in arrogant pride, not from the eye, lest she be wanton-eyed, not from the ear lest she be an eavesdropper, not from the neck lest she be insolent, not from the mouth lest she be a tattler, not from the heart lest she be inclined to envy, not from the hand lest she be a meddler, not from the foot lest she be a gadabout but from the rib, a 'chaste portion of the body.' In spite of all the care and knowledge, something went wrong, and we have been earning our living by the sweat of our brows ever since (Lewontin, 1997, pp. 2).

The third argument one finds for cloning, which I call the hostage argument also occurs across a range of reproductive technologies. John Robertson gives an example of this sort of argument, set in the context of concerns about twinning, but easily extended. Suppose that twins, or triplets, are at a special risk of psychological harm. Still, "it would appear difficult to argue that these disadvantages are so great that the triplet should never have been born. Given that this is the only way for this individual to be born, its birth hardly appears to be a wrongful life that never should have occurred" (Robertson, 1994b, p. 10). This argument also appears in Robertson's discussions of surrogate mothering and other forms of assisted reproduction (Robertson, 1994a).

What happens in the hostage argument is that the future results of cloning, considered as *actual*, are, as it were, put at the mercy of those who wish not to permit cloning: but for cloning, this person would not exist; hence to ban cloning is to effectively will that this person not exist. The future person's existence is thus held hostage to some critic's moral scruples. Despite its popularity, this type of argument appears specious. On this line of thinking, for example, we would have to agree that slavery was legitimate, because the descendants of slaves are better off here in America than they would have been if allowed to remain in Africa (Krimmel, 1992). It is, however, perfectly compatible with being grateful that one is alive, or in America, and so on, that one condemn the circumstances by which one came to be where one was, or indeed came to be at all.

The opposing strand of argument over the ethics of cloning focuses on what an agent is doing in engaging in such activity. In cloning of the sort pioneered by the Scottish geneticist Ian Wilmut, and now carried on by others, the nucleus of an egg is removed, and replaced with the full set of chromosomes of the donor, taken from non-germ cells, e.g., skin cells or udder cells. What is the nature of the agent's doing in this case? It is the proximity of what is done here to technical production which raises concerns in opponents of cloning.

Production as a genus is distinct from other forms of human activity by being concerned with the generation or creation of some form of artifact. Various forms of human action are clearly not so concerned: listening to a symphony, reading Plato, engaging in friendly conversation, and so on. But neither are all forms of production what I am calling *technical* production. We could distinguish the productive activity characteristic of a craftsman, from mere technical making in something like the way we distinguish playing chess from playing tic-tac-toe (MacIntyre, 1982). In the former game, knowledge is essential to successful play, but knowledge does not generate a series of operations sufficient for achieving one's desired outcome: rather, chance, spontaneity and creativity interact with knowledge in determining the overall shape of the outcome; and as one's knowledge increases, the variety of ways one can interact with and integrate chance, spontaneity and creativity also increase. In tic-tac-toe, by contrast, a minimal grasp of the game enables one, with regularity, to determine a specific outcome – a tie, at least – in every game by a series of easily codified rules.

Likewise, the agent involved in technical production differs from a craftsperson. For in technical production, an agent is engaged in an activity in which a series of operations performed upon some set of materials is jointly sufficient, in conjunction with the laws of nature, and to the extent that the materials are adequate, for the creation of the resultant desired end. If this is accepted as a definition of technical production, it allows us to identify cloning as an instance of the technical production of persons, for a series of operations performed upon the egg and the somatic cells is deemed jointly sufficient for the creation of a person, or, in the case of twin-fissure, for the creation of twins.

Two qualifications are important here. One is that the material be adequate: despite the operation of steps deemed sufficient, it is still possible for technical production to fail, specifically when the material is in some way flawed. This, in fact, further differentiates craftsmanship from technical production: a craftsperson's skill is such as to allow her to make the most of flawed material, work around it, even improve her product through integration of the flaw. Technical production fails in the presence of flawed material. For this reason it is not an objection to the identification of cloning with technical production that it often fails.

The claim that the result is a produced *person* might strike some as suspect, if they do not consider, for example, a zygote, to be a person. However, in not all cases of technical production is the product simulateous with the completion of the maker's operations. It is enough that the maker's activity, in conjunction with the normal laws of nature, be sufficient for the desired end to emerge. So even if the zygote is not a person, it will eventually, by the laws of nature, develop into a person in consequence of the cloner's productive activity.

If we identify cloning as technical productive activity resulting in the manufacture of a person, what normative evaluation can we make of this considered as a doing? In attempting to define the proper relation that persons should have to persons, and that persons should have to things, Kant drew a distinction between producible ends and ends in themselves (Kant, 1985; Donagan, 1977a). This notion of an end-in-itself seems best understood as the notion of something of intrinsic, or unconditioned, value. What this amounts to in any given context will require interpretation, but by definition it seems that any case of treating a person as a technically producible end fails to respect her as an end-in-herself, or as of unconditioned worth. For the condition of production is that one have some purpose for that which is produced. Products are not made for their own sake, but for the sake of the producer or, if they are different, the consumer. Hence, as an artifact, a product takes its meaning from the ends of the artificer. It follows that the value of the product is conditioned, relative to the purposes of the artificer; and that hence, where the product is a person, the productive activity fails to respect the person as an unconditioned end, as something of intrinsic worth. Thus Leon Kass, a prominent critic of cloning, writes, "As with any product of our making, no matter how excellent, the artificer stands above it, not as an equal but as a superior, transcending it by its will and creative prowess" (Kass, 1997, pp. 2).

An obvious objection to this is that *of course* the parents (or twin) of the clone will love it, and respect it as an end in itself – in fact, goes the argument, they must want a child more than most people who eventually have children. Nonetheless, for those who accept deontological constraints on action, subsequent good consequences do not render sub-personal treatment legitimate. If purchase of persons, for example, wives, is intrinsically demeaning, it is not made morally acceptable by subsequent good treatment, even love, of the purchased person. If an agent's concern is with the relations she stands in to other persons in her actions, it will not matter that the resultant state of affairs is a good one, if she arrived there through actions which are themselves, and in principle, contrary to the notion of respect for persons as beings of unconditioned worth.

These considerations also help make sense of the debate, in reproductive ethics, over 'slippery slope' arguments. The use of such arguments is common among opponents of various forms of assisted reproduction, who argue that allowing one form – say, IVF – will lead to another – say, cloning. Similar arguments have been proposed with respect to the relationships between contraception and abortion, gene therapy and gene enhancement, or genetic screening and the devaluation of the disabled. The response to such arguments is also common: slippery slope arguments rely upon disputed empirical claims and predictions. Thus, there is no principled objection which the slippery slope argument offers to the possibility that the human community can 'check' itself before it goes too far down one or another reproductive path (Resnick, 1994; Beauchamp and Childress, 1994).

The deontological considerations adduced above, which reflect the concerns of a number of opponents to cloning, and perhaps to other forms of assisted reproduction, suggest an alternative reading to the slippery slope arguments. For they suggest that, in engaging in technical production of persons, one sets one's will in relation to those persons in a certain sort of way, a way already consistent with those relations envisaged as emerging further down the slope. Thus, if IVF is, like cloning, also a form of technical production, then willingness to treat persons as manufacturable items in the former will, by the logic of the will, even if not empirically, lead to the possibility of cloning appearing more and more acceptable.

A question, then, for those who oppose cloning in consequence of its technical character concerns how far back, in the chain of developments in assisted reproduction, this objection applies. Does it, for example, apply to in vitro fertilization? Does it further apply to various forms of artificial insemination? Does the introduction of a consumer relationship into surrogacy arrangements bring surrogacy also closer to the production model (Radin, 1987)? Opponents of cloning, who take it as a paradigm of technical production and for that reason morally suspect, need to address further the question of how far from or near to the paradigm these other forms of technically assisted reproduction are. Central to an evaluation of this issue would be a more fully worked out account of the nature of production or making.

A further question concerns how natural reproduction differs, deontologically, from technical production (Spoerl, 1999), and how these differences fit into answers to the various questions about the family, and the connection between sex and reproduction which the new reproductive technologies raise. These questions will be raised in the next section, as I take them to be connected to questions about the value and place of reproductive liberty and autonomy. Answers to these questions would circle back upon issues raised by forms of assisted reproduction which are not as near to the paradigm of technical production as cloning, such as sperm donation and surrogacy.

IV. PROCREATIVE AUTONOMY AND THE VALUE OF REPRODUCTION

The nature and value of autonomy has been given considerable attention in much recent work in political and ethical theory, even as the value of *reproductive* autonomy has been somewhat undertheorized (Strong, 1997a). Thus, in addressing specifically questions of reproductive autonomy, I will begin by specifying in a broad way three positions about autonomy as such which have been held by various political and ethical theorists. I will then see how these theories might be developed in a context that specifically dealt with reproduction.

One dominant political theory that highlights the importance of autonomy is anti-perfectionism. As put forth by John Rawls, anti-perfectionism as a political theory holds that the state may not choose between competing ideals or thick theories of human goods by promoting some ways of life over others. Rather, the primary norm by which the state operates is that the autonomy of its citizens should be respected in their pursuit of their own life plans, ideals, and thick goods. This autonomy is to be respected up to the point at which an agent's pursuit of their good becomes harmful to others. Self harm, and, as it were, moral harm, do not, however, typically count as reasons for state interference (Rawls, 1974).

As a political view, anti-perfectionist accounts of autonomy have been taken up in the realm of reproductive ethics to justify a lack of state interference in the new reproductive technologies. Permitting abortion would seem to require a judgment of whether the fetus is a person, and thus harmed, but most forms of assisted reproduction seem, as I indicated in the previous section, to leave everyone in some sense *better* off than before – parents now have children, and children now have lives. As long as children are a part, therefore, of some agent or agents' self chosen ideal of the good life, the state would be unjustified in interference in those agents' reproductive choices.

Could such a view operate as a purely moral doctrine? Robert George suggests that David Richards is plausibly interpreted as holding a view of 'moral anti-perfectionism' (George, 1993; Richards, 1986). For the political anti-perfectionist, such as Rawls in *The Theory of Justice*, the state does not arbitrate between competing conceptions of the good, but this does not imply that there are no superior and inferior conceptions of the good. Moral antiperfectionism, by contrast, would hold that "one cannot choose immorally among possible ends...[Hence a] choice which does not violate the rights of others is *morally right*" (George, 1993, pg. 151). Again, 'violates the rights of others' is here not given a moral reading, but is understood in terms of a non-moral notion of harm. As a moral doctrine, then, antiperfectionism licenses all pursuits of ends which do not harm others.

It would seem to follow that there can be no specifically or distinctively *reproductive* version of moral antiperfectionism. Reproductive choices, so long as they do no harm are morally permissible, but not in any way which is distinct from, say, sexual choices, or career choices, or gardening choices. At best we could say that the issue of autonomy is more frequently implicated in the area of reproduction than gardening because of the extent to which human agents are, as a matter of fact, concerned with reproduction.

Are there, nonetheless, defenses of autonomy in the field of reproduction which are morally anti-perfectionist? Such a defense would concede no intrinsic value to reproduction, in any form, but would hold that autonomy in this area was important in order to preserve the opportunity for agents to constitute themselves in this way if they are so inclined subjectively.

It is possible, I think, to read some of what John Roberston says in defense of reproductive autonomy as conforming to this analysis. Procreative liberty, as he calls it, is important because "control over whether one reproduces or not is central to personal identity, to dignity, and to the meaning of one's life" (Robertson, 1994, pg. 24). Such claims are anti-perfectionist, however, only if what matters to one's identity, dignity, and the meaning of one's life does not depend upon any particular value in the pursuits which shape that identity, dignity and meaning, but derive only from the fact that *one has autonomously chosen* those pursuits. Any particular importance in one area over another, would only be contingent upon it's being in fact more frequently chosen and valued than some other pursuit, or in fact more socially valued, etc.

Such a defense does not, however, seem to do justice to the commonly held view that there is something special about reproduction, and the role it plays in human life, that transcends its merely being contingently desired by a large number of people. If all agents simply decided to stop procreating altogether, it would be difficult to accept this as merely one autonomous choice among other possibilities.

A competing conception of autonomy in political philosophy is that of Joseph Raz's 'pluralistic perfectionism' (Raz, 1986). At the level of the state, Raz argues, there is a legitimate interest in promoting some forms of life over others precisely because of their intrinsic and objective value. Moreover, autonomy and self-determination themselves constitute an intrinsic and objective good which it is a legitimate interest of the state to promote. However, Raz also, perhaps paradoxically, holds that autonomy is only an intrinsic and objective good when it is exercised in pursuit of objective and intrinsic goods, not when it is exercised in pursuit of immoral or worthless choices.

Raz's position is already a moral, and not simply, a political account of autonomy. Political consequences, in his view, flow from certain aspects of his moral view. Could there be a specific and distinctive account of reproductive autonomy within the pluralistic perfectionist theory?

In answering this question, let me first address the 'pluralist' aspect of Raz's thought, for it suffices to answer at the start one possible objection to any perfectionist account of reproduction. The objection is that, if reproduction is an objective and intrinsic good, then it becomes somehow mandatory for agents to reproduce; that state coercion is therefore justified; and even, that all the forms of reproductive autonomy so far canvassed would be at least permissible, and perhaps obligatory, as ways of insuring that the good of reproduction was being properly promoted (Atwood, 1986).

A pluralist perfectionist account is not, however, committed to any of these claims, for such an account holds that there are multiple ways of pursuing and promoting goods, not all of which an agent will be able to equally choose. Pluralist accounts of the good, unlike, for example, strict consequentialist accounts, thus allow what some have called 'options,' or the availablility of more than one possible way of pursuing the good.

On a pluralist perfectionist defense of reproductive liberty, then, reproduction will be viewed as one possible good among others. There will thus be a reason for pursuing it, but not an obligation as such (although in some contexts there may be an obligation to pursue it).

Two further requirements for the pluralist perfectionist shape the defense of reproductive autonomy that such a theorist could offer. First, the perfectionist must give some account of the nature of the value that procreation has; second, she must give an account of the moral norms governing procreation, and stipulating ways in which the good of reproduction may and may not be permissibly pursued.

These requirements are necessary for a reason that perhaps undercuts Raz's doctrine. Their necessity derives from Raz's claim that autonomy does *not* have its intrinsic and objective value in situations in which immoral ends are pursued, or in which legitimate ends are immorally pursued. It follows that an account of procreative liberty and autonomy *will* be in some ways distinctive and different from an account of autonomy in some other area. However, as has been argued, Raz's claim that autonomy is itself an intrinsic good seems more difficult to sustain if autonomy has no value in itself to the extent that immoral choices are autonomously made (George, 1993). Autonomy seems to take on an instrumental value, to the extent that it allows us to pursue valuable ends appropriately, but it no longer seems to have intrinsic value. A position much like this may be found in the work of some of the recent writers in the natural law tradition (Finnis, 1980; George, 1993).

The upshot of this line of thought is to move a discussion of the value of procreative liberty in two directions. First, towards the value of liberty and autonomy, considered as of generally instrumental value; and second, towards a discussion of procreation and reproduction as such, and what, if any, objective value they possess, and what, if any, are the moral norms governing pursuit of that value. An account of reproductive autonomy, as a peculiar, or specific type of autonomy legitimately exercised will be the result of intersection of these two lines of consideration.

Part of the above outlined project has already been discussed in this essay. In particular, disputes over the norms governing reproduction have been discussed under two headings: concerns about killing and harming; and concerns about the proper deontic relation to the developing fetus. And underlying this discussion has been a widely, though certainly not universally, shared consensus that at *some* point, norms governing treatment of pre-natal or post-natal humans are determined by the peculiar moral status appropriate to human beings, that of being ends-in-themselves, creatures worthy of respect, beings with dignity, but not price.

If this is so, then it will make sense to see the value of reproduction also, at least in large part, determined by the moral worth of what is reproduced. Persons are not like chairs or cars – their value is independent of any valuing done by any particular agent. And the conclusion which it seems reasonable to draw at this point is that a *primary* source of value in reproduction is the child him or herself.

This would be a conclusion considerably different from the claims made by those whose work skirts the boundaries of anti-perfectionism. In those authors' writings, considerable reference is made to the *experience* of those engaged in reproduction as the grounds for their valuing reproduction. Indeed, this fits with the essentially individualist orientation of much political anti-perfectionism.

An opposing view, by contrast, must see the choice to reproduce as fundamentally guided by the value of the other, for the other's own sake. This, of course, raises fundamental issues of value theory and meta-ethics which cannot be discussed here. [12] It is a position which has also found a home in certain religious doctrines on reproduction and assisted reproduction (Congregation for the Doctrine of the Faith, 1987).

It is apparent, in much recent work in reproductive ethics, that this difference in focus between the objective value of the child and the subjective value of self-determining choice and experience reflects wider differences over the field generally, differences often characterized as conservative and liberal. For reasons that I attempted to make evident in the previous sections, concern for the value of the other from the initial stages of human life tends towards elimination of artificial means of reproduction as insufficiently respectful of human life. On the other hand, an exclusive focus on autonomy, choice, experience, and freely chosen values often accompanies a willingness to embrace nearly all forms of assisted reproduction as widening the scope of our reproductive autonomy.

It is natural, in the face of this divergence, to ask what, if any, connection there is between the value of reproduction as such, and the rather limited range of entirely 'natural' acts which give rise to reproduction on the one hand, and the broader range of 'artificial' acts which can result in the production of offspring on the other.

Again, a range of consequentialist and anti-perfectionist views return similar verdicts about the connection between sexual intercourse and reproduction: some might wish to reproduce in this way, and might find either the reproduction to be benefited by its relation to sex, or the sex to be benefited by its relation to reproduction. However, no necessary connection exists between the two such that moral norms against their separation could make sense.

The more conservative approach tends to see a more necessary connection here between reproduction and intercourse, and, indeed, intercourse within the context of marriage. As Gilbert Meilander has written, "Maintaining the connection between procreation and the sexual relationship of a man and a woman is good both for that relationship and for the children" (Meilaender, 1997, pp. 2).

Such a claim requires, for its defense, a fuller account of the nature and value of marriage than has already, in much of the relevant literature, been provided, and, as well, a more fully articulated view of the nature of human action. For, first, the view of marriage implicit in arguments such as Meilander's seems to require that the unity of a couple in marriage be in some way a unique union, which, in consequence of its special nature, is uniquely valuable. Second, that unity is supposed to be uniquely tied to the possibility of procreation. The connection here appears to be this: the unity of spouses in sexual intercourse is an otherwise impossible unity of function of two bodies, specifically, the reproductive function. And the child that is the offspring of that unity is thus seen as a kind of fruit of the marital union, and, inasmuch as that union is a loving one, the fruit of marital love and friendship.

Thus, in Meilander's account, the possibility of children, hoped for but not chosen as a product, frees spouses from a potential self-absorption; while the offspring is benefited by being the fruit of a loving unity which is uniquely valuable. By contrast, where children are produced, technically, or simply outside a loving union, their being is subordinated to the desires of one or more persons, and they are made objects of choice, rather than hope.

Meilander's claims are set, on his own account, within a specifically religious understanding of marriage and procreation, as are other prominent statements of the conservative position (Congregation for the Doctrine of the Faith, 1987; Grisez, 1993). More work is clearly necessary in articulating what hold on a secular mind such considerations ought to have. Further, even within the religious traditions that have articulated this understanding, there are questions as to how closely tied a stable, loving conjugal union and the generation of children must be (McCormick, 1996).

Why should autonomy be important in this context? And, to the extent that significant strides have been made culturally towards social structures that encourage autonomous choice of marriage and parenthood, why should this be seen as an advance? A possible answer begins with an aspect of the intertwined goods of marriage and procreation which can easily be overlooked in a social context which does not allow for autonomous choice of, say, a marriage partner. Just as there is little, if any, value in religious belief apart from that belief's being voluntary, and in any religious devotion apart from that devotion's being autonomously chosen, so does any form of legitimate friendship – of which marital love seems to be one form – seem to require an element of choice and autonomy as part of its constitution. Friendship, and hence marital love, is a "reflexive good," one into which choice enters as a part of its own actualization (Grisez, Boyle, and Finnis, 1987).

Marriage, as a unique form of friendship, and not, simply, as a coerced physical union, thus depends upon its having been entered into voluntarily, upon the voluntary choice of partner, and, extending now towards its procreative aspect, upon an autonomous openness to new life as the fruit of that autonomously chosen union.

We see, then, that a full understanding of the nature and value of reproductive autonomy requires an account of the nature of human good, and its relationship to choice. Answers which place choice prior to good are likely to have significantly different consequences from answers which place the good prior to choice. Further, adherents to the latter position still need to articulate substantively what goods are at stake in reproduction, and how various forms of human action bear upon those goods.

We have been addressing, in the past few pages, what might be considered 'spill-over' issues surrounding those issues immediately concerned with reproductive ethics – issues of the role of sex, and sexual relationships, of friendship, and marriage, in the broad context of which reproduction is a part. Some similar concerns exist at the other end of reproduction. For it would be surprising, were the conservative account above on the right track, if pregnancy was best characterized as a medical 'problem,' and the delivery room as essentially a theatre for professional medical surgery, from which non-medical personal, such as fathers, were to be excluded. And yet, throughout the twentieth century, such views have not been unknown (Edwards and Waldorf, 1984).

For the anti-perfectionist as well, such an approach might be unsatisfactory, for the medical problem approach tends towards, in certain ways, a limitation of autonomy: when faced with medical problems, difficulties, and diseases, doctors traditionally have made the serious choices. However, it would neither be unsatisfactory as such, nor is there a clear trend towards a wholesale de-medicalization of pregnancy among more 'liberal' thinkers as there is among proponents of a 'thicker' value scheme surrounding reproduction. For inasmuch as pregnancy can sometimes thwart autonomous choices, it may tend to be seen more as a 'disease,' and the language appropriate to a context of medical solution to a health problem may be utilized, e.g., in discussion of abortion options. Likewise, inasmuch as children are conceived of as the products of technological production, a technique oriented model of medicine and surgery might be applied, both in terms of assisted reproductive services, and in terms of delivery room procedure. The emphasis which is placed by some hospitals on caesarian delivery over natural childbirth seems of a piece with this trend.

So again, conceptualizing the medical status of pregnancy, childbirth, and fertility requires an account of the substantive values at stake – are those values limited to the value of autonomy, and if not, what substantive account can be given, and what meta-ethical explanation offered of them?

V. CONCLUSION

It is often said that recent technological developments have radically changed the face of reproductive ethics and left a philosophical void in need of filling. However, I believe that this is an overstatement. The path I have traced in the text above, from issues of personhood and killing, to issues of technical production of children, to issues of autonomy and the relationship between sexual intercourse and children, is a path which has, I believe, been followed before. In his *Republic*, Plato deals with precursor issues to all these problems. We may think, first, of his provisions for infanticide, second, of the eugenic streak which runs through his discussion of the generation of children, and third, of the lack of reproductive and marital autonomy granted to the guardians by the state. While one might certainly disagree with Plato's views on these matters, it is still the case that he began a tradition of philosophizing about such issues, a tradition which was advanced, often in regards to the same issues, by Aristotle, Augustine, Aquinas, Kant, and others. Those working in the field of reproductive ethics must certainly look forward, and I hope I have indicated some topics that require further investigation. But I believe we should also look back, and not forget, in the vertigo induced by the new reproductive technologies, the stability of over two thousand years of philosophical discussion.

Department of Philosophy
University of South Carolina
Columbia, SC 29208, U.S.A.

NOTES

1 Occasionally, one finds the distinction between human and person obscured. Lee Silver, for example, asks whether the conceptus is a human. But he then distinguishes between human in an ordinary sense and human in a special sense (Silver, 1998). The distinction clearly parallels the human/person distinction. On the other hand, recent discussions of individuation appear to raise the possibility that the very early embryo is human, but not *a* human, that is, not an individual human being. I address this issue in the text.

2 'Pre-embryo' is a somewhat vague and politically and ethically charged term. Here, I use it to indicate the embryo prior to the emergence of the primitive streak. Whether the word reflects a genuine distinction in the status of the embryo, or whether the word is itself used to force the appearance of such a distinction is a matter of some debate (Jones and Telfer, 1995; Kischer, 1997; see also references cited in these texts). The issue of whether the word marks a genuine distinction between a stage at which the conceptus is an ontological individual, and a stage at which it is not, is addressed in the text.

3 For example, he points to our beliefs concerning the individuality of other objects which are potentially divisible, such as flatworms (Lee, 1996).

4 Debate over delayed homization may be followed in the following: Ford, 1988; Shannon and Wolter, 1990; Grisez, 1990; Johnson, 1995; Porter, 1995. For an account of the deliberative proceedings of the Human Embryo Research Panel, which accepted the delayed individuation claim, see Tauer, 1997. These texts also contain additional references, which may be of help in working through the various strands of this debate. Recent analytic philosophy has also taken up issues in philosophical embryology: see Howsepian, 1992; Oderberg, 1997; and Tollefsen, 2000.

5 This position, that the capacity to achieve rational agency is what makes something a person, must be distinguished from the view that, in virtue of this capacity, the fetus has the potential to become

a person, and therefore should have the rights of a person. Some arguments about the importance of 'potentiality' confuse the two positions; opponents of abortion are sometimes saddled with the view that the fetus deserves respect because it is a potential person (Benn, 1984; Engelhardt, 1996)

6 The consequence of germ-line therapy and enhancement thus seems to me different from the influence which maternal RNA might have on an embryo. There is a dispute here, as to whether or not (a) a zygote will become a human individual in the absence of maternal genetic information; and (b) whether organizing information is even received from the mother. Some have argued that the zygote will not develop without the maternal information (Bedate and Cefalo, 1989). Antione Suarez has argued that this is false, and has denied that *any* genetic information is received from the mother (Suarez, 1990). However, as Lee argues, "…even if it were true that some information is received from maternal molecules, this would not show that the pre-implantation embryo was not a complete human being. There is no reason to expect that *all* of the future features of the developing organism should be already determined by its internal genetic make-up. Environmental conditions, which could include maternal molecules within the uterus, can determine many of the future characteristics of the developing organism...still, how this information fits within the overall development of this organism is determined from within by the organism's own directed growth...[Such information] does not determine the primary organization and direction of the multitude of cell differentiations and acquisitions and uses of nutrition occurring in this organic system. That primary organization comes from within the embryo itself" (Lee, 1996, pg. 101). My suggestion, made in the text above, is that genetic enhancement performed on a pre-embryo is a restructuring of the embryonic organization from without, rather than an internal reorganization of externally available information, a reorganization attributable to the embryo itself. It is thus *not* like the influence of maternal mitochondrial RNA on the developing embryo.

7 The discussion in the text can hardly be considered a complete discussion of all the issues involved in the problems of genetic therapy and enhancement. The discussion in Section Two of this essay also has ramifications that reach to these issues. A variety of moral, prudential and social concerns, on both sides of the debate, are raised in the recent literature (Suzuki and Knudtson, 1988; Anderson, 1989; Fletcher, 1994; Resnick, 1994; Gardner, 1995; Buchanan, 1996; Engelhardt, 1996; Juengst, 1997; McGee, 1997; Torres, 1997). All the cited works themselves contain valuable references and bibliographies. There is also a journal devoted to these and other issues, *Human Gene Therapy.* The sorts of concerns articulated in these discussions, including, but not limited to, privacy, fairness, scientific advance, beneficence, autonomy, sanctity of human life, artificiality, line drawing, genetic class divisions, eugenics, slippery slopes, and potential longterm hazards, are similar to many of the concerns articulated in discussion of the human genome project (Macer, 1991; Rix, 1991; Skene, 1991; Kitcher, 1994; Rosenberg, 1996; Wiesenthal and Wiener, 1996). Again, references in these works may be consulted for further information. Finally, although in the text I focus on whether genetic enhancement of the pre-embryo constitutes *killing*, it is possible to view the difficulties there raised as problems of *identity* – if, for example, one think that the pre-embryo is an individual but not a person, or simply because one wonders whether enhancement changes the individual the zygote would have been (Zohar, 1991; Persson, 1995; Elliot, 1997).

8 Even so astute an interpreter of traditional morality as Alan Donagan makes this mistake. Donagan suggests that a defender of the view that intention was more important than the wider notion of voluntariness would have to excuse a creditor who foresees but does not intend the ruin of his debtor: "Can it be maintained, as some would like to maintain, that because the creditor's action is not intentional under that description ['ruining his debtor'], then, even if it should be morally impermissible to ruin his debtor, he would not be culpable for doing so?" (Donagan, 1977, pp. 125) The defender of double effect need not claim that *only* the intention is morally relevant, but only that intention is most important in a moral assessment of an agent's acts.

9 In a trolley problem, we are asked to choose which of two tracks to let an out-of-control trolley go down, given that there will be casualties on both sides, or on one but not the other (Foot, 1967). Like the violinist problem, trolley problems elicit intuitions on the differences between what is intended and what is merely accepted as a side effect.

10 Recent discussions of Thomson's argument may be found in Hursthouse, 1988; Kamm, 1992; Lee, 1996. Even more recently David Boonin-Vail has undertaken to defend Thomson's arguments against one of the most prevalent objections to the violinist analogy, viz., that the fetus is present in the mother's womb through the mother's voluntary actions (Boonin-Vail, 1997). Anyone seriously interested in the philosophical, legal, and religious issues surrounding abortion and personhood should see Germain Grisez's early work on the subject (Grisez, 1970).

11 Because of intracellular mitochondrial genes that are shared by identical twins but not clones.

12. E.g., there is a fundamental question concerning the relationship between value and experience in need of settling. The perfectionist view of reproduction must deny the claim that something is valuable merely because it is experienced as such. But then an entire theory of value seems to be required, as well as a theory about the relationship between action and value.

REFERENCES

Anderson, W.F.: 1989, 'Human Gene Therapy: Why Draw a Line?' *The Journal of Medicine and Philosophy* 14, 681-93.

Anderson, W.F.: 1994, 'Human Gene Therapy: Scientific and Ethical Considerations,' 24, 275-291.

Atwood, M.: 1986, *The Handmaid's Tale*, Houghton Mifflin, Boston.

Beauchamp, T. and Childress, J.: 1994, *Principles of Biomedical Ethics*, Fourth Edition, Oxford University Press, New York.

Bedate, C.A. and Cefalo, R.C.: 1989, 'The Zygote: To Be or Not to Be a Person,' *The Journal of Medicine and Philosophy* 14, 641-45.

Benn, S.I.: 1984, 'Abortion, Infanticide, and Respect for Persons,' in Feinberg, J. (Ed.), *The Problem of Abortion* Second Edition, Wadsworth Publishing Co., Belmont.

Boonin-Vail, D.: 1997, 'A Defense of "A Defense of Abortion": On the Responsibility Objection to Thomson's Argument,' *Ethics* 107B, 286-313.

Boyle, J.: 1980, 'Toward Understanding the Principle of Double Effect,' *Ethics* 90, 527-38.

Boyle, J.: 1991, 'Who is Entitled to Double Effect?' *The Journal of Medicine and Philosophy* 16, 475-94.

Buchanan, A.: 1995, 'Equal Opportunity and Genetic Intervention,' *Social Philosophy and Policy* 12, 105-35.

Buchanan, A.: 1996, 'Choosing Who Will Be Disabled: Genetic Intervention and the Morality of Inclusion,' *Social Philosophy and Policy* 13, 18-46.

Buchanan, A., Daniels, D., Wilker, D., and Brock, D.W.: 2000, *From Chance to Choice: Genetics and Justice,* Cambridge University Press, Cambridge.

Clarke, S. and Simpson, E. (eds.):1989, *Anti-Theory in Ethics and Moral Conservatism*, State University of New York Press, Albany.

Clouser, K. and Gert. B.: 1990, 'A Critique of Principlism,' *The Journal of Medicine and Philosophy* 15, 219-36.

Congregation for the Doctrine of the Faith: 1987, *Donum Vitae, Origins* 16, 697-711.

Donagan, A.: 1977a, *The Theory of Morality*, University of Chicago Press, Chicago.

Donagan, A.: 1977b, 'Informed Consent in Therapy and Experimentation,' *The Journal of Medicine and Philosophy* 2, 307-29.

Edwards, M. and Waldorf, M.: 1984, *Reclaiming Birth*, The Crossing Press, Trumansburg, NY.
Elliot, R.: 1997, 'Genetic Therapy, Person-Regarding Reasons and the Determination of Identity,' *Bioethics* 11, 151-69.
Engelhardt, H.T.: 1996a, *Foundations of Bioethics*, Second Edition, Oxford University Press, New York.
Engelhardt, H.T.: 1996b, 'Germ-Line Genetic Engineering and Moral Diversity: Moral Controversies in a Post-Christian World,' *Social Philosophy and Policy* 13, 47-62.
English, J.: 1975, 'Abortion and the Concept of a Person,' *Canadian Journal of Philosophy* 5, 233-43.
Finnis, J.: 1980, *Natural Law and Natural Rights*, The Clarendon Press, Oxford.
Finnis, J.: 1991, *Moral Absolutes: Tradition, Revision, and Truth*, The Catholic University of America Press, Washington, D.C.
Fletcher, J.: 1966, *Situation Ethics*, The Westminster Press, Philadelphia.
Fletcher, J.: 1994, 'Ethical Issues in and Beyond Prospective Clinical Trials of Human Gene Therapy, *The Hastings Center Report* 24, 293-309
Foot, P.: 1967, 'The Problem of Abortion and the Doctrine of Double Effect,' *Oxford Review* 5, 5-15; reprinted in Foot, P.: 1978, *Virtues and Vices, and Other Essays in Moral Philosophy*, Basil Blackwell, Oxford.
Ford, N.: 1988, *When Did I Begin? Conception of the Human Individual in History, Philosophy, and Science*, Cambridge University Press, New York.
Gardner, W.: 1995, 'Can Human Genetic Enhancement be Prohibited?' *The Journal of Medicine and Philosophy* 20, 65-84.
Geach, P.:1956, 'Good and Evil,' *Analysis* 17, 33-42.
George, R.: 1993, *Making Men Moral*, Oxford University Press, New York.
Gilligan, C.: 1982, *In A Different Voice*, Harvard University Press, Cambridge, Massachusetts.
Grisez, G., Boyle, J. and Finnis, J.: 1987, 'Practical Principles, Moral Truth, and Ultimate Ends,' *The American Journal of Jurisprudence* 32, 99-151.
Grisez, G., Boyle, J., Finnis, J., and May, W.: 1988, '"Every Marital Act Ought To Be Open To New Life": Toward a Clearer Understanding,' *The Thomist*, 52, 365-426.
Grisez, G.: 1990, 'When Do People Begin?' *Proceedings of the American Catholic Philosophical Association* 63, 27-47.
Grisez, G.: 1970, *Abortion: The Myths, The Realities, and the Arguments*, Corpus Books, New York.
Grisez, G.: 1978, 'Against Consequentialism,' *American Journal of Jurisprudence* 23, 21-72.
Grisez, G.: 1993, *The Way of Our Lord Jesus Christ, Vol. II*, Franciscan Press, Chicago.
Harris, J.: 1997, 'Goodbye Dolly? The Ethics of Human Cloning,' *The Journal of Medical Ethics* 23, 353-360.
Harris, J.: 1998, *Clones, Genes, and Immortality: Ethics and the Genetic Revolution*, Oxford University Press, Oxford.
Howsepian, A.A.: 1992, 'Who Or What Are We?,' *Review of Metaphysics* 45, 483-502.
Hursthouse, R.: 1988, *Beginning Lives*, Oxford University Press.
Johnson, M.: 1995, 'Questio Disputatia – Delayed Hominization: Reflections on Some Recent Catholic Claims for Delayed Hominization,' *Theological Studies*, 56, 743-63.
Jones, D. and Tefler, B.: 1995, 'Before I Was and Embryo, I Was a Pre-Embryo: Or Was I?' *Bioethics* 9, 32-49.
Juengst, E.: 1997, 'Can Enhancement be Distinguished from Prevention in Genetic Medicine?' *The Medicine and Journal of Philosophy* 22, 125-42.
Kamm, F.: 1992, *Creation and Abortion: A Study in Moral and Legal Philosophy*, Oxford University Press, New York.

Kant, I.: 1959, *Grounding for the Metaphysics of Morals* Lewis White Beck, trans., Indianapolis, Library of the Liberal Arts, Indianapolis.
Kass, L.: 1997, 'Cloning of Human Beings,' Testimony presented to the National Bioethics Advisory Commission, March 14, 1997, *http://www.all.org/abac/clon* sec.htm, 1-3
Khushf, G.: 1997, 'Embryo Research: The Ethical Geography of the Debate,' *The Journal of Medicine and Philosophy* 22, 495-519.
Kischer, C.: 1997, 'The Big lie in Human Embryology: The Case of the Preembryo,' *Linacre Quarterly* 64, 53-61.
Kitcher, P.: 1994, 'Who's Afraid of the Human Genome Project,' *Philosophy of Science Association* 2, 313-21.
Krimmel, H.: 1992, 'Surrogate Mother Arrangements From the Perspective of the Child,' in Alpern, K., ed., *The Ethics of Reproductive Technology*, Oxford University Press, Oxford, pp. 57-70.
Lee, P.: 1996, *Abortion and Unborn Human Life*, The Catholic University of America Press, Washington, D.C
Lewontin, R.: 1997, 'The Confusion Over Cloning,' *The New York Review of Books*, http://www.nybooks.com/nyrev/archives.html, 1-7.
Macer, D.: 1991, 'Whose Genome Project?,' *Bioethics* 5, 183-211.
MacIntyre, A.: 1984, *After Virtue: A Study in Moral Theory*, Second Edition, University of Notre Dame Press, South Bend.
Maynard-Moody, S.: 1995, *The Dilemma of the Fetus*, St. Martin's Press, New York.
McCormick, R.: 1996, 'Human Reproduction: Dominion and Limits,' *The Kennedy Institute of Ethics Journal* 6, 387-392.
McGee, G.: 1997, *The Perfect Baby*, Rowman and Littlefield, New York.
Meilaender, G.: 1997, 'Remarks on Human Cloning,' Testimony presented to the National Bioethics Advisory Commission, March 13, 1997, http://www.all.org/abac/clon-prt.htm, 1-3.
Mirkes, R.: 2001, 'NBAC and Embryo Ethics', *The National Catholic Bioethics Quarterly* 1, 163-187.
Nagel, T.: 1979, 'War and Massacre,' in *Mortal Questions*, Cambridge University Press, Cambridge, pp. 53-74.
Noddings, N.: 1984, *Caring: A Feminine Approach to Ethics and Moral Education*, University of California Press, Berkeley.
Oderberg, D.S.: 1997, 'Modal Properties, Moral Status, and Identity,' *Philosophy and Public Affairs* 26, 259-298.
Persson, I.: 1995, 'Genetic Therapy, Identity, and the Person-Regarding Reasons,' *Bioethics* **9**, 16-31.
Porter, J.: 1995, 'Individuality, Personal Identity, and the Moral Status of the Pre-embryo: A Response to Mark Johnson,' *Theological Studies* 56, 763-770.
Purdy, L.:1996, *Reproducing Persons: Issues in Feminist Bioethics*, Cornell University Press, Ithaca.
Radin, M.: 1987, 'Market-Inalienability,' *Harvard Law Review* 100, 1849-1937
Rawls, J.: 1972, *A Theory of Justice*, Harvard University Press, Cambridge.
Raz, J.: 1986, *The Morality of Freedom*, Oxford University Press, Oxford.
Resnick, D.: 1994, 'Debunking the Slippery Slope Argument Against Human Germ-Line Therapy,' *The Journal of Medicine and Philosophy* 19, 23-40.
Richards, D.A.J.: 1986, *Toleration and the Constitution*, Oxford University Press, 1986.
Rix, B.: 1991, 'Should Ethical Concerns Regulate Science? The European Experience With the Human Genome Project,' *Bioethics* 5, 250-6.
Robertson, J.: 1994a, *Children of Choice*, Princeton University Press, Princeton.
Robertson, J.: 1994b, 'The Question of Human Cloning,' *The Hastings Center Report* 24, 6-14.

Rosenberg, A.: 1996, 'The Human Genome Project: Research Tactics and Economic Strategies,' *Social Philosophy and Policy* 13, 1-17.
Shannon, T. and Wolter, A.: 1990, 'Reflections on the Moral Status of the Pre-embryo,' *Theological Studies* 51, 743-763.
Shannon, T.: 1997a, 'Fetal Status: Sources and Implications,' *The Journal of Medicine and Philosophy* 22, 415-22.
Shannon, T.: 1997b, 'Response to Khushf,' *The Journal of Medicine and Philosophy* 22 (1997) 525-7.
Silver, L.: 1998, *Remaking Eden: Cloning and Beyond in a Brave New World*, Avon Books, New York.
Skene, L.: 1991, 'Mapping the Human Genome: Some Thoughts for Those Who Say "There Ought to be a Law on It",' *Bioethics* 5, 233-49.
Smith, J.: 1993, *Why Humanae Vitae Was Right: A Reader*, Ignatius Press, San Francisco.
Spoerl, J.: '*In Vitro* Fertilization and the Ethics of Procreation,' *Ethics and Medicine* 15, 10-14.
Strong, C.: 1997a, *Ethics in Reproductive and Perinatal Medicine*, Yale University Press, New Haven.
Strong, C.: 1997b, 'The Moral Status of Preembryos, Embryos, Fetuses, and Infants,' *The Journal of Medicine and Philosophy* 22, 457-78.
Strong, C.: 1997c, 'Response to Khushf,' *The Journal of Philosophy and Medicine,* 22 521-23.
Suarez, A.: 1990, 'Hydatidiform Moles and Teratomas Confirm the Human Identity of the Preimplantation Zygote,' *The Journal of Medicine and Philosophy* 15, 627-35.
Tauer, C.: 'Embryo Research and Public Policy: A Philosopher's Appraisal,' *The Journal of Medicine and Philosophy* 22, 423-39.
The Ethics Committee of the American Fertility Society: 1992, 'Ethical Considerations of In Vitro Fertilization,' in Alpern, K., ed., *The Ethics of Reproductive Technology*, Oxford University Press, Oxford, pp. 301-5.
Tollefsen, C.: 2000, 'Embryos, Individuals, and Persons: An Argument Against Embryo Creation and Research,' *Journal of Applied Philosophy* 18, 65-78.
Thomson, J.: 1971, 'A Defense of Abortion,' *Philosophy and Public Affairs* 1, 47-66.
Tong, R.: 1997, *Feminist Approaches to Bioethics*, Westview Press, Boulder, Colorado.
Torres, J.: 1997, 'On the Limits of Enhancement in Human Gene Transfer: Drawing the Line,' *The Journal of Medicine and Philosophy* 22, 43-53.
Warnock, M.: 1985, *A Question of Life: The Warnock Report on Human Fertilisation and Embryology*, Basil Blackwell, Oxford.
Warren, M.: 1973, 'On the Moral and Legal Status of Abortion,' *The Monist* 57, 43-61.
Wiesenthal, D. and Wiener, N.: 1996, 'Privacy and the Human Genome Project,' *Ethics and Behavior* 6, 189-201.
Zohar, N.: 1991, 'Prospects for 'Genetic Therapy' – Can a Person Benefit From Being Altered', *Bioethics* 5, 275-88.

KURT BAYERTZ AND KURT W. SCHMIDT

TESTING GENES AND CONSTRUCTING HUMANS – ETHICS AND GENETICS [1]

Gene technology has had dramatic effects on the field of human medicine. Many of the latest diagnostic and therapeutic procedures are based on gene technological methods, and further progress in this field is expected in the near future. Above all, scientists are convinced that medicine will be revolutionized by the Human Genome Project: both at a theoretical level, involving the understanding of diseases and their geneses, and at a practical level, involving the development of more effective strategies to combat those diseases. They believe us to be at the beginning of a 'molecular medical era' which, as a new 'paradigm', will be analogous in fundamental importance to the 16th century paradigm of anatomy or the 19th century paradigm of cellular pathology.

This far-reaching importance is due to some specific features of gene technology which distinguish it from other technologies, and which are manifested in what may be called its 'depth' and its 'breadth'. On the one hand, gene technology *deeply* invades human nature and living matter generally. On the other hand, it is *broadly* applicable to many different areas of biomedicine and many different aims. Gene technology can be used as a tool or a methodology for (a) biomedical research (basic and applied); (b) diagnostics; (c) therapeutics; (d) industrial or pharmaceutical production. There seems to be no part of the biomedical field for which gene technology is not – at least potentially – of importance.

This 'revolutionary' potential has led gene technology to become a major target for ethical reflection. While there can be no doubts that gene technology will solve many biomedical problems, it is just as transparent that it will also create new ones. This in turn raises many ethical questions, some of which have, in the past few years, already been cause for controversy. These issues cover a broad spectrum which may be divided up into (at least) the following four types:

1. *'Heavy questions', e.g. "Are human beings entitled to uncover the secret of life?" or "Are human beings entitled to 'Play God'?"*
2. *Conceptual problems, e.g. "What meanings may terms like 'disease' or 'health' acquire at a molecular level?" and consequently "Can a precise distinction be made between therapy and enhancement?"*

G. Khushf (ed.), Handbook of Bioethics, 415–438.

3. *Social problems caused by some unwanted side-effects of gene technology, e.g. "How will the proliferation of genetic data influence the self-image of individuals and of the human race in general?" or "Will the widespread use of prenatal diagnostics lead to a decreasing tolerance towards handicapped people?"*
4. *Problems resulting from possible technological risks, e.g. "Should gene therapy be restricted to severe and fatal diseases?" or "How can the reliabilty and validity of genetic tests be guaranteed?"*

It would obviously exceed the limitations of the present paper to analyze or evaluate these different risks and weigh them up against corresponding benefits. Instead, we should like to focus on two of the more fundamental ethical issues surrounding the impact of gene technology on the core of the biomedical enterprise: diagnostics and therapy. Problems of research and of pharmaceutical production will play no role in the following considerations.

In Section I we shall discuss molecular genetic diagnostics and some of its (potential) consequences for the *autonomy* of individuals. The concept and the principle of autonomy are of outstanding importance for the recent bioethics 'paradigm.' But doubts have arisen as to whether additional options gained as a result of biotechnological development can really be seen as benefitting individual autonomy. It is not only a case of individual autonomy *always* being under threat (a principle exists for this very reason, with the explicit intention of protecting it); we will also discuss the hypothesis that the further development and comprehensive application of genetic diagnostics will *structurally* question some of the prerequisites underlying the principle of autonomy: this holds true at the conceptual and social levels.

The main topic of Section II is the *constructive* potential of this new technology. From its beginning, proponents have promised and opponents have warned that gene technology will lead to a deep change in the essence of the medical enterprise, namely a transition from repair to design. This prospect gives rise to a whole series of ethical problems, which we have summarized in the two categories: (1) "Where to draw a line?" and (2) "Why draw a line?". In the final section we return to the concept of autonomy. Although gene technology will provide new options regarding individual self-determination and self-realization, our considerations conclude that this is only one dimension of the development; the other is the price which will inevitably be paid for maximizing the constructive potential of gene technology, namely the autonomy of future generations.

I. GENETIC DIAGNOSTICS AND INDIVIDUAL AUTONOMY

Gene technological diagnostic methods are no longer reserved for the relatively few monogenous hereditary diseases, and are now also being used to detect genetic dispositions for such widespread diseases as cancer or heart disease, as well as some non-hereditary diseases. In so doing, they have expanded beyond the narrower field of human genetics, entering more or less all areas of medicine. In some respects the knowledge gained is similar to that produced by conventional diagnostics, and yet it also differs profoundly. For a long time now there have been intensive and fruitful discussions about the ethical, social and legal implications of this kind of diagnostics; some detailed criteria for its practical application have been drawn up as a result (cf. Andrews *et al.*, 1994; Annas and Elias, 1992; Cook-Degan, 1994; Holtzmann, 1989). We shall not present those results here, but concentrate instead on the implications potentially arising from a structural limitation of individual autonomy.

A. The superindividual character of genetic knowledge and limits of the Lockean paradigm

A special feature of genetic knowledge is that in many cases it is not only knowledge about an individual person, but also knowledge about a more or less large number of other persons. This is due to the fact that genes and genomes are essentially *superindividual* entities. All human individuals share 50 per cent of their genes with each parent, 25 per cent with each grandparent, etc.; human beings even have most of their genes in common with non-related other human beings. This means that the 'object' of a molecular genetic diagnosis cannot always be reduced to the individual person being tested: the diagnosis often amounts to an examination of certain characteristics of an entire group of people, including some not yet born.

If one assumes – as is widely accepted – that genetic knowledge about a particular person has to be viewed as particularly 'sensitive' knowledge, then difficult ethical problems ensue. Take, for example, Mrs. X, who undergoes genetic testing for hereditary breast cancer. Unfortunately the result is positive and she tells her family about it, horrified. In so doing, however, she informs her daughter about the fact that she too is at an increased risk of contracting this disease. The problem being that her daughter had previously had no idea about this risk, maybe not wanting to have been informed. By exercising her right to know, Mrs. X simultaneously violates the right of her daughter not to know. How are the 'autonomies' of the different persons involved – synchronously and diachronously – to be weighted when the field of diagnostics ceases to be limited to individuals? The diversity of appraisals that individuals may have regarding these ethical concerns highlights the values involved in determining an appropriate *standard of care* in the context of genetic diagnostics. Idealization (like tailoring disclosure in informed consent to suit the particular patient) fails to attend to the reality of a need for standardization (Parker and Majeske, 1996).

Of course, the problem of unwanted information is not altogether new. Before genetic testing it was also possible, for some people and in some cases, to deduce health risks from family histories: for example, people with several close relatives who have all died of heart attacks relatively young will necessarily have presumed themselves to be at a significantly higher risk. Nevertheless, the progress of genetic diagnostics is changing the situation both technically and ethically. (a) Risks are rapidly being assessed with increasing accuracy; (b) the knowledge pool is growing in size for two reasons: more and more genetic diseases can be tested for, and the number of persons being tested is increasing; (c) previously, genetic knowledge came to light 'spontaneously' and unavoidably, whereas the information arrived at today through gene technology is derived purposefully and consciously; and (d) modern possibilities of data processing, data transport and data storage render information available anytime, anywhere. This raises the ethical question of whether it is legitimate to perform relevant genetic tests when doing so violates the right of third parties to informational self-determination. Or, to put it another way: how are the rights of the persons involved to be weighed up in such a case?

Let us take a look at another aspect of our above example: Mrs. X hears the news that the results of her genetic tests for hereditary breast cancer are positive and decides *not* to tell her family, in particular her daughter. She asks her doctor to handle the matter confidentially. Let us also assume that her daughter would have been extremely glad to have this information in order to plan her life accordingly. Would withholding this information from her then not amount to an infringement upon her right to self-determination? Furthermore, would the daughter even have had the right to remain ignorant and not want to have this information so central to her life (cf. Rhodes, 1998)?

Whilst some people believe that the superindividual character of genetic information must not be allowed to influence the previously individualistic doctor-patient relationship and that issues surrounding the imparting of disclosure, beneficence and medical secrecy must not be allowed to change that doctor-patient relationship, others propose extending the definition of the term 'patient': whereas in the classic individualistic approach the physician has responsibilities towards the *individual* patient (regarding medical secrecy, etc.), in an alternative framework of genetic consulting the 'patient' could be the entire family environment (cf. Wachbroit, 1993). If, in our example above, this idea of the patient were to include Mrs. X's daughter, then the health professional's informing of the daughter would not constitute a breach of confidentiality. One might even argue that the health professional is not simply *permitted* to inform the daughter, but actually *required* to do so. The moral problem is no longer concern about confidentiality, but now concern about: just who is the patient?

Our interest lies in emphasizing the essentially superindividual character of genetic information and calling to mind the potential limitations which this superindividuality could pose upon individual access to genetic information. According to a widespread view for which John Locke is usually considered to be a source of authority, there is a close connection between the concepts 'person' and 'property'; the 'person' concept includes one's own body, meaning that the relationship between human beings and their bodies may be perceived as a relationship of property: "Every Man has a *Property* in his own *Person*" (Locke, *Second Treatise of Government*, § 27). Genes are beyond doubt part of the human body, and yet in their case one of the central prerequisites underlying Locke's theory is *not* fulfilled: the clear distinction, indeed separateness of individuals and their bodies. Locke could hardly have imagined that access to a part of one's own body could also mean access to parts of another individual's body. It has consequently been suggested that genetic information about any individual be regarded not as personal to that indvidual, but as the *common property* of other people sharing those genes. This would curb individual access to 'one's own' genetic information, thus protecting other individuals.

The Lockean paradigm also reaches its limits at the point where autonomous access of an individual to his or her genetic information affects the rights of an entire group. This problem occurs – whether as part of the *Human Genome Diversity Project* or on the initiative of individual firms – where testing is to be carried out not only on the single genes of single individuals, but on the genomes of entire populations. Some races with genetic peculiarities of interest to researchers decline examination on the grounds of scientific colonialism.[2] Individual race members could almost certainly be won over to participate in such a research program, and the refusal of the majority thus circumnavigated; yet, applying the theory of common property, such use of 'genetic dissidents' would be morally reprehensible. In some countries (e.g. Canada) projects of this kind are therefore only legally permissible if willing individuals can be found to participate *and* if the leaders of the people in question give their consent. Individual rights of access to genetic information are therefore implicitly limited by the essentially superindividual character of that information (Cranor, 1994; Bayertz, 1997).

A similar notion also seems to be behind the *Universal Declaration on the*

Human Genome and Human Rights adopted by UNESCO in November 1997. Here the problem is not discussed in terms of property, however, but in terms of "common heritage." Based on the view that scientific and technological progress plus the new options resulting from it have rendered the human genome a good worthy of protection, this Declaration seeks to establish a path enabling a new balance to be found between the common interests of humanity and individual rights. Article 1 of the Declaration states: "The human genome underlies the fundamental unity of all members of the human family, as well as the recognition of their inherent dignity and diversity. In a symbolic sense, it is the heritage of humanity" (UNESCO, 1997).

This "common hertitage" point of view, which over the last decade has been successful in public policy settings, especially in Europe, is deemed by others to be an outdated body of thought from the 19th century, conceptually flawed and socially dangerous. We cannot conserve this 'natural resource' with declarations (e.g. by forbidding interventions in the human germ line) because, from a scientific point of view, the human genome is not a natural object but an heuristic abstraction, like the anatomist's concept of the human skeleton. Our "worries about preserving its integrity for future generations become concerns about the future of an idea, not a natural resource" (Juengst, 1998b). We should pay more attention *to the social context* of future generations than to their *genetic resources*. In the long run, "we should not be concerned about the ones who benefit from gene therapy or enhancement, rather the preeminent need will be to protect those among the future generations 'unfortunate' enough to enjoy an untampered genetic inheritance from the social discrimination and the unfair disadvantages that they could face in living and working with their genetically engineered neighbours" (Juengst, 1998b).

B. The social context of genetic testing

From a medical point of view, genetic diagnostics has two essential advantages over conventional procedures. Firstly, the *predictive potential* is significantly higher: a genetic test is able to establish the risk of contracting a disease long before first symptoms present themselves. Since detecting a risk early provides opportunity for intervention which are often not there at a later date, this is extraordinarily attractive to the medical profession. Secondly, genetic tests are able to function as a vehicle of medical *individualization*, enabling named persons to be attributed with increased risks for particular diseases to a far greater extent than was ever possible before, even including statements about the expected course and severity of disease (cf. Juengst, 1998a). Seen in this light, individually 'tailored' therapies could well become a future reality.

In many respects this development has to be seen as an achievement: it will provide a chance for medical aid in cases which today are still considered hopeless. And yet the price to be paid for this achievement cannot be ignored for long. Taking the *social context* of this development into account, four implications emerge, all possibly involving limits to individual autonomy.

(1) The burden of knowledge. Within the medical context knowledge is seldom sought for its own sake, but for the sake of the therapeutic options it could provide. This is due to the fact that the goal of medicine is people's health. This goal is (ideally) achieved through therapy; diagnoses are only prerequisites for a correct choice of therapy. In the field of molecular medicine this leads to a serious problem. Knowledge about the genetic (co-)causes of diseases may have increased dramatically in the recent past, and the ability to test for diseases genetically likewise; and yet the increase in therapeutic options has remained modest in comparison. The gulf between the diagnostic and therapeutic possibilities available has continually increased, and this is unlikely to change in the foreseeable future. In many cases where genetic testing is carried out and proves positive, the medical profession can offer no more than that positive result. Sometimes this information alone may be highly valued, and yet this will not be the norm. Although relatively little is known to date about the psychological consequences of being aware of genetic risks, this kind of information is bound to be experienced by most as a burden or even a disaster. A person informed at the age of 18 or 20 about an increased risk of contracting cancer at the age of 50 will live the interim period of three decades under a heavy burden. Under these circumstances, the oracle of Delphi's call for self-knowledge becomes an unreasonable demand; and the notion that expanding one's insight necessarily benefits the autonomous planning of one's life appears naive.

(2) Resurrection of medical priesthood. The only option often remaining in the case of an unfavorable diagnosis is alteration of one's life-plan and adaptation of one's lifestyle to the genetic risk in question. Assuming the persons affected do not sink into bouts of resignation or depression, they will thus attempt to prevent or at least delay the disease they have been predicted to contract. This course of action is not, of course, reprehensible; in some cases it may even be successful. And yet this development will lead to a change in the social role of medicine. Instead of the medical profession working predominantly as a 'repair shop' with a firmly outlined task, an institution will emerge which will also exercise extensive social control over the behavior of entire populations. This can be summed up in two points: (a) The clients of the medical profession will no longer comprise the sick alone, but also and increasingly the *prospectively* sick – i.e. the acutely healthy. (b) The medical profession's original task of providing diagnoses

and therapies for disease will increasingly include prevention, in the sense of influencing lifestyles. 21st century physicians will assume a role occupied in 'primitive' societies by shamans and dominant in the monastic medicine of the Middle Ages: the physician as a priest, holding the secrets to a healthy lifestyle and directing patients or clients towards the 'right' way of living.

A look at the social and political context of this development reveals that shamanism based on genetic diagnostics will not lead to a renaissance of the spiritual dimension of traditional medicine; that it will probably extend far beyond mere consultation and seriously limit individual autonomy. We may assume that medical direction of lifestyle will acquire such authority as a result of various social mechanisms that individuals will be left with no choice. These mechanisms could be direct state coercion, but also and especially economic mechanisms (e.g. exclusion from insurance cover when medical advice is not followed) or numerous forms of indirect social pressure (influence of the media, of the medical profession, etc.). The price for advanced genetic knowledge and its potential benefits will thus be the emergence of a system based on genetic monitoring and behavioral control, assuming the similarly shaped functions of tradition and religion.

(3) Genetic knowledge as a social weapon. One of the characteristics of knowledge is the fact that it is almost impossible to limit. It can be multiplied at will and at little cost, and it can be transported vast distances within a few seconds. In addition, it can be combined and recombined in many different ways (like DNA) and can assume totally new characteristics as a result of these recombinations. Containment of genetic knowledge will therefore be almost impossible. It will possess a "spontaneous" tendency to exceed the boundaries of the medical system and pervade society. The manifold problems surrounding data protection, frequently debated in connection with genetic diagnostics, originate here. They arise from the fact that the data obtained by a genetic diagnosis will often be of interest not only to the individual in question, but also to third parties. Employers, insurance companies and the State immediately spring to mind (Andrews *et al.,* 1994; Draper, 1991; Nelkin and Trancredi, 1989). It is obvious that the interests of these institutions do not always coincide with those of the individuals concerned. For insurance companies, for example, genetic analyses can be a means of parting from genetically burdened persons, thus reducing risks – and costs. Cases such as these have already occurred. It thus seems reasonable to fear that genetic knowledge could become an instrument for discriminating against and disadvantaging persons who – through no fault of their own – are already disadvantaged by their genes. Moreover, this could also hinder or destroy the medical benefits of the technology: members of families with a high risk of contracting certain diseases may refuse to undergo genetic analyses

because they are afraid of subsequently being unable to take out health insurance.

(4) The dialectics of individualization. One of the great attractions of genetic diagnostics is its ability to attribute a particular risk to a particular individual. The resulting opportunities for preventive measures are obvious: a person who is aware of his or her own personal risk is quite likely to adapt his or her lifestyle or take other precautions in order to prevent the onset of the disease. What used to be an unknown and unalterable "fate" can, in the future, increasingly be influenced. In many cases we will thus be able to prevent or reduce suffering. The reverse side of this coin is, of course, that people who know their genetic "fate" and are able to influence it, will suddenly find themselves faced with a previously non-existent responsibility. A cancer or heart attack sufferer will now be forced to hear (from himself or others) that it was all his own fault: a diagnosis was not carried out in time, and no preventive measures were taken. Considering the current financial crisis within the public health care system, it is very improbable that such an attribution of responsibility will remain without social consequences. It will be tempting to laden upon those guilty of omitting to take preventive measures the resulting financial burdens too. Bearing this in mind, further desolidarization is to be feared in countries with public health insurance systems, of which there are many in Europe.

II. From Therapy to Construction

Of course, the expectations raised by gene technology regarding a 'molecular medicine' for the future not only include diagnostics, but also extend to therapy: gene technology will enable diseases not only to be better detected, but also – and especially – to be better treated (OTA, 1984). In this respect, however, little more than hopes or expectations have been realized so far. To date, the most important therapeutic innovations attributed to gene technology have all had to do with the production of medication (human insulin, growth hormones or blood coagulation factors). The long awaited breakthrough involving a *direct* application of gene technological methods to human beings in order to heal disease is still wanting.

Direct application is an idea dating right back to the beginnings of gene technology (Anderson, 1972). It consists of substituting missing or dysfunctional genes in human cells by introducing intact genes from the outside, thus stabilizing, improving or healing the patient's diseased state. Although the first offically authorized gene therapy experiment in 1990 is now a decade old, and although the transfer of genes to human somatic cells should easily have become routine by now judging by the optimistic prognoses around at the time, effective gene therapeutic procedures are still not available. Once the present technical difficulties have been overcome, the great hope is to find not only alternative therapies for diseases already treatable today, but also and especially therapies for (genetic and non-genetic) diseases which are largely untreatable today (Nichols, 1988). Throughout the history of medicine, no other therapeutic procedure has been so intensively discussed before its test phase or so intensively controlled during its entire developmental phase. Somatic gene therapy may therefore be viewed as a model example of bioethically regulated technical innovation (Bayertz *et al.*, 1994). Although it has definitely had its fair share of (technical and ethical) problems and controversies (Anderson and Fletcher, 1980), it is nevertheless internationally believed to represent a desirable addition to previous medical options and one which does not raise any fundamentally new ethical issues (Walters, 1991; Walters and Palmer, 1997).

From its outset, the development of gene technology was surrounded by expectation, discussion and speculation about technical options exceeding far beyond the therapy of individual diseases in individual human beings and the ending of 'reproductive roulette' (Ramsey, 1970; Fletcher, 1974). Why restrict oneself to classical 'conservative' medicine when at some stage gene technology could provide an 'innovative' potential to improve mankind? Eugenic tradition, dating back to Ancient times, suddenly found a new technical ally in gene technology (Duster, 1990). Previously orientated towards the breeding paradigm, eugenics at last seemed capable of breaking through the technical, political and ethical boundaries inherent to it: variation and selection could be renounced in favor of a specific biological engineering of our descendants. To tide us over until the time when a direct controlling of the human gene pool would become possible, technologies such as selective interventions in the germ line, the cloning of human beings or the creation of human-animal hybrids seemingly presented themselves, all of which would allow at least a partial designing of our descendants in the meantime (Humber and Almeder, 1998; Pence, 1998a,b). Thus – so many hoped and at least as many feared – modern biotechnology was well on the way to facilitating the construction of individual human beings and ultimately the reconstruction of the entire human species.

Of course, only a handful have allowed their imaginations to run as far as a complete reconstruction of the human species. And yet it cannot be overlooked that, from its very start, the development of gene technology was linked to the prospect of a new technical and constructive relationship of the human race to itself. This has been emphasized not only by the advocates of gene technology, but also – and maybe even more so – by its critics (cf. Hubbard and Wald, 1993; Nelkin and Lindee, 1995). In all possibility, one group could be just as wrong as the other. Edward O. Wilson (1998, p. 277), for example, is keen to calm anxious souls with his prophecy that future generations will be "genetically conservative," and that the development will never go so far as to apply genetic techniques excessively. This may or may not be true. And yet the crucial issue philosophically speaking is the *evaluation* of this constructive point of view. This is precisely where the ethical controversies are concentrated. Whereas the therapy of disease with the aid of gene technological methods is largely non-controversial, no consensus at all exists about constructive access (however far that may actually go) to the human race to itself. This initially means that it is not the application of gene technology to human beings *per se* which is in need of debate, but the question of which applications are morally permissible and which not. In other words, we have to decide where to draw the line (Anderson, 1989).

A. Where to draw the line?

A line needs to be drawn if, on the one hand, the potential of gene technology is to be exploited for medical (therapeutic) ends and yet, on the other hand, a transition to the 'construction' of human beings is to be avoided. In the discussion two main proposals have been made for drawing such a line:

(1) Somatic cells vs. germline. The abovementioned international consensus about the moral harmlessness of gene therapy refers to gene transfer in human somatic cells. This should be distinguished from gene transfer in germline cells, the difference being that, in the latter case, not only a particular organ of a particular individual is affected by the modification, but all cells plus those of descendants. Manipulations of the human germline have been debated repeatedly and, as a result of the unexpected difficulties encountered by somatic gene therapy, have become the subject of particularly emphatic debate over the last few years. Technical and ethical reservations about this strategy are very strong worldwide, however (Mauron and Thevoz, 1991). In some European countries (including Germany and Switzerland), germline interventions are even illegal, and Article 13 of the European *Convention on Human Rights and Biomedicine* (1997) says:

> An intervention seeking to modify the human genome may only be undertaken for preventive, diagnostic or therapeutic purpose and only if its aim is not to introduce any modification in the genome of any descendants.

There are three - closely connected - reasons for drawing the line between somatic cell therapy (as permitted) and germline cell therapy (as non-permitted). (a) This is not an arbitrary borderline but a biologically given, real difference for which there is empirical confirmation. The biological function of the human genome as a blueprint for the entire organism obviously differs from the genes located in the individual cells which are not totipotent. (b) Whereas gene transfer in somatic cells implies no more than manipulation of a single organ, the scope of germline interventions is far greater: they ultimately change the whole person. (c) And since the latter also affect all descendants of the manipulated individual, their supraindividual effect is potentially infinite.

Ignoring for a moment this totally justified reference to the incomparably grave consequences of intervening in the human germline, and sufficing it to say at this stage that the safety of this procedure would have to meet very high safety standards, this first proposal is convincing only if we assume an identity of genome and person, or at least a very close relationship between the two. And yet genetic determinism of this kind is hardly viable. Although biological relationships of course exist between genome and person, it is at least debatable whether these relationships are as close as the argument presupposes. Just the fact that identical twins possess the same genome and yet are obviously different *people* should be a warning. Neither is it safe to presume that all parts of the human genome are equally relevant to the personality. (After all, only 1% of human DNA basepairs are different from chimpanzee basepairs, rendering 99% identical). A fundamental axiological special position of the human germline is thus hardly plausible.

(2) Therapy vs. enhancement: Other bioethicists have proposed that the line between the permissible and the non-permissible be drawn not on the basis of biological and technical criteria, but with a view to the *goals* aimed at by each intervention. Regardless of whether an intervention were in somatic or germline cells, as long as it were aimed at the therapy or prevention of disease it would be ethically permissible. If it were directed at the enhancement of desirable characteristics, however, then it would be non-permissible (Anderson, 1989). This proposal pays tribute to health as representing a high and generally acknowledged value. The social institution of medicine heeds this value, and it would seem unreasonable to do without relevant technical operations altogether (cf. Harris, 1992; Parens, 1998).

At present it is difficult to imagine realistic indications for therapeutic or preventive interventions in the germline without any possible alternatives. With nearly all genetic diseases there is at least a 50% chance that an embryo will be generated without the diseased gene. With the aid of preimplantation diagnostics, as well as prenatal diagnostics and selective abortion, nearly all genetic defects are avoidable. And yet even if we presuppose the discovery of a therapeutic use for germline interventions which is not only reasonable but also necessary, this approach still poses serious problems. In particular, the division between therapeutic or preventive interventions on the one hand, and enhancement on the other, is far less clear than it may appear at first sight. If we suppose, for example, that a person P has an above average genetic risk of contracting a particular serious disease, and if this risk were to be removed or normalized by means of genetic intervention (whether in somatic or germline cells), then this would clearly constitute prevention. We are aware, however, that (nearly) every human being has an above average risk of contracting several serious diseases. If several or maybe even all of these dispositions were to be eliminated in P, then this would be an obvious case of 'enhancement' or 'eugenics'. But if no clear division exists between prevention and 'enhancement', then this second proposal – based precisely on such a distinction – is only feasible if an arbitrary borderline is accepted.

B. Why draw a line?

The abovementioned difficulties surrounding the drawing of a line are more of a practical than a fundamental nature. In many other areas lacking clear borders we also draw ethical lines. We accept their partially arbitrary character because we are convinced that there have to be such lines. Maybe, instead of debating where to draw the line, we should be asking ourselves why one needs to be drawn at all. The discussion is dominated by four possible answers to this question of 'why'.

(1) Risk and prudence. The most obvious argument for drawing an ethical line between different types of genetic intervention stems from their different levels of immanent risk. We have already seen that germline interventions would have farther-reaching consequences than somatic cell interventions. And clearly the risks connected with the idea of 'enhancing' the human race are completely inestimable. In the foreseeable future the bold ideas of some authors regarding a reformation of human nature will have to remain fantastic:

> Compared with our present-day knowledge of the molecular biology of higher organisms, and our ignorance of the genetics of much of the normal variation in humans, many of these proposals are somewhat analogous to the idea that a boy who has just been given his first electronic set for Christmas, could successfully improve on the latest generation of computers (Vogel and Motulsky, 1996, p. 741).

At each stage in any technical development it is morally imperative to observe justification limits in the light of our ever-limited technical possibilities. In addition, not everything which is technically possible is also in the interests of those involved. Since this insight is often slow in coming, caution and reticence are even more advisable.

Such shrewd considerations do not, however, offer any obvious justification for a fixed and impassable line. It would no doubt be naive to presume that technical limitations will one day disappear *altogether*, and yet it seems safe to predict that they will continue to be pushed and that our possibilities will continue to grow. Shrewd considerations will therefore (very importantly) protect us from hasty steps, but they will not be able to prevent completely the transition to human construction.

(2) The essence of the medical enterprise. Whilst the task of the social institution 'medicine' is to make sick people healthy again or, where this is impossible, to offer them relief in their diseased state, the 'enhancement' of human beings is a different matter altogether. Perceiving a human being as an object in need of perfection ceases to be a medical point of view and becomes a bioengineering project.

This second argument revolves around a generalization of what the medical institution has always believed itself to be – a belief already out of touch with reality. In some of its areas at least, medicine has long been developing in a direction tangibly linked to the idea of human autoevolution using gene technological means. Esthetic surgery (cf. Gilman, 1999), sports medicine and lifestyle drugs are all examples of the departure of medicine, at least partially, from the mere repair of health defects towards a service institution orientated towards the wishes of its customers. Of course this trend may be criticized as negative; many influential authors have done so. For the sociologist Talcott Parsons, esthetic surgical patients are not really patients at all (and this is why many insurance companies refuse to cover them). Leon Kass (1981) is another for whom cosmetic operations do not, strictly speaking, count as 'medical' operations: medicine's only inherent task is to remove physical defects and physically rooted discomfort.

This essentialist concept of medicine overlooks two points. Firstly, it is very difficult to render plausible why the goal of medicine cannot be extended beyond 'health' to include other values. 'Quality of life' could be one such value, especially considering its close connections with the classic value 'health'. Medicine would then emerge as an evolving institution, gradually exceeding its traditional base values and increasingly concerning itself with the fulfilment of all kinds of (morally sound) patient wishes. For H.T. Engelhardt, Jr. (1982), the goals of medical treatment are determined by individuals or groups and cannot be laid down in general terms. Secondly, this essentialist concept overlooks the fact that the term 'health' already contains an uneliminable reference to (extramedical) values, as well as to the relevant (sociohistorical) context. This reference means that constant change within the institution 'medicine' is preprogrammed. In a liberal society granting great scope for the free decisions of individuals, medicine would only be able to distance itself from this individualization trend if it were to adhere to a naturalist definition of 'health', detaching it from the factual wishes and changeable needs of individuals and attaching it to a normatively binding concept of 'human nature'.

(3) The normativity of human nature. This brings us to probably the most fundamental philosophical objection of all regarding the idea of a gene technological reconstruction of the human race. According to this objection, human nature may not be viewed simply as 'biological matter' of neutral worth, which can be modified or optimized at will. Far more, human nature must be seen as an essential part of the human being, necessarily imposing normatively binding limitations. This position may be termed 'substantialist' (Bayertz, 1994) since it presupposes a fixed human substance and rejects as immoral all projects or actions which directly or indirectly call this substance into question. It is of secondary significance whether this substance and its binding nature are justified religiously (e.g. by appealing to God the Creator, the integrity of whose creations are to be respected by man) or with metaphysical and/or natural right arguments (e.g. by appealing to the dignity which human beings have by nature). Even Edward O. Wilson's prophecy that the human beings of the future will be "genetically conservative" is more than a mere (empirical) prediction. Its evocation of human nature causes it to assume an – albeit cloaked – normative dimension.

> Other than the repair of disabling defects, they will resist hereditary change. They will do so in order to save the emotions and epigenetic rules of mental development, because these elements compose the physical soul of the species. (Wilson, 1998, p. 277)

Even if it were possible to 'enhance' a person, according to Wilson that person would then cease to be human: "Neutralize the elements of human nature in favor of pure rationality, and the result would be badly constructed, protein-based computers." Human beings, so his message reads, cannot change themselves without losing themselves. Be it openly or cloaked, all of these positions postulate a categorical duty to preserve human nature or substance.

Any appeal to human nature is also riddled with difficulties. This begins with the fact that it remains – and presumably must remain – notoriously unclear what exactly 'human nature' is. If the legitimacy of technical manipulations in humans depends upon their non-violation of human 'nature' or substance', it becomes hugely significant to have a clear and unambiguous understanding of what exactly this 'nature' is or what exactly constitutes this 'substance'. And yet in place of precise definitions only vague hints are to be found. As neither a clear line 'upwards' nor 'downwards' can be detected, a coincidence seems unlikely. The plasticity and historicity of the human being hinder a clear 'upwards' division between 'nature' and 'culture', whilst the 'downwards' division between human beings and the animal kingdom is being increasingly blurred as science continues to progress (remember, for example, the enormous genetic affinity between us and chimpanzees). Even if a clear definition of human nature were to be found, however, this would not actually help very much: it would still be unclear whether this human nature imposes ethical limitations on human activity. Why should we declare contingent natural facts for non-violable *eo ipso* and deny the realization of any number of human wishes, goals and interests (many of them understandable and legitimate) as a result? In short, the substantialist attitude is wide open to the metaethical accusation of naturalistic fallacy.

(4) Playing God. One categorical objection often raised to any notion of genetic modification or 'improvement' of human beings is that it entails assuming privileges reserved for God. From a religious point of view, this objection seems reasonable since nearly all religions attribute the creation of the human race to gods or god-like beings. Apart from the fact that religious convictions do exist which (under certain circumstances) advocate genetically-based human self-manipulation (cf. Process Theology) or want to leave the door open for further reflection,[3] it is difficult to know where to draw the line with this argumentation. Since not *every* intervention in human nature may be rejected, since we attempt to improve our (phenotypical) nature legitimately in several respects, and since we ultimately influence – albeit relatively non-specifically – the genes of our descendants, criteria must be established to differentiate which interventions are to be deemed illegitimate. In addition, religious faith cannot be taken as an underlying structure of generally binding norms in a pluralistic society. It is possible, however, to shed a very secular light on the "playing God" argument.

> When the objection of playing God is separated from the idea that intervening in this aspect of the natural world is a kind of blasphemy, it is a protest against a particular group of people, necessarily fallible and limited, taking decisions so important to our future. This protest may be on grounds of the bad consequences, such as loss of variety of people, that would come from the imaginative limits of those taking the decisions. Or it may be an expresssion of opposition to such concentration of power, perhaps with the thought: "What right have *they* to decide what kinds of people there should be?" (Glover, 1984, p. 47)

Insofar as this argument draws attention to lacking human wisdom and the fallibility of human decisions, it can only be agreed with. But it certainly does not justify a categorical "No!" to all kinds of intervention in the genetic makeup of human beings; what it justifies is merely a categorical imperative to be extremely cautious in the course of any such undertaking. What carries more weight in this argument is its reference to the power of the manipulators. Even if it were possible to avoid or limit the dangers of a concentration of gene power through a strict individualization of decisions, the decentral power of many individual decision makers would become a new kind of power of (present) human beings over (future) human beings, undermining the autonomy of the latter.

C. Freedom to self-manipulate (subjectivism)

Bearing these considerations in mind, the application of gene technology does not seem bound by any given or fixed moral limitations. This result will please those who believe the specifically human not to be human 'nature', but the ability of human beings to design their lives and world actively and consciously. According to this line of thinking, human beings are different from all other creatures through their free relationship to both the surrounding nature and their own nature, both of which they can change to suit their needs. This position may be termed 'subjectivist' (Bayertz, 1994) because human beings regard themselves primarily as subjects, not bound by nature in their thinking and actions but capable of choosing freely and shaping their world as they see fit. It should be stressed that for subjectivists this is not just a description of the *conditio humana*, but an *evaluative* analysis of the human essence. The subjectivity of the human being is not merely factual but morally decisive: everything in this world with any value at all has that value through human beings and for human beings. Accordingly, nature is 'of neutral worth'; it is 'material' for the human desire to shape things, but it does not possess any value in its own right. This is also true of the human body. It also belongs to this 'outside world' which can be reshaped at will. Nature can and should be 'dominated' for the purpose of self-manipulation and self-realization, both in the human body and in any other parts of nature.

Subjectivism first evolved completely independently of any references to biotechnology or gene technology. It has been the concern of numerous philosophers, especially in the New Age, including – in a particularly extreme form – Jean-Paul Sartre. In his philosophical terminology, human beings 'design' and 'project' themselves, a very literal vocabulary when applied to human plans for genetic self-alteration. The following passage by Sartre comes across as infinite autoevolution devoid of goals, translated into the language of 'phenomenological ontology':

> Since freedom is a being-without-support and without-a-springboard, the project in order to be must be constantly renewed. I choose myself perpetually and can never be merely by virtue of having-been-chosen; otherwise I should fall into the pure and simple existence of the in-itself (...) Our particular projects, aimed at the realization in the world of a particular end, are united in the global project which we are. But precisely because we are wholly choice and act, these partial projects are not determined by the global project. They must themselves be choices; and a certain margin of contingency, of unpredictability, and of the absurd is allowed to each of them, although each project as it is projected is the specification of the global project on the occasion of particular elements in the situation and so is always understood in relation to the totality of my being-in-the-world. (Sartre, 1958, pp. 480f.)

According to Sartre, human free self-design is independent of genetic self-alteration. Human beings have always, and in all circumstances, designed themselves. The fact that human beings had already begun to intervene in the process of reproduction back in primitive societies, whether for contraceptive or for 'proceptive' purposes, is an indication of a continuity of wishes and goals reaching into the present day. Yet if we take a look at the limited and incomplete means available then, and compare them with those possibly available in the future, then it becomes clear how little continuity there is as far as technological means are concerned. The new quality which gene and reproduction technology has lent to human self-determination primarily consists in its technological character, which renders human nature accessible to change and design for subjective purposes in previously unthinkable proportions. Gene technology appears as a huge extension to and reinforcement of subjectivity as far as the biological foundation of human existence is concerned.

There can hardly be any doubt that subjectivism is the currently dominant position in the philosophical evaluation of gene technology. Although in the recent past theoretical efforts to revalidate nature in general (in the context of environmental ethics) and human nature in particular have intensified, and although this expressly happens for the purpose of preventing human subjectivity from going too far, the growing options available to human beings and the consequent ever-growing scope for design of the world and self are generally evaluated positively, even by most bioethicists. Admittedly most of them stop short of Sartre's concept of total freedom; and yet few are prepared to acknowledge categorical limitations to human activity based solely on human nature. The achievements of reproduction technology enable "free individuals to achieve the biological destinies they choose, as, for example, within the area of reproduction" (Engelhardt, 1982, p.72). Why should we refrain from redesigning the world and ourselves in line with our wishes and interests if biotechnology could make it all possible? The corollary of this is: If human beings are autonomous, the moral legitimacy of technological intervention in reproduction may no longer be disputed *a priori*, not even the most fantastic measures within a strategy of genetically improving the human race could be excluded for metaphysical reasons. (This is diametrically opposed to the substantialist standpoint.)

And yet what at first sight may look like a triumph for autonomy, proves at second glance to be a problem of some intricacy. What is increased by the achievements of biotechnology and gene technology is the autonomy of the living, but not that of the future human beings to be 'enhanced' using the new procedures. At least at two levels this leads to limitations which are far graver than they may seem initially.

(1) Risks for others: At all times and in all places, technological activity involves risks. This is also true of biotechnology and gene technology in the future. Matters of risk are usually interpreted as matters of wisdom, and yet that is not the case here. The fundamentally unavoidable risks entered into within a human 'enhancement' project acquire the status of an ethical problem because it is not us entering into them: we burden *other,* future people with them instead. Since these people are not capable of giving their consent, this can only be deemed an ethically permissible course (at a stretch) if the risks are in an appropriate ratio to the expected benefits of the modification for the individuals involved. This would seldom be the case.

(2) Goals for others. This brings us to the goals which enhancement could possibly have. 'Enhancement' is an evaluative term. The further our technical options extend, the more debatable the goals of such modifications become. Two possibilities arise in this context. The first consists in orientating the modifications towards a generally binding ideal or the interests of society. This was the position of classical eugenics. This would obviously increase the power of society over individuals enormously and safeguard it biologically. Far more appropriate in a liberal society is the granting of as much scope as possible for decisions and actions and the promotion of individual multiplicity. In this light, the second possibility seems preferable: to allow parents to decide the goals behind gene technological changes to their offspring. However promising this idea may appear, it would be naive to ignore how tied individuals really are in their decision-making. *De facto* most of them would not reach their decisions autonomously but on the basis of numerous social influences. Precisely because the decisions parents would be making would be important, they would seek orientation from the media, the sports world and relevant 'experts'; analogous to the naming of children[4] and the booms of particular methods of upbringing, fashions and semi-scientific ideologies would play an important role.

The problems facing both of the above options are further aggravated by the temporal distance between the begetters and the begotten: the constructors would opt for those values highly regarded in their *current* social context and try to steer their offspring towards *these* values. And yet what is regarded as positive today will not necessarily be so in our offspring's tomorrow. We never know what characteristics are going to be called for in the future. Just as the science-fictional future usually turns out to be no more than an extrapolation of the present, the 'future human being' will be seen as the current human being with all of its positive characteristics increased. Autonomously speaking this means: the price to be paid for the constructive potential of gene technology is the autonomy of those it would affect. This is markedly different from cosmetic surgery or doping, where the individuals opting for them wish to profit from them themselves and will bear the consequences for doing so during their own lifetimes.

III. ETHICS AND GENETHICS

We pointed out early on that some of the issues raised by GenEthics are very fundamental: 'fundamental' not only in the sense that they deeply affect the future of the human race, but also in the sense that it is difficult to express them in the vocabulary of established ethics, that they even seem to exceed the scope of conventional ethics altogether. Since it has not been possible within the framework of this paper to examine these issues in any depth, we would at least like to conclude with a reference to them, in the hope that they may be paid more attention in the future. This should be the case, at least as long as "bioethics" is interpreted not only as an undertaking which attempts to solve more or less pragmatically the most urgent issues on the agenda at any given time, but also as a genuinely *philosophical* undertaking, going beyond the practical problems of the day in order to examine fundamental metaethical and metaphysical questions as well.

The *first* of these issues stems from the fact that gene technology has (not only, but also) to do with unborn human beings. The bold visions of genetic modification, whether of single individuals or the entire human race, concern the members of future generations. There is a theory that ethical principles developed for actual people cannot just automatically be applied to possible people, and there are arguments to support this. Any attempt to extend the validity of these principles to include possible people seems to lead to paradoxes which in turn signify the end of any meaningful ethical discourse. Unlike other liberal and humanistic expansions of the reference group of these principles (to women, other races, animals)

> sensitivity to the lot of future people cannot be expressed simply by embracing 'them' into the moral community. For it is exactly the indeterminacy of 'them' which makes it impossible to apply contractarian, Kantian, or utilitarian principles to decide 'their' lot. It is not the assumption of timelessness of the moral community which makes theories of ethics incapable of handling genesis problems, but rather the paradox of being expected to provide ethical principles for membership in the community which is the basis of all ethical principles. (Heyd, 1992, pp. 63f.)

If this analysis is correct, we encounter problems here which lie beyond the grip of ethical judgement; we have reached a limit of ethical theory.

Secondly, one of the exceptional qualities of gene technology is its 'synthetic' or 'constructive' potential. At least in principle and in the long-term, it will enable the human genome to be accessed specifically and deliberately. The prospect of a medicine which no longer merely 'repairs' human beings, but also alters and 'improves' them calls into question the moral status of *human nature*. Is it little more than organic matter at stake here, or should an inherent value be attributed to it? Modern ethics has abolished the idea of human nature – and likewise (external) Nature – possessing an inherent value. Accordingly, moral evaluations are to refer exclusively to human wishes, needs or interests. Independently of these instances, Nature herself, or any of her individual states can be neither the source nor the object of such an evaluation. This position, at least with regard to Nature, has been called into question in the light of the ecological crisis (cf. Krebs, 1999). And it is also being increasingly called into question with regard to human nature. This revalidation of human nature can be asserted with 'strong' normative claims or in a moderate sense, comprehending its normativity more in a recommending sense (Siep, 1996). One of the main arguments of the moderate position is the theory that genuine human flourishing may not only be described in subjective termini, but also has an objective and natural dimension which presupposes the recognition of a graduated intrinsic value of Nature (including human nature).

It seems obvious that both of these issues are very fundamental, and that not only our ethical judgment regarding this or that gene technological option depends upon their being answered, but also the structure and content of our ethical thinking altogether. This may be seen as an indication that technological progress not only forces us to assume responsibility for options which continue to extend further and deeper, but also for the ethical categories and principles with which we evaluate these options. This *metaresponsibility* (Bayertz, 1994, pp.181-197) is by no means the smallest problem currently confronting the realm of ethics. Once we begin shaping human nature, we shall also be forced to shape the ethical principles which allow or forbid just this.

Department of Philosophy
University of Münster, Germany

Center for Medical Ethics at the Markus-Hospital
Frankfurt/M., Germany

NOTES

1. Translated by Sarah L. Kirkby (B.A. Hons.)

2. There is no room here to go into detail about the scientific and considerable economic interests involved in uncovering the genetic foundations of ethnic variation. Lucrative diagnostic and therapeutic agents can be developed on the basis of single genes (this is incidentally also true at an individual level).

3. Within "the scope of a theology of creation that emphasizes God's ongoing creative work and that pictures the human being as the created co-creator" the door "to the issue of germ-line intervention for the purpose of therapy and even for enhancing the quality of human life (...) must be kept open" (Peters, 1995, p. 379).

4. Just as social and political trends influence the choice of first names (cf. Wolffssohn and Brechenmacher, 1999).

REFERENCES

Agius, E. and Busuttil, S. (eds.): 1998, *Germ-Line Intervention and our Responsibilities to Future Generations*, Kluwer Academic Publishers, Dordrecht.

Anderson, W.F.: 1972, 'Genetic Therapy', in M.P. Hamilton (ed.), *The New Genetics and the Future of Man,* Eerdmans Publishing Company, Grand Rapids, Mich., pp. 109-124.

Anderson, W.F.: 1989, 'Human Gene Therapy: Why draw a line?', *The Journal of Medicine and Philosophy* 14, 681-693.

Anderson, W.F. and Fletcher, J.C.: 1980, 'Gene Therapy in Human Beings: When is it Ethical to Begin?', *The New England Journal of Medicine* 303 (22), 1293-1297.

Andrews, L.B. *et al.*: 1994, *Assessing Genetic Risks: Implications for Health and Social Policy,* National Academy Press, Washington, D.C.

Annas, G. and Elias, S. (eds.): 1992, *Gene Mapping; Using Law and Ethics as Guides*, Oxford University Press, Oxford.

Bayertz, K.: 1994, *GenEthics. Technological Intervention in Human Reproduction as a Philosophical Problem*, Cambridge University Press, Cambridge.

Bayertz, K.: 1997, 'The Normative Status of the Human Genome: A European Perspective', in K. Hoshino (ed.), *Japanese and Western Bioethics*, Kluwer Academic Publishers, Dordrecht, pp. 167-180.

Bayertz, K. *et al.*: 1994, 'Summary of *Gene Transfer into Human Somatic Cells: State of the Technology, Medical Risks, Social and Ethical Problems: A Report*', *Human Gene Therapy* 5, 465-468.

Chadwick, R.: 1987, *Ethics, Reproduction and Genetic Control*, Croom Helm, London.

Convention for the Protection of Human Rights and Dignity of the Human Being with regard to the Application of Biology and Medicine: Convention on Human Rights and Biomedicine, Oviedo, 4. IV. 1997, European Treaty Series No. 164.

Cook-Degan, R.: 1994, *The Gene Wars: Science, Politics, and the Human Genome*, Norton and Company, New York.

Cranor, C.F. (ed.): 1994, *Are Genes Us? The Social Consequences of the New Genetics,* Rutgers University Press, New Brunswick, NJ.

Duster, T.: 1990, *Backdoor to Eugenics*, Routledge, New York, London.
Draper, E.: 1991, *Risky Business: Genetic Testing and Exclusionary Practices in the Hazardous Workplace*, Cambridge University Press, New York.
Engelhardt, H.T. Jr.: 1982, 'Bioethics in a Pluralist Society', *Perspectives in Biology and Medicine* 26 (1), 64-77.
Engelhardt, H.T. Jr.: 1996, *The Foundations of Bioethics*, Oxford University Press, Oxford, New York, 2nd edition.
Fletcher, J.: 1974, *The Ethics of Genetic Control. Ending Reproductive Roulette,* Anchor Press, Garden City, N.Y.
Gilman, S.L.: 1999, *Making the Body Beautiful. A Cultural History of Aesthetic Surgery*, Princeton University Press, Princeton, NJ.
Glover, J.: 1984, *What Sort of People Should There Be?,* Penguin, Harmondsworth.
Harris, J.: 1992, *Wonderwoman und Superman. The Ethics of Human Biotechnology*, Oxford University Press, Oxford, New York.
Heyd, D.: 1992, *Genethics. Moral Issues in the Creation of People,* University of California Press, Berkeley, Los Angeles, Oxford.
Holtzman, N.A.: 1989, *Proceed with Caution: Predicting Genetic Risks in the Recombinant DNA Era,* Johns Hopkins University Press, Baltimore, MD.
Hubbard, R. and Wald, E.: 1993, *Exploding the Gene Myth. How Genetic Information is Produced and Manipulated by Scientists, Physicians, Employers, Insurance Companies, Educators, and Law Enforcers,* Beacon Press, Boston, Mass.
Humber, J.M. and Almeder, R.F. (eds.): 1998, *Human Cloning*, Humana Press, Totowa, New Jersey.
Juengst, E.T.: 1998a, 'The Ethics of Prediction: Genetic Risk and the Physician-Patient Relationship,' in J.F. Monagle and D.C.Thomasma (eds.), *Health Care Ethics. Critical Issues for the 21st Century,* Aspen Publishers, Gaithersburg, Maryland, pp. 212-227.
Juengst, E.T.: 1998b, 'Should We Treat the Human Germ-Line as a Global Human Resource?', in E. Agius and S. Busuttil (eds.), *Germ-Line Intervention and our Responsibilities to Future Generations*, Kluwer Academic Publishers, Dordrecht, pp. 85-102.
Kass, L.R.: 1981, 'Regarding the End of Medicine and the Pursuit of Health' in A.C. Caplan *et al.* (eds.), *Concepts of Health and Disease*, Addison-Wesley, Reading, Mass., pp. 3-30.
Kevles, D.J.: 1995, *In the Name of Eugenics: Genetics and the Uses of Human Heredity,* Harvard University Press, Cambridge, Mass.
Kevles, D.J. and Hood, L. (eds.): 1992, *The Code of Codes. Scientific and Social Issues in the Human Genome Project*, Harvard University Press, Cambridge, Mass.
Kitcher, P.: 1996, *The Lives to Come. The Genetic Revolution and the Human Possibilities*, Simon & Schuster, New York.
Krebs, A.: 1999, *Ethics of Nature. A Map,* Walter de Gruyter, Berlin, New York.
Locke, J.: [1690], *Two Treatises of Government,* ed. by P. Laslett, Cambridge University Press, Cambridge, 1988.
Mauron, A. and Thévoz, J.M.: 1991, 'Germ-line Engineering: A Few European Voices', *The Journal of Medicine and Philosophy* 16, 649-666.
Murphy, T. and Lappé, M.: 1994, *Justice and the Human Genome Project*, University of California Press, Berkeley.
Nelkin, D. and Lindee, M.S.: 1995, *The DNA Mystique. The Gene as a Cultural Icon,* W.H. Freeman, New York.
Nelkin, D. and Tancredi, L.: 1994, *Dangerous Diagnostics. The Social Power of Biological Information*, Basic Books, New York, 2nd edition.
Nichols, E.K.: 1988, *Human Gene Therapy*, Harvard University Press, Cambridge, Mass.
OTA: 1984, *Human Gene Therapy - A Background Paper*, Office of Technology Assessment, Washington, D.C.

Parens, E.: 1998, 'Is Better Always Good? The Enhancement Project', *Hastings Center Report* 28, (Jan.-Feb.), Special Suppl., S1-S18.

Parfit, D.: 1984, *Reasons and Persons*, Clarendon Press, Oxford.

Parker, L.S. and Majeske, R.A.: 1996, 'Standards of Care and Ethical Concerns in Genetic Testing and Screening,' *Clinical Obstetrics and Gynecology* 39 (4), 873-884.

Pence, G.E.: 1998a, *Flesh of my Flesh. The Ethics of Cloning Humans*, Rowman & Littlefield, Lanham, Oxford.

Pence, G.E.: 1998b, *Who's Afraid of Human Cloning?,* Rowman & Littlefield, Lanham, Oxford.

Peters, T.: 1995, '"Playing God" and Germline Intervention,' *The Journal of Medicine and Philosophy* 20 (4), 365-386.

President's Commission for the Study of Ethical Problems in Medicine and Biomedical and Behavioral Research: 1982, *Splicing Life: A Report on the Social and Ethical Issues of Genetic Engineering with Human Beings*, U.S. Government Printing Office, Washington, D.C.

Ramsey, P.: 1970, *Fabricated Man. The Ethics of Genetic Control,* Yale University Press, New Haven.

Rhodes, R.: 1998, 'Genetic Links, Familiy Ties, and Social Bonds: Rights and Responsibilities in the Face of Genetic Knowledge', *The Journal of Medicine and Philosophy* 23 (1), 10-30.

Sartre, J.-P.: 1958, *Being and Nothingness. An Essay on Phenomenological Ontology*, Trans. by H.E.Barnes. Methuen, London.

Siep, L.: 1996, 'Eine Skizze zur Grundlegung der Bioethik', *Zeitschrift für philosophische Forschung*, Vol. 50, 236-253.

UNESCO: 1997, 'Draft of a Universal Declaration on the Human Genome and Human Rights', Paris, July 25.

Vogel, F. and Motulsky, A.G.: 1996, *Human Genetics. Problems and Approaches*, Springer, Berlin, New York, 3rd edition.

Wachbroit, R.: 1993, 'Rethinking Medical Confidentiality: The Impact of Genetics', *Suffolk University Law Review* XXVII, 1391-1410.

Walters, L.: 1991, 'Ethical issues in human gene therapy', *Journal of Clinical Ethics* 2, 267-274.

Walters, L. and Palmer, J.G.: 1997, *The Ethics of Human Gene Therapy,* Oxford University Press, New York/Oxford.

Weir, R.F. *et al.* (eds.): 1994, *Genes and Human Self-Knowledge. Historical and Philosophical Reflections on Modern Genetics,* Iowa University Press, Iowa City.

Wertz, D.C. and Fletcher, J.C.: 1989, *Ethics and Human Genetics: A Cross-Cultural Perspective*, Springer, Berlin/ New York.

Wilson, E.O.: 1998, *Consilience. The Unity of Knowledge*, Alfred A. Knopf, New York.

Wolffssohn, M. and Brechenmacher, Th.: 1999, *Die Deutschen und ihre Vornamen. Zweihundert Jahre Politik und öffentliche Meinung*, Diana, München.

SECTION V

FOUNDATIONS OF THE HEALTH PROFESSIONS

B. ANDREW LUSTIG

DEATH, DYING, EUTHANASIA, AND PALLIATIVE CARE: PERSPECTIVES FROM PHILOSOPHY OF MEDICINE AND ETHICS

I. INTRODUCTION

The circumstances of dying and death have changed dramatically during the last four decades, especially in developed countries. Those changes have raised both legal and ethical issues concerning the values that should guide end-of-life decision-making and care. At the same time, end-of-life decisions implicate concerns that are central to the philosophy of medicine. This chapter seeks to identify and articulate both sets of issues, and will proceed in three stages. In Part One, I provide a brief overview of the changes in clinical medicine that have led the law and ethics to focus on end-of-life decision making. In Part Two, I identify and articulate a number of philosophical issues that need to be explored if we are appropriately to address end-of-life care. In this section, I seek to make more explicit the richer philosophy of medicine concerns that usually are left unthematized in clinically oriented recommendations for care. Of particular importance here are the conceptual and practical tensions between the objectifying medical language of "disease" and the irreducibly subjective dimensions of the illness experience, tensions likely to be exacerbated in the context of end-of-life decision making. In Part Three, I critically analyze a number of fundamental ethical concerns raised by withholding and withdrawing life-sustaining treatment.

The careful reader will note that Parts Two and Three are, in many respects, quite different vantages from which to view issues in end-of-life clinical settings. Part Two explores fundamental theoretical issues for philosophy of medicine more generally. My working assumption in Part Two is that such broad issues are raised by medicine as a general practice and are of particular relevance to end-of-life care. Part Three surveys problems posed by end-of-life care as they have been addressed in the standard ethics literature. The focus in Part Three is more specific, and less thematic, than the broader discussion in Part Two. The two discussions, while overlapping, remain distinct. As I will suggest in my conclusion (Part Four), the links to be drawn between basic issues in philosophy

G. Khushf (ed.), Handbook of Bioethics, 441–471.

of medicine and standard ethical analyses of issues in end-of-life care emerge as important areas for future research..

II. END-OF-LIFE DECISION MAKING AND THE RISE OF BIOETHICS

In the twentieth century several key factors converged to change dramatically the powers of medicine to respond to life-threatening diseases, to expand significantly the life-spans of individuals, and to increase the range of treatment alternatives available in response to threats to life and well-being. A few examples will suffice to make the point. The development of penicillin, during the early 1940's, and a host of other antibiotics since that time, have allowed modern medicine to treat life-threatening infectious diseases in ways unknown a generation earlier. The development of intensive care units since the 1950's, and in particular the introduction of cardiac monitoring in the 1960's, has raised new possibilities for responding to life-threatening diseases, and has also posed significant questions about the appropriateness of aggressive interventions in particular circumstances and for particular conditions (Bryan-Brown, 1992; Brody, 1995). The development of organ transplant technologies, ventricular assist devices, and a plethora of other medical technologies, has enhanced the possibilities for extending and improving the quality of life for many patients. At the same time, such new technologies have raised questions, at both the micro- and macro-levels, concerning appropriate disposition of resources, according to allocation criteria that are both equitable and likely to be efficacious (Callahan, 1995; Daniels, 1985; Kilner, 1992).

Concerns regarding the protection of human research subjects and debates about the appropriate criteria for allocating new medical technologies are generally deemed the precipitating issues that led to the emergence of modern bioethics as an interdisciplinary field, both historically and conceptually (Rothman, 1992; Jonsen, 1998). But if those issues spawned the "birth of bioethics," the field came of age within the context of several high-profile cases involving decisions about the withholding or withdrawing of life-sustaining interventions, beginning, notably, with the Karen Anne Quinlan case in 1975 (*Quinlan*, 1976), and continuing with Brophy (*Brophy*, 1976), Cruzan (*Cruzan*, 1990), and others (Barber, 1983; Conroy, 1985; *Spring*, 1980; Capron, 1991; Cranford and Ackerman, 1991; McCormick, 1990; Meilaender, 1990; Meisel, 1992; Pence, 1990; Veatch, 1993, Weir, 1989). All such cases involved patients who faced life-threatening conditions and who required significant continuing medical interventions for their survival. The cases generated significant ethical and legal conflicts because they exemplified several values central to both fields for much of the modern era. Those values include: (1) the autonomy of adults (or their surrogates) to control decisions about what constitutes appropriate medical care according to their own values; (2) the commitments of health care professionals to act beneficently toward their patients as a matter of fiduciary obligation; (3) the interest of the state in the preservation of life as a value foundational to the possibility of social flourishing; and (4) the need to protect vulnerable patients.

In a series of landmark decisions, the nature and scope of patient autonomy were clarified. Over time, a working consensus has emerged in secular bioethics that incorporates several core claims. First, in the spirit of *Schloendorff* (*Schloendorff*, 1914) as a long-standing legal precedent, and in accord with an emphasis on autonomy as a central feature of ethics and morality, patients of sound mind are deemed to have decision-making authority over all medical decisions on their own behalf, including those involving life-sustaining treatment at the end of life. Moreover, their wishes are deemed determinative if specified in advance (by so-called advance directives) or if accurately known and articulated by others situated to speak for patients in the event of their incompetence. Second, health care professionals retain their personal and professional integrity in the delivery and management of patient care; thus professionals retain the right to withdraw from or to transfer forms of medical caregiving that they cannot support, for reasons of personal or professional conscience. At the same time, professionals are legally and morally charged not to abandon patients. Third, the doctor-patient relationship is widely discussed as that of a clinical partnership. While clear disparities exist in knowledge and power between physicians and patients, the relationship remains that of moral equality, with both patient and health care professional having specific duties and rights. The doctrine of informed consent is meant to capture the dynamic at work in clinical decision-making, with *both* doctor and patient accountable for a mutually agreed upon decision of what constitutes an appropriate course of care (United States President's Commission, 1983; Faden and Beauchamp, 1986; Wear, 1993).

Despite the importance of the above consensus regarding clinical decision making, the differences between legal and ethical constructs and the realities of clinical care at the end of life remain quite troubling. The Study to Understand Prognoses and Preferences for Outcomes and Risks of Treatments, the so-called "SUPPORT Project," was the first large-scale prospective study of dying patients undertaken in the United States (SUPPORT, 1995). Its results, published in 1995, revealed the stark differences between theoretical discussions about end-of-life care and actual patient experiences. The SUPPORT project, which began in 1989, was conducted in two phases at five major medical centers in the United States. In its first two-year phase, it involved more than 4000 patients, suffering from one of nine conditions with a 50% likelihood of death within six months. The study examined the incidence and quality of discussions between physicians and patients concerning specific medical interventions, as well as the decisions that were reached. Patients and surrogates were interviewed, and their preferences regarding treatment options were determined. The SUPPORT researchers concluded after Phase One that "the hospital experience for many seriously ill patients was unsatisfactory" (SUPPORT, 1995). Although 79% of patients died with a DNR order, 46% of those directives were written within two days of death. Although 31% of patients did not desire CPR, fewer than half their physicians understood (or acknowledged) their wishes. Moreover, 38% of patients spent ten or more days in an intensive care unit before their deaths, and surrogates indicated that 50% of patients experienced moderate to severe pain at least half the time during the last three days before their deaths.

Phase Two of SUPPORT involved the development and dissemination of a clinical intervention designed to enhance the quality of information shared between patients and health care professionals. The second phase included written reports of patients preferences for attending physicians, information concerning advance directives, and training in communication skills for nurse facilitators. However, the intervention led to no significant changes in the timing of DNR orders, physician/patient agreement about DNR orders, the number of undesirable days of treatment, or the prevalence of pain. Hence, despite extensive efforts, the Phase Two attempt to improve the quality of patient care at the end of life appeared to be unsuccessful.

While the remainder of this chapter will be primarily concerned with the theoretical discussion of end-of-life issues, the SUPPORT data are quite sobering. They suggest that significant ongoing efforts to enhance patient and professional education about end-of-life preferences and more effective strategies for improving the quality of clinical communication will be required in order to close the gap between theoretical discussions of doctor-patient relations and the clinical realities of end-of-life care. Moreover, while the hospice setting is often touted as a preferable venue for providing effective palliation for dying patients, the SUPPORT data reveal the obvious need for more effective and appropriate interventions for dying patients, regardless of setting.

III. END-OF-LIFE DECISION MAKING AND THE PHILOSOPHY OF MEDICINE

The consideration of what constitutes appropriate care at the end-of-life raises profound issues about the philosophy of medicine itself, including the following: (1) questions about the goals of medicine; (2) questions about the nature and scope of medicine as a practice; (3) the distinction(s) to be drawn between "caring" and "curing"; (4) the problems posed for patient care by clinical tendencies toward reductionism; (5) the problems generated by medicine's primary, if not exclusive, focus on the physical aspects of disease; (6) tensions posed by the language of "medicalization"; (7) the function of interpretation in medical language; and (8) specific issues posed by the philosophy of palliative care. I will consider each of these topics in turn.

A. The Goals of Medicine

The procedural consensus that supports the right of the competent patient to decide what constitutes appropriate treatment in light of his or her own values does not, of itself, resolve substantive questions concerning the appropriate goals of medicine as a practice in relation to end-of-life care. For example, while patients have a negative right to refuse unwanted treatment, they do not have a positive right to receive whatever treatment they request. In light of the model of clinical partnership, autonomy is a central but not sole value; health care professionals may also invoke beneficence-based judgments to deny forms of care they deem harmful or futile. For example, in debates about medical futility (and as a central motif in debates on physician-assisted suicide), "the integrity of medicine" is often invoked as a crucial countervailing value to the exercise of patient autonomy. But the correlation of medical integrity with judgments of medical futility remains conceptually and procedurally ambiguous, in large measure because the concept of futility is defined and measured according to different criteria by various commentators (Brody and Halevy, 1998).

B. Medicine as a "Practice"

Medicine is a scientifically-based "practice," in Alasdair MacIntyre's sense of that word. According to MacIntyre, there are identifiable goods *constitutive* of any social practice (MacIntyre, 1997). The goods constitutive of medicine are several. They include the promotion of health, the prevention of injury, the cure of disease, and the relief of pain and suffering. In numerous clinical instances, these goods may come into conflict. For example, the effort to cure disease may entail significant pain and suffering in pursuit of that goal. Alternatively, appropriate relief of pain and suffering in terminal patients may requiring that health care professionals modify or forego their usual commitment to "cure" as the anticipated outcome of their ministrations. Moreover, at the macro-level, as is noted by critics of current allocation priorities, the "curative" model tends to emphasize high-technology and aggressive interventions to the detriment of preventive and/or "low technology" approaches that may be of greater value to many patients, especially those with chronically debilitating conditions.

C. "Curing" versus "Caring"

It is no coincidence that medicine, as an increasingly science-based practice, has accorded priority to a "curative" model. In its own history, modern medicine came of age by replacing earlier humoral perspectives, which saw "cure" as a restoration of "balance" in the whole organism, with an approach that views cure as the eradication of specific pathogens as that cause of disease and illness. (By extension, the newly emerging genetic basis of medicine transfers the view of disease pathogens to the genome itself; restoration of genetic normalcy will therefore involve eradication and replacement of "defective" parts of the genome).

There is much to celebrate in the successes of scientific medicine. Its progress during the past five decades has been remarkable in many ways. Yet at the same time, it has posed major challenges to the provision of patient-centered *care.* To make that case, one need not draw some facile dichotomy between caring and curing. Rather, one need only observe a number of troubling tendencies that characterize a primary emphasis on cure-oriented modes of medical intervention (Jecker and Self, 1991; Fox, 1997). As Paul Ramsey reminded us, at the heart of the clinical encounter should be a meeting of health care professional and patient as *persons,* according to the canon of covenant fidelity (Ramsey, 1970). Yet a medicine intent on diagnosis and prognosis of disease may in that very process, fall prey to a number of troubling tendencies. First, a tendency toward reductionism may threaten the commitment of health care professionals to treat patients as persons. Second, an increasingly technologically oriented medicine may exacerbate the strong sense of alienation experienced by patients as part of their subjective experience of illness. Third, by failing to address the tacit dimensions of the patient's illness, such a medicine may, paradoxically, increase the patient's suffering, a result that emerges at odds with one of the goods constitutive of its self-understanding as a practice.

D. Tendencies Toward Reductionism That Thwart Medicine as an Interpersonal Encounter

As noted above, the success of modern medicine is best viewed as a function of the shift from earlier, largely empirically based ministrations, to a scientifically based understanding of disease according to categories of pathophysiology and pathoanatomy (Edelstein, 1977; Khushf, 1992). Rather than the holistic approach that characterized an earlier humoral emphasis on restoring balance, modern medicine's success was achieved largely by an attention to identifying and eradicating specific causes of disease.

As many commentators have observed, certain forms of theoretical and clinical reductionism have followed, with seeming inevitability, from that Copernican shift in perspective on disease (Cassell, 1982; Cassell, 1991; Fox, 1997). According to Cassel, medicine, with its increasing proficiency, has moved from an orientation upon patients as persons to an approach which views patients primarily as bodies. This shift is, of course, understandable as an aspect of medicine's increasing sophistication in identifying the pathogenic bases of disease. In Cassel's judgment, however, it also poses a number of dangers to medicine's traditional ideal of the doctor-patient relationship as an encounter between *persons*.

E. Excessive Focus on Physical Aspects of Disease

Scientifically based medicine exhibits a strong tendency to limit itself to the physical aspects of the patient's experience of illness. Eric Cassel observes a Cartesian dualism prevalent in modern clinical practice, whereby pain is construed only in relation to underlying pathology, and generally delimited as a physical phenomenon (Cassell, 1982). Indeed, the control or relief of pain may or may not be of immediate concern to clinical practice, depending upon the need to identify the underlying pathology that manifests itself in pain's observable symptoms. In a scientifically driven understanding, symptoms, while not epiphenomenal, may provide important clues or cues to an underlying disease or condition, and may, or may not, be seen as appropriate objects of immediate medical management. That is to say, in reducing pain, a science-based understanding of pain as symptom may, in particular circumstances, reduce the ability of clinicians to make an accurate diagnosis.

Cassel notes the paradoxical results that may flow from this reduction of clinical focus to pain as symptomatic. A scientifically based clinical medicine tends to reduce a patient's illness experience to the category of diagnosable disease, and tends to reduce the experience of patient suffering (a holistic notion) to the narrower range of manageable physical symptoms. Both tendencies flow naturally from the self-conscious paradigm shift that occurred in medicine, beginning in the nineteenth century, with a shift toward a scientific model. Much has been gained, but much lost.

F. The Languages of "Disease" and "Illness"

Many commentators have analyzed the different *languages* at work in the clinical experience, depending upon whether the perspective is that of the patient or the medical practitioner (e.g., Engelhardt, 1996, pp. 189-238; McCullough, 1989; Kleinman, 1988). Laurence McCullough notes the difference between the language of "disease," an argot which comports with medicine as a science-based practice, and illness, which is a subjective event or set of events experienced by the patient and interpreted in highly individualized fashion. Disease is a vocabulary based upon general nosological categories. While patients may *typify* a given condition or pathology, the clinical interpretation of sets of symptoms as indicative requires attention to the underlying general categories of diagnosis in terms of which a specific patient complaint is interpreted or construed. Moreover, disease is a *general* language, requiring specific parameters (indeed, perimeters) in order for particular pathophysiological or pathoantomical states to be identified. By contrast, the language of illness is, to significant extent, patient-specific and highly subjective. As Cassell noted the differences between the language of pain and suffering (Cassell, 1982; Cassell, 1991), McCullough discusses the differences between the language of disease and illness (McCullough, 1989). The two sets of distinctions exhibit correlative contrasts. Medicine, in the name of generalizability and greater "objectivity," responds to the physical phenomena of pain, and interprets that pain as physical evidence of disease. The patient's "suffering" is a unitary phenomenon, non-dualistic. The patient, as an embodied self, feels threats to the body holistically, as threats to personal integrity. (These threats are likely to be exacerbated when the patient's prognosis is terminal.) Cassel notes the many experiential aspects of suffering not reducible to medicine's preference for restricting itself to the phenomena of pain. These aspects can include features seldom identified in a medical case history, with its tendency to focus, often quite reductionistically, on the *physical* aspects of disease. Likewise, threats to the integrity of the self, as phenomenologically described in the subjective, non-generalizable, language of illness, do not necessarily correspond, and certainly not isomorphically, to the physical realities of pain (Spicker, 1970; Toombs, 1992). Pain, effectively identified and managed, may well reduce suffering. But just as pain can be minimized, while suffering remains, so pain continues, while suffering is overcome. Neither critical care medicine, nor palliative medicine, can afford to be indifferent to the areas of overlap and difference between the categories of pain and suffering, disease and illness.

G. The Function of Interpretation in Medical Language

There is another significant irony to this discussion. While disease is distinguished from illness by the former's generalizability and predictive value, *both* languages, the quasi-objective argot of disease and the highly subjective language of illness, involve significant degrees of *interpretation* in order to be meaningful descriptions. Despite medicine's efforts at objectivity, even quasi-"objectively" identified pathological states nearly always require interpretation in order for disease, abnormality, or abnormalcy to be diagnosed. McCullough's central example is instructive. He notes that a condition such as pancreatic cancer requires that a threshold be crossed before the "disease" of pancreatic cancer is identified; in this instance, a threshold number of abnormal cells per quadrant of slide must be counted in order for a pathology to be so labeled (McCullough, 1989). Much of the "laboratory turn" in modern medicine requires the same thresholds, and genetic medicine will only intensify such interpretive requirements.

However, the differences between the interpretations of clinical medicine and the existential interpretations of patients remain profound. The interpretations of clinical medicine are, at least in principle, based on a professional base of scientific knowledge, generalizable across patients and populations. The subjective dimensions of illness, however, entail a lack of generalizability; thus the adage, "my cancer is not your cancer." Even more-or-less identical diagnoses and prognoses of disease are therefore quite likely to have very different meanings when experienced as illness by particular patients.

H. The Philosophy of Palliative Care

Much of the emphasis in palliative medicine is a self-conscious effort to overcome the tendencies toward reductionism at work in models of medicine directed primarily at curing disease. The modern hospice movement began in the 1960's in England, and is associated there with the pioneering work of Cicely Saunders (Stoddard, 1992). In the United States, the spread of the modern hospice movement is often associated with the early significant studies of Elisabeth Kubler-Ross, with the Connecticut Hospice the first to be established in this country in 1974 (Stoddard, 1992).

Palliative care in the hospice setting is distinguished by a working philosophy that expressly foregoes further efforts at curative care, and instead focuses on the effective relief of pain and other discomforts associated with terminal diseases. The National Hospice Organization has set forth the standards that are to guide the hospice philosophy of care. They include the following: (1) "support and care for people in the last phases of incurable disease so that they may live as fully and as comfortably as possible;" (2) a recognition of "dying as part of the normal process of living;" (3) a focus on "maintaining the quality of remaining life;" (4) a commitment to "helping patients and families make the transition from health to illness to death to bereavement;" and (5) an emphasis on responding to the "total pain" of patients, which includes attention to physical, emotional, psychological, and spiritual aspects of suffering (American Academy of Hospice and Palliative Medicine, 1998).

As a working philosophy, these hospice commitments embody a "palliative" rather than "curative" approach. While they pose tensions in their own right — for example, the approach to both patient and family as the "unit of care" may generate significant issues with regard to decision making — hospice principles may provide an important corrective to tendencies toward reductionism and alienation discussed above. A more integrated model of end-of-life care in critical care settings may find much that is worthwhile in the working philosophy of hospice care, despite differences in treatment goals and professional ethos.

IV. PARTICULAR ETHICAL ISSUES OF CONCERN POSED BY END-OF-LIFE DECISION MAKING

In end-of-life clinical settings, the commitment to either aggressive interventions or palliative symptom management raises conceptual and practical issues that are of vital ethical concern. While many of these ethical issues call for sustained analysis in light of the foundational philosophy of medicine issues adumbrated in Part II, I will, for the most part, review the standard ethics discussion of these issues on its own terms, that is, as developed in the ethics and policy literature. There is obvious need for constructive engagement between the general philosophy of medicine issues articulated in Part Two with the issues to be discussed in Part Three, and I will, in my conclusion, emphasize the need for such cross-fertilization as an important area for future research. Nevertheless, as I said above, an adequate constructive account of those relations would require a far more extensive discussion than can be ventured here.

The ethical issues raised by end-of-life care in either critical care or hospice settings include the following: (1) the relevance of the distinction between killing and letting die; (2) arguments for and against the legalization of physician-assisted suicide; (3) the normative relevance of dying "naturally," particularly as espoused by hospice principles in the context of high-powered symptom management; (4) the legitimacy of double effect as a moral principle applied to treatment decisions at the end of life; (5) the distinction between withholding and withdrawing treatments; (6) the status of specific treatments, for example, artificial nutrition and hydration, that may or may not be withheld or withdrawn; (7) the distinction between "ordinary" and "extraordinary" means of treatment; and (8) the distinction drawn between "quality of life" and "sanctity of life" as regulative principles. In this section, I will review the ethical discussion of these specific concerns raised by end-of-life care in greater detail.

A. The Distinction Between "Killing" and "Letting Die"

The differences between "killing" and "letting die" have been widely discussed in the clinical and theoretical literature of bioethics. Since 1973, the American Medical Association has upheld both the descriptive and the normative difference between intentional termination of life (an act of killing) and the legitimacy, under conditions generally deemed "terminal," in foregoing further life-sustaining treatment ("letting [a patient] die," American Medical Association, 1988). At the same time, certain features of the conceptual distinctions ordinarily drawn between killing and letting die are problematic and deserving of ongoing scrutiny. Especially problematic, for many commentators, is the difficulty they find in drawing clear conceptual distinctions between "active" measures (such as direct termination of life) and "passive" ones (as in "allowing to die"). Ordinarily, since the time of the President's Commission Report, *Foregoing Life-Sustaining Treatment* (1983), that distinction is framed primarily in terms of causality, and, secondarily, according to judgments about the intentionality of physicians in cases of direct termination of life as compared with cases of "allowing to die" by withholding or withdrawing life-sustaining treatment.

Both the law and morality seek to specify the criteria for discriminating between acceptable and unacceptable actions; the theory of proximate causation is central to such discriminations in both realms of discourse. Nonetheless, theories of causation face difficulties in providing covering models for a range of cases, especially those that may be descriptively overdetermined, hence requiring complex judgments about the relevance of certain "causes" rather than others to the moral and legal evaluation of actions (or inactions) involved. Indeed, much of the literature on assisted suicide and euthanasia finds the ascription of causation central to judgments about the legitimacy of such acts and practices. Much will depend crucially on how one evaluates the distinctions drawn between cases involving separate descriptive categories, for example the descriptive differences between withholding and withdrawing treatment, between omission and commission, and between killing and letting die. Assuming the validity of such descriptive differences between types of actions, what is their normative significance, especially when other features of such differently described actions remain constant? For example, proponents of assisted suicide and even active voluntary euthanasia tend to stress the similarity of intentions in many cases of both "letting die" and "killing." Because they deem such intentions decisive in their evaluations of what constitutes appropriate medical care, they deny the moral force of the descriptive differences that obtain between the two sorts of cases (Battin, 1982; Brock, 1992; Brock, 1993; Levine, 1988; Humphry, 1991; Vaux, 1988). Opponents of assisted suicide and euthanasia, by contrast, argue against those practices on one of two grounds. Either certain forms of actions, such as deliberate killing of the "innocent" are proscribed deontologically, or distinctions are drawn between the justification of individual acts and the justification of social practices, with "slippery slope" concerns invoked as decisive reasons for retaining traditional bans on killing, even though killing may be morally justified in particular cases (Gomez, 1991; Grisez, 1992; Fenigsen, 1989; Hamel, 1991; Jonsen, 1991; Kass, 1991; Wennberg, 1989).

Traditional distinctions between killing and letting die have been subjected to significant criticism, in ways germane to assessing reliance on the distinction in both critical care and hospice settings. James Rachels, in a series of prominent writings, concludes that the "bare difference" between acts of "killing" versus acts of "letting die" is not morally decisive, and elaborates two cases in support of that conclusion. Smith and Jones both desire the death of their six-year old cousins to inherit their money. In one case, Smith drowns the six-year old boy while he's bathing. In the other case, Jones intends to drown his cousin, but the child slips, hits his head on the tub, and drowns; Jones stands idly by and "lets" the boy die. In light of our moral repulsion at both cases, Rachels concludes that the "bare difference" between killing and letting die cannot be morally relevant, much less decisive (Rachels, 1975; Rachels, 1986).

Rachels's contentions can be challenged on several grounds. First, Rachels' cases are imagined in a non-medical context, while the AMA statement focuses on the distinction between killing and letting die in the context of the doctor-patient relationship. Second, both Smith and Jones are morally responsible for the death of their cousins, despite the descriptive difference between "killing" and "letting die." Indeed, the cases illustrate the overlap between the requirements of non-maleficence and limited beneficence in certain paradigmatic cases where rescue is required. Third, Rachels's cases point up a useful feature of moral analysis more generally: viz., some moral distinctions that may be otherwise relevant will be eclipsed by other features of moral choices and actions. Here the motives and intentions of Smith and Jones provoke our moral outrage and decisively influence our moral judgment about the immorality of *both* killing and letting die in such circumstances, whereas the bare difference, descriptively or normatively, between killing and letting die, may not.

Nonetheless, even in the clinical context, Dan Brock suggests that the ease with which most ethicists distinguish withdrawing life-sustaining treatment whereupon death follows shortly from acts of killing is suspect. His example is instructive. In the event that a patient in extremis desires to discontinue a respirator, many ethicists, based on considerations of autonomy and beneficence, conclude that such withdrawal is a clear case of justified "letting die" rather than killing because the relevant *cause* of death is an underlying disease process causing the fatal pathology to the lungs. However, Brock then asks us to consider a change in circumstances. Here a greedy nephew, desirous of his aunt's inheritance, sneaks into her hospital room and discontinues the respirator, resulting in her death. According to Brock, we are likely, in this instance, to find the nephew both morally and legally culpable of "killing" his aunt. The force of Brock's examples should not be lost. While we have other decisive reasons for finding the nephew's action morally and legally reprehensible, it is not obvious that the description of his action as killing distinguishes it from the doctor's action that we generally deem justified. Thus, much like Rachels's famous example, Brock concludes that the *bare* descriptive difference between "killing" and "letting die" does not do obvious moral work in our judgments, at least in ways that distinguish certain cases of withdrawing life-sustaining treatment from our intuitions about other cases we are more likely to describe as acts of killing (Brock, 1993).

Brock's own argument is also open to criticism, for it may be equally plausible to judge the greedy nephew's act as one of culpable letting die rather than killing. If so, the judgments we make to distinguish his action and motive from the action and motive of the physician implicate their very different role-related responsibilities vis-a-vis such action. We may be quite likely to interpret role-related responsibilities of physicians, according to the moral logic of beneficence, in starkly different fashion from the nephew's illicit action, an action which, even without the perverse motive at work in Brock's example, would remain unjustified, given the nephew's lack of professional judgment and medical training. Thus, critics of Rachels's cases, or of Brock's, while not necessarily finding the *bare* moral difference between killing and letting die morally decisive, are less troubled by that unclarity in light of other relevant features, such as role-related responsibilities generated by the doctor-patient relationship. The culpability of moral agents, whatever one's description, will incorporate judgments regarding motives, role-related responsibilities, presence or absence of consent, and warrants for action. Indeed, as we will see below, even in stringently deontological accounts, other aspects of choice and action, including intention, the directness or indirectness of action, or the "innocence" or "non-innocence" of the person toward whom action is taken, often serve to affect, mitigate, or determine judgments of what constitutes proximate causation and moral culpability.

B. Physician-Assisted Suicide and Voluntary Euthanasia

Those who oppose physician-assisted suicide and/or active voluntary euthanasia may base their judgments on either deontological or consequentialist grounds, or both. Deontological bases for opposition often involve appeals to the intrinsic wrongness of killing "innocent" patients, either as a general principle applicable to many settings, or as a role-related, duty-based obligation incumbent upon physicians to uphold a value intrinsic to their practice as oathed professionals dedicated to never directly taking life as a response to suffering or disease. The cogency of these deontological appeals depend upon the warrants that they invoke, be those philosophical, theological, professional, or broadly humanistic. It also depends crucially on the adequacy with which such accounts of moral obligation distinguish between and among different species of actions. As we observed above, it is sometimes difficult to ascribe relevant causation in cases of putative "killing" or "letting die," or perhaps the distinction is overdetermined in ways that may be difficult to disaggregate.

Proponents of assisted suicide and euthanasia often appeal to the moral logic already at work in decisions regarding the withholding or withdrawing of life-sustaining treatment. They emphasize two values as central to the discussion:

autonomy and beneficence. On their reading, the logic of patient autonomy dictates that physician-assisted suicide be legalized as an appropriate option at the end of life. Since patients already have been accorded the legal and moral right to discontinue treatment, with the knowledge that such discontinuance is likely to eventuate in death, proponents argue for broadening that right to legitimate the assistance of a physician to help patients end their lives as a matter of conscious and predetermined choice, rather than as an inevitable outcome of a terminal pathology. In brief, the argument is that "letting die" and "killing," while descriptively different, are both "moral" choices when justified under the rubric of informed and autonomous choice.

This autonomy-based argument is often amplified by a further consideration, since autonomy appears, at first glance, a curious value as a central appeal when arguing for physician assistance in one's suicide, rather than arguing for suicide as a solitary and unaided act (Halevy, 1993). Some commentators speak to the unique conditions within which some terminal patients find themselves. Sometimes (or often), lives may be extended by the very powers of medicine which distinguish recent technologically driven interventions from earlier, largely ineffective, medical ministrations (Brody, 1992). Earlier tableaus of expiring with a "natural" and relatively swift demise have been inexorably altered, in some cases, by the power of medicine to extend patients' lives, though mortality may be forestalled only at the price of significantly greater morbidity. By this reasoning, medicine may appropriately be invoked as an ally to help patients extricate themselves from circumstances that medicine has, to significant degree, served to create or exacerbate.

The second central value to which proponents of assisted suicide appeal is that of beneficence or relief of suffering. The relief of suffering is central to the Hippocratic tradition (Edelstein, 1977). Physicians, as oathed professionals, are obligated, as a matter of professional integrity to respond to the pain and suffering of their patients with compassionate care. Thus, in cases where patients request assistance in terminating a painful state or condition, proponents conclude that the moral logic of beneficence should legally empower the physician who, in good conscience, agrees with that request, to act on the patient's behalf, on grounds of beneficence.

However, there are differences among proponents of physician-assisted suicide about extending the moral logic of beneficence to the practice of voluntary euthanasia. It is difficult to distinguish, on the basis of beneficence, between physician-assisted suicide with (relatively) able-bodied patients and active voluntary euthanasia at the patient's request in certain other cases involving severely disabled patients (for example, late-term ALS patients or terminal quadriplegic). However, many proponents of physician-assisted suicide wish to draw a line between the two practices not on the basis of autonomy or beneficence, but from concern about the dangers of abuse; *i.e.*, a concern based, apparently, in the moral priority they accord non-maleficence (Cassell *et al*, 1992). Thus, they would restrict legalization to the autonomous acts of individual patents, rather than physicians, as the actual agents of deaths by a lethal dose of prescribed drugs. It is important, in this context, therefore, to appreciate that the appeal to the value of beneficence may be limited on other grounds by proponents as well as opponents of physician-assisted suicide.

Clearly, much of the patients' rights movement, in responding to the real and perceived excesses of traditional medical paternalism, is best interpreted as a principled limitation of the imperative of medically driven beneficence on the basis of patient autonomy. Moreover, physician beneficence, at least historically, is also constrained by complementary or alternative professional values espoused within the Hippocratic tradition itself. However, proponents of physician-assisted suicide as a legitimate response to cases involving patients in intractable pain tend to ignore such traditional constraints. Perhaps the basis of Hippocratic constraints emerges to some commentators as an antiquarian oddity. Perhaps, given the spirit of autonomy that has hallmarked the patients' rights movement, the Hippocratic prohibition on assisted suicide is deemed as dispensable as the oath's proscription of abortion. Nonetheless, it remains of theoretical, and, I would contend, practical interest to attend to those constraints historically internal to the profession of medicine, as expressed in the Hippocratic Oath. For the effort to "translate" such traditional constraints into terms palatable to the impulses of modern pluralism would reveal, in foundational ways, how the self-understanding of medicine as a practice may be ineluctably altered by the claims of proponents of assisted suicide for the priority of autonomy. Absent that argument, it is not obvious, either theoretically or practically, that the moral logic of autonomy should assume priority for redefining the appropriate scope of clinical practice in such instances.

Opponents of assisted suicide argue that neither the principle of respect for autonomy nor that of beneficence can be applied as narrowly as proponents of assisted suicide generally suggest without undercutting the logic intrinsic to each value. For example, if autonomy is the primary appeal, it is not at all obvious why provisions restricting the option to patients diagnosed as terminal do not violate autonomy for those non-terminal patients desirous of that option. Nor is it obvious, on grounds of beneficence, that one should restrict physician-assisted suicide to the class of terminal patients, in light of the severe pain and suffering associated with some forms of non-terminal illness, including chronically debilitating conditions.

However, as noted above, many proponents of physician-assisted suicide themselves would limit the logic of beneficence to cases where able-bodied patients are able to terminate their own lives – that is to say, beneficence is limited to cases of patients who can not only autonomously choose to die but can themselves act to effect that end. This appears, at first blush, a curious restriction of autonomy, which otherwise works in tandem with beneficence in the arguments of proponents. Classic proponents of autonomy, for example, find the exercise of will to be decisive: thus, any right, including the right not to be killed, can therefore be waived without injury: *volenti non injuria fit*. In principle, then, proponents of assisted suicide who emphasize autonomy should be disposed to allow voluntary euthanasia as well, so long as clinical and procedural safeguards are put in place. For those who resist the extension of their logic to active euthanasia, a certain incoherence develops. While the autonomous wishes of patients should generally be respected, and while beneficence would seem to dictate relief of suffering for those disabled patients who, while autonomous, cannot effect their own deaths, the concern about potential harm to *others* overcomes the logic of both autonomy and beneficence, which, until this juncture in the argument, have functioned as apparently decisive values.

As the foregoing discussion suggests, then, *both* proponents and opponents of physician-assisted suicide may be troubled by the potential for abuse that may come with legalization. Both camps may stress the need for restricting the claims of autonomy and beneficence because of concerns for non-maleficence. Their differences may well focus on the question of where to draw the appropriate line that works best, as a matter of public policy, to support those several values when they emerge in tension with one another. While opponents and proponents of assisted suicide may disagree profoundly about the basis of their underlying claims, especially when those claims involve differences in basic moral perspective and method, both camps also tend to draw lines that limit the logic of their own conclusions. For example, while opponents of assisted suicide and euthanasia find those options unacceptable as extensions of patient autonomy, few such commentators deny the logic of autonomy central to the bioethics consensus that has developed concerning decisions to withhold or withdraw life-sustaining treatment. It is the additional aspects of assisted suicide and euthanasia that impel them to resist the claims of autonomy being extended to those options. And while some proponents of assisted suicide might support extending the logic of their claims to justifying voluntary euthanasia, many others are not willing to do so, for the reasons discussed above. Moreover, even the latter ordinarily limit the moral logic of their arguments to the class of terminally ill patients, despite the doubtful coherence of that as a principled position.

In addition to deontological bases for opposition to physician-assisted suicide and euthanasia, opposition may also be based on an appeal to what are often called "slippery slope" concerns. Slippery slope concerns may be couched in two ways. In the first version, an appeal is made to extend the justificatory logic at work in one case to another similar case. For example, as we have seen, for some appealing to beneficence as the primary basis for legalizing physician-assisted suicide would dictate legalizing active voluntary euthanasia on the same grounds in certain hard cases.

The second version of slippery slope concerns invokes a range of potentially deleterious consequences on classes of present and future patients that might result from legalizing the practice of physician-assisted suicide. There are a number of these that are often cited, including the following. First, given the emphasis on both autonomy and beneficence, it is very likely, if not inevitable, that certain "hard cases," such as the late stage ALS patient mentioned above, will lead to extending the logic of physician-assisted suicide to active voluntary euthanasia at the patient's request. Second, physician-assisted suicide may lead, over time, to less than subtle pressures upon patients in the affected class of terminal patients who do not wish to avail themselves of that option. It does not require much in the way of imagination to speculate about the way that doctors, especially those who have helped other patients with similar diagnostic and prognostic profiles, bring their own attitudes regarding appropriate care to the bedside of patients who prefer to die "the old fashioned way." Third, there may be subtle or obvious shifts in our societal ethos if legalization creates a new set of expectations and attitudes about "appropriate" choices vis-a-vis particular diagnoses and conditions. Fourth, if the logic of "quality of life" is applied to incompetent or never competent patients, efforts may increase to provide either "substituted judgment" for assisted suicide as an option, or, in keeping with general public attitudes in the wake of legalization, according to a "reasonable person" standard consonant with such attitudes. Fifth, the logic of beneficence, in tandem with assessments about acceptable quality of life, may extend the logic of physician-assisted suicide not only to voluntary but to involuntary euthanasia (Wennberg, 1989).

All such "slippery slope" concerns of the second sort – empirical judgments about the likely social sequelae of legalization as a policy precedent – are problematic. They vary in their force depending upon the anthropological and social assumptions at work in the assessment. Moreover, they require a greater degree of prescience about the future than most prognosticators would claim for themselves. Nor do they, of themselves, override in any obvious way the moral logic at work in basic appeals made to autonomy and beneficence by proponents.

However, a lesser version of slippery slope concerns, while adopting greater humility towards the accuracy of forecasts, appeals to a distinction that is germane to the practice of what has been called "public ethics" (Jonsen and Butler, 1975). There are significant differences, in the societal and policy context, between moral and/or legal justifications of individual *acts* and the justifications of social practices. These differences may involve, as we have noted, appeals to values such as the integrity of medicine as a profession; they also involve "commonsensical" intuitions captured by the useful adage that "hard cases make bad law." As a first approximation in this instance, "emergency or borderline cases" do not possess *general* justificatory force. The differences between the two justificatory processes may be relevant to assisted suicide, as to other areas of bioethics.

C. The Normativity of Appeals to the "Natural" in Discussions of Death and Dying

The normative status of what is deemed "natural" is a deep conceptual issue that cuts across numerous issues in bioethics, from the status of new reproductive technologies and practices at the beginning of life, to issues concerned with appropriate decision making about modes of intervention and care at the end of life. Appeals to dying as a "natural" process are especially important in the philosophy of palliative care. Nonethless, the status of the "natural" as a descriptor with putative moral import remains a contentious question in ethical theory generally, and a brief review of that history sheds light on the problem of such appeals in biomedical ethics. First, appeals to the "natural" as a source of moral duties rely, in one form or another, on some notion of natural teleology, writ large or small. Such reasoning assumes two points that have generated significant critique in recent philosophy. First, such teleology assumes that one can move from description to prescription. That is a controversial claim in two crucial respects. It assumes that "proper" or "appropriate" function can be specified in ways that make its empirical assessment unproblematic. Moreover, the claim implies that, even if such descriptive criteria of appropriateness can be invoked to assess human capacities and functions, that such judgments, of themselves, are generative of moral requirements. It is equally plausible, for example, as critics suggest, that such descriptive generalizations regarding "appropriate" or "desirable" functioning are of merely pragmatic or perhaps aesthetic, rather than moral significance. Or yet again, human modes of action can, simultaneously or at different times, fulfill different functions or aim at different goals. To assume that one criterion for assessment is *either* descriptively or (by some yet to be specified extension) prescriptively determinative of what constitutes the boundaries of the "natural" oversimplifies the complexity of human choices and behaviors. Indeed, "normalcy" is a problematic appeal in putatively empirical realms of discourse, where some effort at observer-neutrality is sought. But the notion of the natural as a *morally* relevant appeal is even more problematic; to assert that relevance itself involves intrinsically a *moral* assessment rather than a matter of scientifically neutral inquiry. As one commentator puts it, "It might yet be possible to find some morally neutral way of examining such notions as human fulfillment in the scientific manner that is required by natural law theories in general. But no such approach has as yet been satisfactorily elaborated, and natural law theories can be regarded as adequate only to the extent that such elaboration is forthcoming" (Hughes, 1986, p. 413).

A more enduring criticism of appeals to the natural as of normative significance derive from G.E. Moore's critique of what he called the "naturalistic fallacy"; viz., the effort to ground ethics in a non-moral account of the nature, capacities, or functions of human beings involves an illicit inference from description to prescription (Hughes, 1986, p. 413). Although Moore's criticism has itself been subject to critique and emendation, it retains its resonance for those aspects of both theoretical and applied ethics that imbue *particular* descriptive accounts with *general* moral significance, or that assume that even *generally* accepted descriptions generate moral duties or imperatives in response to such descriptions.

Such deeply contentious issues involve a vast literature and need not further detain us here. But their relevance to decision-making, in both critical care and hospice contexts, should be briefly noted. Much is made in the literature of palliative care about accepting dying as a "natural" process, with the implications that death should be accepted as the terminus to a process rather than scheduled as an event. This appeal to dying as a "natural" process, while part-and-parcel of popular treatments of death and dying (Kubler-Ross, 1969), emerges as a fairly distinctive professional emphasis in palliative care as a medical specialization, with apparent normative implications in the hospice context. Thus, effective symptom relief is the primary clinical emphasis for hospice clinical care givers, while an understanding of dying as an inevitable, and often quite useful, *stage* with the possibility of growth, integration, and "spiritual" work is specifically embraced by the hospice philosophy.

Despite the excellence of hospice's clinical commitments and philosophy, there may be here a certain theoretical incoherence in its own celebration of dying as "natural." Many palliative interventions are quite aggressive in pursuit of effective symptom management, with largely "unnatural" features assumed to be required as an implication of quality care. Nonetheless, the "naturalness" of dying as a final stage is argued as well, as if the normativity of the natural (or its obverse) will be obvious to caregivers (and philosophers of medicine) in either instance, without the need for extended reflection or argument. While I am largely sympathetic to the intuitions that inform both commitments central to hospice philosophy — to "unnatural" interventions to control symptoms and to the importance of accepting dying as a natural stage, with possibilities for useful completion of life tasks, etc. — the tension between these different moral conclusions raises, without sufficient efforts to resolve them, the general theoretical difficulties already set forth above regarding the normativity of appeals to the natural.

D. The Principle of Double Effect Applied to Treatment Decisions at the End of Life

The principle of double effect, as traditionally espoused in Roman Catholic moral theology, operates within a framework of absolute moral duties, but distinguishes complex acts that involve both positive and negative effects on the basis of what is *intended* by the agent. In the context of decision making at the end of life, double effect has been invoked primarily in cases that involve effective doses of analgesics to relieve pain which may also, in its administration, depress respiration and thereby hasten the process of dying and the onset of death. Double effect has traditionally operated according to a fourfold schema. An act which involves a range of effects, both positive and negative, is evaluated according to the following criteria: (1) the act itself must be intrinsically good, or at least morally neutral; (2) the agent can foresee but not intend the negative effect, and must intend only the good effect; (3) the negative effect cannot be the means to the good effect; and (4) a proportionality must obtain between the good and bad effects.

This four-fold schema, as set forth by a number of commentators (D'Arcy, 1963; Bole, 1991; Mangan, 1949) has recently been reduced to either two or three of its traditional aspects (Boyle, 1980; Boyle, 1991). Some theorists, especially Richard McCormick, have emphasized the fourth element of proportionality as the key to most double effect analysis (McCormick, 1973). These critics of traditional double effect categories conclude that, in many, if not most, actions with a range of effects, the first criterion, which involves judgments about intrinsic evil, cannot be straightforwardly separated from the overall range of effects that follow from the action. Hence, the notion of an "independent moral audit" at work in each of the four criteria of double effect, makes little sense to the proportionalist perspective. For so-called "proportionalists," many goods traditionally judged to be intrinsically immoral are, in fact, "pre-moral goods" or "ontic evils," rather than intrinsically immoral actions. In McCormick's judgment, the assessment of their moral acceptability depends crucially upon an understanding of the way that their "pre-moral" status is justified according to the intrinsically moral values that they do serve, and to which they are inextricably joined. That judgment, in turn, itself depends upon the notion of "proportionate reason," according to which such pre-moral goods are invested, upon appropriate reflection, with moral valence.

Much of this is of theoretical rather than practical interest. Indeed, its theoretical interest may be waning as well, since the discussion often seems laden with somewhat esoteric terminology that is murky, and often only marginally applicable to disputed cases. Nonetheless, the language of intentionality remains crucial to both moral theory and practice. Thus, double effect, as traditionally developed (especially in religious contexts), can be understood as an attempt to accommodate two commitments that may emerge in tension in any deontologically driven world view. First, that perspective sees certain features of some actions as intrinsically wrong-making features; as a result, one is morally constrained from ever directly intending those actions as the means to a desired end, no matter how otherwise laudable that end may be. Second, that perspective appreciates the moral complexity of the world, including, as an often necessary feature of action, the mixed range of effects that may follow from any deliberate choice. Thus, the provisions of double effect serve as criteria for assessing the legitimate scope of actions that we may pursue, given the need for judging such actions by *both* our intentions and the complex effects that such actions cause.

E. The Distinction Between Withholding and Withdrawing Care

Provision of pain relief may well function with other apparently intuitive features of actions to increase the concern of care givers about the morality of their actions. For example, provision of effective analgesics is often described as a "commission" rather than an "omission," with a greater sense of responsibility assumed in the former than the latter. As we have already seen, such intuitions must be analyzed closely, for they may rest on unjustified assumptions about the nature of action and causation. While, admittedly, particular interventions *feel* different to care givers, the moral assessment of such intuitions deserves careful scrutiny.

These same intuitions must be scrutinized when caregivers assume that a moral and/or legal difference obtains between withholding (i.e., not starting) treatment and withdrawing treatment at a later point. In large part, these perceptions may reflect intuitions about some morally relevant difference between acts of "omission" and "commission." As we have seen, such descriptive differences are not, of themselves, morally decisive. Rather, they are contextualized morally by the values and principles at work in particular decisions. Thus, in some cases, withholding treatment may be justified on grounds identical to the withdrawing of treatment in such cases (or others) at a later time.

Here attention must be paid to the practice of medicine itself. Many medical interventions involve tests of therapy. The moral logic of beneficence itself dictates the practical necessity of time-tested trials of therapy, with regular monitoring and evaluation of the efficacy and effectiveness of interventions. Therefore, to insist on some morally relevant distinction attendant to the descriptive difference between withholding and withdrawing may, in numerous cases, lead to the practice of bad medicine either by overtreatment or, in some cases, undertreatment. An unwarranted insistence that treatment, once begun, cannot be withdrawn can lead, obviously, to cases of overtreatment, if caregivers feel compelled to continue treatment once initiated. But insisting on a moral distinction between withholding and withdrawing may also lead to inappropriate undertreatment in cases where caregivers initially forego aggressive interventions for fear of being thereby "locked in."

The consensus of bioethical opinion supports the conclusion that there is no morally relevant difference between decisions to withhold treatment and decisions to withdraw treatment at a later time. In both instances, the relevant standard involves considerations of patient or surrogate consent, a determination of the patient's interests, and an assessment, in patient-centered terms, of the benefits and burdens of either initiating or continuing treatment. In the judgment of two noted commentators, "treatment can always permissibly be withdrawn if it can permissibly be withheld" (Beauchamp and Childress, 1994, p. 199).

F. Distinctions Drawn Among Various Modalities of Treatment

There has been significant debate in the recent literature of bioethics and the law regarding the *range* of interventions that can appropriately be withheld or withdrawn. The status of clinically administered artificial nutrition and hydration (ANH) has been the focus of the greatest controversy, since a significant percentage of commentators endeavor to distinguish ANH as a form of "basic" or "comfort" care that cannot be withdrawn, in distinction from medical interventions that can be appropriately withheld or withdrawn (Carson, 1986; Derr, 1986; Grisez, 1992; Siegler and Weisbard, 1985).

The argument for the distinctiveness of ANH is made on various grounds. For some commentators, the provision of nutrition, hydration, and routine nursing care are requirements of human dignity. For others, ANH should be accorded special status because of the *symbolic* value of nourishment as a fundamental expression of the human effort to heal. The symbolic force of such elemental gestures, on this argument, must be maintained, regardless of the means or context of such provision. For other commentators, the primary concern posed by judgments to withhold and withdraw ANH relate to slippery slope concerns about the plausibly projected deleterious effects of moving from such considerations made in the interests of individuals to those that are made primarily for financial or societal interests, in ways that may threaten the well-being and interests of vulnerable individuals in the future.

It is difficult to know how to measure such slippery slope concerns. Nonetheless, a number of professional bodies, as well as statements from religious groups, have expressed support for assessing ANH in common with other forms of medical intervention. While the provision of food and drink is richly laden with fundamental symbolism, these commentators note the ways that medically administered ANH are disanalogous with ordinarily provided food and drink, noting that ANH may involve significant risks and burdens. Moreover, in some instances, patients die in greater comfort *without* the provision of ANH (Storey, 1992). In addition, there may be other less invasive means for dealing with the discomforts associated with physiological dehydration at the end of life (Lynn and Childress, 1986).

For many commentators, therefore, what emerges as decisive in end-of-life decision making is not the status of any particular form of medical intervention, but the moral logic that should determine what constitutes appropriate treatment in accord with the patient's own values and interests. Thus, neither ANH nor any other form of treatment need *always* be provided. Rather, case-by-case determinations will consider the provision of ANH in light of the benefits and burdens associated with particular interventions. Moreover, as we have already seen, no form of treatment need be continued simply because it has been initiated. Decisions to withhold treatment, and to withdraw treatment at a later time, both should be based on the moral logic of determining the benefits and burdens of continued treatment according to the patient's own values, preferences, and interests.

G. "Ordinary" versus "Extraordinary" Means of Treatment

The distinction between "ordinary" and "extraordinary" means of treatment has a long pedigree. It was originally developed in Roman Catholic moral theological reflections on the nature and scope of one's duty to preserve one's life (Kelly, 1950). As traditionally developed, the distinction served to differentiate forms of treatment that should be deemed obligatory from those that may legitimately be withdrawn. The basis for that discrimination, as applied in the context of modern clinical medicine, has been a patient-centered assessment that focuses on two aspects of any given intervention: (a) its reasonable chance of success; and (b) an assessment, in light of the patient's individual values, of the burdens and benefits induced or sustained by a given medical intervention.

The usage of the language of ordinary and extraordinary means is prone to distortion and misapplication when separated from its historical and systematic context. In its historical context, it developed within a foundationally religious perspective; the benefits and burdens of particular interventions were to be assessed according to one's assessment of their facilitation or thwarting of one's higher or "spiritual" ends; thus, forms of treatment that unduly distract one from the pursuit of those ends are deemed, on that basis, extraordinary (Pius XII, 1992). In Christian terms, life is a fundamental good but not an absolute value. Thus, the duty to preserve one's life is limited by a consideration of the benefits and burdens associated with such efforts at prolongation, and the focus will be, inevitably, subjective: what is deemed by one individual as unduly burdensome may be, on balance, judged to be beneficial by another. Various elements have been central to the discussion of relevant criteria for such assessment; they include the physical, psychological, and emotional pain and suffering associated with particular interventions, financial burdens to oneself or one's family, serious revulsion or aversion to particular interventions, and, more controversially, judgments about the quality of one's life after a given intervention.

The status of artificial nutrition and hydration for patients diagnosed in permanent vegetative state (PVS) has been the focus of extensive recent discussion regarding the meaning and scope of what constitutes ordinary versus extraordinary treatment in this instance. While "burdensomeness" may appear, *strictu sensu*, inapplicable to such patients (since they apparently have no contemporaneous awareness of either burdens or benefits), their prior expressed wishes about such circumstances, including their subjective concerns with the psychological and financial burdens of continued interventions upon *others*, may be quite relevant to judgments made by surrogate decision makers, in accord with the subjective and patient-centered focus of traditional discussion.

On the other hand, if one assumes that "reasonable chance of success" is the primary criterion at work in assessing the PVS patient, a great deal depends on the view of personhood at issue. Central to such judgments are what was traditionally described as a "hope of benefit" *(spes salutis)*. But that judgment itself generates controversy. For some commentators, any treatment that maintains vegetative function without undue burden is deemed beneficial because it is "useful" in maintaining biological function (Meilaender, 1997). Other commentators, however, argue that the moral logic of "usefulness" is ambiguous: it may relate to whether a given treatment is useful in light of its value to patients as persons or to patients solely as the loci for various ongoing biological/vegetative functions. These critics contend that the personalistic focus of traditional language, which subordinates the bodily good of life to the pursuit of higher spiritual ends, is thwarted when the assessment of ordinary and extraordinary reduces to the question of whether particular functions, such as cellular nutrition, can be indefinitely maintained.

H. "Quality of Life" versus "Sanctity of Life"

Much of the discussion of ordinary and extraordinary means trades upon distinctions putatively drawn between assessments of the benefits and burdens associated with an intervention *simpliciter*, and assessments that also include the benefits and burdens of the life subsequent to treatment, in effect, "quality of life' concerns. While "sanctity of life" is often invoked as a deontological bulwark against "quality of life" judgments, its implications for specific treatment choices are seldom specified with rigor or precision, nor does it, in itself, illuminate whether a particular choice should be deemed obligatory or optional (Bayertz, 1996). Moreover, it is difficult to determine what constitutes medically appropriate treatment, especially for patients diagnosed as terminal, without "presupposing some quality-of-life standard and some conception of the life the patient will live after a medical intervention" (Beauchamp and Childress, 1994, p. 216).

Nonetheless, even if "quality of life" judgments are inevitably at work in judgments concerning ordinary versus extraordinary means of treatment, in an expressly religious context, those judgments are not reducible to secular standards of "what makes life worth living." Rather, as we have seen, judgments about the "burdensomeness" of interventions or resultant quality of life are made within an expressly theological framework of stewardship: interventions can be deemed unduly burdensome insofar as they thwart one's pursuit of the "higher ends" of relating to God and preparing for death.

Absent that expressly religious context, the language of ordinary and extraordinary means has been subject, in efforts at secular translation, to significant misinterpretation. In its secular translation, it has often been reduced to a "medical indications" policy, whereby "quasi-objective" considerations of what constitutes the "usual" or ordinary response to a given pathology are viewed as determinative. But that reduction of focus distorts the meaning of the traditional distinction, which maintains a patient-specific focus that is not reducible to general judgments or algorithms.

IV. OVERVIEW AND CONSIDERATIONS FOR FUTURE RESEARCH

This chapter, in summary fashion, has engaged a number of philosophy of medicine issues implicated by clinical medicine generally, as well as particular ethical concerns raised by clinical care at the end of life. As noted throughout this chapter, the complex conceptual dialectic between general issues in the philosophy of medicine and particular ethical issues surrounding end-of-life care emerges as an important area for future research. In addition, there are numerous ethical issues that may be distinctive to hospice care. Among the latter, for example, there may be ethical tensions that arise in hospice situations where an express commitment to forego some forms of aggressive therapies may result in less effective symptom management. (Palliative surgery in some cases is one example.) There also may be tensions about what constitutes appropriate decision-making in the hospice context as hospice and family medicine are the only two specialty areas in medicine expressly committed to viewing the patient *and* family as the "unit of care." In addition, there are general ethical issues, not unique to end-of-life care, where tradeoffs must be made between providing fully effective symptom relief and constraining institutional costs. Finally, there are general justice issues raised in considering the nature and scope of coverage for end-of-life care, issues especially exacerbated by per-diem Medicare reimbursement limits on in-home hospice care.

All of the foregoing issues deserve more extensive scrutiny in their own right. In this chapter, however, I have focused on core philosophy of medicine and ethical concerns raised by end-of-life care. Such concerns are generally relevant to both critical care and hospice settings. While distinctions are appropriately drawn, many of those concerns appear to share greater correspondences than differences, and I have analyzed them accordingly.

By way of summary closing, let me list five areas that deserve greater research and reflection in the future. First, regarding issues in the philosophy of medicine, more extensive attention should be paid to the relations between the goods constitutive of medicine as a practice and the goals of medicine as professionally elaborated, as well as to the relevance of both discussions to debates about physician-assisted suicide and euthanasia. Second, greater dialogue should be fostered between critical care and hospice practitioners, in keeping with a commitment to the ideal of providing a continuum of care for patients. There are strong tendencies in medicine toward depersonalization and alienation, as we have observed. The philosophy of palliative care, as developed in hospice settings, may, with careful attention to the differences between those settings and critical care contexts, offer a useful corrective to such tendencies. Third, greater scrutiny should be given to the nature and scope of the concept of causation in order to determine its relevance to the putative moral difference between "killing" and "letting die." Fourth, the logic of double effect merits continuing attention, since that principle, in detailing the relations between the structure of acts and the intentions of moral agents, may help to clarify the relevance and force of traditional deontological constraints upon certain choices and actions, including decisions about end-of-life decision making. Finally, the principle of "sanctity of life" should be specified with greater precision in order to overcome its vagueness as a moral appeal meant to illuminate treatment choices.

Program on Biotechnology, Religion and Ethics
Rice University
Houston, TX, U.S.A.

REFERENCES

American Academy of Hospice and Palliative Medicine: 1998, *Unipac One: The Hospice/Palliative Medicine Approach to End-of-Life Care*, Kendall/Hunt Publishing Company, Dubuque, Iowa.

American Medical Association, Council on Ethical and Judicial Affairs: 1988, *Euthanasia: Report C* in *Proceedings of the House of Delegates,* American Medical Association, Chicago (June).

Barber v. Superior Court, 147 Cal. App. 3d 1006, 195 Cal. Rptr. 484 (1983).

Battin, M.P.: 1982, *Ethical Issues in Suicide*, Prentice-Hall, Englewood Cliffs, New Jersey.

Bayertz, K (ed.): 1996, *Sanctity of Life and Human Dignity*, Kluwer Academic Publishers, Dordrecht.

Beauchamp, T.L. and Childress, J.F.: 1994, *Principles of Biomedical Ethics* (4th ed.), Oxford University Press, New York.

Bole, T.J. (ed.): 1991, 'Double Effect: Theoretical Function and Bioethical Implications,' *Journal of Medicine and Philosophy* 16:5, 467-585.

Boyle, J.: 1989, 'Sanctity of Life and Suicide: Tensions and Developments within Common Morality,' in B. Brody (ed.), *Suicide and Euthanasia: Historical and Contemporary Perspectives*, Kluwer Academic Publishers, Dordrecht, 221-250.

Boyle, J.: 1980, 'Toward Understanding the Principle of Double Effect,' *Ethics* 90, 527-38.

Boyle, J.: 1991, 'Who is Entitled to Double Effect?' *The Journal of Medicine and Philosophy* 16, 475-94.

Brock, D.W.: 1993, *Life and Death: Philosophical Essays in Biomedical Ethics*, Cambridge University Press, Cambridge.

Brock, D.W.: 1992, 'Voluntary Active Euthanasia,' *Hastings Center Report*, 22:2, 11-12.

Brody, B.: 1988, *Life and Death Decision Making*, Oxford University Press, New York.

Brophy v. New England Sinai Hospital, Inc., 398 Mass. 417, 497 N.E. 2d 626 (1986).

Bryan-Brown, C.W.: 1992, 'Pathway to the Present: A Personal View of Critical Care,' in J. Civetta, R. Taylor, and R. Kirby (eds.), *Critical Care*, 2nd ed., J.P. Lippincott Company, Philadephia, 5-11.

Buchanan, A. and Brock, D.W.: 1989, *Deciding for Others: The Ethics of Surrogate Decision-Making*, Cambridge University Press, New York.

Callahan, D.: 1989, 'Can We Return Death to Disease?,' *Hastings Center Report* 19:1, 5.Callahan, D.: 1995, *What Kind of Life: The Limits of Medical Progress*, Georgetown University Press, Washington, D.C.

Capron, A.M.: 1991, '*In re* Helga Wanglie,' *Hastings Center Report* 21:5, 26-28.

Capron, A.M.: 1992, 'Euthanasia and the Netherlands: American Observations,' *Hastings Center Report* 22:2, 30-33.

Carson, R.: 1986, 'The Symbolic Significance of Giving to Eat and Drink,' in J. Lynn (ed.), *By No Extraordinary Means*, Indiana University Press, Bloomington, IN.

Cassell, E.: 1982, 'The Nature of Suffering and the Goals of Medicine,' *NEJM* 306, 639-645.

Cassell, E.: 1991, *The Nature of Suffering and the Goals of Medicine*, Oxford University Press, New York.

Cranford, R.E., Rie, M.A., and Ackerman, F.: 1991, 'Helga Wanglie's Ventilator,' *Hastings Center Report* 21:4, 23-29.

Cruzan v. Director, Missouri Department of Health, 110 S.Ct. 2841 (1990).

Daniels, N.: 1985, *Just Health Care*, Cambridge University Press, Cambridge.

D'Arcy, E.: 1963, *Human Acts: An Essay in Their Moral Evaluation*, Clarendon Press, Oxford.

Derr, P.: 1986, 'Why Food and Fluids Can Never Be Denied,' *Hastings Center Report* 16:1 28-30.

De Wachter, M.A.: 1989, 'Active Euthanasia in the Netherlands,' *Hastings Center Report* 22:2, . 23-30.

Dresser, R.S., and Robertson, J.: 1989, 'Quality of Life and Non-Treatment Decisions for Incompetent Patients: A Critique of the Orthodox Approach,' *Law, Medicine, and Health Care* 17:3, 234-44.

Edelstein, L.: 1977, 'From 'The Professional Ethics of the Greek Physician,' in S. Reiser, A. Dyck, and W. Curran (eds.), *Ethics in Medicine: Historical Perspectives and Contemporary Concerns*, MIT Press, Cambridge, Ma., 40-51.

Engelhardt, H.T.: 1989, 'Death by Free Choice,' in B. Brody (ed.), *Suicide and Euthanasia: Historical and Contemporary Perspectives*, Kluwer Academic Publishers, Dordrecht, pp. 251-280.

Engelhardt, H.T.: 1996, *The Foundations of Bioethic*, 2nd edition, Oxford University Press, New York.

Faden, R., and Beauchamp, T.: 1986, *A History and Theory of Informed Consent*, Oxford University Press, New York.

Fenigsen, R.: 1989, 'A Case Against Dutch Euthanasia,' *Hastings Center Report* 19 (Supp. 1), 22-29.

Fox, E.: 1997, 'Predominance of the Curative Model of Medical Care: A Residual Problem,' *JAMA* 278 (9), 761-63.

Gomez, C.F.: 1991, *Regulating Death: Euthanasia and the Case of the Netherlands*, The Free Press, New York.

Grisez, G.: 1992, 'Should Nutrition and Hydration Be Provided to Permanently Unconscious and Other Mentally Disabled Persons?' *Issues in Law and Medicine* 5:2, 165-179.

Gruzalski, B.: 1988, 'Death by Omission,' in B. Brody (ed.), *Moral Theory and Moral Judgment*, Kluwer Academic Publishers, Dordrecht, 75-85.

Hamel, R., (ed.): 1991, *Active Euthanasia, Religion, and the Public Debate*, The Park Ridge Center, Chicago.

Hastings Center: 1987, *Guidelines on the Termination of Life-Sustaining Treatment and the Care of the Dying*, Indiana University Press, Bloomington, In..

Humphry, D.: 1991, *Final Exit: The Practicalities of Self-Deliverance and Assisted Suicide for the Dying*, The Hemlock Society, Eugene, OR.

In re Conroy, 486 A. 2d 1209 (N.J. 1985).

In re Quinlan, 70 N.J. 10, 355 A.2d 647 (1976).

In the matter of Spring, Mass. 405 N.E. 2d 115 (1980), at 488-89.

Jecker, N.S. and Self, D.J.: 1991, 'Separating Care and Cure: An Analysis of Historical and Contemporary Images in Nursing and Medicine,' *The Journal of Medicine and Philosophy* **16**,. 285-306.

Jennings, B., Callahan, D., and Caplan, A.L.: 1988, 'Ethical Challenges of Chronic Illness,' *Hastings Center Report* **18**:1 (Supp. 1), 1-16.

Jonsen, A. And Butler, L.: 1975, 'Public Ethics and Policy Making,' *Hastings Center Report* 5:4, 19-31.

Jonsen, A.: 1998, *The Birth of Bioethics*, Oxford University Press, New York.

Jonsen, A.: 1991, 'What is at Stake?,' *Commonweal* 118:14 (Supp. 1), 2-4.

Kass, L.:1991, 'Why Doctors Must Not Kill,' *Commonweal* 118: 14 (Supp. 1), 10.

Kelly, G.: 1950, "The Duty of Using Artificial Means of Preserving Life," *Theological Studies* 11, 203-20.

Keown, J.: 1991, 'On Regulating Death,' *Hastings Center Report* 22: 2, 39-43.

Khuse, H.: 1987, *The Sanctity-of-Life Doctrine in Medicine: A Critique*, Oxford University Press,Cambridge.

Khushf, G.: 1992, 'Post-Modern Reflections on the Ethics of Naming,' in J. Peset and D. Gracia (eds.), *The Ethics of Diagnosis*, Kluwer Academic Publishers, Dordrecht, 275-300.

Kilner, J.: 1990, *Who Lives? Who Dies? Ethical Criteria in Patient Selection*, Yale University Press, New Haven.

Kleinman, A.: 1988, *The Illness Narratives: Suffering, Healing, and the Human Condition*, Basic Books, New York.

Kohl, M. (ed.): 1975, *Beneficent Euthanasia*, Prometheus Books, Buffalo, New York.

Levine, M.P.: 1988, 'Coffee and Casuistry: It Doesn't Matter Who Caused What,' in B. Brody (ed.), *Moral Theory and Moral Judgment*, Kluwer Academic Publishers, Dordrecht, 87-98.

Lynn, J., and Childress, J.F.: 1986, 'Must Patients Always Be Given Food and Water?,' *Hastings Center Report* 13:5, 17-21.

Lynn, J. (ed.): 1986, *By No Extraordinary Means: The Choice to Forgo Life-Sustaining Food and Water*, Indiana University Press, Bloomington, In.

McCarrick, P.M.:1992, *Active Euthanasia and Assisted Suicide*, Scope Note 18, The Kennedy Institute of Ethics at Georgetown University, Washington, D.C.

McCormick, R.: 1990, 'Clear and convincing: The Case of Nancy Cruzan,' *Midwest Medical Ethics* **6**:4, 10-12.

McCormick, R.: 1973, *Ambiguity in Moral Choice*, Marquette University Press, Milwaukee, Wisconsin.

McCullough, L.: 1989, 'The Abstract Character and Transforming Power of Medical Language,' *Soundings* 72, 111-125.

MacIntyre, A.: 1997, *After Virtue: A Study in Moral Theory* 2nd ed., University of Notre Dame Press, Notre Dame, In.

Mangan, J.: 1949, 'An Historical Analysis of the Principle of Double Effect,' *Theological Studies* **10**, 41-61.

Meilaender, G.: 1990, 'The Cruzan Decision: 95 Theses for Discussion,' *Midwest Medical Ethics* 6:4, 1, 3-5.
Meisel, A.: 1992, "The Legal Consensus about Foregoing Life-Sustaining Treatment: Its Status and Prospects,' *Kennedy Institute of Ethics Journal* 2:4, 309-345.
Micetich, K., Steicker, P., and Thomasma, D.: 1983, 'Are Intravenous Fluids Morally Required for Dying Patients?' *Archives of Internal Medicine* 143 (May), 975-978.
O'Rourke, K.: 1989, 'Should Nutrition and Hydration Be Provided to Permanently Unconscious and Other Mentally Disabled Persons?,' *Issues in Law and Medicine* 5:2, 181-196.
Pence, G. E.: 1990, *Classic Cases in Medical Ethics*, McGraw Hill, New York.
Pius XII: 1992, 'The Prolongation of Life,' in K. Wildes, F. Abel, J. Harvey (eds.), *Birth, Suffering, and Death: Catholic Perspectives at the Edges of Life*, Kluwer Academic Publishers, Dordrecht, 209-215.
Post, S.: 1988, 'Family Casemaking: Moral Commitments and the Burden of Care,' *Second Opinion* 8, 114-27.
Rachels, J.: 1975, 'Active and Passive Euthanasia,' *NEJM* 292:2, 78-80.
Rachels, J.: 1986, *The End of Life: Euthanasia and Morality*, Oxford University Press, New York.
Ramsey, P.: 1970, *The Patient as Person*, Yale University Press, New Haven.
Schloendorff v. Society of New York Hospital. 1914. 211 N.Y. 125, 105 N.E. 92. 95.
Siegler, M. and Weisbard, A.: 1985, 'Against the Emerging Stream: Should Fluids and Nutritional Support Be Discontinued?' *Archives of Internal Medicine* 145 (January), 129-32.
Spicker, S. (ed.):1970, *The Philosophy of the Body: Rejections of Cartesian Dualism*, Quadrangle/New York Times Publishing Company, New York.
Steinbock, B. (ed.):1980, *Killing and Letting Die*, Prentice-Hall, Englewood Cliffs, New Jersey.
Stoddard, S.: 1992, *The Hospice Movement: A Better Way of Caring for the Dying*, Vintage Books, New York.
Storey, P.: 1992, 'Artificial Feeding and Hydration in Advanced Illness,' in K. Wildes, F. Abel, J. Harvey (eds.), *Birth, Suffering, and Death: Catholic Perspectives at the Edges of Life*, Kluwer Academic Publishers, Dordrecht, 67-75.
SUPPORT Principal Investigators, The: 1995, 'A Controlled Study to Improve Care for Seriously Ill Hospitalized Patients: The Study to Understand Prognoses and Preferences for Outcomes and Risks of Treatments (SUPPORT),' *JAMA* 274: 1591-98.
Toombs, K.: 1992, *The Meaning of Illness*, Kluwer Academic Publishers, Dordrecht.
United States President's Commission for the Study of Ethical Problems in Medicine and Biomedical and Behavioral Research: 1983, *Deciding to Forego Life-Sustaining Treatment: A Report on the Ethical, Medical, and Legal Issues in Treatment Decisions*, Washington, D.C.
Vaux, K.L.: 1988, 'Debbie's Dying: Mercy Killing and the Good Death,' *JAMA* 259:14, 2410-2411.
Veatch, R.M.: 1993, 'Forgoing Life-Sustaining Treatment: Limits to the Consensus,' *Kennedy Institute of Ethics Journal*, 3:1, 1-19.
Wanzer, S., Adelstein *et al.*: 1984, "The Physician's Responsibility toward Hopelessly Ill Patients,' *NEJM* 310:15, 955-959.
Weir, R.F.: 1989, *Abating Treatment with Critically Ill Patients: Ethical and Legal Limits to the Medical Prolongation of Life*, Oxford University Press, New York.
Wennberg, R.: 1989, *Terminal Choices: Euthanasia, Suicide, and the Right to Die*, Eerdmans Publishing Company, Grand Rapids, Mi.
Wolf, S.: 1989, 'Holding the Line on Euthanasia,' *Hastings Center Report* 19:1 (Suppl.), 13-15.

OSBORNE P. WIGGINS AND MICHAEL ALAN SCHWARTZ

PHILOSOPHICAL ISSUES IN PSYCHIATRY

I. INTRODUCTION: MENTAL DISTURBANCES AND WESTERN MEDICAL SCIENCE

In earlier periods of human history some members of society were certainly seen as mentally disturbed or at least as behaving in remarkably exceptional ways. This persistent, unusual behavior was often interpreted in moral or religious terms: the person was viewed as violating the established morality or as introducing a higher morality, as being demon-possessed or divinely inspired. These extraordinary modes of behavior were thus conceived through schemes of interpretation that were already operative in the society.

Today these modes of behavior would be *medically* conceived. The person would no longer be morally condemned or subjected to exorcism. He or she would now be seen as suffering from an "illness" somewhat similar to other (i.e., physical) illnesses, with uniform symptomatologies and supposedly uniform etiologies. Since these are *medical* problems, they must be understood through scientific concepts and treated with scientifically established techniques.

Only in the past two decades have signs appeared that this scientific-medical approach to mental disturbances is on the right track. With the increasing "medicalization" of psychiatry during these decades, psychiatry has sought to model its treatment procedures, theoretical concepts, and research methods on those of biomedicine. This modeling has produced significant advances. We shall mention only three of the most prominent. (1) The official diagnostic manual published by the American Psychiatric Association, *Diagnostic and Statistical Manual of Mental Disorders (DSM),* has continued to undergo refinement and is more and more viewed as authoritative. This manual has established criteria of diagnosis that have high degrees of reliability, and now efforts are underway to determine their validity. (2) Proof of the efficacy of several different kinds of pharmacotherapy has given psychiatry chemical tools for alleviating suffering that rival the effective medications of other medical specialties. These advances in psychopharmacology have been accompanied by a growing suspiciousness about the scientific credentials and therapeutic efficacy of Freudian and other forms of psychotherapy. (3) Related to pharmacotherapy

G. Khushf (ed.), Handbook of Bioethics, 473–488.

have been studies of brain chemistry and structure that spur the hope that the nueuromechanisms of mental disorders will be increasingly understood and therefore controlled. Hence neuroscience, with its solid experimental foundation, plays a larger and larger role in the basic conception of mental disorders and in strategies for their treatment.

As a result of these recent advances, there are those who think that psychiatry has at last entered upon the sure path of a successful medical specialty. For example, here is how Joseph T. Coyle and Richard Mollica, both of Harvard Medical School, depict the capabilities of present-day psychiatry:

> Psychiatry, fortunately, has developed highly reliable, phenomenologically based diagnostic instruments (DSM-IV and ICD-10) that are easy to apply for the diagnosis of common psychiatric disorders. Furthermore, a new generation of psychotherapeutic drugs has been developed that are more effective and exhibit fewer and less serious side effects than the previous generation of psychotropic drugs. The symptoms of the most common disorders, including depression, post-traumatic stress disorder, anxiety disorders, schizophrenia and bipolar disorder, can be very effectively managed in most patients with psychotropic drugs (Coyle et al, p. 494).

These advances have seemed to many, moreover, to confirm the *superiority* of the scientific approaches that produced them. Hence certain spokespersons for these approaches have boasted that the future of psychiatry lies with them and that the fruitlessness of other approaches will become increasingly apparent. It will be noted that these "successful" approaches – *DSM,* psychopharmacology, and neuroscience – are ones closely associated with the natural sciences, and this too seems to confirm psychiatry's close kinship with biomedicine. The approaches toward which a growing skepticism has been directed are the ones akin to the social and psychological disciplines.

Not everyone believes, however, that the "successful" approaches alone will yield a fully adequate understanding of mental disorders and their effective treatment. Many psychiatrists feel that other perspectives too need to be included, but there remains significant confusion about just *how* they should be included. This confusion is so great that most practitioners are wary of any "comprehensive framework" that might be put forward as the systematic and unified discipline of psychaitry. Any such comprehensive framework, it is rightly said, must be "philosophical," and widespread suspicion exists regarding the demonstrability of large philosophical frameworks. Otherwise stated, philosophical systems are feared to be too subjective. Nevertheless, the confusion in psychiatry to which we have alluded concerns some basic philosophical problems. Among those problems are the following. First, because of the growth of neuroscience, the mind/body question has re-appeared in a slightly new form: Can mental disorders be conceived exclusively as "diseases of the brain" without bothering with a careful understanding of the psychopathology? And if both neurobiology and psychopathology are to be considered, how does one combine these two? Second, the nature/nurture problem has resurfaced: To what extent should mental disorders be conceived as influenced by social environment and to what extent can they be approached as biologically conditioned, ultimately even genetically conditioned? Third, what sorts of methods are most appropriate for research and what sorts of procedures are most effective in treatment? Also among the philosophical problems are, fourth, the ethical problems: What is the proper ethical treatment of persons with mental disorders?

At present little agreement exists on how to answer these questions. Faced with these apparently unanswerable philosophical questions, therefore, only a few options remain available for psychaitrists to decide how to understand their professional field and daily work.

One option would be that of *pure pragmatism:* one adopts "what works" without too much worry about why or how it works. One despairs of being able to make comprehensive coherent sense of psychiatric practice and tries simply "to do what's effective."

Another option is to settle for one's own *personally satisfying world-view,* remaining convinced that nothing more rationally defensible can be attained. This world-view may contain a variety of elements from different domains, neuroscience, diverse psychotherapies, religion, common sense morality, etc. Indeed, this collection of items may even be fitted into a roughly unified whole. But one sees it as one's *personal* perspective and therefore incapable of proof – indeed incapable of defense against other contrasting and competing "perspectives."

Both of the above options are, of course, permeated by skepticism. Rationality is deemed incapable of supplying and grounding psychiatry as a whole, and consequently only a non-rational ground is available. The kind of pragmatism that pervades all of medicine, however, may offer consolation in the midst of this intellectual resignation: you don't have to understand as long as the patient got better.

II. THE NEED TO GO BEYOND EMPIRICAL SCIENCE TO A PHILOSOPHY OF THE PERSON

It would appear, then, that the scientific and biomedical approaches which psychiatry has adopted and which have produced numerous successes remain insufficient for answering several larger questions which those very scientific approaches raise. Those questions can be addressed, we think, only through a philosophical inquiry.

The question of mind and body – or, more precisely stated, of mind and neurobiological organism – poses the question of the nature of the *human being* who is this mixture of mind and body (or biological organism). Likewise, the question of nature and nurture in their influence on mental disturbances points to the question of the *person* who is to some extent a product of his or her nature or nurture. Moreover, questions about appropriate research methods and therapeutic procedures in psychiatry must face the fact that the "mental disorders" which are the topics of research and therapy are not separable from the *persons* whose mental disorders they are.

Of course, once it is recognized that these puzzles that arise *within* psychiatry can be resolved only with an adequate philosophy of the person, it will be quickly perceived that this philosophy will have to reach far beyond what is now deemed the province of psychiatry. Any conception of the person which can deal adequately with the nature/nurture perplexity will have to encompass normal as well as abnormal human conditions. Moreover, it will have to clarify the nature of society and culture sufficiently to show how these large realities continually shape personhood. Hence this philosophy of the person must discuss the cultural, social, and natural constituents of personhood as these manifest themselves in both normal and abnormal experiences.

"Person," moreover, is the concept that can unify the disparate principles of psychiatric ethics. This is already evinced in the formulation of the ethical principle of autonomy as "respect for persons." The Kantian notion of person out of which the formulation originated is no longer accepted, but some understanding of persons is still assumed when ethicists speak of dignity and freedom. Clearly the assumption is that dignity and autonomy are basic constituents of personhood that deserve respect. But present-day psychiatric ethics, despite its distance from the Kantian "person," never replaces it with an alternative conception that would demonstrate the moral value of those constituents. Similarly, the other principles such as non-maleficence and beneficence can be more clearly connected with autonomy as soon as we comprehend how the ethical imperatives not to harm or to remove harm presuppose values of persons which deserve not to be violated. Similarly, the principle of justice assumes values in persons which are violated when there is discrimination against them or they do not receive the goods and services they deserve. As long as this unifying concept of persons is not clarified, the various ethical principles individually hang in the air without any relation except that implicitly acknowledged when we apply them.

We will seek now to move toward a philosophical understanding of persons as the presupposed reality of psychiatric thought and practice. In view of the present state of psychiatry as we have outlined it above, however, any conception of person that will serve psychiatry today will have to take fully into account the knowledge of persons we have in the sciences. Any notion of persons that does not include what we know about them from neuroscience or experimental psychology, for example, will remain alien to present-day psychiatry. We must therefore encompass the findings of the sciences in the general conception we wish to develop. But it will nonetheless remain necessary to go beyond these findings, i.e., to move at a genuinely philosophical level, because, as we have seen, the sciences, when considered alone, raise questions for psychiatrists which those sciences prove incapable of answering.

III. SCIENCE AS A HUMAN INTELLECTUAL ACHIEVEMENT

Psychiatry has taken over many thought patterns from Western science in general and from Western scientific medicine in particular. Nevertheless, if it is to be effective, psychiatry must adapt these thought patterns to its own peculiar subject matter, human mental disturbances. And consequently the question arises of whether this adaptation is adequate. In other words, the question arises of whether psychiatry has succeeded in doing justice to its subject matter. To be more precise, the question is whether psychiatry has done *interpretive* and *clinical* justice to its peculiar subject matter, a subject matter related to but still different from those of other sciences and medical specialties. In order to address this question of the adequacy of present-day psychiatry to its subject matter, we shall clarify some of the basic features of Western science. Clarifying these features will permit us to critically judge this adequacy and to suggest approaches that will render psychiatry more adequate.

We can today no longer speak of "science" as a univocal term. There rather exists a plurality of sciences employing a variety of methods. The sciences are differentiated either by their *subject matters,* the realities they study, or by the *methods* they use to study them. In the developed sciences, the subject matters are defined by the concepts and theories dominant in those sciences. In other words, the subject matter *is* what the theories about the subject matter conceive it as being. Of course, in any particular science there may exist competing theories about the subject matter. And moreover, the adequacy of the theories and concepts can be tested by appealing to evidence of the subject matter. But it is still the competing and empirically testable theories that determine for us what we take the subject matter of the science to be.

If we are concerned then about how adequate sciences may be as conceptualizations of their subject matters, we need to understand the conditions necessary for constructing scientific theories. We shall concentrate on only one such condition for theory construction here: the need to define concepts by abstracting from certain features of the reality under study. In 1903, Johann

Theodore Merz surveyed the different aspects of nature portrayed in the different sciences of his time. In the midst of this survey he comments,

> The different aspects of nature which I have reviewed in the foregoing chapters and the various sciences which have been elaborated by their aid, comprise what may appropriately be termed the abstract study of natural objects and phenomena. Though all the methods of reasoning with which we have so far become acquainted originated primarily through observation and the reflection over things natural, they have this in common that they – for the purpose of examination – remove their objects out of the position and surroundings which nature has assigned them: that they *abstract* them. This process of abstraction is either literally a process of removal from one place to another, from the great work- and storehouse of nature herself to the small workroom, the laboratory of the experimenter; or – where such removal is not possible – the process is carried out merely in the realm of contemplation; one or two special properties are noted and described, whilst the number of collateral data are for the moment disregarded. (A third method, not developed at the time, is the creation of "unnatural" conditions and, thereby, the production of "unnatural" phenomena [Paul Feyerabend's comment].)
>
> There is, moreover, in addition to the aspect of convenience, one very powerful inducement for scientific workers to persevere in their process of abstraction... This is the practical usefulness of such researches in the arts and industries... The wants and creations of artificial life have thus proved the greatest incentives to the abstract and artificial treatment of natural objects and processes for which the chemical and electrical laboratories with the calculating room of the mathematician on the one side and the workshop and factory on the other, have in the course of the century become so renowned...(cited in Feyerabend, p. 153).

Merz thus portrays "the spirit of abstraction" as necessary for scientific conceptualization. He characterizes the process of abstraction this way: "one or two special properties are noted and described, whilst the number of collateral data are for the moment disregarded." The "collateral data" disregarded by one science may be precisely the "one or two special properties" to which another science is devoted. But still what one science examines carefully another science disregards.

Merz then immediately notes an intellectual interest moving in the opposite direction:

> There is, however, in the human mind an opposite interest which fortunately counteracts to a considerable extent the one-sided working of the spirit of abstraction in science... This is the genuine love of nature, the consciousness that we lose all power if, to any great extent, we sever or weaken that connection which ties us to the world as it is – to things real and natural: it finds its expression in the ancient legend of the mighty giant who derived all his strength from his mother earth and collapsed if severed from her... In the study of natural objects we meet (therefore) with a class of students who are attracted by things as they are... Their sciences are the truly descriptive sciences, in opposition to the abstract ones (Feyerabend, p. 153).

We shall pursue below this concern, expressed by Merz, for "truly descriptive sciences" which "fortunately" counteract "the one-sided working of the spirit of abstraction in science." But first we would like to emphasize an implication of Merz's view of abstraction as necessary for scientific conceptualization. This is the implication that the sciences are unavoidably *many.* The sciences are many because each one studies only "one or two special properties" of nature and disregards others. Hence other sciences are needed to examine the other properties. As John Ziman writes,

> There is no simple "scientific" map of reality – or if there were, it would be much too complicated and unwieldy to be grasped or used by anyone. But there are many different maps of reality, from a variety of scientific viewpoints (Feyerabend, p. 154).

But ultimately it must be said that the sciences are many because each is a one-sided conceptualization of a many-sided whole. As Paul Feyerabend has said, "The world is a complex and many-sided thing" (Feyerabend, pp. 151-152).

In psychiatry this spirit of abstraction is today under the guidance of the medicalization. The basic concepts and procedures of biomedicine determine what tends to be thematized and what disregarded in psychiatry. Viewing human mental problems as "mental disorders" is already an abstraction: it is a one-sided focus on certain features of these problems while setting aside others. Moreover, conceiving these problems as mental disorders entails that our way of practically dealing with them will proceed through differential diagnosis, treatment plans, and assessment of improvement – all modeled on biomedicine.

Moreover, the spirit of abstraction takes a further step when the features selected by the theory are assigned *metaphysical priority:* these features are alone seen as the "really real" ones. All other features are disregarded as not as real or even as unreal – as "purely subjective" or epiphenomenal. Physicalism (or materialism) is one example of this sort of metaphysical step. In physicalism it is claimed that the only truly real properties of things are their physical properties, and "physical properties" is usually taken to mean those properties depicted by modern natural science, physics, chemistry, and biology. Hence the yellow color of the object is merely "subjective" because the only "objective" properties of the object are its atomical and molecular properties. Such a physicalism requires that any non-physical properties be somehow "reduced to" physical ones or disregarded altogether. In the philosophy of mind today there are strong tendencies toward physicalism: many philosophical arguments are mounted in order to prove that it is possible to explain all forms of human behavior by referring solely to physical causes, and again "physical" means those properties explicable in terms from the natural sciences.

If each of the sciences has been able to secure its own special subject matter by abstractly separating it from other aspects of reality, then we might ask what it is that each science "abstracts from." What is the "reality" out of which the sciences have been constructed through processes of abstraction? Whatever this reality is it is clear that it must harbor *in nucleo all* the realities which the various sciences and disciplines will abstractively delineate. Of course, prior to the abstractions and the development of the sciences of them these various realities lie more or less latent and opaque in the fundamental reality. But this does allow us to say that this fundamental reality must be a multifarious and complicated one: it must contain a "plenitude of being" such that the many different sciences can separate out sectors of it and study them. As Feyerabend says in the quote cited above, "The world is a complex and many-sided thing."

What then is the fundamental reality out of which all the sciences abstract and delimit their greatly differing subject matters? We submit that the fundamental reality is what Husserl has called "the lifeworld," the world as we directly encounter it in everyday perception and action. It is not possible here to explicate the historical origin of the Western sciences. We would like only to suggest that that origin can be found in the thought of Xenophanes and other Pre-Socratic philosophers who, when confronted with the relativity of their own religious, moral, and political beliefs – their own sociocultural world-view – conceded this relativity and then asked about the "being" beyond their own culture, i.e., non-relative, true being.

IV. THE LIFEWORLD

The lifeworld is the world as given in pre-scientific experience. The lifeworld is therefore the world as we directly encounter it in our everyday perception and bodily action. The saw in the workshop is not an object composed of atoms and molecules; it is rather the tool with which I cut pieces of wood, the object I manipulate in a certain bodily way in order to actualize a certain goal. This knowledge we have of things and people at the level of ordinary perception and action is a pre-linguistic knowledge. The action of sawing is an action whose meaning I learn through watching someone else saw or sawing myself. I may, of course, learn to associate the sound with that meaningful action, and the word may then form part of my native language. The word, however, originally derived its meaning from perception and bodily activity. At the pre-linguistic level the saw is perceived by me as having a typical meaning: it is the *kind* of object that one can bodily manipulate in a typical way in order to achieve a typical purpose. Of course, the object is also perceived as a particular thing. But it is perceived as a *particular* thing with a *generic* meaning. Natural language, which also forms part of the lifeworld, derives much of its meaningfulness from these typical meanings of things and events that are familiar to us pre-linguistically through perception and bodily action (Husserl, 1970; Gurwitsch, 1974).

These generic senses of things that pervade perception and action also inform our interactions with other people. When I encounter the mailperson and she hands me my mail, I have a *typical* understanding of her action and motivation: her intention in behaving as she does is to deliver my mail to me. Indeed my act of taking the mail from her makes sense to me only because I typify her motivation in this generic way. In our interaction with other people in the lifeworld we are constantly typifying the meanings of their actions; i.e., we are understanding the subjective experiences of what they are doing, at least in a typical and generic fashion.

Karl Jaspers introduced into psychiatry the notion of "understanding" *(Verstehen)* that he had learned from Wilhelm Dilthey, Georg Simmel, and others (Jaspers, 1963). "Understanding" in this technical sense connotes making sense of other people's subjective experiences through perceiving their overt behaviors, i.e., their verbal utterances, bodily gestures, and facial expressions. This making sense of a person's subjective experiences through his or her overt expressions does not involve inference or projection. Understanding another person is simply a unified act of *perceiving* this person. In such perception this person is given to me as a psychophysical whole. Or rather, my pre-scientific perception of other people is "psychophysically neutral": my perception does not distinguish between the person's experiences and his or her behavior.

Such understanding of people is operative daily in our interaction with others in the common sense lifeworld. Psychiatric understanding of patients goes beyond such common sense understanding, but it always draws on vast portions of this pre-scientific knowledge of human beings. Psychiatric understanding, no matter how technical it becomes, must always be supplemented by the understanding of other people that the psychiatrist possesses by virtue of being a human being existing with other people in the common sense lifeworld. Most often this supplementation of scientific understanding by pre-scientific understanding goes unstated: it is silently presupposed. But pre-scientific *Verstehen* is, even in its implicitness, operative in informing the scientific interpretation of patients.

To say this is not to diminish the limitations and relativity of pre-scientific knowledge, however. Each real lifeworld is culturally relative. In other words, common sense understanding of human beings is ethnocentric. And it is precisely in order to overcome such relativity and ethnocentricity that science arises. Scientists strive for *universal knowledge,* i.e., a knowledge not limited to or biased by the cultural meanings of any particular lifeworld. Notice, however, that we designate this universality as the "striving" of science. It is an *aim* of scientific activity. This does not assure that it is ever attained. But even if it is not fully attained, it does serve as a guiding ideal which directs and informs scientific activity. This striving for universal knowledge is one of the defining features of science as a human project different from others (Gurwitsch, 1974).

V. PHENOMENOLOGY

The explication we have just furnished above of our everyday experience in the lifeworld has been a phenomenological explication. Phenomenology is thus the study of human experiences and the objects of those experiences precisely as those experiences are lived through by the subjects whose experiences they are. In brief, phenomenology is a description of experience and the world as experienced. Phenomenology, therefore, does not study the world "as it exists in itself"; it rather studies *the world as experienced* by the subjects under study, and correlatively it studies the *psychological experiences* of those subjects. This correlation between experience and the object of experience phenomenologists call "intentionality."

The natural sciences abstract from the world as it appears in the everyday experience of the subjects who inhabit that world. Natural science, in other words, abstracts from the lifeworld. One example of natural scientific abstraction is found in the way in which it disregards the person's daily experience of his or her own *lived body.* Natural science rather focuses on the *biological organism* as conceived through its theories and concepts. Phenomenology performs no such abstractions. One of its main topics is the lived body as experienced by the subject whose body it is. Another central topic of phenomenology is different sociocultural lifeworlds as they are experienced by the people who live their daily lives in them. Thus phenomenology is a more concrete discipline than natural science in the sense that phenomenology does not disregard what the natural sciences do systematically disregard.

As Jaspers pointed out, phenomenology employs the method of "understanding" *(Verstehen):* it seeks to makes sense of what the other person is experiencing by interpreting the words, expressions, and behaviors of the person. This sort of understanding can be rendered scientific, however, if it always remains critically based on this *evidence,* namely, the words, expressions, and behaviors of the subjects (Jaspers, 1963).

In order to characterize the experiences of subjects, phenomenology must, of course, develop its own technical vocabulary. But this vocabulary aims merely to accurately capture the features of the subject's experiences precisely as the subject lives through these experiences. Hence technical phenomenological language remains very close to the concrete experiences of people.

Phenomenology is crucial for psychiatry, we believe, because phenomenology can portray the patient's mental problems *as the patient experiences them.* It depicts the mental illness *as it is subjectively lived by the patient.* Phrased somewhat differently, it describes mental illness *as a lived human experience.* And it also explicates the world of the patient, *the world as experienced by the patient in his or her mental illness.*

Phenomenology can thus serve as an underlying science for psychopathology. It is imperative, however, that psychopathology be based on a more inclusive phenomenology. Psychopathology should comprise only a part of the more general discipline of phenomenology. The more general discipline must study normal mental life as well as abnormal. Psychopathology needs phenomenology because the latter can provide a systematic conception of mental life as a whole as well as basic concepts of the different kinds of mental processes and their properties that the former can then apply to mental disorders.

Moreover, phenomenology can assist psychiatry in additional ways. (1) A phenomenological understanding can help in making sense of the brain scans and other sources of evidence for neurobiological hypotheses. Phenomenological knowledge can guide neurobiological thinking. (2) Phenomenological understanding can help make sense of the symptomatological manifestations of the illness. If we know what the patient's experience is like, we can make better sense of why she behaves and presents as she does. (3) Phenomenological understanding also helps the clinician make sense of the patient's response to treatment once therapy is underway.

Experimental psychology is related to phenomenology through its subject matter, just as experimental psychology is related to phenomenological psychopathology through its subject matter. The differences betweeen the experimental and phenomenological approaches lie in their methods. It is important to recognize, however, that experimental psychology and psychopathology, insofar as their subject matters are human experience, must also utilize the method of scientific *Verstehen.* The evidence gathered by the experimental scientist must be interpreted so that it is evidence for certain features of mental processes. And the reasoning from the experimental evidence to those mental features – as well as the reasoning from the supposed mental features back to the evidence – involves scientific *Verstehen.*

Any understanding of other people, however, must draw on the ways in which we understand other people pre-scientifically. It must thus draw on the common-sensical understanding of people we already have as participants in a sociocultural lifeworld. Here we possess both a pre-linguistic (typifications) and a linguistic way of understanding people. Through clinical experience, however, the psychiatrist develops psychiatric typifications and psychiatric concepts.

VI. A NOTE ON "PHENOMENOLOGY" AS USED IN PRESENT-DAY PSYCHIATRY

DSM-IV, by appearing to be "phenomenological," has brought about the neglect of a more thorough phenomenological psychopathology. It has brought about this neglect because too many people in the field have been persuaded that the restriction of *DSM* "phenomenology" to what can be "observed," "operationalized," and established with "reliability" are necessary for a "scientific" psychiatry. But this *DSM* "phenomenology" differs greatly from the phenomenology that Karl Jaspers pioneered with this 1911 book, *General Psychopathology. DSM* "phenomenology" is strongly influenced by the logical empiricist philosophy of Carl Hempel. Jaspers' phenomenology is defined by the methods which he found in the thought of Wilhelm Dilthey, Georg Simmel, Max Weber, and Edmund Husserl. In another essay we have shown in detail that Jaspers' methods involve reflection, description, intuitive representation, as well as the kind of understanding *(Verstehen)* that we pinpointed above (Wiggins et al, 1997). We are convinced that a return to and systematic development of Jaspersian phenomenological psychopathology would provide a very helpful supplement to the other methods and concepts available in psychiatry today.

VII. THE SOCIAL SCIENCES AND HUMANITIES

The social sciences and humanities remain close to the concreteness of the lifeworld. They do this by explicating the ways in which the subjects who inhabit the various sociocultural lifeworlds subjectively experience those worlds. This they seek to do for *all* social lifeworlds, throughout history and across the globe. The social sciences and the humanities thus start from the phenomenological viewpoint that we sketched above: they start from the world as it appears to experiencing subjects. They then seek to provide *explanations* of the origin, nature, and persistence of these cultural lifeworlds and the subjects' experiences of them. Such explanations refer to factors that, as such, are not directly part of the subjects' experiences. But the point is to explain the existence of these lifeworlds and the subjective experiences (Gurwitsch, 1974).

Social scientists can make sense of the historical lifeworlds they study only because these scientists, as sociocultural subjects, inhabit their own lifeworld and know at the pre-scientific level how to make sense of this, their native lifeworld. In other words, again pre-scientific understanding of people and the world make scientific understanding of other people and their worlds possible. But scientific understanding goes beyond pre-scientific by developing its own scientific concepts and theories. In this way scientific understanding can transcend the relativity and one-sidedness of the scientist's own culture while still arising out of it and being nourished by it.

For the psychiatrist, the social sciences and humanities can provide a variety of different perspectives on the patient and his or her mental problems. Multiple perspectives are required for an adequate view of human beings because human beings are multifaceted realities. The great variety of the "human sciences" gives testimony to the multiple dimensions of and possibilities for human life. The social sciences and humanities can thus break down assumptions about human beings that remain too narrow or one-sided. They allow one to place one's own lifeworld "in perspective." And moreover they provide concepts that allow us to make sense of how culture and society shape "human nature" into a variety of historical forms.

VIII. THE NATURAL SCIENCES

All the sciences, as we have claimed, have been constructed through processes of abstraction. The natural sciences, however, in addition to employing extremely high levels of abstraction, use two other intellectual operations, formalization and idealization.

Formalization is the intellectual operation whereby a term with meaningful content is replaced by a mere placeholder, as, for example, when "The snow is white" is reformuated as *"S is w."* Formalization is therefore necessary for logic and mathematics. Since natural scientific accounts of reality are largely mathematical or at least aspire to be mathematical, natural scientific thinking systematically formalizes its subject matter: the "nature" of the natural sciences is a mathematized nature, or at least a nature on the way to mathematization. Such a nature differs greatly, of course, from the nature that we directly perceive in the lifeworld (Husserl, 1970).

Idealization is the intellectual operation whereby things are imagined in their perfect, flawless, or "ideal" form. Thus idealization too is required for mathematics. For example, the circle of geometry is the perfect circle, the circle all of whose points are exactly equidistant from a center point. It is clear that such a geometrical circle can exist only in imagination because all "real circles" are imperfect. The high level theories and "laws of nature" which the natural sciences develop are idealizations, and they thus depict an idealized nature (Husserl, 1970).

Idealization functions, however, at even a more fundamental level in natural scientific thinking. For example, the theories of the natural sciences are undergirded by assumptions regarding universal causal determinism. The idea of universal causal determinism is itself an *idealization.* It is the idea that every event has a cause and that, moreover, that cause can be determined with exactitude. The intellectual roots of such an idealized notion lie in the rough and ready causation that we experience everyday among events in the lifeworld. Such lifeworld causation is not entirely regular and exact, but is rather typical or "more or less" regular. Through our attempts to control events, however, we achieve some success, and we thereby render events more regular and predictable. From these successes, we obtain the notion of the "more regular." Pursuing this notion in imagination, we can conceive of consecutive events which are *more and more regular.* Taking such imaginings far enough, we arrive at the notion of the perfectly regular, the precisely regular. In other terms we reach the idea of *a law-governed sequences of events.* Notice that what imagination does in such cases is *idealize* certain experiences: events that are experienced in a certain way are imagined as *more and more* like that. At the limit point of such idealizations lies the perfectly law-governed universe. Such *perfect* laws of events can be expressed in their absolute precision only in a *mathematical* language. And thus idealization requires mathematization just as the latter requires the former.

This idealization of universal causal determinism has not and cannot be proven to be true. But it can and does serve to guide natural scientific thinking. It thus functions as a "regulative idea" for the natural scientific mind. And, as we have noted, the presupposition of the mathematizability of all natural occurrences is a similar idealization that functions as a regulative idea. These idealizations are components of the "cloak of ideas" that has been thrown over nature (Husserl, 1970) and that thereby constitute the "nature" studied by modern natural science. Scientists then assume that they must pinpoint the precise cause of an event and that this relationship between cause and effect must be expressed in a mathematical formula. While these aims are not always reached, they nevertheless function as "ideals" that guide thought.

IX. CONCLUSION: THE "GAPS" AND (AGAIN) THE NEED FOR A PHILOSOPHY OF THE PERSON

There is thus, as Fred Kersten has pointed out, a definite "gap" between the lifeworld and the universe as depicted in natural scientific theories: i.e., there exists very little similarity between the lifeworld and the universe of natural science, or, stated the other way around, there is a sizable conceptual difference between the lifeworld and the scientific universe (Kersten, 1997).

With the increasing specialization of the various sciences, different scientific sub-universes have been constructed. Sometimes there is very little similarity between scientific sub-universes. The most significant gap exists, of course, between the sub-universes of natural science and those of social science. The social sciences bear a greater verisimilitude to the pre-scientific social lifeworld than the natural sciences do to the natural lifeworld. This is because the social sciences have been developed more through inductive generalization from the lifeworld than the natural sciences have. The social sciences are in this sense, then, "less abstract," i.e., "less abstract" in the sense of disregarding fewer qualities of the various cultural lifeworlds. Many generalizations of the social sciences can be understood as simply higher level (i.e., more exact and more accurate) conceptualizations of generalizations that are already meaningful in the native languages of the lifeworlds. Hence the sciences and the humanities present no system or integration. We are left with a disparate array of disciplines which exhibit little, if any, versimilitude.

This conceptual gap between the natural sciences and the social sciences and humanities, of course, generates the infamous mind/body problem which has proven to be so intractable despite widespread acknowledgment of the need to "solve" it. Related to this is the gap between "nature" and "nuture." Hence we encounter again the current problems for psychiatry that we mentioned earlier, and psychiatry is faced with the difficulty of deciding what to do about these gaps. We suggest again that what is needed is a unified philosophy of the person. This philosophy must take into account the various findings and claims of the natural sciences, social sciences, and humanities. But it needs to develop an integral theory of the person which delineates the connections among the diverse disciplines. In another place we have tried to outline such a philosophy of the person (Schwartz et al., 2000). Without such a philosophical theory psychiatry will continue in the confusion that results when the pieces of what is vaguely known to be a whole nevertheless refuse to fit.

Department of Philosophy
University of Louisville
U.S.A.

Department of Psychiatry
Tufts University School of Medicine
U.S.A.

REFERENCES

Diagnostic and Statistical Manual of Mental Disorders, Fourth Edition, (DSM-IV): 1994, American Psychiatric Assocation, Washington, D.C.

Coyle, JT., and Mollica, R: 1998, 'Forum – Psychiatry in Medical Education, Comments,' *Current Opinion in Psychiatry* 11:5, 493-495.

Feyerabend, P: 1999, *Conquest of Abundance: A Tale of Abstraction versus the Richness of Being*, edited by Bert Terpstra, pp. 153, 154; © The University of Chicago Press. Reprinted with permission of The University of Chicago Press, Chicago.

Graham, G, and Stephens, GL (eds): 1994, *Philosophical Psychopathology*, The MIT Press, Cambridge.

Gurwitsch, A: 1974, *Phenomenology and the Theory of Science*, Northwestern University Press, Evanston.

Husserl, E.: 1970, *The Crisis of European Sciences and Transcendental Phenomenology*, D. Carr (trans.), Northwestern University Press, Evanston.

Jaspers, K: 1963, *General Psychopathology*, The University of Chicago Press, Chicago.

Kersten, F: 1997, *Galileo and the 'Invention' of Opera: A Study in the Phenomenology of Consciousness*, Kluwer Academic Publishers, Dordrecht.

McHugh, PR., and Slavney, PR: 1986, *The Perspectives of Psychiatry*, The Johns Hopkins University Press, Baltimore.

Sadler, JZ, Wiggins, OP, and Schwartz, MA (eds.): 1994, *Philosophical Perspectives on Psychiatric Diagnostic Classification*, The Johns Hopkins University Press, Baltimore, MD.

Schwartz, MA, and Wiggins, O: 1985, 'Science, Humanism, and the Nature of Medical Practice: A Phenomenological View,' *Perspectives in Biology and Medicine* 28:3, 331-361.

Schwartz, MA, and Wiggins, OP: 1986, 'Logical Empiricism and Psychiatric Classification,' *Comprehensive Psychiatry*, No. 2 (March/April), pp. 101-114.

Schwartz, MA., Wiggins, OP, and Norko, MA: 1995, 'Prototypes, Ideal Types, and Personality Disorders: The Return to Classical Phenomenology,' in *The DSM-IV Personality Disorders*, ed by W. Livesley, John, Guilfod Publications, Inc. New York, pp. 417- 432.

Schwartz, MA, and Wiggins, OP: 2000, 'Pathological Selves,' in *Exploring the Self: Philosophical and Psychopathological Perspectives on Self-Experience*, ed. by D. Zahavi, John Benjamins Publishing Company, Amsterdam.

Slavney, PR, and McHugh, PR: 1987, *Psychiatric Polarities: Methodology and Practice*, The Johns Hopkins University Press, Baltimore.

Spitzer, M, Uehlein, F, Schwartz, M, and Mundt, C (eds.): 1992, *Phenomenology, Language, and Schizoprhenia*, Springer Verlag, New York.

Spitzer, M: 1999, *The Mind within the Net: Models of Learning, Thinking, and Acting*, The MIT Press, Cambridge.

White, KL: 1988, *The Task of Medicine: Dialogue at Wickenburg*, The Henry J. Kaiser Family Foundation, Menlo Park.

Wiggins, OP, and Schwartz, MA: March 1997, 'Edmund Husserl's Influence on Karl Jaspers's Phenomenology,' *Philosophy, Psychiatry, and Psychology – PPP*, 4:1 (March), pp. 15-36.

SARA T. FRY

NURSING ETHICS

Throughout the nursing profession's short history, nursing ethics has been primarily concerned with two issues: (1) describing the characteristics of the "good" nurse, and (2) identifying nurses' ethical practices. In this chapter, the evolution of nursing ethics' treatment of these two issues during the 20th century will be described. I will argue that nursing ethics needs to continue to address these issues in the 21st century and that how they are understood will be strongly influenced by changes in health care delivery and the workplaces of nurses.

I. DEFINING THE "GOOD" NURSE AND ETHICAL PRACTICE

A. The "Good" Nurse as an Obedient, Cooperative, and Dutiful Helper

During the early days of the 20th century, nursing ethics was understood as the articulation of the customs, habits, and moral rules that nurses follow in the care of the sick (Robb, 1921; Aikens, 1931). Robb (1921), for example, defined ethics as "the science that treats of human actions from a standpoint of right and wrong" (p. 13). Nursing ethics therefore concerned "the rules of conduct adapted to the many diverse circumstances attending the nursing of the sick" (Robb, 1921, p. 16). Early nurse leaders believed that the science of ethics must be learned by the nurse so that competency in nursing practice could be attained and ethical behaviors assured. By learning ethics, the nurse learned the moral duties and rules expected and could be relied upon to engage in certain ethical behaviors. When such behaviors were performed, ethical practice resulted and the nurse was considered competent or a "good" nurse.

This view of the close connection between ethical behavior and nursing competence was repeated in early texts on nursing ethics. Aikens (1931) described nursing ethics as the "ideals, customs, and habits . . accumulating around the . . . name and general characteristics of the . . . trained nurse" (p. 36). The nurse was admonished to follow "patterns of ethical behavior" in order to demonstrate "keenness of ethical duty" for the practice of nursing (Aikens, 1931, p. 19). The "good" nurse was a nurse who acted out of duty and who consistently engaged in ethical behaviors. Such a nurse was also considered competent.

G. Khushf (ed.), Handbook of Bioethics, 489–505.

Nursing ethics thus described the "good" nurse as an individual who was virtuous and who followed certain rules in caring for the sick. Ethical practice was the performance of ethical behavior as a moral duty.

Ethical behaviors that were expected of the nurse included loyalty (to the physician and also to the training school), modesty, sobriety, honesty, truthfulness, trustworthiness, obedience, promptness, quietness, cheerfulness, and deference to authority figures (Nutting and Dock, 1907). Besides being a "good woman" (which apparently meant being morally pure and of good breeding), the nurse was to act dignified, cultured, courteous, and reserved (Robb, 1921). The nurse was also expected to cooperate with the health care institution or agency and with the physician in providing care to the patient.

Today, we recognize these behaviors as nursing etiquette or polite forms of behavior associated with the social role of the nurse of the early 20th century. However, the performance of these behaviors was essential to being considered a "good" nurse. Any nurse could follow rules and be of a virtuous character. But without the practice of certain behaviors, a nurse could not be considered a "good" nurse. The competence of the nurse was thus a reflection of her moral conduct, personal characteristics, inner strength, and the performance of ethical behaviors.

B. The "Good" Nurse as an Accountable and Principled Independent Practitioner

After World War II, the role of the nurse changed from being an obedient, cooperative helper to the physician, to being an independent practitioner. In this new role, the nurse was held morally (and legally) accountable for what had been done (or not done) in providing nursing care. While the nurse was still expected to be cooperative, the new expectation of accountability created changes in how nurses' ethical duties and behaviors were understood. Guidelines for nurses' ethical behaviors were now described in codes of ethics. The International Council of Nurses (ICN) *Code of Ethics for Nurses* (2000) was first accepted in 1953 while the American Nurses Association (ANA) *Code for Nurses* (1985) was accepted in 1950. It is no surprise that early versions of both codes used language that closely mirrored the language of physicians' codes of ethics of the same time period. However, nursing codes of ethics also described nurses' ethical responsibilities in terms of health outcome – the promotion of health, the prevention of illness, the restoration of health, and the alleviation of suffering (ICN, 2000). Ethical behaviors were still expected of the nurse but there was less emphasis on the performance of these behaviors as individual moral duties. Instead, the nurse's primary moral duty was to be accountable to the patient, the profession, and the public for how the nurse fulfilled his/her obligations to achieve health outcomes. The standards for legal accountability appeared in nurse practice acts and licensing regulations administered by state-governed Boards of Nursing. The standards for moral accountability were defined in codes of ethics created by members of the profession.

The profession's requirements for competent and ethical practices also appeared in statements called "standards of nursing practice" (ANA, 1998). These standards were established by specialty practice groups such as maternal/child nursing, psychiatric nursing, and medical/surgical nursing. They listed the moral and legal requirements for safe and competent practice within the specialty and defined what was considered competent nursing practice. Thus, the "good" nurse was defined in terms of technical competence rather than in terms of personal moral characteristics or even the performance of ethical behaviors. Ethical behaviors were still expected of the nurse but these behaviors were secondary to competent practice. If a nurse practiced competently, his/her practice was usually considered ethical.

In the late 1970s, nursing ethics began to use the language, concepts, and principles of biomedical ethics in defining the "good" nurse and the characteristics of ethical practice (Fry, 1986; Lumpp, 1979; Stenberg, 1979). For example, the Preamble of the ANA *Code for Nurses* (1985) defined the ethical framework for nurses' decision making in terms of principles of biomedical ethics such as beneficence, nonmaleficence, justice, and autonomy. Being morally accountable meant that the nurse justified his/her ethical decision making and ethical actions in terms of these principles. Ethical practice was providing ethical reasons for nursing actions in patient care while the "good" nurse met standards for moral and legal accountability in patient care.

At the same time, nursing ethics was also influenced by the language and principles of cognitive psychology (Kohlberg, 1976; Rest, 1986). For example, the relationship between the nurse's ethical reasoning and ethical behavior was explained by psychological processes or structures of moral reasoning (Crisham, 1981; Ketefian, 1981; Parker, 1991). The result was that moral psychology and moral development theory both influenced how ethical practices in nursing was conceptualized – ethical practice was the application of moral reasoning skills, and ethical principles and rules. Thus, ethics was no longer taught in nursing education so that the nurse would learn moral rules and duties. It was taught so that the nurse could learn ethics as a cognitive and analytical skill that could be applied in patient care.

The growing body of nursing ethics literature at this time demonstrates the influence of both cognitive psychology and biomedical ethics on how the "good" nurse and ethical practice were defined (Ketefian and Ormond, 1988). The nurse was no longer expected to develop virtues or to practice certain ethical behaviors. The nurse applied ethical reasoning and the language of biomedical ethics as skills that were learned in the classroom. As Penticuff (1991) points out, the non-reflective use of these theoretical frameworks had an unfortunate effect on the development of nursing ethics. Their use limited knowledge of the variables that influence ethical practice, impeded theory development in nursing ethics, and prevented a clear articulation of what is distinctive about nursing ethics and nurses' ethical practices.

The clinical judgment model (Gordon, et al., 1994), in particular, clearly demonstrates the limiting influences of these theoretical frameworks on the conceptualization of nurses' ethical practices. Clinical judgment is essentially a reasoning process that is deliberate, analytical, and conscious, and has three dimensions – diagnostic, ethical, and therapeutic. Based on information processing theory, decision theory, and cognitive psychology, the clinical judgment model views the nurse as a rational, impartial person who gathers critical information about the patient, identifies problems, and states them in a specific language or "nursing diagnosis" (Gordon, 1994). The nurse then weighs options in a disciplined and systematic manner, and chooses a nursing intervention based on its potential to resolve the problem or to minimize adverse outcomes (Gordon, Murphy, et al., 1994). Ethical reasoning in the model is a cognitive and problem-solving process guided by utility-based or duty-based theories that results in therapeutic and presumably ethical nursing actions. These actions are supported by nursing theory and justified by universal ethical principles.

This strong, influential view of clinical judgment has unfortunately created a view of nurses' ethical practice that is not congruent with the reality of that practice. It has promoted a prescriptive and process-oriented approach that is rarely experienced in nursing practice (Yeo, 1989), while the theoretical structures of utility and duty may not even be accurate representations of the foundations for nurses' ethical decision making (Fry, 1989). Furthermore, important aspects of the nurse-patient relationship – especially, the moral elements of caring and connectedness – are virtually overlooked in this model of clinical judgment (Benner, 1990; Benner, et al., 1996).

C. The "Good" Nurse as a Caring Professional in Co-Presence with the Patient

During the past few years, a number of nursing scholars have suggested approaches to move beyond the above described "impoverished" view of nurses' ethical practices. One approach adopts a richer view of nursing practice where health and illness, "the nature of nursing, what it means to be a nurse, and essential features of nurse-patient transactions are examined" (Penticuff, 1991, p. 246). After all, the goals of nurses' ethical practice concern human needs which "are either met or not met within nurse-patient transactions" (p. 246). Penticuff is supported in this approach by other nurses who are focusing on the specific nature of the nurse-patient relationship (Bishop and Scudder, 2001; Gadow, 1999; Hess, 1996; Liaschenko, 1997) and the ethical goals of nursing care (Gastmans, et al., 1998; Taylor, 1998).

Another approach has focused on how skilled nurses make judgments (including ethical judgments) in their everyday practice. This approach has led to a view of clinical judgment involving engaged, practical reasoning that is quite different from the earlier view that nursing judgment is deliberative and consciously analytical. According to Benner, Tanner, and Chesla (1996),

> Experienced nurses reach an understanding of a person's experience with an illness, and . . . their response to it, not through abstract labeling such as nursing diagnoses, but rather through knowing the particular patient, his typical pattern of responses, his story, and the way in which illness has constituted his story .... (p. 1).

This view of clinical judgment refers to the ways in which nurses (a) understand patient problems, issues, and concerns, (b) deal with salient information, and (c) respond in concerned and deliberate ways. It includes decision-making, discrimination, and intuitive expertise on the part of the nurse. The "good" nurse is one who uses this type of clinical judgement to produce good outcomes for the patient.

Essential to ethical practice is the nurse's ability to "know" the patient or being attuned to the patient in such a way that the nurse recognizes and interprets patient cues that usually occur before they are noted by monitors or laboratory tests (Benner, et al., 1996). It is not merely knowing various diagnostic labels that might be applied to the changes in the patient, but responding to the changes expertly and ethically. Ethical practice is "skilled ethical comportment," the embodied skilled know-how of relating to patients in ways that are respectful, responsive, and supportive.[1] In this view of ethical practice, the words, intents, beliefs and values of the nurse are encased in his/her stance, touch, and orientation – "thoughts and feelings are fused with physical presence and action" (Benner, et al., 1996). It is a type of practice that develops over time, is not rule-governed or procedural, and is guided by an understanding of particular human concerns in particular contexts.

Being a caring nurse is, of course, important to this view of ethical practice. People with health care needs are often vulnerable and nurses are expected to use caring behaviors in addressing these needs. But what is caring's status in nursing ethics? Is caring a moral duty of the nurse or an ethical principle that guides nursing actions? Is caring a special virtue of the nurse, as some have argued (Brody, 1988; Knowlden, 1990)? Or is caring simply valued in patient care because it supports human concern for people with health care needs?

Nursing ethics has been interested in caring as a moral foundation for nurses' ethical practice because traditional ethical theories (and the ethics of duty or obligation) have not proved adequate to address the reality of health care relationships, in general, and the nurse-patient relationship, in particular. In the nurse-patient relationship, the nurse is involved with the patient in such a way that concern is created about how the patient experiences his/her world. Traditional ethical theories, principles, and rules have insufficient content to address this kind of involvement. They do have content to address the ends (outcomes) of this form of involvement (increased patient good or welfare) and perhaps the structure of the involvement (duty or obligation), but they cannot address the nature of the involvement itself. As a result, nurses have found traditional ethical theories conceptually inadequate to characterize the moral dimensions of the nurse-patient relationship. They have focused on caring as an essential characteristic of ethical practice and redefined the "good" nurse as the nurse who demonstrates caring attitudes and behaviors in patient care. Yet, caring's status as a marker for ethical practice remains ambiguous.

Bishop and Scudder (2001) suggest that caring is really a "presence" – a therapeutic presence that includes an attitude of personal concern as well as skill and knowledge about caring. Caring is not emotion or sentimental. Rather it is a way of being with others that assures them of personal concern for their well being. Such presence fosters the well being of individuals by transforming how they experience their world and ultimately fosters the healing process. It lets patients know that they are not only being cared for, but that the one providing care, really does care.

This view of caring is repeated by others (Benner, et al., 1996) and defined as "the alleviation of vulnerability; the promotion of growth and health; the facilitation of comfort, dignity, or a good or peaceful death; . . . and the preservation and extension of human possibilities" (p. 233). Caring is a behavior and stance that is learned and then demonstrated within the context of good nursing practice (Jacques, 1993).

I have argued elsewhere that caring is simply a value and a behavior (Fry, 1989) and not a virtue, as some have charged (Curzer, 1993). What this means is that caring behaviors as well as the attitudes and feelings that underlie these behaviors are valued in patient care. Caring is a necessary but not a sufficient condition for ethical practice. However, others strongly disagree with this position. Several ethicists (Curzer, 1993; Nelson, 1992) and nurses (Crigger, 1997; Phillips, 1993; Warelow, 1996) argue against the use of caring in conceptualizing nurses' ethical practices. Kuhse (1997), in particular, voices concern that an emphasis on caring in nursing to the exclusion of traditional ethical principles will "reinforce the traditional subordinate roles of nurses and facilitate the continued exclusion of nurses from ethical discourse" (p. x).

Rather than characterizing caring as a special virtue of the nurse or as a transcendental co-presence of the nurse with the patient, Kuhse urges a "dispositional notion of care" (p. 150) where caring is understood as a willingness and openness to apprehend the health concerns of the patient. This notion of caring means to be receptive to the needs of patients as particular persons with special needs, beliefs, and desires. Such caring, according to Kuhse (1997), should be a component of nurses' ethical practices and, indeed, of all health care practices.

There is considerable support for Kuhse's view of caring in the nurse-patient relationship.[1] Some argue that caring needs to be conceptualized as a central concept of nurses' ethical practices (Dyson, 1997; Khushf, 1997) and not just as a value, principle or virtue. Others claim that descriptions of caring need to be theoretically adequate in that the necessary conceptual features of the properties of caring are made explicit (Gaut, 1983; Holden, 1991; Smerke, 1990). A few argue that caring should not be isolated from traditional bioethical principles, such as justice, autonomy, and beneficence (Fry and Johnstone, 2001; Omery, 1995).

By the end of the 20th century, there is a general consensus in the nursing ethics literature that the "good" nurse is a caring nurse and that ethical practice necessarily requires caring behaviors and attitudes towards patients on the part of the nurse. But how does caring fit into the moral foundations for nursing practice? Are there influences on abilities of a nurse to care and on the amount of time it takes to be caring toward patients? Do changes in health care delivery affect the ability of the nurse to provide caring behaviors and ultimately ethical practices? These are questions that need to be addressed in the 21st century by modern nursing ethics.

II. NURSING ETHICS IN THE 21ST CENTURY

A. Reflecting on What Is Already Known

Over the past 100 years, several moral concepts of nursing practice have emerged. These concepts support a moral foundation for nursing practice, help define the "good" nurse, and have important implications for understanding what characterizes nurses' ethical practices. The concept of *cooperation* stems from early conceptions of the relationship of the nurse to the physician and other health care workers. The concept of *accountability* developed out of the moral responsibility of the nurse as an independent nurse practitioner after World War II. The concept of *caring* developed from the perceived need for the nurse to perform caring behaviors as health care delivery became a highly impersonalized and complex system of multiple health care workers. The nurse is the one health care worker who still maintains a close relationship with the patient despite these changes. Thus, caring behaviors, while always expected of the nurse, have a new importance and meaning in modern health care delivery. One new concept, *advocacy*, has emerged only in recent years in relation to trends in the protections of patients' rights in health care and the roles that nurses have in helping patients to exercise these rights. Each of these concepts are briefly described because together, they provide a starting point for nursing ethics inquiry in the 21st century and contribute to our understanding of the "good" nurse and nurses' ethical practices.

Cooperation. Historically, cooperation was viewed as a type of special loyalty shared by members of a professional group (Robb, 1921). Nightingale also wrote of cooperation as "human combination" where individuals maintain and strengthen a community (of nurses) by working toward a common goal (Nutting and Dock, 1907). However, contemporary views of cooperation consider it a multidimensional concept that includes *active participation* with others to obtain quality care for patients, *collaboration* in designing approaches to nursing care, and *reciprocity* with those with whom nurses professionally identify (Fry, 1994). It means to consider the values and goals of those with whom one works as one's own values and goals thus fostering supportive networks and close working relationships among members of the health care team.

The concept of cooperation provides support for nursing actions such as working with others toward shared goals, keeping promises, and forgoing personal interests in order to maintain the health care team's integrity. These actions express human interactions traditionally valued by people and support the ideal of professional collaboration in designing patient care (Jameton, 1984). The moral role for cooperation is the maintenance of working relationships and conditions that express obligations toward the patient and are mutually agreeable (Corser, 1998). The "good" nurse cooperates and collaborates with other health care workers to bring about patient good and to improve health care practices.

Accountability. The concept of accountability became morally important to nursing when nurses began to practice independently of the physician and take personal responsibility for nursing actions. The concept of accountability has two major attributes – answerability and responsibility – that define accountability as being answerable for how one has carried out his/her responsibility (Fry, 1994). The responsibility of the nurse is universally understood as (a) promoting health, (b) preventing illness, (c) restoring health, and (d) alleviating suffering (ICN, 2000). A nurse is accountable when he/she explains how this responsibility has been carried out, justifying the choices and actions according to accepted moral standards or norms. The nurse is always accountable to the patient and the profession, and may be accountable to the employer and to society for what has been done (or not done) in providing nursing care (Hilbig and Manning, 1999; Pyne, 1991; Rowe, 2000). Public trust and confidence in the nursing profession depends, in part, on nurses being viewed as responsible and answerable for their actions. The "good" nurse is an accountable nurse and only when a nurse meets the requirements of moral accountability can his/her practices be considered ethical.

Caring. The concept of caring supports the moral obligation of the nurse to provide nursing care to those who need it. Since the relationship is created by the patient's need for nursing care, the nurse provides care (cares for the patient) and also demonstrates caring behaviors toward the patient because doing so promotes patient good or welfare (Smerke, 1990). Thus, in nursing, caring is defined as a form of involvement with patients that creates concern about how they experience their world (Benner and Wrubel, 1989). While all communities and cultures practice caring behaviors that preserve health and life, nurse caring is specifically directed toward the vulnerability of the patient with health care needs and is expressed in caring attitudes and behaviors. It requires "sentiment and skills of connection and involvement, as well as caregiving, knowledge and skills" (Benner and Gordon, 1996, p. 44). Because nurse caring, in part, concerns patient good, it is morally significant in health care and not simply a human behavior.

There is little consensus on the necessary conditions for a nurse's behavior to be considered caring (Fry, 1991; Shiber and Larson, 1991). Likewise, there is little consensus on how much caring the nurse is obligated to exhibit in the nurse-patient relationship (Jacques, 1993; Valentine, 1991). Yet caring attitudes and behaviors are expected of the nurse and are valued by the public (Benner and Wrubel, 1989). Caring is considered an important moral concept with implications for the nature and quality of the nurse-patient relationship. Although the attributes and ethical dimensions of nurse caring have not yet been adequately described, most nurses agree that the "good" nurse is a caring nurse and that caring is a necessary condition for ethical practice.

Advocacy. The concept of advocacy has been recently considered morally important to nursing practice. With changes in the delivery of health care and the growing number of professional and non-professional groups who are involved in providing health care services, nurses have added an advocacy role to the nurse-patient relationship (Gadow, 1989; Gates, 1994; Mallik, 1997). This advocacy role does not mean that the nurse is the primary defender of patient rights within the health care system, the person who informs the patient of his rights, and prevents violations of the patient's rights. This is a legalistic view of advocacy that does not really fit with nurse advocacy as it is actually practiced in health care.

Most nurses view their advocacy role as one of negotiation on behalf of the patient. This means that the nurse assists the patient to discuss his needs, interests, and choices consistent with his values and lifestyles. The nurse also helps the patient examine the advantages and disadvantages of various health options in order to make decisions most consistent with personal beliefs and values (Fry, 1994). The patient is regarded as a fellow human being who is entitled to respect. The nurse negotiates with the health care insurer, the health care system at large, with other health care providers, and with the patient and his family in order to protect the patient's values, human dignity, physical integrity, privacy, and choices. When the patient is not competent to make choices, the nurse advocates for the patient's welfare as defined by the patient before he became ill or as defined by his family members or substitute decision-makers. If no one else is available to define the welfare of the patient, the nurse promotes the best interests of the patient to the best of his/her nursing ability.

This view of advocacy is most consistent with the tenets of the International Council of Nurses' (ICN) *Code of Ethics for Nurses* (2000) and the American Nurses Association (ANA) *Code for Nurses* (1985). It characterizes the moral responsibility of the nurse to protect the patient from harm. While some question whether nurses can be effective patient care advocates (Abrams, 1978; Annas, 1974), others clearly affirm advocacy's role as a moral concept for nursing practice (Gates, 1994; Gaylord and Grace, 1995; Sellin, 1995; Snowball, 1996). The "good" nurse is one who advocates for patients and advocacy-motivated actions contribute to ethical nursing practices.

B. Researching What is Not Known

Despite renewed interest in nursing ethics over the past 20 years, the ethical issues that nurses experience in providing patient care are just beginning to be identified. In addition, very little is known about the essential elements of nurses' ethical practices. In particular, influences on nurses' ethical practices have not been clearly identified (Penticuff, 1991) and nurses' ethical practices are not well conceptualized or described (Benner, et al., 1996; Gastmans, et al., 1998; Taylor, 1998). One of the major tasks for nursing ethics in the 21st century will be to carefully identify what is known and not known about nurses' ethical practices in a changing health care environment. Identifying the "good" nurse in the 21st century may well depend on the results of this effort.

Ethical Issues in Nursing Practice. Surprising as it may seem, there has not been an accurate record of the ethical issues that nurses encounter in their practices over the past 65 years. As a result, little is known about the resources that nurses actually use to deal with ethical issues or what they need in order to practice ethically.

The earliest study of ethical issues in nursing practice was a content analysis of the diaries of 95 nurses who recorded the ethical problems they encountered in nursing practice over a three-month period (Vaughan, 1935). The diary entries were analyzed according to Augustino Lehmkuhl's classifications of moral problems with slight modifications. [3] Vaughan's analysis identified a total of 2265 ethical issues, 67 problems of etiquette, and 110 questions about ethical behavior. This amounted to an average of 23.4 moral problems for each nurse participant in the study.

The ethical issue encountered most often and reported by a majority of the respondents (88%) was a lack of cooperation between nurses and physicians, and between nurses in general. Other frequently encountered issues included duties to patients, lying (including dishonest charting), lust, and alcohol use among nurse colleagues. A lack of cooperation between nurses and physicians (and physician/nurse conflict) is a recurring theme in studies of ethical issues experienced by practicing nurses, despite the recognition that cooperation is an important moral concept of nursing practice.

More recently, nurses have studied how frequently nurses in different practice environments encounter specific ethical issues in their practices, how disturbed they are by them, and the influence of demographic and work-related variables on the frequency and the disturbance levels of ethical issues (Berger, et al., 1991; Omery, et al., 1995; Scanlon, 1994;). The results differ among the study findings but issues such as inadequate staffing, difficulties in the nurse/physician relationship, pain relief and management, inappropriate allocation of resouces, end-of-life decisions, cost-containment issues that affect patient welfare, and incompetent and irresponsible colleagues were more frequently experienced than other issues in all studies.

Two methodologically similar studies have yielded the most useful information about ethical issues in nursing practice, how frequently they are experienced, how nurses handle the issues, and the ethics education needed by nurses in order to practice ethically (Fry and Damrosch, 1994; Fry and Riley, 1999). Surveying a large probability sample of nurses working in Maryland and six New England states, the surveys found that protecting patients' rights and human dignity, respecting/non-respecting informed consent to treatment, staffing patterns that limit patient access to nursing care, quality of care, and prolonging the living/dying process with inappropriate measures were ethical issues frequently experienced by nurses. All of these issues concern the nurse's advocacy role with the patient. The nurses also rated these issues as highly disturbing to them when they occurred. The majority of the New England nurse respondents (83.8%) reported that they handled ethical issues by discussing them with a nurse peer while less than 14% consulted with an Ethics Committee. Only a few nurses (11.6%) thought their workplace resources were adequate to help nurses deal with ethical issues. With these findings, it was no surprise that the majority of the New England nurses (59%) reported that they had a moderate to great need for ethics education in order to practice ethically.

While these recent studies have identified ethical issues in nursing practice, it is still not known how nurses respond to particular issues when experienced, and how they use resources in the workplace to handle specific issues. Furthermore, it is not known the extent to which workplace factors influence the abilities of nurses to handle issues and which ethics resources in the workplace are most helpful to nurses. Clearly, further research needs to be conducted as changes occur in health care delivery and nurses are presented with new and more difficult ethical issues in providing patient care. The advocacy role of the nurse also needs further clarification and support. The "good" nurse is a nurse who protects the rights of patients, especially where patients' privacy, quality of care, choices, and access to beneficial health care services are concerned. More information about how nurses practice ethically as an advocate for the patient is necessary.

Influences on Nurses' Ethical Practices. Several influences on nurses' ethical practices were identified in early studies about the ethical decision making of nurses. Moral reasoning abilities, decision-making styles, education, attitudes, and values were identified in studies about the moral reasoning abilities of nurses to make ethical judgments (Crisham, 1981; Ketefian, 1981a, 1981b; Munhall, 1980; Murphy, 1976), the attitudes and values of nurses making ethical decisions (Davis and Slater, 1989), and nurses' perceptions of ethical problems (Gramelspacher, Howell, and Young, 1986).

Other studies have indicated that perceived powerlessness and role confusion significantly reduce nurses' abilities to resolve ethical conflicts and practice ethically (Erlen and Frost, 1991; Prescott and Dennis, 1985). One other study has indicated that nurses' sensitivity to the ethical aspects of a patient care situation may influence their ethical practice (Oddi, et al., 1995). Beyond the results of these studies, however, little is known about the influences on nurses' ethical practice.

As the delivery of health care has changed, the roles of nurses with patients, the amount of time that nurses spend with patients, and the context for the nurse-patient relationship have all changed. But little is known about how and to what extent these changes, other elements in the workplace, and health care organizations themselves, affect nurses' ethical practice (Chambliss, 1996). Clearly, such knowledge is needed for the development of nursing ethics in the 21st century.

III. CONCLUSION

Nursing ethics in the 21st century will continue to be concerned with describing and communicating the characteristics of the "good" nurse, and describing nurses' ethical practices. However, there is a growing concern that what constitutes nurses' ethical practices is changing as patients are experiencing, by virtue of reduced reimbursements for health care services, limited time to be in a nurse-patient relationship and to receive nursing care services. Nurses' ethical practices have always been centered on the nurse-patient relationship and if the opportunity for that relationship changes, then what constitutes ethical practice may change, as well. Nursing ethics in the 21st century must be concerned with defining and describing nurses' ethical practices, raising questions about what such practice requires, and the issues that nurses confront in their advocacy roles with patients. The continued and accurate identification of ethical issues that nurses experience, how nurses handle ethical issues, and the resources nurses need to practice ethically will contribute to this effort. The "good" nurse of the 21st century will be a composite ideal derived from new conceptualizations of nursing practice and nurses' ethical practice, and shaped by empirical evidence about the ethical reality of nurses' practices.

Boston College School of Nursing
Chestnut Hill, MA, USA

NOTES

1. Benner (2000) has recently argued that good nursing practice minimally requires seven moral sources and skills: (1) relational skills in meeting the other, drawing on life-manifestations of trust, mercy and openness of speech; (2) perceptiveness or recognizing when a moral principle such as injustice is at stake; (3) skilled know-how that allows for ethical comportment and action in particular encounters; (4) moral deliberation and communication skills that allow for justification of and experiential learning about actions and decisions; (5) an understanding of the goals or ends of good nursing practice; (6) participation in a practice community that allows for character development to actualize and extend good nursing practice; and (7) the capacity to love ourselves and our neighbors, and the capacity to be loved. Benner asserts that all seven aspects of moral life are required for good and ethical nursing practice.

2. Edwards (2001) argues that nursing might be better served by recognizing two types of caring. *Intentional care* involves a particular kind of mental attitude in response to human vulnerability and suffering which supports caring's moral foundation for nursing practice. This kind of caring is very similar to Kuhse's views on caring in nursing. Edwards goes a little further than Kuhse, however, when he discusses another type of caring, *ontological care*, which is "an inescapable feature of leading a human life" (p. 124). This notion of caring, based in part on Heidegger's sense of care, concerns *deep care* and *identity constituting care.* He argues that these two senses of ontological care are relevant to nursing in that in order for the nurse to perform acts of intentional care, the nurse must first consider care at the ontological level. The "good" nurse is a nurse who understand ontological care and is thus better able to provide intentional care.

3. Lehmkuhl's classifications were a central component of theological ethics in seminary education during the 1920s. Vaughan omitted Lehmkuhl's "duties to the state" and added "duties to the patient," "duties to the hospital," and "duties to the profession." The modification of Lehmkuhl's classifications yielded 33 groups of moral problems for the analysis of the data.

REFERENCES

Abrams, N.: 1978, 'A contrary view of the nurse as patient advocate,' *Nursing Forum* 17, 258-267.

Aikens, C.: 1931, *Studies in Ethics for Nurses,* W.B. Saunders, Philadelphia.

American Nurses Association: 1985, *Code for Nurses with Interpretative Statements,* ANA, Washington, DC.

American Nurses Association: 1998, *Standards of Clinical Nursing Practice,* 2nd ed., ANA, Washington, DC.

Annas, G.: 1974, 'The patient rights advocate: Can nurses effectively fill the role?', *Supervisor Nurse,* 5, 20-23, 25.

Benner, P.: 1990, 'The moral dimensions of caring,' in J. Stephenson (Ed.), *Care, Research and State of Art,* American Academy of Nursing, Kansas City, MO, pp. 5-17.

Benner, P.: 2000, 'The role of embodiment, emotion, and lifeworld in nursing practice,' *Nursing Philosophy* 1, 5-19.

Benner, P. and Gordon, S.: 1996, 'Caring practice,' in S. Gordon, P. Benner, and N. Noddings (Eds.), *Caregiving: Readings in Knowledge, Practice, Ethics, and Politics,* University of Pennsylvania Press, Philadelphia, PA, pp. 40-55.

Benner, P. and Wrubel, J.: 1989, *The Primacy of Caring: Stress and Coping in Health and Illness*, Addison-Wesley, Menlo Park, CA.

Benner, P., Tanner, C.A., and Chesla, C.: 1996, *Expertise in Nursing Practice: Caring, Clinical Judgment and Ethics,*, Springer Publishing, New York.

Berger, M.C., Seversen, A., and Chvatal, R.: 1991, 'Ethical issues in nursing,' *Western Journal of Nursing Research* 13, 514-521.

Bishop, A.H. and Scudder, J.R.: 2001, Nursing Ethics: Holistic Caring Practice. 2nd Ed., Jones and Bartlett Publishers, Boston, MA.

Brody, J.: 1988, 'Virtue ethics, caring, and nursing,' *Scholarly Inquiry for Nursing Practice* 2, 90-97.

Chambliss, D.F.: 1996, *Beyond Caring,*. University of Chicago Press, Chicago.

Corser, W.D.: 1990, 'A conceptual model of collaborative nurse-physician interactions: The management of traditional influences and personal tendencies,' *Scholarly Inquiry for Nursing Practice* 12, 325-41, 343-346.

Crigger, N.J.: 1997, 'The trouble with caring: A review of eight arguments against an ethic of care', *Journal of Professional Nursing* 13, 217-221.

Crisham, P.: 1981, 'Measuring moral judgment in nursing dilemmas,' *Nursing Research* 30, 104-110.

Curzer, H.J.:1993, 'Fry's concept of care in nursing ethics,' *Hypatia* 8, 174-183.

Davis, A.J. and Slater, P.V.: 1989, 'U.S. and Australian nurses' attitudes and beliefs about the good death', *Image: The Journal of Nursing Scholarship* 21, 34-39.

Dyson, L.: 1997, 'An ethic of caring: Conceptual and practical issues,' *Nursing Inquiry* 4, 196-201.

Edwards, S.D.: 2001, *Philosophy of Nursing: An Introduction,* , Palgrave, Hampshire, Great Britain.

Erlen, J. A. and Frost, B.: 1991, 'Nurses' perceptions of powerlessness in influencing ethical decisions,' *Western Journal of Nursing Research* 13, 397-407.

Fry, S.T.: 1986, 'Ethical inquiry in nursing: The definition and method of biomedical ethics,' *Perioperative Nursing Quarterly* , 2, 108.

Fry, S.T.: 1989, 'Toward a theory of nursing ethics,' *Advances in Nursing Science* 11(4), 9-22.

Fry, S.T.: 1991, 'A theory of caring: Pitfalls and promises,' in D. Gaut and M. Leininger (Eds.), *Caring: The Compassionate Healer*, National League for Nursing, New York, pp. 161-172.

Fry, S.T.: 1994, *Ethics in Nursing Practice: A Guide to Ethical Decision Making,* International Council of Nurses, Geneva.

Fry, S.T. and Damrosch, S.: 1994, 'Ethics and human rights issues in nursing practice: A survey of Maryland nurses', *The Maryland Nurse* 13, 11-12.

Fry, S.T. and Riley, J.M.: 1999, Ethics and Human Rights in Nursing Practice: A Multi-State Study of Registered Nurses (abstract), The Nursing Ethics Network, Boston, MA [http://www.nursingethicsnetwork.org]

Fry, S.T. and Johnstone, M.J.: 2001: Ethics in Nursing Practice: A Guide to Ethical Decision Making, 2nd Ed., Blackwell Publishing, London

Gadow, S.: 1989, 'Clinical subjectivity: Advocacy for silent patients,' *Nursing Clinics of North America* 24, 535-541.

Gadow, S.: 1999, 'Relational narrative; The postmodern turn in nursing ethics,' Scholarly Inquiry for Nursing Practice 13, 57-70.

Gastmans, C., Dierckz deCasterle, B., and Schotsmans, P.: 1998, 'Nursing considered as moral practice: A philosophical-ethical interpretation of nursing,' *Kennedy Institute of Ethics Journal* 8(1), 43-69

Gates, B.: 1994, *Advocacy: A Nurse's Guide..* Scutari Press, London.

Gaut, D.A.: 1983, 'Development of a theoretically adequate description of caring,' *Western Journal of Nursing Research* 5, 313-323.

Gaylord, N. and G race, P.: 1995, 'Nursing advocacy: An ethic of practice,' *Nursing Ethics* 2, 11-18.

Gordon, M.: 1994, *Nursing Diagnosis: Process and Application,* , Mosby Books, St. Louis, MO.

Gordon, M., Murphy, C.P., Candee, D., Hiltunen, E.: 1994, 'Clinical judgment: An integrated model,' *Advances in Nursing Sciences* 16, 55-70.

Gramelspacher, G.P., Howell, J.D., and Young, M.J., 'Perceptions of ethical problems by nurses and doctors,' *Archives of Internal Medicine 146*, 577-578.

Hess, J. D.: 1996, 'the ethics of compliance: A dialectic,' *Advances in Nursing Science* 19, 18-27.

Hilbig, J.I. and Manning, J.: 1999, 'Accountability and responsibility underpins successful pain management,' *Journal of Perianesthesia Nurses* 14, 390-392.

Holden, R.: 1991, 'An analysis of caring: Attributions, contributions, and resolutions,' *Journal of Advanced Nursing* 16, 893-898.

International Council of Nurses: 2000, *Code of Ethics for Nurses*, ICN, Geneva.

Jacques, R.: 1993, 'Untheorized dimensions of caring work: Caring as a structural practice and caring as a way of seeing,' *Nursing Administration Quarterly* 17, 1-10.

Jameton, A.: 1984, *Nursing Practice: The Ethical Issues,* Prentice-Hall, Englewood Cliffs, NJ.

Ketefian, S.: 1981, 'Moral reasoning and moral behavior among selected groups of practicing nurses,' *Nursing Research* 30, 171-176.

Ketefian, S. and Ormond, I.: 1988, *Moral Reasoning and Ethical Practice in Nursing: An Integrative Review,* National League for Nursing, New York.

Khushf, G.: 1997, 'Nursing ethics at the juncture of two kinds of care,' *The South Carolina Nurse* (July, August, September), 3-4.

Knowlden, V.: 1990, 'The virtue of caring in nursing', in M. M. Leininger (ed.), *Ethical and Moral Dimensions of Care,* Wayne State University Press, Detroit, MI, pp. 211-267.

Kohlberg, L.: 1976. 'Moral stages and moralizaton: The cognitive-developmental approach,' in T. Lickona (ed.), *Moral Development and Behavior: Theory, Research, and Social Issues,* Holt, Rinehart and Winston, New York.

Kuhse, H.: 1997, *Caring: Nurses, Women and Ethics,* Blackwell, Maldon, MA

Lehmkuhl, A.: 1910, *Theologia Moralis* Vol. I-II, Friburg; Brisgoviae, Herder..

Liaschenko, J.: 1997, 'Ethics and the geography of the nurse-patient relationship: Spatial vulnerabilities and gendered space,' *Scholarly Inquiry for Nursing Practice: An International Journal* 11, 45-59.

Lumpp, F.: 1979, 'the role of the nurse in the bioethical decision-making process,' *Nursing Clinics of North America* 14, 13-21.

Mallik, M.: 1997, 'Advocacy in nursing: Perceptions of practicing nurses,' *Journal of Clinical Nursing* 6, 303-313.

Munhall, P.: 1980, 'Moral reasoning levels of nursing students and faculty in a baccalaureate nursing program,' *Image: The Journal of Nursing Scholarship* 12, 57-61.

Murphy, C.P.: 1976, *Levels of Moral Reasoning in a Selected Group of Nursing Practioners,* Ph.D. Dissertation, Teachers College, Columbia University.

Nelson, H.L.: 1992, 'Against caring,' *The Journal of Clinical Ethics* 3, 8-15.

Nutting, M.A. and Dock, L.L.: 1907, *A History of Nursing,*, Vol 2, G.P. Putnam's and Sons, New York.

Oddi, L.F., Cassidy, V.R., and Fisher, C.: 1995, 'Nurses' sensitivity to the ethical aspects of clinical practice,' *Nursing Ethics* 2, 197- 209.

Omery, A.: 1995, 'Care: The basis for a nursing ethics?', Journal of Cardiovascular Nursing 9(3), 1-10.

Omery, A., Henneman, E., Billet, B., Luna-Raines, M. and Brown-Saltzman, K.: 1995, 'Ethical issues in hospital-based nursing practice,' *Journal of Cardiovascular Nursing* 9, 42-53.

Parker, R.S.: 1991, 'Measuring nurses' moral judgments,' *Image: Journal of Nursing Scholarship* 22, 213-218.

Penticuff, J.H.: 1991, 'Conceptual issues in nursing ethics research,' *The Journal of Medicine and Philosophy* 16, 235-258.

Phillips, P.: 1993, 'A deconstruction of caring,' *Journal of Advanced Nursing 18,* 1554-1558.

Prescott, P.A. and Dennis, K.E.: 1985, 'Power and powerlessness in hospital nursing departments,' *Journal of Professional Nursing* 1, 348-355.

Pyne, R.: 1991, 'Accountability,' *Nursing Times* 87, 25-28.

Rest, J.R.: 1986, *Moral Development: Advances in Researach and Theory.* Praeger, New York.

Robb, I. H.: 1921, *Nursing Ethics: For Hospital and Private Use,* E.C. Loeckert, Cleveland.

Rowe, J.A.: 2000, 'Accountability: A fundamental component of nursing practice,' *British Journal of Nursing* 9, 549-552.

Scanlon, C.: 1994, 'Ethics survey looks at nurses experiences,' *The American Nurse* (November-December), 22.

Sellin, S.C.: 1995, 'Out on a limb: A qualitative study of patient advocacy in institutional nursing,' *Nursing Ethics* 2, 19-29.

Shiber, S. and Larson, E.: 1991, 'Evaluating the quality of caring: Structure, process and outcome,' *Holistic Nursing Practice* 5, 57-66.

Smerke, J.: 1990, 'Ethical components of caring,' *Critical Care Nursing Clinics of America 2,* 509-513.

Snowball, J.: 1996, 'Asking nurses about advocating for patients: Reactive and proactive accounts,' *Journal of Advanced Nursing* 24, 67-75.

Stenberg, M.J.: 1979, 'The search for a conceptual framework as a philosophic basis for nursing ethics: An examination of code, contract, context, and covenant,' *Military Medicine* 144, 9-22.

Taylor, C.R.: 1998, 'Reflections on "nursing considered as moral practice",' *Kennedy Institute of Ethics Journal* 8, 71-82.

Valentine, K.: 1991, 'Comprehensive assessment of caring and its relationship to outcome measures', *Journal of Nursing Quality Assessment* 6, 59-68.

Vaughan, R.H.: 1935, *The Actual Incidence of Moral Problems in Nursing: A Preliminary Study in Empirical Ethics*, Studies in Nursing Education, Vol II, Fascicle 2. Washington, D.C., The Catholic University of America).

Warelow, P.J.: 1996, 'Is caring the ethical ideal?', *Journal of Advanced Nursing* 24, 655-661.

Yeo, M.: 1989, 'Integration of nursing theory and nursing ethics,' *Advances in Nursing Science* 11(3), 33-42.

LAURENCE McCULLOUGH

GEROETHICS

I. INTRODUCTION

Geriatrics comprises a medical and nursing specialty concerned with the health care of elderly patients. By convention, this is the population of patients 65 years of age and older. In the United States, geriatrics is now a recognized subspecialty of the specialty of internal medicine, with its own professional association of physicians, the American Geriatrics Society, and journal, *Journal of the American Geriatrics Society*. Gerontology is the interdisciplinary field of study of the elderly, using a biopsychosocial approach to study aging from the perspectives of biology, medicine, nursing, social work, the social sciences, and the humanities. In the United States, the Gerontological Society of America is the interdisciplinary association for such study of aging and publishes several scholarly journals. Bioethics plays a prominent role in both geriatrics and gerontology – in teaching as well as in research.

Bioethics in geriatrics/gerontology – or geroethics – concerns ethical issues in aging and encompasses a broad range of issues. From the late 1960s, when bioethics emerged as a field (Jonsen, 1998), until the early 1980s, the agenda of ethics and aging was dominated by issues at the end of life, especially death and dying, advance directives, and hospice care. Indeed, these end-of-life issues became synonymous with ethics and aging. The dominance of these issues had a distorting effect on the literature, leading to the neglect of less glamorous and exciting issues, especially the challenges of informed consent in the elderly population and long-term care. In the current volume these "four alarm" topics are addressed elsewhere; namely, in the chapters on advance directives (Cohen); issues in death and dying and palliative care and hospice (Lustig). I will, therefore, not address these issues here. I will focus instead on two ethical issues: (a) a preventive ethics approach to decision making in the care of geriatric patients; and (b) ethical issues on long-term care decision making that confront many elders, their families, health care professionals, and health care institutions. In my treatment of both, I will identify and discuss both current issues and future challenges.

G. Khushf (ed.), Handbook of Bioethics, 507–523.

II. A PREVENTIVE ETHICS APPROACH TO DECISION MAKING IN GERIATRICS

A. *Preventive Ethics*

During its first two decades bioethics in general and its sub-field, clinical ethics, came to be dominated by an acute-care, reactive approach. This reflected, in considerable measure, the need of clinicians to respond to ethical conflicts when they occurred in patient care. Moreover, solving ethical conflicts engaged and excited those trained in ethics. This reactive approach to clinical ethics also reflected, unwittingly, the reactive approach of acute medical care, which is designed precisely to respond to clinical problems *after they occur*. Preventive medicine does not have much influence on clinical judgment, decision making, and practice – at least in American hospitals, in critical care units, where the reactive approach both to medical practice and to ethical issues in the clinical setting continues to dominate. Clinical ethics consultation, for example, appears very much to be dominated by a reactive approach. It should therefore come as little surprise that preventive ethics has been slow to develop in the context of American hospital-based medicine and health care, dominated, as it surely still is, by the reactive approach to patients' problems.

Preventive ethics emphasizes clinical strategies that are prospective in their orientation. These strategies anticipate ethical conflicts, the potential for which is taken to be recognizable and manageable *before* such conflicts actually occur. As in so many aspects of American bioethics, a preventive ethics approach was introduced by the law, in the form of advance directives, which are precisely designed to project patients' preferences into the future when they will be needed to guide the medical and nursing care of patients who have lost the capacity to participate in the informed consent process. The concept of preventive ethics as such was first introduced into the bioethics literature in the early 1990s (Chervenak and McCullough, 1990a; 1990b). Preventive ethics strategies have been proposed for the management of ethical issues in obstetrics and gynecology (McCullough and Chervenak, 1994), for the management of futility by primary care physicians (Doukas, et al., 1996), for the provision of screening tests in the primary care setting (Doukas, et al, 1997), for percutaneous endoscopic gastrostomy tube placement (Rabeneck, et al., 1997), for the care of mentally ill and depressed women in reproductive decisions (Coverdale, et al. 1995; Coverdale, et al., 1997) and in pregnancy (Coverdale et al., 1996), and geriatrics (McCullough, et al., 1999).

Preventive ethics has crucial applications in the care of geriatric patients because, well before they become terminally ill, elderly patients are at risk for diminished or even lost capacity to participate in the informed consent process, especially from dementing disorders and other neurological diseases and injuries such as strokes. In the absence of knowledge of their prior values and preferences, clinical decision making for such patients can pose significant ethical challenges, captured under the rubric of surrogate decision making (Buchanan and Brock, 1989). The occurrence of such challenges can be minimized by a preventive ethics approach.

B. *The Informed Consent Process*

The purpose of the informed consent process is to form a therapeutic alliance between the physician and other health care professionals, on the one hand, and the patient, on the other. In this alliance the professional and patient work together to protect and promote the health-related interests of the patient. The health care professional brings to this alliance the fiduciary obligation to care for the patient as the professional's primary concern, with the self-interest of the professional a systematically secondary concern. The patient brings to this alliance his or her informational needs and desire to participate in the decision-making process, both of which can vary.

C. *Ageism in the Informed Consent Process*

Ageism, a bias against the ability of elders to make their own decisions, has long been recognized in both geriatrics and gerontology, as unacceptable among health care professionals. This reflects the consensus opposition in the bioethics literature to paternalism on the part of health care professionals. The professional-patient relationship should be based on respect for the patient's person, dignity, and autonomy. This concept of the professional-patient relationship is not peculiarly American, as evidenced by its adoption in the literature in other countries (Gillon and Lloyd, 1994) and by national associations of physicians in other countries (Fineschi, 1997).

It is worth noting, however, that elderly patients, when they were younger, experienced health care institutions and professionals in the past, who were more paternalistic than they may be now. Elders can and do experience health care institutions and professionals who are indeed ageist in their attitudes, practices, and policies. As a consequence, some elders have absorbed ageist attitudes towards themselves – they may not genuinely appreciate that they, and not health care professionals, do indeed have the right to decide on the course of their health care. This attitude should, respectfully and sensitively, be challenged, by inviting such patients to take seriously the prerogative to play major role in the decision making process and not defer that role to others, especially the physician (Wear, 1993). The first preventive ethics task in geriatrics is therefore make clear to elderly patients the role that they should play in the clinical decision-making process: the role that the elder wants to play – which is a larger role than some elders may think to be the case. It is therefore crucial that clinicians ascertain the amount of information that elders actually want and the extent of decisional authority that they actually want in the informed consent process. Empirical research needs to be undertaken to develop validated clinical tools for doing so. In the meantime, clinicians should assure elderly patients that they can play whatever decision-making role that they prefer.

The philosophical point here is that respect for autonomy in the informed consent process imposes fairly uniform obligations of disclosure of information on the part of the clinician. In American bioethics, this disclosure requirement should conform to the reasonable person standard: the clinician should disclose to the patient salient aspects of clinical judgment so that the patient can have the information that the patient needs to replicate the clinician's judgment (Wear, 1993). This has also been described as the "transparency" standard (Brody, 1989). Respect for autonomy does *not,* however, mean that the patient has to play some stipulated role in the consent process, especially the role of exercising autonomy *in full* and taking *complete* power to govern the professional-patient relationship. Instead, the patient should be encouraged to take the decision-making role that the patient wants to take – a role what can vary considerably. Elderly patients should be no exception to this concept of respect for the autonomy of the patient. This is an especially important consideration in multicultural contexts, e.g., when the geriatric patient's adult children request that information be withheld from the patient and given to them instead so that they can make decisions for their parent (Freedman, 1993).

D. *Elders with Diminished Decision-Making Capacity*

As a rule, dementing disorders and neurological diseases and injuries that can cause diminished decision-making capacity occur with greater frequency among the elderly than with younger patients. These disorders, diseases, and injuries display biologic variability and thus have a variable effect on decision-making capacity. This variation occurs across patients with a similar diagnosis, e.g., in thesuccessive stages of Alzheimer's disease. This variation can also occur within a single patient over time, e.g., the waxing and waning of decision-making capacities at different times of the day or in response to environmental stressors. Moreover, diminished decision-making capacity can occur as a chronic phenomenon, and decline over time, as occurs in dementing disorders such as Alzheimer's disease and with successive strokes.

This chronically and variably diminished decision-making capacity (Coverdale, et al., 1996) can variously affect the steps in which patients exercise their autonomy in the informed consent process. These steps include the following:

1. Attending to, registering, and recalling the information provided by the clinician.
2. Reasoning from present events, alternative strategies for the clinical management of the patient's problem(s), to their future consequences, or cognitive understanding (White, 1994).
3. Evaluating those consequences in terms of one's values and beliefs or evaluative understanding (White, 1994), which includes appreciation that those consequences could indeed happen to oneself (Grisso and Appelbaum, 1998).
4. Reaching a judgment about which alternative(s) to accept, based on the preceding steps.
5. Expressing this judgment in meaningful terms, i.e., in terms of cognitive and evaluative understanding and *not* in terms of the values and preferences of the clinician (Coverdale, et al. 1996).

It is now accepted, at least in American law and bioethics, that patients should be presumed to be capable of completing these steps and thus participating in the informed consent process. Each of these steps builds on and therefore presumes its predecessors. The mere presence of chronic or variable impairment of the ability to complete one or more of these steps, however, is not by itself evidence of either significant or irreversible impairment of them. One approach in geroethics is to insist that such impairment be clinically evaluated and attempts made to reverse that impairment (McCullough, et al., 1999). Only if such attempts fail and the patient exhibits significantly impaired capacity should surrogate decision making be employed. Patients must be provided sufficient opportunity to exercise their autonomy in the informed consent process, even and especially if they need the clinician's assistance to do so (McCullough, 1988).

In the primary care setting, an additional preventive ethics strategy is to explore over time the patient's values, so that a record of these can be compiled in the patient's chart. That is, the development and expression of the basis of each patient's evaluative understanding in various clinical decisions should be recorded by clinicians. Disease-specific approaches have been proposed in the literature for accomplishing this important goal of preventive ethics (Singer, et al., 1994; 1995; 1997; Berry and Singer, 1998) and also for end-of-life decision making (Doukas and McCullough, 1999).

This information can then guide the substituted judgment approach to surrogate decision making, which is the ethically preferred approach (Buchanan and Brock, 1989). On this standard, the surrogate decision maker should not be asked for his or her preference from among the medically reasonable alternatives. Instead, the surrogate should be asked to provide information about the patient's values and beliefs. The surrogate can then be provided with information about the patient's value history from the patient's chart. This information should then be used to form *substituted evaluative understanding,* which can then guide surrogate decision making. Limitations to this approach have been identified in the literature, e.g., projecting past decisions about the management of one type of disease to different types of disease (Reilly, et al., 1995). Substituted judgment should therefore be undertaken with attention to those limits.

Sometimes no one has recorded such information (which, in the author's judgment, should no longer be accepted in clinical practice) and no one has the relevant information about the significantly impaired elder's values and preferences. If preventive ethics becomes uniformly adopted in primary-care geriatrics, these cases should become very rare indeed, especially for patients whose chronic illnesses require regular out-patient management. In these cases the surrogate decision maker relies on the clinician's judgment about which alternatives protect and promote the patient's health-related interests. The best-interest standard instructs the surrogate to elect a course of clinical management from among these alternatives.

E. *Diseases of Aging: Challenging Orthodoxy in Geriatrics and Gerontology*

It is a commonplace of geriatrics and gerontology that there are diseases associated *with* aging, but not diseases *of* aging. This distinction was developed to counter intellectually, scientifically, and clinically unfounded prejudices against the elderly in the many forms that ageism once took. This distinction has done important intellectual, moral, clinical, educational, and political work, to be sure. Nonetheless, I want to challenge it.

I do so, in part, from the appropriate skepticism that we learn to cultivate, from science and philosophy alike, in response to orthodoxy, especially when orthodoxy becomes unquestioning or even unquestionable and thus assumes the status of ideology. I do so from recent work with geropsychiatric colleagues in my institution on ethical implications of impaired executive control functions that can be caused by, among other factors, frontal lobe disorders (Workman, et al., 2000; Grimes et al. ,2000).

Impaired executive control functions lead, on our analysis of them, to impairments of all three components of autonomy – intentionality, understanding, and non-control – identified in the now classic work (in the field of bioethics) of Faden and Beauchamp (1986) on informed consent. These patients experience distinctive deficits, especially in intentionality – the ability to make and adapt plans – and in non-control – the ability to resist environmental cues that are detrimental to one's plans. As a consequence, they experience significantly impaired autonomy, requiring surrogate decision making about their care subsequent to hospitalization for health problems that themselves are the sequellae of impaired intentionality and non-control. The ethical issue concerns whether the substituted judgment standard, i.e., decision making based on the patient's prior values such as independence, or the best interest standard, i.e., decision making based on prudent judgments about how best to protect the patient's health, functional status, and remaining independence and autonomy, should guide surrogate decision making for these patients.

The importance of impaired executive control functions secondary to frontal lobe disorders for the orthodoxy described above is that, if investigators such as Royall (1997) are correct, then there is a strong correlation between an autonomy-disabling pathology, executive dysfunction, and aging. Such disorders may well indeed be diseases *of* the aging brain, not diseases *associated with but not caused by aging*. This is, to be sure, a gerontologically heterodox – even heretical – view, but this fact by itself does not make the view scientifically unsound.

This gerontological heterodoxy has important implications for the ethics and law of aging. There may be diseases of aging that significantly impair the autonomy of elders and may do so with increasing frequency in the oldest old. If this is the case, should there be routine screening of elders for these deficits of autonomy? If such deficits are discovered and significantly impair the elder's ability to complete the five steps of decision making described above, and if there is also no way presently to treat such deficits, should we discount the autonomy of affected elders? If so, to what degree?

Consider, for example, the impact answers to these questions could have on crucial, everyday decisions such as driving and the quotidian agenda of long-term care decisions (Kane and Caplan, 1990, 1993, see, also, below) or more momentous decisions to complete an advance directives, in which decision – by definition – a great deal is at stake. As a stable matter of both ethics and law in the United States, both advance directive and long-term care decision making involves and therefore presumes intact autonomy. On the Faden-Beauchamp (1986) analysis, autonomy includes intentionality, the ability to make and adapt plans. Executive dysfunctions impair this ability, frequently significantly. Moreover, autonomy on the Faden-Beauchamp (1986) analysis includes non-control, the freedom from substantially controlling influences. Frontal lobe disorders can impair impulse control, making the elder more susceptible to substantially controlling environmental cues, such as the strongly stated and directive recommendations of involved family members and health care professionals. Should the "plans" of affected elders at hospital discharge planning be respected and implemented? How should power over patients with impaired non-control be exercised by clinicians and family members?

In most American jurisdictions witnesses to last wills and testaments and to durable powers of attorney attest to soundness of mind of the person executing these documents. Should witnesses do so in the absence of evaluation of executive control functions? Most statutes governing living wills require witnesses only to attest that the patient did indeed make the decision made orally or in writing, but not to attest to competence. Should this policy be changed, given the apparent prevalence of frontal lobe disorders with increasing age? Clinical decision making by elders, if it involves any assessment at all by clinicians, usually involves only cognitive assessment via such clinical tools as the Mini-Mental Status Examination. But this tool evaluates only cognitive functions that Faden and Beauchamp (1986) call understanding; this examination does not evaluate intentionality and non-control. Given the stakes for elders and involved family members, and the importance of at once respecting the elder's exercise of autonomy and the reasonable limits of family members on the direct and indirect burdens of post-hospitalization care, should the evaluation of decision-making capacity for the elder's role in decision making about discharge planning be more thorough?

Is even raising these questions an unwarranted moral, social, and legal threat to elders? I think that raising these questions does indeed pose a threat to the autonomy of affected elders, but it may be warranted scientifically, clinically, and ethically. It is not unreasonable to be disturbed by this judgment, creating an important and compelling item for the future agenda of law and ethics in aging.

F. *Summary of a Preventive Ethics Approach*

The preventive ethics approach described above creates a new set of obligations for clinicians to take a prospective approach to clinical decision making with and for elderly patients. Obviously, preventive ethics appeals to respect for the patient's autonomy – both when the patient can exercise it in the informed consent process and in advance of the times that the patient may not be able to express it. This approach assumes that there is sufficient continuity in personal identity over time to provide a reliable basis for the approach.

There is an interesting debate in the bioethics literature (see, e.g., Dresser and Robertson, 1989; and the chapter by Cohen in this volume) about just which concept of personal identity is presupposed by advance decision making, one important tool of preventive ethics, as noted above. There is, of course, a rich literature in the history of philosophy on this topic, which has not been drawn upon sufficiently in the current debates. In my own view, complete or highly coherent personal identity over time, however, is not required by preventive ethics in general and advance decision making in particular. More demanding concepts of personal identity have been used in the geroethics literature to question the concept of projected decision making that is at the heart of the preventive ethics approach described above and to advance directives, in particular (Dresser and Robertson, 1989). These more demanding concepts appear to require a metaphysics of personal unity or near-complete coherence. There is an interesting and relevant historical precedent for this, namely, of Leibniz's monads. The monad generates, by appetition, all of its perceptions (which are its properties and provide for its identity over time). Because each perception is contained in its predecessor and contains its successor, *personal identity* does indeed become equivalent to *personal unity* over time.

This concept of personal identity as personal unity demands too much because it is not consistent with lived experience and what I take to be the instruction of philosophy by contemporary clinical experience of brain function. Clinical experience teaches that an individual's current cognitive and evaluative functioning is variably coherent with that individual's past such functioning and that individuals can get along well enough even when that coherence is low. Moreover, the brain does not function as a single organ, but rather as a more or less well associated collection of quasi-organs that have evolved in sufficient association – but not unity – over time.

This concept of personal identity as personal unity also demands too much in that it violates the presumption of autonomy that is usually taken to set low and readily achievable thresholds for performance of the five steps of patient decision making described above. The concept of personal identity proposed by such critics of projected decision making as Dresser and Robertson (1989) sets those thresholds considerably higher, thus denying autonomy to many and therefore subtly but powerfully undermining the presumption of autonomy that is supposed to govern the ethics and law of the informed consent process.

III. LONG-TERM CARE DECISION MAKING

A. *Long-Term Care and its Demands on Families*

Long-term care has been defined as "a set of health, personal care and social services delivered over a sustained period of time to persons who have lost or never acquired some degree of functional capacity" (Kane and Kane, 1987, p. 4). Long-term care needs trigger responses within the elder's family to meet those needs (Kane and Caplan, 1990; 1993). The typical pattern of response is that the female members of the family provide so-called "informal" long-term care services, in the sense that such work is unpaid and provided by individuals not trained in the health professions. In contemporary American society, women tend increasingly to be in the work place and so, when a woman makes the decision to meet the long-term care needs of an elderly family member, she does so either by adding hours of "informal" work at the beginning and end of her work-day (thus imposing psychosocial costs on her family as well as opportunity costs on herself) or at the expense to herself and her family of lost income from less hours at work, sometimes including no hours at work. Reducing one's work hours, taking an extended and, in the United States, unpaid leave (as now permitted under federal law), or quitting paid work altogether impose psychosocial costs on women who make these difficult decision, as well as opportunity costs.

Through such heroic efforts and non-trivial financial, opportunity-cost, physical, and psychosocial sacrifice the long-term care needs of millions of older Americans are met every day. Some states have experimented with ways to help families cope with these caregiving burdens, while others have done little – for the simple, and correct, reason that doing so is probably not cost-beneficial. These expenditures will probably not save that public source any significant amount of money. Any costs that do occur show up, as it were, on someone else's books. Payers, private and public alike, have an incentive only to avoid cost-non-beneficial strategies when they themselves incur the unnecessary added cost. The non-system of American health care – with a mix of private and public payers and private and public providers with no central management or guiding policies that shows no signs of moving toward a system – perpetuates the shifting of care burdens onto families, in good measure, because there is no consistent economic incentive for a more rational alternative.

B. *Policy Assumptions Regarding the Family's Role in Long-Term Care*

Medicare has been taking advantage of the non-system of health care in the United States for more than a decade with its prospective payment system under Diagnostic Related Groups (DRGs). DRGs-based payment provides hospitals – for-profit and not-for-profit alike – with an incentive to shorten length of stay as a major means to control costs and thus continue to profit from Medicare. Medicare has benefitted by being able to slow the rate of inflation for its hospital expenditures, thus postponing financial and political crisis for the Medicare Trust Fund. In addition, more and more geriatric medical care is being shifted to the out-patient setting, following the general trend in medical care of the past decade. As a consequence, patients admitted to hospitals tend to be sicker and frailer than they were in the past. Also as a consequence, geriatric patients are frequently discharged from the hospital with new or increased long-term care needs. The Veterans Health Affairs system is now rapidly following suit, e.g., in profiling of physicians that focuses intensely on utilization of hospitalization and hospital resources such as the formulary as a measure of quality.

In this environment it seems plain that there is an implicit policy assumption in Medicare's DRGs-based payment system and the VA shift to controlling utilization of hospital admissions and days: Families will pick up where hospital admissions used to continue. Put in a more unfriendly way, implicit Medicare and VA policy is to take full advantage of the sense of moral obligation on which so many American families act to meet the long-term care needs of their loved ones. This can become exploitation when the burdens of long-term care become unreasonably excessive, e.g., when they result in significant financial loss, compromised health of caregivers, or fractured families. Medicaid and private insurance, in my experience teaching in a major children's hospital, have adopted the same implicit policy for meeting the long-term care needs of disabled and seriously ill children – an experience from which geriatrics and gerontology need to learn.

This exploitation of families has occurred without any explicit public policy debate about whether such policies are consistent with social justice. We know that most of the "informal" long-term care is provided by women, raising the justice-based issue of whether the exploitation of these women unfairly builds on and sustains gender bias in our society. Men in the household, and children, need to work more hours to replace needed, lost income. It does not take much reflection to appreciate that Americans – through our elected representatives in the Congress – have decided that it is just fine to impose these burdens on families and to do so with little or no public debate about the justice of such policies.

As a consequence, we put families between the rock of doing everything and the hard place of doing less than everything or sometimes nothing and often judging themselves selfish or morally inadequate for doing the latter. There is a middle ground: setting reasonable limits on one's caregiving obligations. Identifying this middle ground can be undertaken within the larger context of identifying the values and preferences of elders and involved family members and negotiating caregiving obligations, shares, burdens, and limits in a values-rich discourse (McCullough and Wilson, 1995). The health policy challenge will be how to adjust policy constraints to accommodate families that, on careful and informed reflection, reach the well-reasoned judgment that they can no longer provide "informal" long-term care at their current levels.

Now, accommodating such families will almost certainly mean an increase in taxes, both state and federal. We live in an America in which the old concept of citizenship, built around virtues of self-sacrifice and compassion for the unfortunate, has become attenuated, perhaps destroyed (Holstein and Cole, 1995). This language needs to be revived, although without the special pleading that advocates for the elderly sometimes employ in our public discourse.

C. *Long-Term Care and Bioethics*

Inquiry into long-term care has, in important ways, enriched bioethics itself. It now seems plain that the ethical principle of respect for autonomy and the concept of autonomy of patients was developed in and for the acute-care setting. The decision-making group in this setting was understood to be the physician and health care team with the patient, with the patient's family regarded as third parties to the decision-making process (Beauchamp and McCullough, 1984). Ethical issues concerned decisions made in the hospital about hospital-based care, especially admission to and discharge from the critical care unit. Patient autonomy was understood to involve the exercise of rights by the patient about hospital-based clinical management of the patient's problems. Discharge planning did not get much attention in the bioethics literature; this setting of decision-making, where families are vitally involved, is still largely ignored in the bioethics literature.

Recent work on ethics and aging has called this narrow focus of decision making and narrow concept of autonomy into question. George Agich has forcefully argued that long-term care requires us to rethink the eviscerated concept of autonomy that developed in the acute-care setting, a setting that ignores the life-world of the elderly, a concept that Agich takes from the phenomenologist Alfred Schütz (Agich, 1993). Based on such a phenomenological analysis of the life-world of the elderly, Agich draws a useful and very powerful distinction between nodal and interstitial autonomy and autonomous decision making. Nodal decisions involve the "either/or" choices, such as to continue to live at home or to move to congregate housing or even a nursing home. Interstitial autonomy concerns the everyday decisions we all make about people with whom we want to spend time, having some quiet time, the objects of art or decoration with which we want to be surrounded, and the like. Agich notes that there are indeed powerful policy constraints on nodal decisions in long-term care, for example, those created by income eligibility for various publicly funded services. These constraints can and do restrict the exercise of autonomy in nodal long-term care decisions. Once nodal decisions have been made, however, many opportunities for the exercise of interstitial autonomy remain and the ethics of long-term care decision making is incomplete if it is omits this varied and large domain of decision making and meaning for elders. Agich's innovative and original analysis of autonomy in long-term care expands the agenda of ethics in long-term care to include both nodal and interstitial decision making. Because the latter is not restricted by policy constraints, it should receive greater attention and emphasis, especially in long-term care institutions. These, therefore, need not be "total" institutions (Lidz, et al, 1993). On this account, too, we can see that acute-care, hospital-based bioethics has been impoverished by failing to acknowledge the importance of interstitial autonomy, which is exercise chiefly in the discharge planning process.

Nodal decisions in long-term care frequently involve trade-offs between safety and independence. Bart Collopy (1995) has recently provided a provocative analysis about these trade-offs, arguing that they invoke a false dichotomy. Safety, Collopy argues, has been construed narrowly to mean physical safety only, e.g., from falls or wandering off grounds onto a busy thoroughfare. If safety is rethought, Collopy argues, in psychosocial terms, then personal and psychological and social safety become just as important as physical safety. Independence, on this account, involves psychosocial safety, which is now seen as one end of the continuum of biopsychosocial safety, with the other end concerning physical safety. Within this biopsychosocial concept of safety physical safety does not automatically receive priority, as it often, in fact, does, especially in the judgments of health care professionals and concerned family members. Instead, it may be obligatory to respect the elder's person and autonomy, to risk physical safety in order to preserve a larger, more meaningful, and therefore more morally compelling domain of psychosocial safety.

Long-term care decision making needs to take account of these remarkable conceptual advances. Family members may place a priority on physical safety of an elderly loved one partly to avoid or reduce caregiving burdens on themselves, including the psychosocial burden of worry or the disruption of phone calls from parents' neighbors about dad wandering yet again in the neighborhood in the wee hours of the morning. Indeed, family members may place such a priority on physical safety that they describe the elder's circumstances in ways that differ markedly from the description the elder offers, itself shaped by psychosocial safety. A small fire in the kitchen from an untended pot on the stove thus becomes an unacceptable emergency for family members but a minor nuisance to the elder who cannot countenance leaving home and who seeks to sustain her sense of personal identity – a core concern of biopsychosocial safety.

This problem has been termed "contested reality" in the recent literature (McCullough, et al., 1995). The contest does not arise from some value-neutral account of the facts but from value-laden accounts of the elder's life world. An emphasis on physical safety links to a concern with nodal decisions – dad just cannot be left at home alone any longer, according to an exhausted and concerned daughter – to reach one description. An emphasis on psychosocial safety links to a concern with interstitial decisions – what's the fuss about; it wasn't a big fire and was out as soon as the fire department put water on it, dad says. Long-term care decision making that ignores the problem and challenges of value-laden contested reality will probably only make interpersonal and intrapsychic conflicts among elders and family members worse in an already stressful decision-making process.

D. *Future Challenges*

In long-term care, the stage has been set to build on these innovative conceptual analyses and ethical arguments. It is time to undertake clinical investigation of experimental long-term care decision-making processes developed in response to this original and important work. This reflects a broad trend in clinical bioethics: in many areas the conceptual groundwork has been undertaken and has stabilized and this conceptual groundwork needs to be transformed into innovative and proven clinical practice. For example, it would be worth studying how frequently patients with significant executive dysfunction are presumed to be capable of making their own decisions and then make decisions that they are unable to implement, compromising their subsequent health status and ability to exercise whatever level of capacity of "autonomous" decision making that is left to them. On the basis of data generated from such a study, one could go on to devise experimental interventions in the decision making process, to attempt to compensate for the deficits of executive function and measure the effect on outcomes such as successful out-patient care. In such a combination of ethics and health services research is to be found the future of geroethics. Ethical analysis generates the research questions and also helps to identify relevant end-points for measurement. Accepted methods of health services research are used in the design and analysis of such research.

E. *Summary of Long-Term Care*

Long-term care decisions involve everyday, low-profile ethical issues and concerns. They therefore do not attract the attention of bioethics in the way that ethical issues in the reactive approach do, e.g., physician-assisted suicide. Nonetheless, long-term care decisions prompt scholars and teachers of bioethics to explore important conceptual matters, such as whether personal identity should be understood as a form of personal unity, and important ethical questions, such as the need and legitimacy of family members to place limits on their obligation to provide long-term care to frail elders (McCullough and Wilson, 1995).

IV. CONCLUSION

Bioethics in the clinical setting has now reached a stage of conceptual development that leads to proposals for how to implement those concepts in clinical practice (McCullough, et al., 1995; Kane and Caplan, 1990; 1993). The proposed clinical strategies now must be put to the test. Thus, one can look to the future of geroethics and expect to see increased work on the design, implementation, and clinical investigation of such strategies. Outcomes-based geroethics research will become the norm. Such research should be expected not only to improve patient care but also to enrich the ongoing conceptual development of the field.

Center for Medical Ethics and Health Policy
Baylor College of Medicine
Houston, TX, U.S.A.

REFERENCES

Agich, G.J.: 1993, *Autonomy and Long-term Care*, Oxford University Press, New York.

Beauchamp, T.L., McCullough, L.B.: 1984, *Medical Ethics: The Moral Responsibilities of Physicians*, Prentice- Hall, Inc, Englewood Cliffs, NJ.

Berry, S.R., Singer, P.A.: 1998, 'The cancer specific advance directive,' *Cancer* 82, 1570-1577.

Brody, H.: 1989, 'Transparency: informed consent in primary care," *Hastings Center Report* 19, 5-9.

Buchanan, A.E., Brock, D.W.: 1989, *Deciding for Others: The Ethics of Surrogate Decision Making*, Cambridge University Press, Cambridge.

Chervenak, F.A., McCullough, L.B.: 1990a, Clinical guides to preventing ethical conflicts between pregnant women and their physicians,' *American Journal of Obstetrics and Gynecology*, 162, 303- 307.

Chervenak, F.A., McCullough, L.B.: 1990b, 'Preventive ethics strategies for drug abuse during pregnancy,' *TheJournal of Clinical Ethics* 1, 157-158.

Collopy, B.: 1995, 'Safety and independence: rethinking some basic concepts in long-term care,' in McCullough, L.B., Wilson, N.L., eds., *Long-Term Care Decisions: Ethical and Conceptual Dimensions*, Johns HopkinsUniversity Press, pp. 137-152.

Coverdale, J.H., Bayer, T,L., McCullough, L.B., Chervenak, F.A.: 1995, 'Sexually transmitted disease prevention services for female chronically mentally ill patients,' *Community Mental Health Journal* 31, 303-315.

Coverdale, J.H., Chervenak, F.A., McCullough, L.B., Bayer, T.L.: 1996, 'Ethically justified clinically comprehensive guidelines for the management of the depressed pregnant patient,' *American Journal of Obstetrics and Gynecology* 171, 169-173.

Coverdale, J.H., McCullough, L.B., Chervenak, F.A., Bayer, T.L., Weeks, S.: 1997, 'Clinical implications of respect for autonomy in the psychiatric treatment of pregnant patients with depression,' *Psychiatric Services* 48, 209-212.

Doukas, D.J., McCullough, L.B.: 1996, A preventive ethics approach to counseling patients about clinical futility in the primary care setting,' *Archives of Family Medicine* 5, 589-592.

Doukas, D.J., McCullough, L.B.:1999, 'The values history in well elder care,' in Gallo, J., et al., eds., *Reichel's Care of the Elderly: Clinical Aspects of Aging*, 5th ed., in press.

Dresser, R, Robertson, J.: 1989, 'Quality of life and treatment decisions for incompetent patients: a critique of the orthodox approach,' *Law, Medicine, and Health Care*, 17, 234-268.

Faden, R.R., Beauchamp, T.L.: 1986, *A History and Theory of Informed Consent*, Oxford University Press, New York.

Fineschi V., Turillazi, E., Cateni, C.: 1997, 'The new Italian code of medical ethics,' *Journal of Medical Ethics*, 23, 239-244.

Freedman, B.: 1993. 'Offering truth: One ethical approach to the uninformed cancer patient,' *Archives on Internal Medicine* 153, 572-576.

Gillon, R., Lloyd, A.: 1994, *Principles of Health Care Ethics*, John Wiley & Sons, Chichester, England.

Grisso T., Appelbaum, P.S.: 1998, *Assessing Competence to Treatment: A Guide for Physicians and other Health Professionals*, Oxford University Press, New York.

Grimes, A.L., McCullough, L.B., Kunik, M.E., Molinari, V., Workman, R.: 2000, 'Informed consent and neuroanatomic correlates of intentionality and voluntariness among psychiatric patients,' *Psychiatric Services* 51, 1561-1567.

Jonsen, A.R.: 1998, *The Birth of Bioethics*, Oxford University Press, New York.

Holstein, M., Cole, T.R.: 1995, *Long-Term Care: A Historical Reflection*, in McCullough, L.B., Wilson, N.L., eds., *Long-Term Care Decisions: Ethical and Conceptual Dimensions*, Johns Hopkins University Press, pp. 15-34.

Kane, R.A., Caplan, A.L., eds.: 1990, *Ethical Issues in the Everyday Lives of Nursing Home Residents*, Springer Publishing Co., New York.

Kane, R.A., Caplan, A.L., eds.: 1993, *Ethical Conflicts in the Management of Home Care: The Case Manager's Dilemma*, Springer Publishing Co., New York.

Kane, R.A., Kane, R.L.: 1987, *Long-Term Care: Principles, Programs, and Policies*, Springer Publishing Co., New York.

Lidz, C.W., Fischer, L., Arnold, R.M.: 1993, *The Erosion of Autonomy in Long-Term Care*, Oxford University Press, New York.

McCullough, L.B.: 1988, 'An ethical model for improving the patient-physician relationship,' Inquiry 25, 454-468.

McCullough, L.B., Chervenak, F.A.: 1994, *Ethics in Obstetrics and Gynecology*, Oxford University Press, NewYork.

McCullough, L.B., Wilson, N.L., eds.: 1995, Long-Term Care Decisions: Ethical and Conceptual *Dimensions,* Johns Hopkins University Press, Baltimore, MD.

McCullough, L.B., Wilson, N.L., Rhymes, J.A., Teasdale, T.A.: 1995, 'Managing the conceptual and ethical dimensions of long-term care decision making: a preventive ethics approach,' in McCullough, L.B., Wilson, N.L., eds., *Long-Term Care Decisions: Ethical and Conceptual Dimensions*, Johns Hopkins University Press, pp. 221-240.

Rabeneck, L., McCullough, L.B., Wray, N.P.: 1997, Ethically justified, clinically comprehensive guidelines for percutaneous endoscopic gastrostomy (peg) tube placement,' *The Lancet* 349, 496- 498.

McCullough, L.B., Rhymes, J.A., Teasdale, T.A., Wilson, N.L.: 1999, 'Preventive ethics in geriatric practice,' in Gallo, J., et al., eds., *Reichel's Care of the Elderly: Clinical Aspects of Aging*, 5th ed., in press.

Reilly, R.B., Teasdale, T.A., McCullough, L.B.: 1995, 'Projecting patients' preferences from living wills: an invalid strategy for management of dementia with life-threatening illness,' Journal of the *American Geriatrics Society* 42, 997-1003.

Royall, D.R., Cordes, J., Polk, M.: 1997, 'Executive control and the comprehension of medical information by elderly retirees,' *Experimental Aging Research* 23, 301-313.

Singer, P.: 1994, 'Disease-specific advance directives,' *Lancet* 344: 594-596.

Singer, P.A., Thiel, E.C., Naylor, D., Richardson, R.M.A., Llewellyn-Thomas, H., et al.: 1995, 'Life-sustaining treatment preferences of hemodialysis patients: implications for advance directives,' *Journal of the American Society of Nephrology* 6, 1410-1417.

Singer, P.A., Thiel, E.C., Salit, I., Flanagan, W., Naylor, C.D.: 1997, 'The HIV-specific advance directive', *Journal of General Internal Medicine* 12, 729-735.

Wear, S.: 1993, *Informed Consent: Patient Autonomy and Physician Beneficence within Clinical Medicine*, Kluwer Academic Publishers, Dordrecht, The Netherlands.

White, B.C.: 1994, *Competence to Consent*, Georgetown University Press, Washington, DC.

Workman, R,H., McCullough, L.B., Molinari, V., Kunik, M.E., Orengo, C., Khalsa, D.K., Rezabek, P.: 2000, 'Clinical ethical implications of impaired executive control functions for patient autonomy,' *Psychiatric Services* 51, 359-363.

DOUGLAS L. WEED

ETHICS AND PHILOSOPHY OF PUBLIC HEALTH

I. MAJOR THEMES AND BASIC QUESTIONS

To review the ethics and philosophy of public health from what scholars have written to date – a review of the 'landscape' as revealed in the intellectual maps of book chapters and journal articles – is not a simple task. Many details of this intellectual journey are missing. Nevertheless, in this same sparse literature lie hints of several major themes, which if further developed, could be woven together like threads of a great tapestry into an overarching and comprehensive philosophical foundation for public health. Such a feat is clearly beyond the scope of this paper. Here I hope only to begin to point the way for those who can develop these themes more fully in the future. In the order in which they will be described, the main themes of the ethics and philosophy of public health include: the relationship between public health and medicine, and ontological matters including the nature of health and disease as public phenomena as well as the nature of disease causation. Perhaps most relevant to the readers of this volume are the ethics of public health interventions, but it is difficult to separate the nub of that issue – when and for whom shall we intervene? – from the complex nature of evidence-based scientific judgments made in search of the causes of disease. Because so little has been written and because I believe that to understand the ethics of public health – and especially the ethics of intervention – one must also understand the ontological and epistemological frameworks that impact on decision making, I begin with the broader issue regarding the very existence of the topic at hand.

II. IS THERE A PHILOSOPHY OF PUBLIC HEALTH?

It may seem reasonable to create a catalog of possible applications of current ideas in bioethics and philosophy of science to issues in public health, along the lines suggested by Pellegrino (1986) and others (Engelhardt, 1986) for an analogous situation in medicine. Such a compilation we could call a philosophy *in* (meaning, "applied to") public health, but such an approach leaves open the question of whether a philosophy *of* public health exists at all.

G. Khushf (ed.), Handbook of Bioethics, 525–547.

By a philosophy *of* public health I mean a general theory of the nature and practice of public health encompassing ontological, epistemological, and ethical concerns (Weed, 1999).

Not everyone, of course, might place ethics snugly within the philosophy of a discipline such as public health or medicine. Caplan (1992), for example, clearly distinguishes between the ethics of medicine and its philosophy, preferring to align the latter with ontological and epistemological concerns typically found in the philosophy of science. Others, including Engelhardt and Wildes (1995) and Pellegrino and Thomasma (1981) include ontological, epistemological, and ethical concerns within a philosophy of medicine.

In the rare papers that address the "philosophy of public health" this broader view has been adopted; see, for example, the piece by Beauchamp in the Encyclopedia of Bioethics (Beauchamp, 1995) and a more recent effort in the Encyclopedia of Public Health (Weed, 2001). Both accounts note that a philosophy of public health must acknowledge the central place of community, scientific method, prevention, and the limits placed on individual autonomy in the name of the common good. The more recent account (Weed, 2001) recognizes the constant pressure of scientific uncertainty on efforts to acquire knowledge about health in human populations and notes the important role that values – of communities and of public health professionals—will play in tempering judgments about what should be done in the name of the common good.

With so little written on the topic, it is easy to answer the question "is there a philosophy of public health?" negatively, similar to the way in which Lindahl (1990) concluded perhaps a bit prematurely that a philosophy of medicine does not exist because so little has been written about it. Indeed, the difference between the quantities of published literature on the two topics is so remarkable – somewhere on the order of Mount Medicine and a molehill – that if a philosophy of medicine doesn't make the grade, then philosophy of public health cannot possibly. But others argue that a philosophy of medicine does not exist because it is not a distinct philosophical subdiscipline, too easily relabeled bioethics and philosophy of science (Caplan, 1992) or alternatively a subdiscipline of medicine itself (Wulff, 1992). Engelhardt and Wildes (1995) describe the philosophy of medicine as an ambiguous enterprise comprised of four activities: speculative medicine, the logic of medicine, philosophy of science in medicine, and bioethics. Pellegrino, who with Thomasma strongly advocates the need for and existence of a philosophy of medicine (1981), would likely argue that the key criteria for answering questions concerning the philosophy of any discipline include (1986): whether (in this case) public health is a distinct activity and whether a philosophy of public health differs from public health itself. Accepting even the first of these criteria (and given that a philosophy of public health is in such an embryonic stage that answering the second question is impossible for now), the key initial consideration is whether public health is distinct from other practices. Medicine, one of public health's closest scholastic and professional sisters, seems an obvious choice for contrast.

Is public health distinct from medicine?

In contemporary American practice, medicine primarily involves the treatment of disease in individual patients by individual licensed practitioners whereas public health primarily involves the prevention of disease in populations by organized community efforts. The two appear to differ, therefore, in terms of the basic activity or goal (treatment versus prevention), in terms of who is practiced upon (individuals versus populations), and in terms of who is doing the practicing. It follows that the relationship between practitioner and participant in these two activities will likely also differ. Historically, differences between medicine and public health have not always been so prominent, and yet today they strike many commentators as profound, as if a schism has emerged between two not-altogether-friendly professional cultures (Kerr-White, 1991).

Putting aside possible solutions to this difficult historical and social problem (Reiser, 1996), we may more closely examine the extent to which any of these differences – some apparent and some real – provide interesting starting points for this journey into the landscape of the philosophy of public health. The distinction between treatment and prevention seems fundamental and suggests two very different approaches to the corresponding professional practices and, by extension, differences if not in their philosophical foundations then in the application of philosophical concepts. If treatment is the goal, then the key question for the medical practitioner becomes: "is there a disease here to treat and what should be done to alleviate the suffering and signs it has caused in this patient?" If prevention is the goal, then the key question for the public health practitioner becomes: "what causes disease and what can be done to prevent disease from occurring in this population or community?" At least three issues of philosophical interest arise:

(1) what counts as disease matters; medicine presumes the existence of disease whereas public health presumes its absence

(2) the cause of disease (in public health) can be distinguished from disease as cause (of symptoms and signs)

(3) preventive interventions in public health are primarily directed at populations within which some individuals are disease-free but at risk for disease, whereas medicine primarily intervenes on sick individuals.

Clearly it will be necessary to examine the ontological issues surrounding the nature of health and disease, disease causation, and the distinction between what some have dubbed "sick individuals and sick populations" (Rose, 1985).

Epistemological and ethical issues are also important. Interestingly, interventions in medicine and public health arise from the same large pot of scientific research in biological, epidemiological, and behavioral disciplines, and with rare exceptions the conceptual and methodologic issues are so similar that it is very difficult to discern whether or not substantive differences exist between epistemological frameworks for medicine on the one hand and public health on the other. And yet the types of interventions and especially the ethical conditions under which their use are justified may be quite different. A therapeutic intervention in medicine, for instance, can require considerable risk on the part of the patient; life-saving craniotomies and immune-suppressing chemotherapies are only two of many possible examples. A preventive intervention, on the other hand, is less likely to incur significant risk for the individual and more likely to involve structural changes in the social and physical environment of communities; fluoridation of water supplies to prevent dental caries, sewage treatment systems, and anti-smoking billboards are a few examples.[1]

The ethical conditions under which interventions are used and justified appear also to differ for medicine and public health.[2] For example, some have argued that interventions are required in medicine, whereas in public health – wherein interventions are typically applied to so-called 'healthy' or 'disease-free' populations – interventions may be ethically justified only if scientific research has demonstrated them to be effective (Charlton, 1993). Furthermore, the vast majority of medical interventions are undertaken with the knowledge and consent of those practiced upon, whereas in public health, as Beauchamp reminds us (1995), the individual may not ever be consulted about community-level interventions about which he or she has little direct say: premarital blood tests, immunization requirements, seatbelt and no-smoking regulations are prime examples.[3]

Distinguishing between individuals and populations, either as professional practitioners or as recipients of medical and public health interventions, seems an obvious approach to convincingly separate medicine from public health. But even this time-honored distinction may be in a state of flux with the two "sides" moving closer together rather than farther apart. The rise of managed care and group practices has convincingly changed the way medicine is practiced in the United States, with less emphasis on the individual decisions made by the practitioner and more emphasis on group (if not corporate) decisions. Furthermore, there are signs that a population perspective is gaining or perhaps regaining prominence in medicine (Greenlick, 1992) inasmuch as 'clinical epidemiology' (Mackenbach, 1995) and 'public health medicine' (Charlton, 1993; Leck, 1993) appear to be reinvigorating both the evidence-based approach to clinical decisionmaking as well as what has long been called preventive and community medicine. Likewise, in public health circles, there are those who complain that public health is far too oriented towards a "public" comprised of individuals rather than a "public" comprised of communities. More on this issue will emerge in a later section but it is clear that population-based thinking (including concepts, research methods, and data) has become increasingly important in the practice of medicine and in public health. (Nijhuis & van der Maesen, 1994; Diez-Roux, 1998)[4]

Nevertheless, a key distinction between medicine and public health remains the nature of the relationship between practitioner and the recipient(s) of that practice. In medicine, the patient typically seeks out the practitioner and asks for her help. In public health, the link between community and practitioner is less clear. Although there are examples of communities seeking help for what they perceive to be public health problems (e.g. the inhabitants of Love Canal, the residents of Long Island, New York and of North Karelia in Finland, and numerous groups of picnic-goers complaining of acute gastrointestinal distress), these may not be the most common scenarios and the person(s) or institutions to whom these requests are made may not be responsible for any subsequent interventions. Indeed, what counts as a public health practitioner at the level of community intervention seems best described as a complex social arrangement including in various proportions: governmental institutions, educational and scientific institutions, and private industry, with public interest in the environment, health promotion, and disease prevention an important partner. Occasionally, single prominent individuals – e.g. former Surgeon General Everett Koop – have an extraordinary influence on public health. These various participants merge together to form public health movements or campaigns that have a considerable impact on our culture as well as on the incidence, prevalence, and mortality of disease. The "War on Cancer," the HIV/AIDS phenomenon, and the efforts (largely Australian and North American) to limit public tobacco consumption, are a few recent examples. Vaccination programs to prevent and control the spread of influenza, poliomyelitis, and smallpox are historical examples. Many of the same factors found in these campaigns are relevant to the practice of medicine – the war on cancer, for example, involves as much the search for better treatments as better preventive interventions –

the key difference here between public health and medicine, at least in terms of the relationship between the participant and practitioner, is that medicine can define for itself a focal point for practice, what Pellegrino and Thomasma (1981) call the 'clinical event' – between clinician and patient – an event which does not appear to have a clear analogy in public health. A recent conversation in the public health literature has focused upon the extent to which schools of public health and the profession of epidemiology have become disengaged from public health practice at the community level (Institute of Medicine, 1988; Terris, 1992; Shy, 1997) but the nature of the relationship between practitioner (whether individual professional or institution) and the community remains unclear. Within epidemiology, there are those who would leave the practice of public health – the intervention programs – to others, so that their lives as professional scientists would remain unsullied (Walker, 1997; Rothman and Poole, 1985) whereas there are others who believe that epidemiologists and other public health professionals have a duty at the level of the individual practitioner to act in the best interests of communities, including but not limited to those which participate in research studies (Shy, 1997). We are left, therefore, with an interesting problem: what obligations do public health professionals have to intervene in communities?

From this rather cursory exposition, it is clear that public health cannot be completely dissected away from medicine. To be sure, there are important differences as discussed above, but overlap also exists. Medical practitioners practice public health (often called preventive medicine or less frequently health promotion) when they advise their individual patients to refrain from tobacco or other self-imposed disease-causing hazards (Lawrence, 1990). Public health practitioners in turn design and manage screening programs that not only identify disease in its early stages but include treatments that hopefully are effective in reducing mortality (so-called secondary prevention). Furthermore, in both medicine and public health there is increasing attention placed on the scientific basis of judgments and decisions, what amounts to an emphasis on evidence-based interventions, whether therapeutic or preventive, with the evidence often methodologically indistinguishable. Indeed, there are some ensuing philosophical problems (e.g. how one judges scientific evidence to be sufficient to make a claim about the efficacy of an intervention) that appear to be shared by the practice of public health and medicine. A sharp distinction between the two disciplines and a clear border between the landscape of the philosophy of public health and that of medicine may therefore not be possible.

Examining the differences and similarities between these disciplines has provided the groundwork upon which an interesting set of problems can be examined. If I were to choose the two most important and most controversial

– keystone issues, in other words – for a philosophy of public health, they would be:

Keystone Issue #1: What justifies the decision to implement a preventive intervention, to move, in other words, from scientific evidence to public action?

This problem is certainly at the center of many current controversies in public health practice, whether involving primary prevention (i.e. the removal or reduction of harmful disease-causing factors) or screening for early disease (i.e., secondary prevention). Examples of the former include the historical debates on smoking and lung cancer and a host of more recent debates too numerous to mention in detail such as Agent Orange, radon gas, the so-called Gulf War Syndrome, and exposure to electromagnetic fields. A recent example of the justification problem in secondary prevention (i.e., screening) is the prominent controversy surrounding the use of mammography to detect and treat early breast cancer in women under age 50 (Fletcher, 1997).

Keystone Issue #2: Are public health professionals obliged to participate in preventive interventions? To whom are they so obliged?

As noted above, this problem emerges in an environment in which many academic departments in public health schools across the U.S. have become increasingly isolated from the communities they serve. In addition, there is a sentiment among some senior epidemiologists that policymaking and other aspects of preventive decisionmaking are inappropriate professional activities.

These keystone issues are only the tip of a very large iceberg. And so, rather than first focusing in upon their analysis, I take a couple steps backwards to sketch out some of the ontological and epistemological frameworks of contemporary public health. Ontological concerns include the nature of public health, with its distinction, introduced earlier, between individuals and communities, the nature of health, of disease and of disease causation. Epistemological concerns emphasize the problem of making causal claims and recommendations about the need for preventive interventions in the face of empiric uncertainty or what some philosophers call underdetermination (McMullin, 1995).

My intent throughout the remainder of this chapter is not to analyze but rather to provide descriptive stepping-stones for those who would examine the philosophy of public health in greater detail.

III. ONTOLOGICAL CONCERNS: PUBLIC HEALTH AND DISEASE CAUSATION

In order to move forward with a discussion of core ethical and philosophical issues in public health, it will be important to examine the multiple meanings of some key concepts: public, health, public health, disease, and cause. It has been recognized that different definitions for these concepts may underlie and ultimately influence whether a preventive intervention is employed (Nijhuis and van der Maesen, 1994), leading some to suggest that public health practitioners should at the least disclose their definitions. In the context of prostate cancer screening it has be shown that this claim has practical relevance; there are indications that different published opinions about this controversial public health screening approach can be traced to different philosophically relevant although typically implicit assumptions (Weed, 1999).

Indeed, public health practitioners may not only hold different meanings of "public health" but these may be different enough and non-negotiable enough to have emerged from partially incommensurable worldviews, not necessarily from different scientific paradigms per se but rather from different social and professional traditions, and heavily influencing the decisions made by their respective adherents. Under these circumstances, it seems prudent to carefully lay out what can be meant by the term: "public health." There are several different possibilities. The "public" can spring from within an individualistically oriented philosophy in which the total (the community) is the sum of the actions and motivations of distinct individuals. A rather different notion of the "public" can emerge from a more collectivistically oriented social philosophy in which social institutions and constructs comprise the community, still containing individuals but not reducible to them. In this broader view, communities can have characteristics relevant to their collective health; herd immunity against infectious diseases is an example.

Similarly, the concept of "health" can be described as either a natural scientific (mechanistic) view in which health is non-disease and disease in turn is brought about by causal mechanisms affecting each individual's biophysiological and neurophysiological system – a reductionist view – or it can be seen as a more holistic phenomenon reflecting a dynamic equilibrium that includes not only the natural mechanistic view but also a social systemic perspective. And it is interesting to observe that the current practice of public health research – largely but not exclusively maintained conceptually by the discipline of epidemiology – reflects two ultimately complementary but currently disparate schools of thought each of which is gathering increasingly strident voices (Rothman et al., 1998). One vows allegiance to the need for biological (especially molecular) research to understand the nature of disease etiology – a mechanistic and reductionist view of disease primarily affecting individual – and another emphasizes social (or so-called contextual) factors including ethnicity, social class, income and income inequality, to help causally explain disease phenomena as public or community events (Diez-Roux, 1998; Gori, 1998; Pearce and McKinlay, 1998). That there is such a debate about the meanings of public health is a good sign for the discipline because its fundamentally humanistic nature requires some conceptual inquiry (Weed, 1995), although differing views clearly make for a more complex landscape.

Another dimension to the ontology of public health is reflected in conceptual concerns about the nature of health and disease. At issue is a longstanding philosophical problem: that of distinguishing between facts and values, asking whether a value-free (or so-called objective) account of disease and especially health is possible, independent of the subjective preferences of the community. For a disease-specific and historical example, see Saraf's (1998) recent paper on anorexia nervosa, generally recognized today as a psychological disorder with potentially profound pathophysiological effects, and considered a faith inspired sacrifice of devout Christians (holy anorexia) in 14th Century Europe. The problem is typically framed in terms of a debate between reductionists and relativists. The reductionist view separates fact from value and provides a statistical account of disease that is presumably value-free (Boorse). Relativists promote the value-laden and culturally-sensitive view of health and disease (Khushf, 1995). Neither perspective seems adequate (Khushf) and a middle-ground has emerged in which a biological basis of value is substituted for subjective values (Lennox, 1995). These objectivist theories rest on an understanding of value as grounded in the desirability of an objective goal, with the ultimate goal as life itself. Disease remains classifiable by a taxonomy of biological dysfunctions (Sade, 1995).[5]

In public health, the great bulk of concern lies in preventing the occurrence of such biological dysfunctions by intervening at various, typically prediagnostic, stages in the natural history of the disease process. Primary prevention, for example, intervenes at a very early stage, and is classically described as the removal or reduction of exposure to that which causes disease. A philosophy of public health therefore requires careful consideration of the question: what is a cause? And can causality effectively capture the primary relationship between disease and exposure factor events across the broad spectrum of scientific knowledge from the innermost workings of the cell to the broad expanse of social and ecologic events that may influence the occurrence of disease in populations? Clear answers to these questions are unavailable in the current literature on the philosophy of public health.

Occasionally commentators espouse definitions of disease causation, with overly simplistic deterministic versions predominating among clinicians and epidemiologists (Rothman, 1976; Stehbens, 1985), probabilistic or statistical accounts the most highly developed (Olsen, 1993; Koopman and Weed, 1990; Elwood, 1988; Lilienfeld and Stolley, 1994), and counterfactual accounts predominating among the few statisticians who have discussed causation (Rubin, 1974; Holland, 1986). A comprehensive discussion of disease causation seems an important component of a philosophy of public health because identifying manipulable causal factors is central to the goals of prevention.

Such a strong emphasis on the manipulability of causes may sound rather like an appeal to pragmatism, but before that label can be resoundingly stuck onto the enterprise, it is important to point out that some causal factors may not be manipulable but still remain important to the explanation of etiology (e.g., gender and ethnicity) and that causes may not be sufficient to explain the occurrence of disease in populations. For those public health practitioners primarily involved in searching for factors that affect the biologic functions of individual organisms – the so-called reductionist view – causes are but one part of a more general notion of disease mechanisms. Likewise, for those investigators primarily involved in searching for the social origins of disease, cause may not be the only (or even the most prominent) form of explanation. Metaphorical approaches, e.g., the "spider on the web of causation" (Krieger, 1994), and systems theory (Koopman, 1996; Weed, 1998) approaches have been proposed as complementary forms of explanation, as have more general forms of determination (Weed, 1986).[6]

Nevertheless, causes remain an important locus of philosophical inquiry as well as scientific research because they carry with them the promise of "making a difference" if removed in time to affect subsequent disease incidence (Parascandola and Weed, submitted). How one goes about judging cause from scientific evidence designed to discover its whereabouts is another foundational problem in the philosophy of public health, largely described in epistemological terms.[7]

IV. EPISTEMOLOGICAL CONCERNS: LOGIC, UNDERDETERMINATION, JUDGMENT, AND VALUES

Public health practitioners and especially academics have, for at least fifty years, been interested in the interpretation of scientific evidence with special emphasis on judging cause from observational and experimental studies (Weed, 1995). Occasionally this discussion takes on a decisively philosophical air (Weed, 1986; Rothman, 1988; Susser, 1991; Karhausen, 1996), but the choice of philosophical schools and concepts in these efforts has often been highly selective, depending in large part upon the rather colloquial interests of those in public health who, after realizing that philosophy of science might have something to offer, jumped into that literature, found some ideas worthy of discussion and application, and wrote a few papers on such topics, thereby creating a focus from perspectives that the professional philosopher could easily consider passe if not old-fashioned. The epidemiologists' emphasis on Popperian falsification and refutation is a good example (Buck, 1975; Jacobsen, 1976; Maclure, 1985; Weed, 1988), with papers contrasting the relative merits of induction and deduction still making their voices known in the professional literature (Greenland, 1998).

Only recently have more contemporary epistemological bridges been built between public health and the philosophy of science. An excellent example is that of underdetermination (McMullin, 1995) which, for the purposes here, can be defined as a lack of proof or disproof in the relationship between observations and the theories proposed to explain the occurrence of disease. Typically, public health investigators examine hypotheses that are often rather loosely connected to vaguely stated causal theories. The implications of underdetermination for appraising these hypotheses emerges when attempting to solve the problem of judging whether observations – for example, the mix of epidemiologic and biologic studies collected and summarized in reviews of evidence – represent causation or its alternatives, including most prominently, chance or bias (Weed, 1997). When faced with this problem, investigators have traditionally invoked what are called causal inference methods which involve causal "criteria" comprising a mix of qualitative and quantitative concepts with which one judges the evidence. These so-called criteria represent prominent scientific values; the consistency of the findings (a variant of the notion of replicability), the magnitude of the measures observed, the existence of a dose-response curve, and the extent to which the findings are biologically plausible are almost universally employed (Weed and Gorelic, 1996) with others, such as the temporal relationship between purported cause and its effect, added when deemed relevant by the individual judging the evidence. Indeed, the process of judging evidence in contemporary public health science is clearly if not extraordinarily value-laden, with these scientific (or constitutive) values – the criteria and their accompanying rules of inference – playing alongside extrascientific (or contextual) values (Longino, 1990). Clear differences in these and other values appear to adequately explain prominent differences in causal judgments in controversial cases, such as the purported relationship between induced abortion and breast cancer (Weed, 1997). Put another way, the current landscape of causal judgment in contemporary public health seems best described as a decisionmaking process that takes place after scientific evidence accumulates but does not (indeed perhaps cannot) decidedly determine the best choice of hypothesis – the phenomenon of underdetermination – coupled with a retreat to value-laden inferential methods containing both qualitative and quantitative components.

The extent to which other even more prominent themes in contemporary epistemological circles – such as pragmatism and realism – could form at least a cornerstone in the foundation of a philosophy of public health is a matter of considerable importance. The value-ladeness of contemporary causal inference as practiced by public health professionals seems to represent what Beauchamp (1995) refers to as pragmatism, with its emphasis on what "works" and its appeal to consensus – in this case a methodological consensus – or as Rorty might put it, solidarity within the communities calling themselves scientists for a methodology in which an individualistic interpretation of central features of the method are not so much tolerated as encouraged (Rorty, 1995). Indeed, as noted earlier, the emphasis on identifying causal factors which, once removed, may reduce the occurrence of disease adds fuel to the pragmatists' fires. But among the few who attempt to bridge the rather large gaps between contemporary philosophy of science and what would count as a philosophy of public health, there are some who wonder (Weed, 1997) and others who claim (Renton, 1994) that objective accounts of public health science – e.g. a realist interpretation of causation and causal inference in epidemiology – are also important. Finding what common ground exists between the camps of pragmatists and realists and applying those claims to the theory and practice of public health seems a reasonable activity for future inquiry.

Before moving on to the topic of the ethics of public health, and especially of public health interventions, it should be noted that causal inference is not the sole epistemologically-oriented issue in public health, although it lies very near if not at the center of any such discussion. Public health practice also involves the *application* of interventions and the assessment of those actions. Distinctions between the "efficacy" and the "effectiveness" of interventions (Last, 1988), for example, may therefore have implications for the epistemological component of a philosophy of public health. Similarly, it would be unfair to conclude from this presentation that randomized trials, whether clinically or community-based, do not have some special epistemological concerns for the philosopher involved in public health. Indeed, causal inference methodology would be influenced if more randomized prevention trials were carried out.

V. ETHICS

The final leg of a journey through the landscape of the philosophy of public health involves ethics. As noted above, two key concerns will provide a focus for this discussion: (1) what justifies a public health intervention? and (2) what obligations do public health practitioners – whether individuals or institutions – have to intervene, and for whom do they do so?

A. The Ethics of Intervention

The question about what justifies a public health intervention, is key not only because it lies at the center of so many specific controversies, but also because interventions in public health may prominently restrict autonomy, because these same interventions may have prominent risks as well as benefits, because they are often applied to communities whose individual members are not diseased and some who will never be, and finally, because public health interventions at the community level can involve considerable costs to society. Put another way, the decision to go forward with a public health intervention clearly involves every one of the battered-but-not-beaten principles of bioethics (Beauchamp and Childress, 1994): autonomy, beneficence, nonmaleficence, and justice.

Having mentioned these principles, it is probably necessary to briefly refer to the multiplicity of ethical theories and to some extent, ethical decisionmaking models, that have been featured and debated in the bioethics literature (Weed, 1996), although their impact on the theory and practice of public health to date has been relatively small. Certainly these "four principles" are the most familiar ethical framework among public health practitioners, although citing them – even using them predominantly in analyses – does not preclude a consideration of other views. Indeed, virtue theory and casuistry have clear connections to public health decisionmaking (Weed, 1996; Weed and McKeown, 1998), and others, such as feminist theory, are also likely to be applicable. And perhaps one or another of these theories or methods will afford a philosophy of public health some special measure in the long term, although I rather doubt a convincing case can be made to exclude from consideration all but one. A complementary approach to principles, used recently in a discussion of ethics and epidemiology(Weed, 1996), is to put forward a list of moral rules, what can be considered a locus of moral consensus. Not surprisingly, within this list are found concerns about autonomy, about doing good and preventing harm, and about fairness.

Surveying the literature on the ethics of public health, one finds considerable concern about how to justify interventions that impinge on the rights of the individual because they smack of paternalism (Last, 1987; Wooley, 1990; Last, 1992) or because the public is inadequately informed about the expected benefits and potential harms (Skrabanek, 1990; Gillon, 1990; Horner, 1992; Calman, 1993). Typically legislative or regulatory in nature, involving such issues as mandatory safety devices (seat-belts and motorcycle helmets), limits on personal behaviors (e.g. public tobacco use), and required activities such as immunizations or premarital blood tests, these public health interventions require careful ethical analysis, with the primary concern whether the common good (i.e., the good of individuals taken together as a group or as communities) is sufficiently met by the intervention to warrant an abrogation of individual liberties. As Beauchamp notes, a philosophy of public health must adequately define this notion of a common good, because it is so important to an assessment of the ethics of interventions that tread on claims of individual autonomy (Beauchamp, 1995).

Cole (1995) argues that some regulations result from a failure on the part of the state (although other avenues exist such as public health professionals) to convince the public through less obtrusive means, such an education, that interventions are beneficial. He believes that education, however formal, has a clear moral basis because it is intended to enhance the will of the individual and because no penalty is imposed on those who do not accept the message. One might argue, however, that such a "clear moral basis" seems to emerge directly from the perspective of the individual than rather from the community at large. To put it another way, if the individual's autonomy (free choice) is paramount, then the intervention most easily justified is education because it appears to have no down side (at least from the individual's perspective). The community, on the other hand, whether considered as a group of individuals or as an identifiable social construct comprised of individuals but not reducible to them, may clearly not benefit at all from an education intervention if no one (at the individual level) chooses to change his or her behavior. Note how critical one's ontological commitment (i.e., to individuals or to communities) matters in this arrangement.

Similarly, epistemological commitments have a strong role to play in the assessment of the ethics of intervention. This is most prominently demonstrated if it can be (very easily) assumed that public health interventions have both benefits and risks. Note that the "benefit" here is the reduction in risk for an untoward event (typically disease or injury) at the individual level and a reduction in the rates of disease at the population or community level. A "risk" in this same context involves an increase in the opportunity of a side-effect that can be biological (e.g., immunizations may involve a small risk of untoward neurological effects), economic (e.g., screening for early disease involves a risk of medical expenses if disease is found), and social (e.g., screening may also stigmatize an individual or an entire community if disease – say, HIV or cancer – is uncovered). Characterizing the risk/benefit ratio of potential public health interventions, therefore, is a complex affair, involving assessments of scientific evidence that, as discussed above, are as much qualitative as quantitative, and perhaps as subjective as they are objective.

Some public health interventions, especially those involving early detection (screening for early disease), have some special ethical concerns because they can be applied to individuals who are at extraordinarily low risk of the disease and therefore stand to gain very little or no benefit from the intervention and yet are exposed to the risks that such an intervention entails. Or, to put the same problem into the language of communities, interventions that look for evidence of early disease so that it may be treated may offer more risks to some groups than benefits. Putting aside the difficult problem of the assessment of risks and benefits and the certainty with which they can be described, the problem of exposing what would otherwise be considered healthy individuals or communities (with respect to the disease in question) to potentially harmful interventions deserves special mention. Screening for prostate cancer with prostate specific antigen (PSA) and vaccinations for acute infectious diseases are just two of many possible examples (Weed, 1999; Lachmann, 1998).

B. Who Intervenes in Public Health and for Whom?

Those who participate in the contemporary practice of public health – at the level of the individual practitioner especially but also at the level of institutions – seem adrift in a sea of indecision about what role these potential "practitioners" play in the decisions about and implementations of interventions. To put this issue in terms analogous to those used in the practice of contemporary clinical medicine: what is the nature of the relationship between the practitioner of public health and the recipients of that practice? What are its ethical dimensions?

The problem exists in large part because it is not easy to focus upon who is 'doing' (i.e. performing or being responsible for) the various public health interventions that may impinge on individual autonomy, that typically involve benefits and risks, and may involve considerable costs to individuals and communities. Some lack of clarity is due to the fact that there are many different kinds of public health interventions, reaching individuals and communities in different ways.

For some public health disciplines, especially epidemiology, this issue is reaching a critical point. Some professional epidemiologists are extraordinarily uncomfortable about participating in any activity beyond traditional scientific research practice. For them, public health practice – whether it arrives as policymaking or as direct interventions – is either too complicated or too imbued with the bad taste of advocacy to be acceptable. On the other side of the debate are those who cite historical precedents for participating in public health (Terris, 1992); for them, epidemiologists began their search for etiology because they needed information to support and to participate directly in interventions and policymaking. And finally, there is a case to be made for professional epidemiologists to embrace an active role in both scientific research and public health interventions policy based ultimately upon the notion that being a 'professional' means professing something to someone: in this case, professing to prevent disease to the public (Weed, 1994).

Epidemiologists who make such a public pronouncement – a 'profession' – may then join in the practice of public health along with physicians and other clinicians, other public health professionals, legislative bodies, private firms, non-profit organizations, and private citizens. Even so, the relationship of these practitioners to the community remains elusive. Indeed, in public health the public itself can act as its own practitioner, an idea that appears to fly in the face of traditional models if it were not for the extraordinary popularity of alternative and complementary medicine. Indeed, we might propose that public health has both a traditional component (representing preventive interventions sanctioned by official governmental and community organizations, by the public health professional community, by private industry, and by medical practitioners) and an alternative component (representing the community's efforts to prevent disease and promote health on its own terms).

VI. RESEARCH FRONTIERS

It seems fitting to end this account of the ethics and philosophy of public health somewhere along the road to our future (Holland and Stewart, 1998; Levy, 1998), where we, as public health professionals and as individuals and communities comprising the 'public' itself, negotiate the risks, intervene when and where it is most appropriate, and – not always with the same voice – decide which direction to take at the next crossroads (Beaglehole and Bonita, 1997). For those who in the tradition of philosophy seek to better understand

– to illuminate – the intellectual problems that undergird public health, I offer the following.

Public health is fundamentally a practical endeavor. Its philosophers, in my view, should therefore set their sights on the problems that matter to public health practice and work to provide solutions to those problems as best they can. Blackburn calls this a "middle-ground" approach to philosophy (Blackburn, 1999); philosophical reflection that is continuous with the practice of public health and thus matters to the ways in which public health professionals, the public, and all other involved institutions go about their daily business. This approach does not deny the importance nor necessity of reflection at a somewhat higher level of understanding, at that rarefied level of thought characteristic of academic philosophers, or what Blackburn calls "high ground philosophy." Certainly there should be as many connections as possible laid between those whose training and experience springs from public health practice and seek to "philosophize" about public health and those whose backgrounds begin with the great themes of philosophy – knowledge, mind, reasoning, free will, the self and the world – to name a few, and seek ways to turn their considerable intellectual talents towards the ethics and philosophy of public health. How to make those connections, for example by teaching philosophers and ethicists about public health practice and vice versa, by training public health practitioners in the traditions and techniques of ethics and philosophy, is beyond the scope of this paper. I close with a list of topics – see Table 1 – that sit somewhere in-between the daily practice of public health and the starry heights of philosophical reflection. It is my hope that these may prove useful for those who would undertake the journey.

Table 1: Public Health's Philosophical Issues

This list is not complete, but hopefully gives a sense of the complexity, breadth, and depth of the issues that can be connected to public health's philosophical foundations. No special significance should be given to the order in which the individual items appear.

Ontological Concerns:
Concepts of Public and Health and Disease
Individuals and Populations
Reductionism and a Systems View
Health and Disease as Concepts
Definitions, Models, and Theories of Cause: Determinism,
Probabilism, and Counterfactuals

What Counts as Prevention?
Sustainability

Epistemological Concerns:
Causal and Preventive Knowledge and Inference
Complexity of Evidence: Biology to Society
Underdetermination, Refutation, and Verification
Objectivity and Values in Methods
Theory and Practice in Methods of Interpretation and Inference
Uncertainty, Proof, and Weighing Evidence
Realism and the Weight of Paradigms
Chaos and Complexity

Professional Ethics:
Code of Ethics for Public Health Professions
Duties, Obligations, and Core Values
Individuals and their Professional Character: Virtue
Who is Practiced Upon?
Who is Responsible for Practicing Prevention? Who Decides?
Science, Policymaking, and Advocacy

Ethics of Public Health:
Obligations, Responsibility and Commitment
The Precautionary Principle
Human Rights
Animal Rights (Experimentation)
Informed Consent
Confidentiality
Autonomy and the Public or Common Good
Misconduct, Integrity, and Public Trust
Communitarian and Virtue Ethics
Community Participation in Public Health Decisionmaking
Health Disparities

National Cancer Institute
Washington, DC, U.S.A.

NOTES

1. There are, of course, therapeutic interventions that incur very little risk (e.g. swallowing two capsules of a mild analgesic) and preventive interventions that are hardly risk-free (e.g., a screening colonscopy).
2. There are other situations, however, for which finding a sharp distinction between medicine and public health on ethical grounds may be difficult. Both practices rely primarily upon a basic sense of beneficence. Thus the medical practitioner(s) asks, "what can I (we) do to help this individual ill patient?" and the public health practitioner asks, "what can we (I) do to help this population avoid disease?" Both are concerned with the risks and benefits of intervention, with issues of cost and economics and therefore justice, and with issues of autonomy, although as Beauchamp (1995) noted, distinguishing between respect for persons and respect for communities seems worthy of more attention.
3. In both medicine and public health there is a considerable amount of self-directed intervention that takes place without the direct knowledge of any practitioner. Indeed, for primary prevention in public health involving lifestyle choices, the intent of effective public health may well be that individuals make their own decisions with very little if any intervention on the part of a professional practitioner. Choosing whether to smoke or not, eat several or no servings of fruits and vegetables, exercise regularly, or drink moderate versus excessive amounts of alcoholic beverages is something that individuals and therefore populations make without necessarily consulting with a public health professional. In medicine, the same sort of self-directed intervention also occurs, not only for the small aches and injuries of everyday existence well cared for by two analgesics and a bandage, but also for more serious conditions in the form of alternative therapies. Some alternative medical practices exist, but for the most part, patients self-medicate.
4. Likewise, the genuine explosion of progress in studying and understanding the biology of the cell in molecular and genetic terms has had important effects on both the individual patient in clinical medicine and on the community in public health.
5. The extent to which this definition handles situations in which individuals or society reject life as the ultimate goal, e.g. in cases of holy anorexia and (for that matter) anorexia nervosa (Saraf, 1998), is an interesting issue worthy of more attention than can be given here.
6. Although epistemological concerns will be examined later, it is interesting to point out that contemporary public health decisionmakers are now faced with the problem of understanding the etiology and natural history of disease conditions and health in terms of a very broad structure of scientific knowledge, encompassing many different levels ranging from molecular through cellular and on to the organismic level and beyond that to social interactions, the environment, and group phenomena (e.g. herd immunity) that cannot be captured as the sum of the individuals making up the group. The nature of this structure, the interconnectedness of its layers or levels or substructures, how causation acts within and between levels, and the types of conceptual frameworks (e.g. systems analysis) that may be fruitful for navigating its innards has been a subject of a lively controversy in the public health literature, the so-called "black box debate" pitting biomolecular adherents against their socially-oriented counterparts (Weed, 1997; Diez-Roux 1998; Gori 1998; Pearce & McKinlay 1998; Gori 1998).
7. Indeed, one might posit that to best judge whether a particular factor is a cause from evidence collected in epidemiologic and other studies one might need a handy if not clear definition of cause from which one's criteria for judgment would emerge (influenced, of course, by one's epistemological framework), but the current practice of judging cause in public health is interestingly devoid of careful descriptions of the nature of causation. Rather, the current practice is comprised of guidelines called 'causal criteria' that may assume one form of causal hypothesis or another, although the assumption is not made explicit. There is, in other words, no clear link

between ontological and epistemological frameworks and yet it seems reasonable that there should or could be such.

REFERENCES

Beaglehole, R. and Bonita, R.: 1997, *Public Health at the Crossroads: Achievements and Prospects*, Cambridge University Press, New York.

Beauchamp, D.E.: 1995, 'Philosophy of Public Health,' in W. Reich (ed), *Encyclopedia of Bioethics,* Simon and Schuster MacMillan , New York, 4, pp. 2161-2166.

Beauchamp, T.L. and Childress, J.F. : 1994, *Principles of Biomedical Ethics*, Oxford University Press, New York.

Beauchamp, T.L. : 1996, 'Moral Foundations,' in Coughlin, S.S. and Beauchamp, T.L., *Ethics and Epidemiology,* Oxford, New York, pp. 24-25.

Blackburn, S.: 1999, *Think*, Oxford University Press, New York.

Blaney, R.: 1987, 'Why prevent disease?,' in S. Doxiadis (ed.), *Ethical Dilemmas in Health Promotions,* John Wiley & Sons, pp. 47-56.

Boorse, C. :1977, 'Health as a theoretical concept,' *Philosophy of Science* 44, 542-573.

Buchanan, D. R.: 2000, *An Ethic for Health Promotion*, Oxford University Press, New York.

Buck, C.: 1975, 'Popper's philosophy for epidemiology,' *International Journal of Epidemology* 4, 159-168.

Calman, K.C. : 1993, 'Ethics and the public health,' *Medico-Legal Journal* 62, 190-203.

Caplan, A.L.: 1992, 'Does the philosophy of medicine exist?,' *Theoretical Medicine* 13, 67-77.

Charlton, B.G.: 1993, 'Public health medicine--a different kind of ethics?,' *Journal of the Royal Society of Medicine* 86, 194-195.

Cole, P.: 1995, 'The moral bases for public health interventions,' *Epidemiology* 6, 78-83.

Diez-Roux, A.V.: 1998, 'On genes, individuals, society and epidemiology,' *American Journal of Epidemiology* 148, 1027-1032.

Diez-Roux, A.V.: 1998, 'Bringing context back into epidemiology: variables and fallacies in multilevel analysis,' *American Journal of Public Health* 88, 216-222.

Elwood, M.: 1988, *Casual Relationships in Medicine: a Practical System for Critical Appraisal*, Oxford, New York.

Engelhardt, H.T.: 1986, 'From philosophy *and* medicine to philosophy *of* medicine,' *Journal of Medicine and Philosophy* 11, 3-8.

Engelhardt, H.T. and Wildes, K.W.: 1995, 'Philosophy of Medicine,' in W. Reich (ed), *Encyclopedia of Bioethics,* Simon & Schuster MacMillan , New York, pp. 1680-1684

Fletcher, S.W.: 1997, 'Whether scientific deliberation in health policy recommendations? Alice in the Wonderland of breast-cancer screening,' *New England Journal of Medicine* 336, 1180-1183.

Gillon, R.: 1990, 'Ethics in health promotion and prevention of disease,' *Journal of Medical Ethics* 16, 171-172.

Gori, G.B.: 1998, 'Epidemiology and public health: is a new paradigm needed or a new ethic?,' *Journal of Clinical Epidemiology* 51, 637-641.

Greenlick, M.R.: 1992, 'Educating physicians for population-based clinical practice,' *Journal of the American Medical Association* 267, 1645-1648.

Holland, W.W. and Steward, S.: 1998, 'Public health: where should we be in 10 years?,' *Journal of Epidemiology and Community Health* 52, 278-279.

Holland, P.W.: 1986, 'Statistics and causal inference,' *Journal of American Statistical Association* 81, 945-960.

Horner, J.S.: 1992, 'Medical ethics and public health,' *Public Health* 106, 185-192.

Institute of Medicine: 1988, *The Future of Public Health*, National Academy Press, Washington, D.C.
Jacobsen, M.: 1976, 'Against Popperized epidemiology,' *International Journal of Epidemology* 5, 9-11.
Karhausen, L.R.: 1996, 'The logic of causation in epidemiology,' *Scandinavian Journal of Social Medicine* 24, 8-13.
Khushf, G.: 1995, 'Expanding the horizon of reflection on health and disease,' *Journal of Medicine and Philosophy* 20, 461-473.
Koopman, J.S. and Weed, D.L.: 1990, 'Epigenesis theory: a mathematical model relating causal concepts of pathogenesis in individuals to disease patterns in populations,' *American Journal of Epidemiology* 132, 366-390.
Koopman, J.S.: 1996, 'Comment: emerging objectives and methods in epidemiology,' *American Journal of Public Health* 39, 277-285.
Krieger, N.: 1994, 'Epidemiology and the web of causation: has anyone seen the spider?', *Social Sciences of Medicine* 39, 887-903.
Lachmann, P.J.: 1998, 'Public health and bioethics,' *Journal of Medicine and Philosophy* 23, 297-302.
Lake, L.T.: 1997, 'A partnership for public health,' *Minnesota Medicine* 80, 20-24.
Last, J.M.: 1988, *A Dictionary of Epidemiology*, Oxford University Press, New York
Last, J.M.: 1992, 'Ethics and public health policy,' in Maxcy, Rosenau, Last (13th ed.), *Public Health and Preventive Medicine*, pp. 1187-1196.
Last, J.M.: 1987, 'The ethics of paternalism in public health,' *Canadian Journal of Public Health* 78, 3-4.
Lawrence, R.S.: 1990, 'The role of physicians in promoting health,' *Health Affairs* 9, 122-132.
Leck, I.: 1993, 'Clinical and public health ethics: conflicting or complementary?,' *Journal of the Royal College of Physicians (London)* 27, 161-168.
Lennox, J.G.: 1995, 'Health as an objective value,' *Journal of Medicine and Philosophy* 20, 499-511.
Levy, B.S.: 1998, 'Creating the future of public health: values, vision, and leadership,' *American Journal of Public Health* 88, 188-192.
Lilienfeld, D.E. and Stolley, P.D. (3rd ed): 1988, *Foundation of Epidemiology*, Oxford, New York.
Lindahl, B.I.B.: 1990, 'Editorial,' *Theoretical Medicine* 11, 1-3.
Longino, H.E.: 1990, *Science as Social Knowledge*, University Press, Princeton, NJ.
Loomis, D. and Wing, S.: 1990, 'Is molecular epidemiology a germ theory for the end of the twentieth century?,' *International Journal of Epidemiology* 19, 1-3.
Mackenbach, J.P.: 1995, 'Public health epidemiology,' *Journal of Epidemiology and Community Health* 49, 333-334.
Maclure, M. : 1985, 'Popperian refutation in epidemiology,' *American Journal of Epidemiology* 121, 343-350.
McMullin, E.: 1995, 'Underdetermination,' *Journal of Medicine and Philosophy* 20, 233-252.
Nijhuis, H.G.J. and Van der Maesen, L.J.G.: 1994, 'The philosophical foundations of public health: an invitation to debate,' *Journal of Epidemiology and Community Health*_ 48, 1-3.
Olsen, J.: 1993, 'Some consequences of adopting a conditional deterministic causal model in epidemiology,' *European Journal of Public Health* 3, 204-209.
Parascandola, M. and Weed, D.L. 'Causation in epidemiology', submitted for publication
Pearce, N. and McKinlay, J.B. : 1998, 'Back to the future in epidemiology and public health: response to Dr. Gori,' *Journal of Clinical Epidemiology* 51, 643-646
Pellegrino, E.D.: 1986, 'Philosophy *of* medicine: towards a definition,' *Journal of Medicine and Philosophy* 11, 9-16.

Pellegrino, E.D, and Thomasma, D.C.: 1981, *A Philosophical Basis of Medical Practice: Toward a Philosophy and Ethic of the Healing Professions*, Oxford, New York.

Reiser, S.J.: 1996, 'Medicine and public health: pursuing a common destiny,' *Journal of the American Medical Association* 276, 1429-1430.

Renton, A. : 1994, 'Epidemiology and causation: a realist view,' *Journal of Epidemiology and Community Health* 48, 79-85.

Rorty, R.: 1995, *Objectivity, Relativism, and Truth*, Cambridge University Press, New York.

Rose, G.: 1985, 'Sick individuals and sick populations,' *International Journal of Epidemiology* 14, 32-38.

Rothman, K.J.: 1976, 'Causes,' *American Journal of Epidemiology*, 104, 587-592.

Rothman, K.J. and Poole, C.: 1985, 'Science and policymaking,' *American Journal of Public Health* 75, 340-341.

Rothman, K.J., Adami, H.O., and Trichopoulos, D.: 1998, 'Should the mission of epidemiology include the eradication of poverty?,' *Lance*t 352, 810-813.

Rothman, K.J.: 1988, 'Causal inference,' *Chestnut Hill*, ERI, MA.

Rubin, D.B.: 1974, 'Estimating causal effects of treatments in randomized and nonrandomized studies,' *Journal of Educational Phychology* 66, 688-701

Sade, R.M.: 1995, 'A theory of health and disease: the objectivist-subjectivist dichotomy,' *Journal of Medicine and Philosophy* 20, 513-525.

Saraf, M.: 1998, 'Holy anorexia and anorexia nervosa: society and concept of disease,' *Pharo*s 61, :2-4.

Shy, C.M.: 1997, 'The failure of academic epidemiology,' *American Journal of Epidemiology* 145, 479-484

Skrabanek, P.: 1990, 'Why is preventive medicine exempted from ethical constraints?,' *Journal of Medical Ethics* 16, 187-190.

Stehbens, W.E.: 1985, 'The concept of cause in disease,' *Journal of Chronic Disease* 38, 947-950.

Susser, M.: 1991, 'What is a cause and how do we know one? A grammar for pragmatic epidemiology,' *American Journal of Epidemiology* 133, 635-648.

Terris, M.: 1992, 'The Society of Epidemiologic Research (SER) and the future of epidemiology,' *American Journal of Epidemiology* 36, 909-915.

Vandenbroucke, J.P.: 1988, ' "The Causes of Cancer" a miasma theory for the end of the twentieth century?,' *International Journal of Ejpidemiology* 17, 708-709.

Walker, A.M.: 1997, ' "Kangaroo Court": Invited commentary to Shy's "The Failure of Academic Epidemiology: witness for the prosecution",' *American Journal of Epidemiology* 145, 485-456

Weed, D.L.: 1995, 'Causal and Preventive Inference,' in Greenwald, P., Kramer, B.S. and

Weed, D.L. *Cancer Prevention and Control*, Marcel Dekker, New York.

Weed, D.L.: 1997, 'Underdetermination and incommensurability in contemporary epidemiology,' *Kennedy Institute on Ethics Journal* 7, 107-127.

Weed, D.L. and Gorelic, L.S.: 1996, 'The practice of causal inference in cancer epidemiology,' *Cancer Epidemiology, Biomarkers and Prevention* 5, 303-311.

Weed, D.L.: 1994, 'Science, ethics guidelines, and advocacy,' *Annals of Epidemiology* 4, 166-171.

Weed, D.L.: 1996, 'Epistemology and ethics,' in Coughlin, S.S. and Beauchamp, T.L., *Ethics and Epidemiology*, pp. 76-94, Oxford, New York.

Weed, D.L. and McKeown, R.E.: 1998, 'Epidemiology and virtue ethics,' *International Journal of Epidemiology* 27, 343-348.

Weed, D.L.: 1995, 'Epidemiology, humanities, and public health,' *American Journal of Public Health* 85, 914-918.

Weed, D.L.: 1986, 'On the logic of causal inference,' *American Journal of Epidemiology* 123, 965-979.

Weed, D.L.: 1999, 'Towards a philosophy of public health,' *Journal of Epidemiology and Community Health* 53, 99-104.

Weed, D.L.: 1998, 'Beyond black box epidemiology,' *American Journal of Public Health* 88, 12-14.

White, K.L.: 1991, *Healing the Schism: Epidemiology, Medicine, and the Public's Health*, Springer-Verlag, New York .

Whitbeck, C.: 1977, 'Causation in medicine: the disease-entity model,' *Philosophy of Science* 44, 619-637.

Woolley, F.R.: 1990, 'Medical ethics, technology and public health,' *Asia-Pacific Journal of Public Health* 4, 228-233.

Wulff, H,R.: 1992, 'Philosophy of medicine--from a medical perspective,' *Theoretical Medicine* 13, 79-85.

NOTES ON CONTRIBUTORS

Kurt Bayertz, Ph.D., is Professor of Philosophy at the University of Munster, Germany.

Tom L. Beauchamp, Ph.D., is Professor of Philosophy and Senior Research Scholar, Georgetown University, Washington, DC., U.S.A.

Joseph Boyle, Ph.D., is Professor of Philosophy, St. Michaels' College, University of Toronto, Toronto, Canada.

Dan W. Brock, Ph.D., is Senior Scientist and a member of the Department of Clinical Bioethics at the National Institutes of Health, Bethesda, MD, U.S.A.

Baruch Brody, Ph.D., is Professor, Center for Medical Ethics and Health Policy, Baylor College of Medicine, and Professor, Department of Philosophy, Rice University, Houston, TX, U.S.A.

K. Danner Clouser,[†] Ph.D., was University Professor of Humanities Emeritus at Penn State University College of Medicine, Hershey, PA, U.S.A.

Cynthia Cohen, J.D., is Senior Research Fellow, Kennedy Institute of Ethics, Georgetown University, Washington, D.C., U.S.A.

David DeGrazia, Ph.D., is Associate Professor of Philosophy, George Washington University, Washington, DC, U.S.A.

Sara T. Fry, R.N., Ph.D., is the Henry Luce Professor of Nursing Ethics, Boston College School of Nursing, Chestnut Hill, MA, U.S.A.

Bernard Gert, Ph.D., is Eunice and Julian Cohen Professor for the Study of Ethics and Human Values, Dartmouth College and Adjunct Professor of Psychiatry at Dartmouth Medical School, Hanover, New Hampshire, U.S.A.

Albert R. Jonsen, Ph.D. is Professor of Ethics in Medicine Emeritus, Department of Biomedical History, University of Washington, Seattle, Washington, U.S.A.

George Khushf, Ph.D., is the Humanities Director of the Center for Bioethics and Medical Humanities, and Associate Professor, Department of Philosophy, University of South Carolina, Columbia, SC, U.S.A.

Mark G. Kuczewski, Ph.D., is the Director of the Neiswanger Institute for Bioethics and Health Policy, Stritch School of Medicine, Loyola University of Chicago, Maywood, IL, U.S.A.

B. Andrew Lustig, Ph.D., is the Director of the Program in Bioetechnology, Religion and Ethics, Rice University, Houston, TX, U.S.A.

Laurence McCullough, Ph.D., is Professor of Medicine and Medical Ethics in the Center for Medical Ethics and Health Policy, Baylor College of Medicine, Houston, TX, U.S.A.

Hilde Lindemann Nelson, Ph.D., is Associate Professor in the Department of Philosophy at Michigan State University, East Lansing, MI, U.S.A.

Lennart Nordenfelt, Ph.D., is Professor of Philosophy of Medicine in the Department of Health and Society, Linkoping University, Linkoping, Sweden.

Edmund D. Pellegrino, M.D., is Professor Emeritus of Medicine and Medical Ethics, a Senior Research Scholar of the Kennedy Institute of Ethics, and Adjunct Professor of Philosophy at Georgetown University, Washington, D.C., U.S.A.

Kurt Schmidt, Ph.D., is Director of the Center for Medical Ethics at the St. Markus Hospital, Frankfurt am Main, Germany.

Michael Alan Schwartz, M.D., is Professor and Vice Chairman, Department of Psychiatry, Case Western University School of Medicine, Cleveland, OH, U.S.A.

David C. Thomasma†, Ph.D., was The Fr. Michael I. English, S.J., Professor of Medical Ethics and the Director of Medical Humanities at the Stritch School of Medicine, Loyola University of Chicago, Maywood, IL, U.S.A.

Christopher Tollefsen, Ph.D., is Associate Professor in the Department of Philosophy, University of South Carolina, Columbia, SC, U.S.A.

Rosemarie Tong, Ph.D., is Distinguished Professor in Health Care Ethics, Department of Philosophy, and Director, Center for Professional and Applied Ethics, at University of North Carolina, Charlotte, North Carolina, U.S.A.

Stephen Wear, Ph.D., is Clinical Associate Professor, Department of Medicine, and Co-Director of the Center for Clinical Ethics and Humanities in Health Care, University of Buffalo, Buffalo, NY, U.S.A.

Douglas L. Weed, M.D., M.P.H., Ph.D., is Chief, Office of Preventive Oncology and Dean of Education and Training in the Division of Cancer Prevention at the National Cancer Institute, Bethesda, MD, U.S.A.

Osborne Wiggins, Ph.D., is Professor of Philosophy at the University of Louisville, Louisville, KY, U.S.A.

Richard M. Zaner, Ph.D., is Professor Emeritus of Medicine and Philosophy, Vanderbilt University, Nashville, TN, U.S.A.

INDEX

Philosophy and Medicine

1. H. Tristram Engelhardt, Jr. and S.F. Spicker (eds.): *Evaluation and Explanation in the Biomedical Sciences.* 1975 ISBN 90-277-0553-4
2. S.F. Spicker and H. Tristram Engelhardt, Jr. (eds.): *Philosophical Dimensions of the Neuro-Medical Sciences.* 1976 ISBN 90-277-0672-7
3. S.F. Spicker and H. Tristram Engelhardt, Jr. (eds.): *Philosophical Medical Ethics.* Its Nature and Significance. 1977 ISBN 90-277-0772-3
4. H. Tristram Engelhardt, Jr. and S.F. Spicker (eds.): *Mental Health.* Philosophical Perspectives. 1978 ISBN 90-277-0828-2
5. B.A. Brody and H. Tristram Engelhardt, Jr. (eds.): *Mental Illness.* Law and Public Policy. 1980 ISBN 90-277-1057-0
6. H. Tristram Engelhardt, Jr., S.F. Spicker and B. Towers (eds.): *Clinical Judgment.* A Critical Appraisal. 1979 ISBN 90-277-0952-1
7. S.F. Spicker (ed.): *Organism, Medicine, and Metaphysics.* Essays in Honor of Hans Jonas on His 75th Birthday. 1978 ISBN 90-277-0823-1
8. E.E. Shelp (ed.): *Justice and Health Care.* 1981 ISBN 90-277-1207-7; Pb 90-277-1251-4
9. S.F. Spicker, J.M. Healey, Jr. and H. Tristram Engelhardt, Jr. (eds.): *The Law-Medicine Relation.* A Philosophical Exploration. 1981 ISBN 90-277-1217-4
10. W.B. Bondeson, H. Tristram Engelhardt, Jr., S.F. Spicker and J.M. White, Jr. (eds.): *New Knowledge in the Biomedical Sciences.* Some Moral Implications of Its Acquisition, Possession, and Use. 1982 ISBN 90-277-1319-7
11. E.E. Shelp (ed.): *Beneficence and Health Care.* 1982 ISBN 90-277-1377-4
12. G.J. Agich (ed.): *Responsibility in Health Care.* 1982 ISBN 90-277-1417-7
13. W.B. Bondeson, H. Tristram Engelhardt, Jr., S.F. Spicker and D.H. Winship: *Abortion and the Status of the Fetus.* 2nd printing, 1984 ISBN 90-277-1493-2
14. E.E. Shelp (ed.): *The Clinical Encounter.* The Moral Fabric of the Patient-Physician Relationship. 1983 ISBN 90-277-1593-9
15. L. Kopelman and J.C. Moskop (eds.): *Ethics and Mental Retardation.* 1984 ISBN 90-277-1630-7
16. L. Nordenfelt and B.I.B. Lindahl (eds.): *Health, Disease, and Causal Explanations in Medicine.* 1984 ISBN 90-277-1660-9
17. E.E. Shelp (ed.): *Virtue and Medicine.* Explorations in the Character of Medicine. 1985 ISBN 90-277-1808-3
18. P. Carrick: *Medical Ethics in Antiquity.* Philosophical Perspectives on Abortion and Euthanasia. 1985 ISBN 90-277-1825-3; Pb 90-277-1915-2
19. J.C. Moskop and L. Kopelman (eds.): *Ethics and Critical Care Medicine.* 1985 ISBN 90-277-1820-2
20. E.E. Shelp (ed.): *Theology and Bioethics.* Exploring the Foundations and Frontiers. 1985 ISBN 90-277-1857-1

Philosophy and Medicine

21. G.J. Agich and C.E. Begley (eds.): *The Price of Health.* 1986
ISBN 90-277-2285-4
22. E.E. Shelp (ed.): *Sexuality and Medicine.* Vol. I: Conceptual Roots. 1987
ISBN 90-277-2290-0; Pb 90-277-2386-9
23. E.E. Shelp (ed.): *Sexuality and Medicine.* Vol. II: Ethical Viewpoints in Transition. 1987 ISBN 1-55608-013-1; Pb 1-55608-016-6
24. R.C. McMillan, H. Tristram Engelhardt, Jr., and S.F. Spicker (eds.): *Euthanasia and the Newborn.* Conflicts Regarding Saving Lives. 1987
ISBN 90-277-2299-4; Pb 1-55608-039-5
25. S.F. Spicker, S.R. Ingman and I.R. Lawson (eds.): *Ethical Dimensions of Geriatric Care.* Value Conflicts for the 21th Century. 1987 ISBN 1-55608-027-1
26. L. Nordenfelt: *On the Nature of Health.* An Action-Theoretic Approach. 2nd, rev. ed. 1995 SBN 0-7923-3369-1; Pb 0-7923-3470-1
27. S.F. Spicker, W.B. Bondeson and H. Tristram Engelhardt, Jr. (eds.): *The Contraceptive Ethos.* Reproductive Rights and Responsibilities. 1987
ISBN 1-55608-035-2
28. S.F. Spicker, I. Alon, A. de Vries and H. Tristram Engelhardt, Jr. (eds.): *The Use of Human Beings in Research.* With Special Reference to Clinical Trials. 1988
ISBN 1-55608-043-3
29. N.M.P. King, L.R. Churchill and A.W. Cross (eds.): *The Physician as Captain of the Ship.* A Critical Reappraisal. 1988 ISBN 1-55608-044-1
30. H.-M. Sass and R.U. Massey (eds.): *Health Care Systems.* Moral Conflicts in European and American Public Policy. 1988 ISBN 1-55608-045-X
31. R.M. Zaner (ed.): *Death: Beyond Whole-Brain Criteria.* 1988
ISBN 1-55608-053-0
32. B.A. Brody (ed.): *Moral Theory and Moral Judgments in Medical Ethics.* 1988
ISBN 1-55608-060-3
33. L.M. Kopelman and J.C. Moskop (eds.): *Children and Health Care.* Moral and Social Issues. 1989 ISBN 1-55608-078-6
34. E.D. Pellegrino, J.P. Langan and J. Collins Harvey (eds.): *Catholic Perspectives on Medical Morals.* Foundational Issues. 1989 ISBN 1-55608-083-2
35. B.A. Brody (ed.): *Suicide and Euthanasia.* Historical and Contemporary Themes. 1989 ISBN 0-7923-0106-4
36. H.A.M.J. ten Have, G.K. Kimsma and S.F. Spicker (eds.): *The Growth of Medical Knowledge.* 1990 ISBN 0-7923-0736-4
37. I. Löwy (ed.): *The Polish School of Philosophy of Medicine.* From Tytus Chałubiński (1820–1889) to Ludwik Fleck (1896–1961). 1990
ISBN 0-7923-0958-8
38. T.J. Bole III and W.B. Bondeson: *Rights to Health Care.* 1991
ISBN 0-7923-1137-X

39. M.A.G. Cutter and E.E. Shelp (eds.): *Competency*. A Study of Informal Competency Determinations in Primary Care. 1991 ISBN 0-7923-1304-6
40. J.L. Peset and D. Gracia (eds.): *The Ethics of Diagnosis*. 1992 ISBN 0-7923-1544-8
41. K.W. Wildes, S.J., F. Abel, S.J. and J.C. Harvey (eds.): *Birth, Suffering, and Death*. Catholic Perspectives at the Edges of Life. 1992 [CSiB-1] ISBN 0-7923-1547-2; Pb 0-7923-2545-1
42. S.K. Toombs: *The Meaning of Illness*. A Phenomenological Account of the Different Perspectives of Physician and Patient. 1992 ISBN 0-7923-1570-7; Pb 0-7923-2443-9
43. D. Leder (ed.): *The Body in Medical Thought and Practice*. 1992 ISBN 0-7923-1657-6
44. C. Delkeskamp-Hayes and M.A.G. Cutter (eds.): *Science, Technology, and the Art of Medicine*. European-American Dialogues. 1993 ISBN 0-7923-1869-2
45. R. Baker, D. Porter and R. Porter (eds.): *The Codification of Medical Morality*. Historical and Philosophical Studies of the Formalization of Western Medical Morality in the 18th and 19th Centuries, Volume One: Medical Ethics and Etiquette in the 18th Century. 1993 ISBN 0-7923-1921-4
46. K. Bayertz (ed.): *The Concept of Moral Consensus*. The Case of Technological Interventions in Human Reproduction. 1994 ISBN 0-7923-2615-6
47. L. Nordenfelt (ed.): *Concepts and Measurement of Quality of Life in Health Care*. 1994 [ESiP-1] ISBN 0-7923-2824-8
48. R. Baker and M.A. Strosberg (eds.) with the assistance of J. Bynum: *Legislating Medical Ethics*. A Study of the New York State Do-Not-Resuscitate Law. 1995 ISBN 0-7923-2995-3
49. R. Baker (ed.): *The Codification of Medical Morality*. Historical and Philosophical Studies of the Formalization of Western Morality in the 18th and 19th Centuries, Volume Two: Anglo-American Medical Ethics and Medical Jurisprudence in the 19th Century. 1995 ISBN 0-7923-3528-7; Pb 0-7923-3529-5
50. R.A. Carson and C.R. Burns (eds.): *Philosophy of Medicine and Bioethics*. A Twenty-Year Retrospective and Critical Appraisal. 1997 ISBN 0-7923-3545-7
51. K.W. Wildes, S.J. (ed.): *Critical Choices and Critical Care*. Catholic Perspectives on Allocating Resources in Intensive Care Medicine. 1995 [CSiB-2] ISBN 0-7923-3382-9
52. K. Bayertz (ed.): *Sanctity of Life and Human Dignity*. 1996 ISBN 0-7923-3739-5
53. Kevin Wm. Wildes, S.J. (ed.): *Infertility: A Crossroad of Faith, Medicine, and Technology*. 1996 ISBN 0-7923-4061-2
54. Kazumasa Hoshino (ed.): *Japanese and Western Bioethics*. Studies in Moral Diversity. 1996 ISBN 0-7923-4112-0

Philosophy and Medicine

55. E. Agius and S. Busuttil (eds.): *Germ-Line Intervention and our Responsibilities to Future Generations.* 1998 ISBN 0-7923-4828-1
56. L.B. McCullough: *John Gregory and the Invention of Professional Medical Ethics and the Professional Medical Ethics and the Profession of Medicine.* 1998 ISBN 0-7923-4917-2
57. L.B. McCullough: *John Gregory's Writing on Medical Ethics and Philosophy of Medicine.* 1998 [CiME-1] ISBN 0-7923-5000-6
58. H.A.M.J. ten Have and H.-M. Sass (eds.): *Consensus Formation in Healthcare Ethics.* 1998 [ESiP-2] ISBN 0-7923-4944-X
59. H.A.M.J. ten Have and J.V.M. Welie (eds.): *Ownership of the Human Body.* Philosophical Considerations on the Use of the Human Body and its Parts in Healthcare. 1998 [ESiP-3] ISBN 0-7923-5150-9
60. M.J. Cherry (ed.): *Persons and Their Bodies.* Rights, Responsibilities, Relationships. 1999 ISBN 0-7923-5701-9
61. R. Fan (ed.): *Confucian Bioethics.* 1999 [APSiB-1] ISBN 0-7923-5853-8
62. L.M. Kopelman (ed.): *Building Bioethics.* Conversations with Clouser and Friends on Medical Ethics. 1999 ISBN 0-7923-5853-8
63. W.E. Stempsey: *Disease and Diagnosis.* 2000 PB ISBN 0-7923-6322-1
64. H.T. Engelhardt (ed.): *The Philosophy of Medicine.* Framing the Field. 2000 ISBN 0-7923-6223-3
65. S. Wear, J.J. Bono, G. Logue and A. McEvoy (eds.): *Ethical Issues in Health Care on the Frontiers of the Twenty-First Century.* 2000 ISBN 0-7923-6277-2
66. M. Potts, P.A. Byrne and R.G. Nilges (eds.): *Beyond Brain Death.* The Case Against Brain Based Criteria for Human Death. 2000 ISBN 0-7923-6578-X
67. L.M. Kopelman and K.A. De Ville (eds.): *Physician-Assisted Suicide.* What are the Issues? 2001 ISBN 0-7923-7142-9
68. S.K. Toombs (ed.): *Handbook of Phenomenology and Medicine.* 2001 ISBN 1-4020-0151-7; Pb 1-4020-0200-9
69. R. ter Meulen, W. Arts and R. Muffels (eds.): *Solidarity in Health and Social Care in Europe.* 2001 ISBN 1-4020-0164-9
70. A. Nordgren: *Responsible Genetics.* The Moral Responsibility of Geneticists for the Consequences of Human Genetics Research. 2001 ISBN 1-4020-0201-7
71. J. Tao Lai Po-wah (ed.): *Cross-Cultural Perspectives on the (Im)Possibility of Global Bioethics.* 2002 ISBN 1-4020-0498-2
72. P. Taboada, K. Fedoryka Cuddeback and P. Donohue-White (eds.): *Person, Society and Value.* Towards a Personalist Concept of Health. 2002 ISBN 1-4020-0503-2
73. J. Li: *Can Death Be a Harm to the Person Who Dies?* 2002 ISBN 1-4020-0505-9

Philosophy and Medicine

74. H.T. Engelhardt, Jr. and L.M. Rasmussen (eds.): *Bioethics and Moral Content: National Traditions of Health Care Morality*. Papers dedicated in tribute to Kazumasa Hoshino. 2002 ISBN 1-4020-6828-2
75. L.S. Parker and R.A. Ankeny (eds.): *Mutating Concepts, Evolving Disciplines: Genetics, Medicine, and Society*. 2002 ISBN 1-4020-1040-0
76. W.B. Bondeson and J.W. Jones (eds.): *The Ethics of Managed Care: Professional Integrity and Patient Rights*. 2002 ISBN 1-4020-1045-1
77. K.L. Vaux, S. Vaux and M. Sternberg (eds.): *Covenants of Life. Contemporary Medical Ethics in Light of the Thought of Paul Ramsey*. 2002 ISBN 1-4020-1053-2
78. G. Khushf (ed.): *Handbook of Bioethics: Taking Stock of the Field from a Philosophical Perspective*. 2004 ISBN 1-4020-1870-3; Pb: 1-4020-1893-2
79. A. Smith Iltis (ed.): *Institutional Integrity in Health Care*. 2003 ISBN 1-4020-1782-0
80. R.Z. Qiu (ed.): *Bioethics: Asian Perspectives A Quest for Moral Diversity*. 2003 ISBN 1-4020-1795-2
81. M.A.G. Cutter: *Reframing Disease Contextually*. 2003 ISBN 1-4020-1796-0

KLUWER ACADEMIC PUBLISHERS – DORDRECHT / BOSTON / LONDON